T0331594

Introduction to Random Graphs

From social networks such as Facebook, the World Wide Web and the Internet, to the complex interactions between proteins in the cells of our bodies, we constantly face the challenge of understanding the structure and development of networks. The theory of random graphs provides a framework for this understanding, and in this book the authors give a gentle introduction to the basic tools for understanding and applying the theory. Part one includes sufficient material, including exercises, for a one-semester course at the advanced undergraduate or beginning graduate level. The reader is then well prepared for the more advanced topics in Parts two and three. A final part provides a quick introduction to the background material needed.

All those interested in discrete mathematics, computer science or applied probability and their applications will find this an ideal introduction to the subject.

Alan Frieze is a Professor in the Department of Mathematical Sciences at Carnegie Mellon University. He has authored more than 300 publications in top journals and was invited to be a plenary speaker at the Seoul ICM 2014. In 1991 he received the Fulkerson prize in discrete mathematics.

Michał Karoński is a Professor in the Departments of Mathematics and Computer Science at Adam Mickiewicz University and Emory University. He is founder of the Discrete Mathematics group in Poznań and since 1990 has served as co-Editor-in-Chief of *Random Structures and Algorithms*.

Introduction to Random Graphs

ALAN FRIEZE
Carnegie Mellon University

MICHAŁ KAROŃSKI
Adam Mickiewicz University and Emory University

CAMBRIDGE
UNIVERSITY PRESS

University Printing House, Cambridge CB2 8BS, United Kingdom

One Liberty Plaza, 20th Floor, New York, NY 10006, USA

477 Williamstown Road, Port Melbourne, VIC 3207, Australia

314-321, 3rd Floor, Plot 3, Splendor Forum, Jasola District Centre, New Delhi - 110025, India

79 Anson Road, #06-04/06, Singapore 079906

Cambridge University Press is part of the University of Cambridge.

It furthers the University's mission by disseminating knowledge in the pursuit of education, learning and research at the highest international levels of excellence.

www.cambridge.org
Information on this title: www.cambridge.org/9781107118508

First published 2016

A catalogue record for this publication is available from the British Library

Library of Congress Cataloging in Publication data
Frieze, Alan, 1945–
Introduction to random graphs / Alan Frieze, Carnegie-Mellon University, Pennsylvania, Michał Karoński, Universytet im. Adama Mickiewiczowa w Poznaniu, Poland.
pages cm
Includes bibliographical references and index.
ISBN 978-1-107-11850-8 (hardback)
1. Random graphs. 2. Combinatorial probabilities. 3. Probabilities.
I. Karonski, Michal. II. Title.
QA166.17.F75 2015
511′.5–dc23 2015022579

ISBN 978-1-107-11850-8 Hardback

To Carol and Jola

Contents

Preface

Our purpose in writing this book is to provide a gentle introduction to a subject that is enjoying a surge in interest. We believe that the subject is fascinating in its own right, but the increase in interest can be attributed to several factors. One factor is the realization that networks are "everywhere." From social networks such as Facebook, the World Wide Web and the Internet to the complex interactions between proteins in the cells of our bodies, we face the challenge of understanding their structure and development. By and large natural networks grow in an unpredictable manner and this is often modeled by a random construction. Another factor is the realization by Computer Scientists that NP-hard problems are often easier to solve than their worst-case suggests and that an analysis of running times on random instances can be informative.

History

Random graphs were used by Erdős [274] to give a probabilistic construction of a graph with large girth and large chromatic number. It was only later that Erdős and Rényi began a systematic study of random graphs as objects of interest in their own right. Early on they defined the random graph $\mathbb{G}_{n,m}$ and founded the subject. Often neglected in this story is the contribution of Gilbert [367] who introduced the model $\mathbb{G}_{n,p}$, but clearly the credit for getting the subject started goes to Erdős and Rényi. Their seminal series of papers [275], [277], [278], [279] and, in particular, [276] on the evolution of random graphs laid the groundwork for other mathematicians to become involved in studying properties of random graphs.

In the early eighties the subject was beginning to blossom and it received a boost from two sources. First was the publication of the landmark book of Béla Bollobás [130] on random graphs. Around the same time, the

Discrete Mathematics group at Adam Mickiewicz University began a series of conferences in 1983. This series continues biennially to this day and is now a conference attracting more and more participants.

The next important event in the subject was the start of the journal *Random Structures and Algorithms* in 1990 followed by *Combinatorics, Probability and Computing* a few years later. These journals provided a dedicated outlet for work in the area and are flourishing today.

Scope of the book

We have divided the book into four parts. Part one is devoted to giving a detailed description of the main properties of $\mathbb{G}_{n,m}$ and $\mathbb{G}_{n,p}$. The aim is not to give best possible results, but instead to give some idea of the tools and techniques used in the subject, as well as to display some of the basic results of the area. There is sufficient material in Part one for a one-semester course at the advanced undergraduate or beginning graduate level. Once one has finished the content of the first part, one is equipped to continue with material of the remainder of the book, as well as to tackle some of the advanced monographs such as Bollobás [130] and the more recent one by Janson, Łuczak and Ruciński [432].

Each chapter comes with a few exercises. Some are fairly simple and these are designed to give the reader practice with making some of the estimations that are so prevalent in the subject. In addition, each chapter ends with some notes that lead through references to some of the more advanced important results that have not been covered.

Part two deals with models of random graphs that naturally extend $\mathbb{G}_{n,m}$ and $\mathbb{G}_{n,p}$. Part three deals with other models. Finally, in Part four, we describe some of the main tools used in the area along with proofs of their validity.

Having read this book, the reader should be in a good position to pursue research in the area and we hope that this book will appeal to anyone interested in Combinatorics or Applied Probability or Theoretical Computer Science.

Acknowledgements

Several people have helped with the writing of this book and we would like to acknowledge their help. First there are the students who have sat in on courses based on early versions of this book and who helped to iron out the many typo's etc.

We would next like to thank the following people for reading parts of the book before final submission: Andrew Beveridge, Deepak Bal, Malgorzata Bednarska, Patrick Bennett, Mindaugas Bloznelis, Antony Bonato, Boris Bukh, Fan Chung, Amin Coja-Oghlan, Colin Cooper, Andrzej Dudek, Asaf Ferber, Nikolas Fountoulakis, Catherine Greenhill, Dan Hefetz, Paul Horn, Hsien–Kuei Hwang, Jerzy Jaworski, Tony Johansson, Mihyun Kang, Michael Krivelevich, Tomasz Łuczak, Colin McDiarmid, Andrew McDowell, Hosam Mahmoud, Mike Molloy, Tobias Müller, Rajko Nenadov, Wesley Pegden, Boris Pittel, Dan Poole, Pawel Prałat, Andrzej Ruciński, Katarzyna Rybarczyk, Matas Šileikis, Greg Sorkin, Joel Spencer, Dudley Stark, Angelika Steger, Prasad Tetali, Andrew Thomason, Linnus Wästlund, Nick Wormald and Stephen Young.

Thanks also to Béla Bollobás for his advice on the structure of the book.

Conventions/Notation

Often in what follows, we give an expression for a large positive integer. It might not be obvious that the expression is actually an integer. In which case, the reader can rest assured that he/she can round up or down and obtained any required property. We avoid this rounding for convenience and for notational purposes.

In addition we list the following notation:

Mathematical relations

- $f(x) = O(g(x))$: $|f(x)| \leq K|g(x)|$ for some constant $K > 0$ and all $x \in \mathbf{R}$.
- $f(x) = \Theta(g(x))$: $f(n) = O(g(x))$ and $g(x) = O(f(x))$.
- $f(x) = o(g(x))$ as $x \to a$: $f(x)/g(x) \to 0$ as $x \to a$.
- $A \ll B$: A/B is sufficiently small for the succeeding arguments.
- $A \gg B$: A/B is sufficiently large for the succeeding arguments.
- $A \approx B$: $A/B \to 1$ as some parameter converges to 0 or ∞ or another limit.
- $[n]$: This is $\{1,2,\dots,n\}$. In general, if $a < b$ are positive integers, then $[a,b] = \{a, a+1, \dots, b\}$.
- If S is a set and k is a non-negative integer then $\binom{S}{k}$ denotes the set of k-element subsets of S. In particular, $\binom{[n]}{k}$ dnotes the set of k-sets of $\{1,2,\dots,n\}$.

Graph Notation

- $G = (V,E)$: $V = V(G)$ is the vertex set and $E = E(G)$ is the edge set.
- $e(G)$: $|E(G)|$.

- $N(S) = N_G(S)$ where $S \subseteq V(G)$. $\{w \notin S : \exists v \in S \text{such that } \{v,w\} \in E\}$.
- For a graph H, $aut(H)$ denotes the number of automorphisms of H.

Random Graph Models

- $[n]$: The set $\{1,2,\ldots,n\}$.
- $\mathcal{G}_{n,m}$: The family of all labeled graphs with vertex set $V = [n] = \{1,2,\ldots,n\}$ and exactly m edges.
- $\mathbb{G}_{n,m}$: A random graph chosen uniformly at random from $\mathcal{G}_{n,m}$.
- $E_{n,m} = E(\mathbb{G}_{n,m})$.
- $\mathbb{G}_{n,p}$: A random graph on vertex set $[n]$ where each possible edge occurs independently with probability p.
- $E_{n,p} = E(\mathbb{G}_{n,p})$.
- $\mathbb{G}_{n,m}^{\delta \geq k}$: $G_{n,m}$, conditioned on having minimum degree at least k.
- $\mathbb{G}_{n,n,p}$: A random bipartite graph with vertex set consisting of two disjoint copies of $[n]$ where each of the n^2 possible edges occurs independently with probability p.
- $\mathbb{G}_{n,r}$: A random r-regular graph on vertex set $[n]$.
- $\mathcal{G}_{n,\mathbf{d}}$: The set of graphs with vertex set $[n]$ and degree sequence $\mathbf{d} = (d_1, d_2, \ldots, d_n)$.
- $\mathbb{G}_{n,\mathbf{d}}$: A random graph chosen uniformly at random from $\mathcal{G}_{n,\mathbf{d}}$.
- $\mathbb{H}_{n,m;k}$: A random k-uniform hypergraph on vertex set $[n]$ and m edges of size k.
- $\mathbb{H}_{n,p;k}$: A random k-uniform hypergraph on vertex set $[n]$ where each of the $\binom{n}{k}$ possibles edge occurs independently with probability p.
- $\vec{\mathbb{G}}_{k-out}$: A random digraph on vertex set $[n]$ where each $v \in [n]$ independently chooses k random out-neighbors.
- $\mathbb{G}_{k-out}$: The graph obtained from $\vec{\mathbb{G}}_{k-out}$ by ignoring orientation and coalescing multiple edges.

Probability

- $\mathbb{P}(A)$: The probability of event A.
- $\mathbb{E}\,Z$: The expected value of random variable Z.
- $h(Z)$: The entropy of random variable Z.
- $Po(l)$: A random variable with the Poisson distribution with mean l.
- $N(0,1)$: A random variable with the normal distribution, mean 0 and variance 1.
- $\text{Bin}(n,p)$: A random variable with the binomial distribution with parameters n, the number of trials and p, the probability of success.
- $EXP(l)$: A random variable with the exponential distribution, mean l i.e. $\mathbb{P}(EXP(l) \geq x) = e^{-lx}$. We sometimes say *rate* $1/l$ in place of mean l.

- w.h.p.: A sequence of events $\mathscr{A}_n, n = 1, 2, \ldots$, is said to occur *with high probability* (w.h.p.) if $\lim_{n\to\infty} \mathbb{P}(\mathscr{A}_n) = 1$.
- $\stackrel{D}{\to}$: We write $X_n \stackrel{D}{\to} X$ to say that a random variable X_n *converges in distribution* to a random variable X, as $n \to \infty$. Occasionally we write $X_n \stackrel{D}{\to} N(0,1)$ (resp. $X_n \stackrel{D}{\to} Po(l)$) to mean that X has the corresponding normal (resp. Poisson) distribution.

PART I

Basic Models

1
Random Graphs

Graph theory is a vast subject in which the goals are to relate various graph properties i.e. proving that Property A implies Property B for various properties A,B. In some sense, the goals of Random Graph theory are to prove results of the form "Property A almost always implies Property B." In many cases Property A could simply be "Graph G has m edges." A more interesting example would be the following: property A is "G is an r-regular graph, $r \geq 3$" and Property B is "G is r-connected." This is proved in Chapter 10.

Before studying questions such as these, we need to describe the basic models of a random graph.

1.1 Models and Relationships

The study of random graphs in their own right began in earnest with the seminal paper of Erdős and Rényi [276]. This paper was the first to exhibit the threshold phenomena that characterize the subject.

Let $\mathcal{G}_{n,m}$ be the family of all labeled graphs with vertex set $V = [n] = \{1,2,\ldots,n\}$ and exactly m edges, $0 \leq m \leq \binom{n}{2}$. To every graph $G \in \mathcal{G}_{n,m}$, we assign a probability

$$\mathbb{P}(G) = \binom{\binom{n}{2}}{m}^{-1}.$$

Equivalently, we start with an empty graph on the set $[n]$, and insert m edges in such a way that all possible $\binom{\binom{n}{2}}{m}$ choices are equally likely. We denote such a random graph by $\mathbb{G}_{n,m} = ([n], E_{n,m})$ and call it a *uniform random graph*.

We now describe a similar model. Fix $0 \leq p \leq 1$. Then for $0 \leq m \leq \binom{n}{2}$, assign to each graph G with vertex set $[n]$ and m edges a probability

$$\mathbb{P}(G) = p^m (1-p)^{\binom{n}{2}-m},$$

where $0 \leq p \leq 1$. Equivalently, we start with an empty graph with vertex set $[n]$ and perform $\binom{n}{2}$ Bernoulli experiments inserting edges independently with probability p. We call such a random graph, a *Binomial random graph* and denote it by $\mathbb{G}_{n,p} = ([n], E_{n,p})$. This was introduced by Gilbert [367].

As one may expect there is a close relationship between these two models of random graphs. We start with a simple observation.

Lemma 1.1 *A random graph $\mathbb{G}_{n,p}$, given that its number of edges is m, is equally likely to be one of the $\binom{\binom{n}{2}}{m}$ graphs that have m edges.*

Proof Let G_0 be any labeled graph with m edges. Then since

$$\{\mathbb{G}_{n,p} = G_0\} \subseteq \{|E_{n,p}| = m\}$$

we have

$$\begin{aligned}
\mathbb{P}(\mathbb{G}_{n,p} = G_0 \mid |E_{n,p}| = m) &= \frac{\mathbb{P}(\mathbb{G}_{n,p} = G_0, |E_{n,p}| = m)}{\mathbb{P}(|E_{n,p}| = m)} \\
&= \frac{\mathbb{P}(\mathbb{G}_{n,p} = G_0)}{\mathbb{P}(|E_{n,p}| = m)} \\
&= \frac{p^m (1-p)^{\binom{n}{2} - m}}{\binom{\binom{n}{2}}{m} p^m (1-p)^{\binom{n}{2} - m}} \\
&= \binom{\binom{n}{2}}{m}^{-1}.
\end{aligned}$$

□

Thus $\mathbb{G}_{n,p}$ conditioned on the event $\{\mathbb{G}_{n,p}$ has m edges$\}$ is equal in distribution to $\mathbb{G}_{n,m}$, the graph chosen uniformly at random from all graphs with m edges.

Obviously, the main difference between those two models of random graphs is that in $\mathbb{G}_{n,m}$ we choose its number of edges, while in the case of $\mathbb{G}_{n,p}$ the number of edges is the Binomial random variable with the parameters $\binom{n}{2}$ and p. Intuitively, for large n random graphs $\mathbb{G}_{n,m}$ and $\mathbb{G}_{n,p}$ should behave in a similar fashion when the number of edges m in $\mathbb{G}_{n,m}$ equals or is "close" to the expected number of edges of $\mathbb{G}_{n,p}$, i.e. when

$$m = \binom{n}{2} p \approx \frac{n^2 p}{2}, \tag{1.1}$$

or, equivalently, when the edge probability in $\mathbb{G}_{n,p}$

$$p \approx \frac{2m}{n^2}. \tag{1.2}$$

Throughout the book, we use the notation $f \approx g$ to indicate that $f = (1+o(1))g$, where the $o(1)$ term depends on some parameter going to 0 or ∞.

We next introduce a useful "coupling technique" that generates the random graph $\mathbb{G}_{n,p}$ in two independent steps. We then describe a similar idea in relation to $\mathbb{G}_{n,m}$. Suppose that $p_1 < p$ and p_2 is defined by the equation

$$1 - p = (1 - p_1)(1 - p_2), \tag{1.3}$$

or, equivalently,

$$p = p_1 + p_2 - p_1 p_2.$$

Thus an edge is not included in $\mathbb{G}_{n,p}$ if it is not included in either of $\mathbb{G}_{n,p_1}$ or $\mathbb{G}_{n,p_2}$.

It follows that

$$\mathbb{G}_{n,p} = \mathbb{G}_{n,p_1} \cup \mathbb{G}_{n,p_2},$$

where the two graphs $\mathbb{G}_{n,p_1}, \mathbb{G}_{n,p_2}$ are independent. So when we write

$$\mathbb{G}_{n,p_1} \subseteq \mathbb{G}_{n,p},$$

we mean that the two graphs are *coupled* so that $\mathbb{G}_{n,p}$ is obtained from $\mathbb{G}_{n,p_1}$ by superimposing it with $\mathbb{G}_{n,p_2}$ and replacing eventual double edges by a single one.

We can also couple random graphs $\mathbb{G}_{n,m_1}$ and $\mathbb{G}_{n,m_2}$ where $m_2 \geq m_1$ via

$$\mathbb{G}_{n,m_2} = \mathbb{G}_{n,m_1} \cup \mathbb{H}.$$

Here $\mathbb{H}$ is the random graph on vertex set $[n]$ that has $m = m_2 - m_1$ edges chosen uniformly at random from $\binom{[n]}{2} \setminus E_{n,m_1}$.

Consider now a graph property $\mathcal{P}$ defined as a subset of the set of all labeled graphs on vertex set $[n]$, i.e. $\mathcal{P} \subseteq 2^{\binom{n}{2}}$. For example, all connected graphs (on n vertices), graphs with a Hamiltonian cycle, graphs containing a given subgraph, planar graphs, and graphs with a vertex of given degree form a specific "graph property."

We state below two simple observations that show a general relationship between $\mathbb{G}_{n,m}$ and $\mathbb{G}_{n,p}$ in the context of the probabilities of having a given graph property $\mathcal{P}$.

Lemma 1.2 *Let $\mathcal{P}$ be any graph property and $p = m/\binom{n}{2}$ where $m = m(n), \binom{n}{2} - m \to \infty$. Then, for large n,*

$$\mathbb{P}(\mathbb{G}_{n,m} \in \mathcal{P}) \leq 10m^{1/2}\,\mathbb{P}(\mathbb{G}_{n,p} \in \mathcal{P}).$$

Proof By the law of total probability,

$$\mathbb{P}(\mathbb{G}_{n,p} \in \mathscr{P}) = \sum_{k=0}^{\binom{n}{2}} \mathbb{P}(\mathbb{G}_{n,p} \in \mathscr{P} \mid |E_{n,p}| = k)\, \mathbb{P}(|E_{n,p}| = k)$$
$$= \sum_{k=0}^{\binom{n}{2}} \mathbb{P}(\mathbb{G}_{n,k} \in \mathscr{P})\, \mathbb{P}(|E_{n,p}| = k)$$
$$\geq \mathbb{P}(\mathbb{G}_{n,m} \in \mathscr{P})\, \mathbb{P}(|E_{n,p}| = m).$$

Recall that the number of edges $|E_{n,p}|$ of a random graph $\mathbb{G}_{n,p}$ is a random variable with the Binomial distribution with parameters $\binom{n}{2}$ and p. Applying Stirling's formula:

$$k! = (1+o(1))\left(\frac{k}{e}\right)^k \sqrt{2\pi k}, \tag{1.4}$$

and putting $N = \binom{n}{2}$, we get

$$\mathbb{P}(|E_{n,p}| = m) = \binom{N}{m} p^m (1-p)^{\binom{n}{2}-m}$$
$$= (1+o(1)) \frac{N^N \sqrt{2\pi N}\, p^m (1-p)^{N-m}}{m^m (N-m)^{N-m}\, 2\pi \sqrt{m(N-m)}} \tag{1.5}$$
$$= (1+o(1)) \sqrt{\frac{N}{2\pi m(N-m)}}.$$

Hence,

$$\mathbb{P}(|E_{n,p}| = m) \geq \frac{1}{10\sqrt{m}},$$

so

$$\mathbb{P}(\mathbb{G}_{n,m} \in \mathscr{P}) \leq 10 m^{1/2}\, \mathbb{P}(\mathbb{G}_{n,p} \in \mathscr{P}).$$

□

We call a graph property $\mathscr{P}$ *monotone increasing* if $G \in \mathscr{P}$ implies $G + e \in \mathscr{P}$, i.e. adding an edge e to a graph G does not destroy the property. For example, connectivity and Hamiltonicity are monotone increasing properties. A monotone increasing property is *non-trivial* if the empty graph $\bar{K}_n \notin \mathscr{P}$ and the complete graph $K_n \in \mathscr{P}$.

A graph property is *monotone decreasing* if $G \in \mathscr{P}$ implies $G - e \in \mathscr{P}$, i.e. removing an edge from a graph does not destroy the property. Properties of a graph not being connected or being planar are examples of monotone decreasing graph properties. Obviously, a graph property $\mathscr{P}$ is monotone increasing if and only if its complement is monotone decreasing. Clearly not all graph properties are monotone. For example, having at least half of the vertices having a given fixed degree d is not monotone.

From the coupling argument it follows that if $\mathcal{P}$ is a monotone increasing property then, whenever $p < p'$ or $m < m'$,

$$\mathbb{P}(\mathbb{G}_{n,p} \in \mathcal{P}) \leq \mathbb{P}(\mathbb{G}_{n,p'} \in \mathcal{P}), \tag{1.6}$$

and

$$\mathbb{P}(\mathbb{G}_{n,m} \in \mathcal{P}) \leq \mathbb{P}(\mathbb{G}_{n,m'} \in \mathcal{P}), \tag{1.7}$$

respectively.

For monotone increasing graph properties we can get a much better upper bound on $\mathbb{P}(\mathbb{G}_{n,m} \in \mathcal{P})$, in terms of $\mathbb{P}(\mathbb{G}_{n,p} \in \mathcal{P})$, than that given by Lemma 1.2.

Lemma 1.3 *Let $\mathcal{P}$ be a monotone increasing graph property and $p = \frac{m}{N}$. Then, for large n and p such that $Np, N(1-p)/(Np)^{1/2} \to \infty$,*

$$\mathbb{P}(\mathbb{G}_{n,m} \in \mathcal{P}) \leq 3\,\mathbb{P}(\mathbb{G}_{n,p} \in \mathcal{P}).$$

Proof Suppose $\mathcal{P}$ is monotone increasing and $p = \frac{m}{N}$, where $N = \binom{n}{2}$. Then

$$\begin{aligned}\mathbb{P}(\mathbb{G}_{n,p} \in \mathcal{P}) &= \sum_{k=0}^{N} \mathbb{P}(\mathbb{G}_{n,k} \in \mathcal{P})\mathbb{P}(|E_{n,p}| = k) \\ &\geq \sum_{k=m}^{N} \mathbb{P}(\mathbb{G}_{n,k} \in \mathcal{P})\mathbb{P}(|E_{n,p}| = k).\end{aligned}$$

However, by the coupling property we know that for $k \geq m$,

$$\mathbb{P}(\mathbb{G}_{n,k} \in \mathcal{P}) \geq \mathbb{P}(\mathbb{G}_{n,m} \in \mathcal{P}).$$

The number of edges $|E_{n,p}|$ in $\mathbb{G}_{n,p}$ has the Binomial distribution with parameters N, p. Hence,

$$\begin{aligned}\mathbb{P}(\mathbb{G}_{n,p} \in \mathcal{P}) &\geq \mathbb{P}(\mathbb{G}_{n,m} \in \mathcal{P}) \sum_{k=m}^{N} \mathbb{P}(|E_{n,p}| = k) \\ &= \mathbb{P}(\mathbb{G}_{n,m} \in \mathcal{P}) \sum_{k=m}^{N} u_k,\end{aligned} \tag{1.8}$$

where

$$u_k = \binom{N}{k} p^k (1-p)^{N-k}.$$

Now, using Stirling's formula,

$$u_m = (1+o(1)) \frac{N^N p^m (1-p)^{N-m}}{m^m (N-m)^{N-m} (2\pi m)^{1/2}} = \frac{1+o(1)}{(2\pi m)^{1/2}}.$$

Furthermore, if $k=m+t$ where $0\le t\le m^{1/2}$ then

$$\frac{u_{k+1}}{u_k}=\frac{(N-k)p}{(k+1)(1-p)}=\frac{1-\frac{t}{N-m}}{1+\frac{t+1}{m}}$$
$$\ge \exp\left\{-\frac{t}{N-m-t}-\frac{t+1}{m}\right\}=1-o(1),$$

after using Lemma 21.1(a), (b) to obtain the first inequality and our assumptions on N,p to obtain the second.

It follows that

$$\sum_{k=m}^{m+m^{1/2}} u_k \ge \frac{1-o(1)}{(2\pi)^{1/2}}$$

and the lemma follows from (1.8). □

Lemmas 1.2 and 1.3 are surprisingly applicable. In fact, since the $\mathbb{G}_{n,p}$ model is computationally easier to handle than $\mathbb{G}_{n,m}$, we repeatedly use both lemmas to show that $\mathbb{P}(\mathbb{G}_{n,p}\in\mathcal{P})\to 0$ implies that $\mathbb{P}(\mathbb{G}_{n,m}\in\mathcal{P})\to 0$ when $n\to\infty$. In other situations we can use a stronger and more widely applicable result. The theorem below, which we state without proof, gives precise conditions for the asymptotic equivalence of random graphs $\mathbb{G}_{n,p}$ and $\mathbb{G}_{n,m}$. It is due to Łuczak [535].

Theorem 1.4 *Let* $0\le p_0\le 1$, $s(n)=n\sqrt{p(1-p)}\to\infty$, *and* $\omega(n)\to\infty$ *arbitrarily slowly as* $n\to\infty$.

(i) *Suppose that* $\mathcal{P}$ *is a graph property such that* $\mathbb{P}(\mathbb{G}_{n,m}\in\mathcal{P})\to p_0$ *for all*

$$m\in\left[\binom{n}{2}p-\omega(n)s(n),\binom{n}{2}p+\omega(n)s(n)\right].$$

Then $\mathbb{P}(\mathbb{G}_{n,p}\in\mathcal{P})\to p_0$ *as* $n\to\infty$.

(ii) *Let* $p_-=p-\omega(n)s(n)/n^3$ *and* $p_+=p+\omega(n)s(n)/n^3$. *Suppose that* $\mathcal{P}$ *is a monotone graph property such that* $\mathbb{P}(\mathbb{G}_{n,p_-}\in\mathcal{P})\to p_0$ *and* $\mathbb{P}(\mathbb{G}_{n,p_+}\in\mathcal{P})\to p_0$. *Then* $\mathbb{P}(\mathbb{G}_{n,m}\in\mathcal{P})\to p_0$, *as* $n\to\infty$, *where* $m=\lfloor\binom{n}{2}p\rfloor$.

1.2 Thresholds and Sharp Thresholds

One of the most striking observations regarding the asymptotic properties of random graphs is the "abrupt" nature of the appearance and disappearance of certain graph properties. To be more precise in the description of this phenomenon, let us introduce *threshold functions* (or just *thresholds*) for monotone graph properties. We start by giving the formal definition of a threshold for a monotone increasing graph property $\mathscr{P}$.

Definition 1.5 A function $m^* = m^*(n)$ is a *threshold* for a monotone increasing property $\mathscr{P}$ in the random graph $\mathbb{G}_{n,m}$ if

$$\lim_{n\to\infty} \mathbb{P}(\mathbb{G}_{n,m} \in \mathscr{P}) = \begin{cases} 0 & \text{if } m/m^* \to 0, \\ 1 & \text{if } m/m^* \to \infty, \end{cases}$$

as $n \to \infty$.

A similar definition applies to the edge probability $p = p(n)$ in a random graph $\mathbb{G}_{n,p}$.

Definition 1.6 A function $p^* = p^*(n)$ is a *threshold* for a monotone increasing property $\mathscr{P}$ in the random graph $\mathbb{G}_{n,p}$ if

$$\lim_{n\to\infty} \mathbb{P}(\mathbb{G}_{n,p} \in \mathscr{P}) = \begin{cases} 0 & \text{if } p/p^* \to 0, \\ 1 & \text{if } p/p^* \to \infty, \end{cases}$$

as $n \to \infty$.

It is easy to see how to define thresholds for monotone decreasing graph properties and therefore we leave this to the reader.

Notice also that the thresholds defined above are not unique, since any function which differs from $m^*(n)$ (resp. $p^*(n)$) by a constant factor is also a threshold for $\mathscr{P}$.

A large body of the theory of random graphs is concerned with the search for thresholds for various properties, such as containing a path or cycle of a given length, or, in general, a copy of a given graph, or being connected or Hamiltonian, to name just a few. Therefore the next result is of special importance. It was proved by Bollobás and Thomason [150].

Theorem 1.7 *Every non-trivial monotone graph property has a threshold.*

Proof Without loss of generality assume that $\mathscr{P}$ is a monotone increasing graph property. Given $0 < \varepsilon < 1$ we define $p(\varepsilon)$ by

$$\mathbb{P}(\mathbb{G}_{n,p(\varepsilon)} \in \mathscr{P}) = \varepsilon.$$

Note that $p(\varepsilon)$ exists because

$$\mathbb{P}(\mathbb{G}_{n,p} \in \mathscr{P}) = \sum_{G\in\mathscr{P}} p^{|E(G)|}(1-p)^{N-|E(G|}$$

is a polynomial in p that increases from 0 to 1. This is not obvious from the expression, but it is obvious from the fact that $\mathcal{P}$ is monotone increasing and that increasing p increases the likelihood that $\mathbb{G}_{n,p} \in \mathcal{P}$.

We will show that $p^* = p(1/2)$ is a threshold for $\mathcal{P}$. Let $G_1, G_2, \ldots, G_k$ be independent copies of $\mathbb{G}_{n,p}$. The graph $G_1 \cup G_2 \cup \ldots \cup G_k$ is distributed as $\mathbb{G}_{n,1-(1-p)^k}$. Now $1-(1-p)^k \leq kp$, and therefore by the coupling argument

$$\mathbb{G}_{n,1-(1-p)^k} \subseteq \mathbb{G}_{n,kp},$$

and so $\mathbb{G}_{n,kp} \notin \mathcal{P}$ implies $G_1, G_2, \ldots, G_k \notin \mathcal{P}$. Hence,

$$\mathbb{P}(\mathbb{G}_{n,kp} \notin \mathcal{P}) \leq [\mathbb{P}(\mathbb{G}_{n,p} \notin \mathcal{P})]^k.$$

Let ω be a function of n such that $\omega \to \infty$ arbitrarily slowly as $n \to \infty$, $\omega \ll \log\log n$. (We say that $f(n) \ll g(n)$ or $f(n) = o(g(n))$ if $f(n)/g(n) \to 0$ as $n \to \infty$. Of course in this case we can also write $g(n) \gg f(n)$.) Suppose also that $p = p^* = p(1/2)$ and $k = \omega$. Then

$$\mathbb{P}(\mathbb{G}_{n,\omega p^*} \notin \mathcal{P}) \leq 2^{-\omega} = o(1).$$

On the other hand for $p = p^*/\omega$,

$$\frac{1}{2} = \mathbb{P}(\mathbb{G}_{n,p^*} \notin \mathcal{P}) \leq \left[\mathbb{P}(\mathbb{G}_{n,p^*/\omega} \notin \mathcal{P})\right]^{\omega}.$$

So

$$\mathbb{P}(\mathbb{G}_{n,p^*/\omega} \notin \mathcal{P}) \geq 2^{-1/\omega} = 1 - o(1).$$

□

In order to shorten many statements of theorems in the book we say that a sequence of events $\mathcal{E}_n$ occurs *with high probability* (w.h.p.) if

$$\lim_{n\to\infty} \mathbb{P}(\mathcal{E}_n) = 1.$$

Thus the statement that says p^* is a threshold for a property $\mathcal{P}$ in $\mathbb{G}_{n,p}$ is the same as saying that $\mathbb{G}_{n,p} \notin \mathcal{P}$ w.h.p. if $p \ll p^*$, while $\mathbb{G}_{n,p} \in \mathcal{P}$ w.h.p. if $p \gg p^*$.

In many situations we can observe that for some monotone graph properties more "subtle" thresholds hold. We call them "*sharp thresholds.*" More precisely,

Definition 1.8 A function $m^* = m^*(n)$ is a *sharp threshold* for a monotone increasing property $\mathcal{P}$ in the random graph $\mathbb{G}_{n,m}$ if for every $\varepsilon > 0$,

$$\lim_{n\to\infty} \mathbb{P}(\mathbb{G}_{n,m} \in \mathcal{P}) = \begin{cases} 0 & \text{if} \quad m/m^* \leq 1-\varepsilon \\ 1 & \text{if} \quad m/m^* \geq 1+\varepsilon. \end{cases}$$

A similar definition applies to the edge probability $p = p(n)$ in the random graph $\mathbb{G}_{n,p}$.

Definition 1.9 A function $p^* = p^*(n)$ is a *sharp threshold* for a monotone increasing property $\mathcal{P}$ in the random graph $\mathbb{G}_{n,p}$ if for every $\varepsilon > 0$

$$\lim_{n\to\infty} \mathbb{P}(\mathbb{G}_{n,p} \in \mathcal{P}) = \begin{cases} 0 & \text{if} \quad p/p^* \leq 1-\varepsilon \\ 1 & \text{if} \quad p/p^* \geq 1+\varepsilon. \end{cases}$$

We illustrate both types of threshold in a series of examples dealing with very simple graph properties. Our goal at the moment is to demonstrate basic techniques to determine thresholds rather than to "discover" some "striking" facts about random graphs.

We start with the random graph $\mathbb{G}_{n,p}$ and the property

$$\mathcal{P} = \{\text{all non-empty (non-edgeless) labeled graphs on } n \text{ vertices}\}.$$

This simple graph property is clearly monotone increasing and we show below that $p^* = 1/n^2$ is a threshold for a random graph $\mathbb{G}_{n,p}$ of having at least one edge (being non-empty).

Lemma 1.10 *Let $\mathcal{P}$ be the property defined above, i.e. stating that $\mathbb{G}_{n,p}$ contains at least one edge. Then*

$$\lim_{n\to\infty} \mathbb{P}(\mathbb{G}_{n,p} \in \mathcal{P}) = \begin{cases} 0 & \text{if } p \ll n^{-2} \\ 1 & \text{if } p \gg n^{-2}. \end{cases}$$

Proof Let X be a random variable counting edges in $\mathbb{G}_{n,p}$. Since X has the Binomial distribution, then $\mathbb{E}X = \binom{n}{2}p$, and $\mathrm{Var}X = \binom{n}{2}p(1-p) = (1-p)\mathbb{E}X$.

A standard way to show the first part of the threshold statement, i.e. that w.h.p. a random graph $\mathbb{G}_{n,p}$ is empty when $p = o(n^{-2})$, is a very simple consequence of Markov's inequality, called the First Moment Method, see Lemma 20.2. It states that if X is a non-negative integer valued random variable, then

$$\mathbb{P}(X > 0) \leq \mathbb{E}X.$$

Hence, in our case

$$\mathbb{P}(X > 0) \leq \frac{n^2}{2}p \to 0$$

as $n \to \infty$, since $p \ll n^{-2}$.

On the other hand, if we want show that $\mathbb{P}(X > 0) \to 1$ as $n \to \infty$ then we cannot use the First Moment Method and we should use the Second Moment Method, which is a simple consequence of the Chebyshev inequality, see Lemma 20.3. We use the inequality to show *concentration around the mean*. By this we mean that w.h.p. $X \approx \mathbb{E}X$. The Chebyshev inequality states that if

X is a non-negative integer valued random variable then

$$\mathbb{P}(X > 0) \geq 1 - \frac{\operatorname{Var} X}{(\mathbb{E} X)^2}.$$

Hence, $\mathbb{P}(X > 0) \to 1$ as $n \to \infty$ whenever $\operatorname{Var} X/(\mathbb{E}X)^2 \to 0$ as $n \to \infty$. (For proofs of both of the above Lemmas see Section 20.1 of Chapter 20.)

Now, if $p \gg n^{-2}$ then $\mathbb{E}X \to \infty$ and therefore

$$\frac{\operatorname{Var} X}{(\mathbb{E}X)^2} = \frac{1-p}{\mathbb{E}X} \to 0$$

as $n \to \infty$, which shows that the second statement of Lemma 1.10 holds, and so $p^* = 1/n^2$ is a threshold for the property of $\mathbb{G}_{n,p}$ being non-empty. □

Let us now look at the degree of a fixed vertex in both models of random graphs. One immediately notices that if $\deg(v)$ denotes the degree of a fixed vertex in $\mathbb{G}_{n,p}$, then $\deg(v)$ is a binomially distributed random variable, with parameters $n-1$ and p, i.e. for $k = 0, 1, 2 \ldots, n-1$,

$$\mathbb{P}(\deg(v) = d) = \binom{n-1}{d} p^d (1-p)^{n-1-d},$$

while in $\mathbb{G}_{n,m}$ the distribution of $\deg(v)$ is Hypergeometric, i.e.

$$\mathbb{P}(\deg(v) = d) = \frac{\binom{n-1}{d}\binom{\binom{n-1}{2}}{m-d}}{\binom{\binom{n}{2}}{m}}.$$

Consider the monotone decreasing graph property that a graph contains an isolated vertex, i.e. a vertex of degree zero:

$$\mathscr{P} = \{\text{all labeled graphs on } n \text{ vertices containing isolated vertices}\}.$$

We show that $m^* = \frac{1}{2} n \log n$ is the sharp threshold function for the above property $\mathscr{P}$ in $\mathbb{G}_{n,m}$.

Lemma 1.11 *Let $\mathscr{P}$ be the property that a graph on n vertices contains at least one isolated vertex and let $m = \frac{1}{2}n(\log n + \omega(n))$. Then*

$$\lim_{n\to\infty} \mathbb{P}(\mathbb{G}_{n,m} \in \mathscr{P}) = \begin{cases} 1 & \text{if } \omega(n) \to -\infty \\ 0 & \text{if } \omega(n) \to \infty. \end{cases}$$

Proof To see that the second statement of Lemma 1.11 holds we use the First Moment Method. Namely, let $X_0 = X_{n,0}$ be the number of isolated vertices in the random graph $\mathbb{G}_{n,m}$. Then X_0 can be represented as the sum of indicator random variables

$$X_0 = \sum_{v \in V} I_v,$$

where

$$I_v = \begin{cases} 1 & \text{if } v \text{ is an isolated vertex in } \mathbb{G}_{n,m} \\ 0 & \text{otherwise.} \end{cases}$$

So

$$\begin{aligned} \mathbb{E}X_0 &= \sum_{v \in V} \mathbb{E}I_v = n \frac{\binom{\binom{n-1}{2}}{m}}{\binom{\binom{n}{2}}{m}} \\ &= n\left(\frac{n-2}{n}\right)^m \prod_{i=0}^{m-1}\left(1 - \frac{4i}{n(n-1)(n-2) - 2i(n-2)}\right) \\ &= n\left(\frac{n-2}{n}\right)^m \left(1 + O\left(\frac{(\log n)^2}{n}\right)\right), \end{aligned} \tag{1.9}$$

assuming that $\omega = o(\log n)$.

Hence,

$$\mathbb{E}X_0 \le n\left(\frac{n-2}{n}\right)^m \le ne^{-\frac{2m}{n}} = e^{-\omega},$$

for $m = \frac{1}{2}n(\log n + \omega(n))$. ($1 + x \le e^x$ is one of the basic inequalities stated in Lemma 21.1.)

So $\mathbb{E}X_0 \to 0$ when $\omega(n) \to \infty$ as $n \to \infty$ and the First Moment Method implies that $X_0 = 0$ w.h.p.

To show that Lemma 1.11 holds in the case when $\omega \to -\infty$ we first observe from (1.9) that in this case

$$\begin{aligned} \mathbb{E}X_0 &= (1 - o(1))n\left(\frac{n-2}{n}\right)^m \\ &\ge (1 - o(1))n\exp\left\{-\frac{2m}{n-2}\right\} \\ &\ge (1 - o(1))e^{-\omega} \to \infty, \end{aligned} \tag{1.10}$$

The second inequality in the above comes from Lemma 21.1(b), and we have once again assumed that $\omega = o(\log n)$ to justify the first equation.

We caution the reader that $\mathbb{E}X_0 \to \infty$ does not prove that $X_0 > 0$ w.h.p. Chapter 5 has an example of a random variable X_H, where $\mathbb{E}X_H \to \infty$ and yet $X_H = 0$ w.h.p.

We now use a stronger version of the Second Moment Method (for its proof see Section 20.1 of Chapter 20). It states that if X is a non-negative integer valued random variable then

$$\mathbb{P}(X > 0) \ge \frac{(\mathbb{E}X)^2}{\mathbb{E}X^2} = 1 - \frac{\operatorname{Var} X}{\mathbb{E}X^2}. \tag{1.11}$$

Notice that

$$\begin{aligned}
\mathbb{E}X_0^2 &= \mathbb{E}\left(\sum_{v\in V} I_v\right)^2 = \sum_{u,v\in V} \mathbb{E}(I_u I_v) \\
&= \sum_{u,v\in V} \mathbb{P}(I_u = 1, I_v = 1) \\
&= \sum_{u\neq v} \mathbb{P}(I_u = 1, I_v = 1) + \sum_{u=v} \mathbb{P}(I_u = 1, I_v = 1) \\
&= n(n-1)\frac{\binom{\binom{n-2}{2}}{m}}{\binom{\binom{n}{2}}{m}} + \mathbb{E}X_0 \\
&\le n^2\left(\frac{n-2}{n}\right)^{2m} + \mathbb{E}X_0 \\
&= (1+o(1))(\mathbb{E}X_0)^2 + \mathbb{E}X_0.
\end{aligned}$$

The last equation follows from (1.9).

Hence, by (1.11),

$$\begin{aligned}
\mathbb{P}(X_0 > 0) &\ge \frac{(\mathbb{E}X_0)^2}{\mathbb{E}X_0^2} \\
&= \frac{(\mathbb{E}X_0)^2}{(1+o(1))((\mathbb{E}X_0)^2 + \mathbb{E}X_0)} \\
&= \frac{1}{(1+o(1)) + (\mathbb{E}X_0)^{-1}} \\
&= 1 - o(1),
\end{aligned}$$

on using (1.10). Hence, $\mathbb{P}(X_0 > 0) \to 1$ when $\omega(n) \to -\infty$ as $n \to \infty$, and so we can conclude that $m = m(n)$ is the sharp threshold for the property that $\mathbb{G}_{n,m}$ contains isolated vertices. □

For this simple random variable, we worked with $\mathbb{G}_{n,m}$. We will, in general, work with the more congenial independent model $\mathbb{G}_{n,p}$ and translate the results to $G_{n,m}$ if so desired.

For another simple example of the use of the Second Moment Method, we will prove

Theorem 1.12 *If $m/n \to \infty$ then w.h.p. $\mathbb{G}_{n,m}$ contains at least one triangle.*

Proof Because having a triangle is a monotone increasing property we can prove the result in $\mathbb{G}_{n,p}$ assuming that $np \to \infty$.

Assume first that $np = \omega \leq \log n$ where $\omega = \omega(n) \to \infty$ and let Z be the number of triangles in $\mathbb{G}_{n,p}$. Then

$$\mathbb{E}Z = \binom{n}{3} p^3 \geq (1 - o(1))\frac{\omega^3}{6} \to \infty.$$

We remind the reader that simply having $\mathbb{E}Z \to \infty$ is not sufficient to prove that $Z > 0$ w.h.p.

Next let $T_1, T_2, \ldots, T_M, M = \binom{n}{3}$ denote the triangles of K_n. Then

$$\begin{aligned}
\mathbb{E}Z^2 &= \sum_{i,j=1}^{M} \mathbb{P}(T_i, T_j \in \mathbb{G}_{n,p}) \\
&= \sum_{i=1}^{M} \mathbb{P}(T_i \in \mathbb{G}_{n,p}) \sum_{j=1}^{M} \mathbb{P}(T_j \in \mathbb{G}_{n,p} \mid T_i \in \mathbb{G}_{n,p}) \qquad (1.12) \\
&= M\,\mathbb{P}(T_1 \in \mathbb{G}_{n,p}) \sum_{j=1}^{M} \mathbb{P}(T_j \in \mathbb{G}_{n,p} \mid T_1 \in \mathbb{G}_{n,p}) \qquad (1.13) \\
&= (\mathbb{E}Z) \sum_{j=1}^{M} \mathbb{P}(T_j \in \mathbb{G}_{n,p} \mid T_1 \in \mathbb{G}_{n,p}).
\end{aligned}$$

Here (1.13) follows from (1.12) by symmetry.

Now suppose that T_j, T_1 share σ_j edges. Then

$$\begin{aligned}
\sum_{j=1}^{M} \mathbb{P}(T_j \in \mathbb{G}_{n,p} \mid T_1 \in \mathbb{G}_{n,p}) &= 1 + \sum_{j:\sigma_j=1} \mathbb{P}(T_j \in \mathbb{G}_{n,p} \mid T_1 \in \mathbb{G}_{n,p}) \\
&\quad + \sum_{j:\sigma_j=0} \mathbb{P}(T_j \in \mathbb{G}_{n,p} \mid T_1 \in \mathbb{G}_{n,p}) \\
&= 1 + 3(n-3)p^2 + \left(\binom{n}{3} - 3n + 8\right)p^3 \\
&\leq 1 + \frac{3\omega^2}{n} + \mathbb{E}Z.
\end{aligned}$$

It follows that

$$\operatorname{Var} Z \leq (\mathbb{E}Z)\left(1 + \frac{3\omega^2}{n} + \mathbb{E}Z\right) - (\mathbb{E}Z)^2 \leq \frac{10\omega^5}{n}.$$

Applying the Chebyshev inequality we get

$$\mathbb{P}(Z = 0) \leq \mathbb{P}(|Z - \mathbb{E}Z| \geq \mathbb{E}Z) \leq \frac{\operatorname{Var} Z}{(\mathbb{E}Z)^2} \leq \frac{60\omega^5}{n\omega^3} = o(1).$$

This proves the theorem for $p \leq \frac{\log n}{n}$. For larger p we can use (1.6). □

We can in fact use the Second Moment Method to show that if $m/n \to \infty$ then w.h.p. $\mathbb{G}_{n,m}$ contains a copy of a k-cycle C_k for any fixed $k \geq 3$. See Theorem 5.3, see also Exercise 1.4.7.

1.3 Pseudo-Graphs

We sometimes use one of the two following models that are related to $\mathbb{G}_{n,m}$ and have a little more independence. (We will use Model A in Section 7.3 and Model B in Section 6.4.)

Model A: We let $\mathbf{x} = (x_1, x_2, \ldots, x_{2m})$ be chosen uniformly at random from $[n]^{2m}$.

Model B: We let $\mathbf{x} = (x_1, x_2, \ldots, x_{2m})$ be chosen uniformly at random from $\binom{[n]}{2}^m$.

The (multi-)graph $\mathbb{G}_{n,m}^{(X)}, X \in \{A, B\}$ has vertex set $[n]$ and edge set $E_m = \{\{x_{2i-1}, x_{2i}\} : 1 \leq i \leq m\}$. Basically, we are choosing edges with replacement. In Model A we allow loops and in Model B we do not. We get simple graphs by removing loops and multiple edges to obtain graphs $\mathbb{G}_{n,m}^{(X*)}$ with m^* edges. It is not difficult to see that for $X \in \{A, B\}$ and conditional on the value of m^* that $\mathbb{G}_{n,m}^{(X*)}$ is distributed as $\mathbb{G}_{n,m^*}$, see Exercise (1.4.10).

More importantly, we have that for $G_1, G_2 \in \mathcal{G}_{n,m}$,

$$\mathbb{P}(\mathbb{G}_{n,m}^{(X)} = G_1 \mid \mathbb{G}_{n,m}^{(X)} \text{ is simple}) = \mathbb{P}(\mathbb{G}_{n,m}^{(X)} = G_2 \mid \mathbb{G}_{n,m}^{(X)} \text{ is simple}), \tag{1.14}$$

for $X = A, B$.

This is because for $i = 1, 2$,

$$\mathbb{P}(\mathbb{G}_{n,m}^{(A)} = G_i) = \frac{m!2^m}{n^{2m}} \text{ and } \mathbb{P}(\mathbb{G}_{n,m}^{(B)} = G_i) = \frac{m!2^m}{\binom{n}{2}^m 2^m}.$$

Indeed, we can permute the edges in $m!$ ways and permute the vertices within edges in 2^m ways without changing the underlying graph. This relies on $\mathbb{G}_{n,m}^{(X)}$ being simple.

Secondly, if $m = cn$ for a constant $c > 0$ then with $N = \binom{n}{2}$, and using Lemma 21.2(d),

$$\mathbb{P}(\mathbb{G}_{n,m}^{(X)} \text{ is simple}) \geq \binom{N}{m} \frac{m!2^m}{n^{2m}} \geq$$

$$(1 - o(1)) \frac{N^m}{m!} \exp\left\{-\frac{m^2}{2N} - \frac{m^3}{6N^2}\right\} \frac{m!2^m}{n^{2m}}$$

$$= (1 - o(1)) e^{-(c^2 + c)}. \tag{1.15}$$

It follows that if $\mathscr{P}$ is some graph property then

$$\mathbb{P}(G_{n,m} \in \mathscr{P}) = \mathbb{P}(\mathbb{G}^*_{n,m} \in \mathscr{P} \mid \mathbb{G}^*_{n,m} \text{ is simple}) \leq \tag{1.16}$$

$$(1+o(1))e^{c^2+c}\,\mathbb{P}(\mathbb{G}^{(X)}_{n,m} \in \mathscr{P}). \tag{1.17}$$

Here we have used the inequality $\mathbb{P}(A \mid B) \leq \mathbb{P}(A)/\mathbb{P}(B)$ for events A, B.

We will use this model a couple of times and (1.16) shows that if $\mathbb{P}(\mathbb{G}^{(X)}_{n,m} \in \mathscr{P}) = o(1)$ then $\mathbb{P}(\mathbb{G}_{n,m} \in \mathscr{P}) = o(1)$, for $m = O(n)$.

Model $\mathbb{G}^{(A)}_{n,m}$ was introduced independently by Bollobás and Frieze [140] and by Chvátal [187].

1.4 Exercises

We point out here that in the following exercises, we have not asked for best possible results. These exercises are for practise. You will need to use the inequalities from Section 21.1.

1.4.1 Suppose that $p = d/n$ where $d = o(n^{1/3})$. Show that w.h.p. $G_{n,p}$ has no copies of K_4.

1.4.2 Suppose that $p = d/n$ where $d > 1$. Show that w.h.p. $G_{n,p}$ contains an *induced* path of length $(\log n)^{1/2}$.

1.4.3 Suppose that $p = d/n$ where $d = O(1)$. Prove that for all $S \subseteq [n], |S| \leq n/\log n$, we have $e(S) \leq 2|S|$, where $e(S)$ is the number of edges contained in S.

1.4.4 Suppose that $p = \log n/n$. Let a vertex of $G_{n,p}$ be small if its degree is less than $\log n/100$. Show that w.h.p. there is no edge of $G_{n,p}$ joining two small vertices.

1.4.5 Suppose that $p = d/n$ where d is constant. Prove that w.h.p. no vertex belongs to more than one triangle.

1.4.6 Suppose that $p = d/n$ where d is constant. Prove that w.h.p. $G_{n,p}$ contains a vertex of degree at least $(\log n)^{1/2}$.

1.4.7 Suppose that $k \geq 3$ is constant and that $np \to \infty$. Show that w.h.p. $\mathbb{G}_{n,p}$ contains a copy of the k-cycle, C_k.

1.4.8 Suppose that $0 < p < 1$ is constant. Show that w.h.p. $G_{n,p}$ has diameter two.

1.4.9 Let $f : [n] \to [n]$ be chosen uniformly at random from all n^n functions from $[n] \to [n]$. Let $X = \{j : \nexists i \text{ s.t. } f(i) = j\}$. Show that w.h.p. $|X| \approx e^{-1}n$.

1.4.10 Prove Theorem 1.4.

1.4.11 Show that conditional on the value of m^{X*} that $\mathbb{G}_{n,m}^{X*}$ is distributed as $\mathbb{G}_{n,m^*}$, where $X = A, B$.

1.5 Notes

Friedgut and Kalai [316] and Friedgut [314] and Bourgain [154] and Bourgain and Kalai [153] provide much greater insight into the notion of sharp thresholds. Friedgut [315] gives a survey of these aspects. For a graph property $\mathscr{A}$ let $\mu(p,\mathscr{A})$ be the probability that the random graph $\mathbb{G}_{n,p}$ has property $\mathscr{A}$. A threshold is *coarse* if it is not sharp. We can identify coarse thresholds with $p\frac{d\mu(p,\mathscr{A})}{dp} < C$ for some absolute constant $0 < C$. The main insight into coarse thresholds is that to exist, the occurrence of $\mathscr{A}$ can in the main be attributed to the existence of one of a bounded number of small subgraphs. For example, Theorem 2.1 of [315] states that there exists a function $K(C,\varepsilon)$ such that the following holds. Let $\mathscr{A}$ be a monotone property of graphs that is invariant under automorphism and assume that $p\frac{d\mu(p,\mathscr{A})}{dp} < C$ for some constant $0 < C$. Then for every $\varepsilon > 0$ there exists a finite list of graphs $G_1, G_2, \ldots, G_m$ all of which have no more than $K(\varepsilon, C)$ edges, such that if $\mathscr{B}$ is the family of graphs having one of these graphs as a subgraph then $\mu(p, \mathscr{A}\Delta\mathscr{B}) \leq \varepsilon$.

2

Evolution

Here begins our story of the typical growth of a random graph. All the results up to Section 2.3 were first proved in a landmark paper by Erdős and Rényi [276]. The notion of the *evolution* of a random graph stems from a dynamic view of a *graph process*: viz. a sequence of graphs:

$$\mathbb{G}_0 = ([n], \emptyset), \mathbb{G}_1, \mathbb{G}_2, \ldots, \mathbb{G}_m, \ldots, \mathbb{G}_N = K_n,$$

where $\mathbb{G}_{m+1}$ is obtained from $\mathbb{G}_m$ by adding a random edge e_m. We see that there are $\binom{n}{2}!$ such sequences and $\mathbb{G}_m$ and $\mathbb{G}_{n,m}$ have the same distribution.

In the process of the evolution of a random graph we consider the properties possessed by $\mathbb{G}_m$ or $\mathbb{G}_{n,m}$ w.h.p., when $m = m(n)$ grows from 0 to $\binom{n}{2}$, while in the case of $\mathbb{G}_{n,p}$ we analyze its typical structure when $p = p(n)$ grows from 0 to 1 as $n \to \infty$.

In this chapter we mainly explore how the typical component structure evolves as the number of edges m increases.

2.1 Sub-Critical Phase

The evolution of Erdős–Rényi type random graphs has clearly distinguishable phases. The first phase, at the beginning of the evolution, can be described as a period when a random graph is a collection of small components that are mostly trees. Indeed the first result in this section shows that a random graph $\mathbb{G}_{n,m}$ is w.h.p. a collection of tree-components as long as $m = o(n)$ or, equivalently, as long as $p = o(n^{-1})$ in $\mathbb{G}_{n,p}$. For clarity, all results presented in this chapter are stated in terms of $\mathbb{G}_{n,m}$. Due to the fact that computations are much easier for $\mathbb{G}_{n,p}$ we first prove results in this model and then the results for $\mathbb{G}_{n,m}$ follow by the equivalence established either in Lemmas 1.2 and 1.3

or in Theorem 1.4. We also assume, throughout this chapter, that $\omega = \omega(n)$ is a function growing slowly with n, e.g. $\omega = \log\log n$ will suffice.

Theorem 2.1 *If $m \ll n$, then $\mathbb{G}_m$ is a forest w.h.p.*

Proof Suppose $m = n/\omega$ and let $N = \binom{n}{2}$, so $p = m/N \le 3/(\omega n)$. Let X be the number of cycles in $\mathbb{G}_{n,p}$. Then

$$\begin{aligned}\mathbb{E}X &= \sum_{k=3}^{n} \binom{n}{k} \frac{(k-1)!}{2} p^k \\ &\le \sum_{k=3}^{n} \frac{n^k}{2k} \frac{3^k}{\omega^k n^k} \\ &= O(\omega^{-3}) \to 0.\end{aligned}$$

Therefore, by the First Moment Method, (see Lemma 20.2),

$$\mathbb{P}(\mathbb{G}_{n,p} \text{ is not a forest}) = \mathbb{P}(X \ge 1) \le \mathbb{E}X = o(1),$$

which implies that

$$\mathbb{P}(\mathbb{G}_{n,p} \text{ is a forest}) \to 1 \text{ as } n \to \infty.$$

Notice that the property that a graph is a forest is monotone decreasing, so by Lemma 1.3

$$\mathbb{P}(\mathbb{G}_m \text{ is a forest}) \to 1 \text{ as } n \to \infty.$$

(Note that we have actually used Lemma 1.3 to show that $\mathbb{P}(\mathbb{G}_{n,p}$ is not a forest) $= o(1)$ implies that $\mathbb{P}(G_m$ is not a forest) $= o(1)$.) □

We next examine the time during which the components of G_m are isolated vertices and single edges only, w.h.p.

Theorem 2.2 *If $m \ll n^{1/2}$ then $\mathbb{G}_m$ is the union of isolated vertices and edges w.h.p.*

Proof Let $p = m/N$, $m = n^{1/2}/\omega$ and let X be the number of paths of length two in the random graph $\mathbb{G}_{n,p}$. By the First Moment Method,

$$\mathbb{P}(X > 0) \le \mathbb{E}X = 3\binom{n}{3} p^2 \le \frac{n^4}{2N^2\omega^2} \to 0,$$

as $n \to \infty$. Hence

$$\mathbb{P}(\mathbb{G}_{n,p} \text{ contains a path of length two}) = o(1).$$

Notice that the property that a graph contains a path of length two is monotone increasing, so by Lemma 1.3,

$$\mathbb{P}(\mathbb{G}_m \text{ contains a path of length two}) = o(1),$$

and the theorem follows. □

Now we are ready to describe the next step in the evolution of $\mathbb{G}_m$.

Theorem 2.3 *If $m \gg n^{1/2}$, then $\mathbb{G}_m$ contains a path of length two w.h.p.*

Proof Let $p = \frac{m}{N}, m = \omega n^{1/2}$ and X be the number of paths of length two in $\mathbb{G}_{n,p}$. Then

$$\mathbb{E}X = 3\binom{n}{3}p^2 \approx 2\omega^2 \to \infty,$$

as $n \to \infty$. This, however, does not imply that $X > 0$ w.h.p.! To show that $X > 0$ w.h.p. we apply the Second Moment Method.

Let $\mathscr{P}_2$ be the set of all paths of length two in the complete graph K_n, and let $\hat{X}$ be the number of *isolated* paths of length two in $\mathbb{G}_{n,p}$, i.e. paths that are also components of $\mathbb{G}_{n,p}$. We show that w.h.p. $\mathbb{G}_{n,p}$ contains such an isolated path. Now,

$$\hat{X} = \sum_{P \in \mathscr{P}_2} I_{P \subseteq_i \mathbb{G}_{n,p}}.$$

We always use $I_{\mathscr{E}}$ to denote the indicator for an event $\mathscr{E}$. The notation $\subseteq_i$ indicates that P is contained in $\mathbb{G}_{n,p}$ as a component (i.e. P is isolated). Having a path of length two is a monotone increasing property. Therefore we can assume that $m = o(n)$ and so $np = o(1)$ and the result for larger m follows from monotonicity and coupling. Then

$$\begin{aligned}\mathbb{E}\hat{X} &= 3\binom{n}{3}p^2(1-p)^{3(n-3)+1} \\ &\geq (1-o(1))\frac{n^3}{2}\frac{4\omega^2 n}{n^4}(1-3np) \to \infty,\end{aligned}$$

as $n \to \infty$.

In order to compute the second moment of the random variable $\hat{X}$ notice that,

$$\hat{X}^2 = \sum_{P \in \mathscr{P}_2}\sum_{Q \in \mathscr{P}_2} I_{P \subseteq_i \mathbb{G}_{n,p}} I_{Q \subseteq_i \mathbb{G}_{n,p}} = \sum\nolimits^*_{P,Q \in \mathscr{P}_2} I_{P \subseteq_i \mathbb{G}_{n,p}} I_{Q \subseteq_i \mathbb{G}_{n,p}},$$

where the last sum is taken over $P, Q \in \mathscr{P}_2$ such that either $P = Q$ or P and Q are vertex disjoint. The simplification that provides the last summation is precisely the reason that we introduce path-components (isolated paths). Now

$$\mathbb{E}\hat{X}^2 = \sum_P \left\{ \sum_Q \mathbb{P}(Q \subseteq_i \mathbb{G}_{n,p} | \, P \subseteq_i \mathbb{G}_{n,p}) \right\} \mathbb{P}(P \subseteq_i \mathbb{G}_{n,p}).$$

The expression inside the brackets is the same for all P and so

$$\mathbb{E}\hat{X}^2 = \mathbb{E}\hat{X} \left(1 + \sum_{Q \cap P_{(1,2,3)} = \emptyset} \mathbb{P}(Q \subseteq_i \mathbb{G}_{n,p} | \, P_{(1,2,3)} \subseteq_i \mathbb{G}_{n,p}) \right),$$

where $P_{(1,2,3)}$ denotes the path on vertex set $[3] = (1,2,3)$ with middle vertex 2. By conditioning on the event $P_{(1,2,3)} \subseteq_i \mathbb{G}_{n,p}$, i.e. assuming that $P_{(1,2,3)}$ is a component of $\mathbb{G}_{n,p}$, we see that all of the nine edges between Q and $P_{(1,2,3)}$ must be missing. Therefore

$$\mathbb{E}\hat{X}^2 \leq \mathbb{E}\hat{X} \left(1 + 3\binom{n}{3} p^2 (1-p)^{3(n-6)+1} \right) \leq \mathbb{E}\hat{X} \left(1 + (1-p)^{-9} \mathbb{E}\hat{X} \right).$$

So, by the Second Moment Method (see Lemma 20.5),

$$\begin{aligned} \mathbb{P}(\hat{X} > 0) &\geq \frac{(\mathbb{E}\hat{X})^2}{\mathbb{E}\hat{X}^2} \geq \frac{(\mathbb{E}\hat{X})^2}{\mathbb{E}\hat{X}\left(1 + (1-p)^{-9}\mathbb{E}\hat{X}\right)} \\ &= \frac{1}{(1-p)^{-9} + [\mathbb{E}\hat{X}]^{-1}} \to 1 \end{aligned}$$

as $n \to \infty$, since $p \to 0$ and $\mathbb{E}\hat{X} \to \infty$. Thus

$$\mathbb{P}(\mathbb{G}_{n,p} \text{ contains an isolated path of length two}) \to 1,$$

which implies that $\mathbb{P}(\mathbb{G}_{n,p}$ contains a path of length two$) \to 1$. As the property of having a path of length two is monotone increasing it in turn implies that

$$\mathbb{P}(\mathbb{G}_m \text{ contains a path of length two}) \to 1$$

for $m \gg n^{1/2}$ and the theorem follows. □

From Theorems 2.2 and 2.3 we obtain the following corollary.

Corollary 2.4 *The function $m^*(n) = n^{1/2}$ is the threshold for the property that a random graph $\mathbb{G}_m$ contains a path of length two, i.e.*

$$\mathbb{P}(\mathbb{G}_m \textit{ contains a path of length two}) = \begin{cases} o(1) & \textit{if } m \ll n^{1/2}. \\ 1 - o(1) & \textit{if } m \gg n^{1/2}. \end{cases}$$

As we keep adding edges, trees on more than three vertices start to appear. Note that isolated vertices, edges and paths of length two are also trees on one, two and three vertices, respectively. The next two theorems show how long we have to "wait" until trees with a given number of vertices appear w.h.p.

Theorem 2.5 *Fix $k \geq 3$. If $m \ll n^{\frac{k-2}{k-1}}$, then w.h.p. $\mathbb{G}_m$ contains no tree with k vertices.*

Proof Let $m = n^{\frac{k-2}{k-1}}/\omega$ and then $p = \frac{m}{N} \approx \frac{2}{\omega n^{k/(k-1)}} \leq \frac{3}{\omega n^{k/(k-1)}}$. Let X_k denote the number of trees with k vertices in $\mathbb{G}_{n,p}$. Let $T_1, T_2, \ldots, T_M$ be an enumeration of the copies of k-vertex trees in K_n. Let

$$A_i = \{T_i \text{ occurs as a subgraph in } \mathbb{G}_{n,p}\}.$$

The probability that a tree T occurs in $\mathbb{G}_{n,p}$ is $p^{e(T)}$, where $e(T)$ is the number of edges of T. So,

$$\mathbb{E}X_k = \sum_{t=1}^{M} \mathbb{P}(A_t) = Mp^{k-1}.$$

But $M = \binom{n}{k}k^{k-2}$ since one can choose a set of k vertices in $\binom{n}{k}$ ways and then by Cayley's formula choose a tree on these vertices in k^{k-2} ways. Hence

$$\mathbb{E}X_k = \binom{n}{k}k^{k-2}p^{k-1}. \tag{2.1}$$

Noting also that (see Lemma 21.1(c))

$$\binom{n}{k} \leq \left(\frac{ne}{k}\right)^k,$$

we see that

$$\begin{aligned}
\mathbb{E}X_k &\leq \left(\frac{ne}{k}\right)^k k^{k-2} \left(\frac{3}{\omega n^{k/(k-1)}}\right)^{k-1} \\
&= \frac{3^{k-1}e^k}{k^2\omega^{k-1}} \to 0,
\end{aligned}$$

as $n \to \infty$, seeing as k is fixed.

Thus we see by the First Moment Method that,

$$\mathbb{P}(\mathbb{G}_{n,p} \text{ contains a tree with } k \text{ vertices}) \to 0.$$

This property is monotone increasing and therefore

$$\mathbb{P}(\mathbb{G}_m \text{ contains a tree with } k \text{ vertices}) \to 0.$$

□

Let us check what happens if the number of edges in $\mathbb{G}_m$ is much larger than $n^{\frac{k-2}{k-1}}$.

Theorem 2.6 *Fix $k \geq 3$. If $m \gg n^{\frac{k-2}{k-1}}$, then w.h.p. $\mathbb{G}_m$ contains a copy of every fixed tree with k vertices.*

Proof Let $p = \frac{m}{N}, m = \omega n^{\frac{k-2}{k-1}}$, where $\omega = o(\log n)$ and fix some tree T with k vertices. Denote by $\hat{X}_k$ the number of *isolated* copies of T (T-components) in $\mathbb{G}_{n,p}$. Let $aut(T)$ denote the number of automorphisms of a graph H. Note that there are $k!/aut(T)$ copies of T in the complete graph K_k. To see this, choose a copy of T with vertex set $[k]$. There are $k!$ ways of mapping the vertices of T to the vertices of K_k. Each map f induces a copy of T and two maps f_1, f_2 induce the same copy if and only if $f_2 f_1^{-1}$ is an automorphism of T.

So,

$$\mathbb{E}\hat{X}_k = \binom{n}{k} \frac{k!}{aut(T)} p^{k-1} (1-p)^{k(n-k)+\binom{k}{2}-k+1} \tag{2.2}$$

$$= (1+o(1)) \frac{(2\omega)^{k-1}}{aut(T)} \to \infty.$$

In (2.2) we have used the fact that $\omega = o(\log n)$ to show that $(1-p)^{k(n-k)+\binom{k}{2}-k+1} = 1+o(1)$.

Next let $\mathscr{T}$ be the set of copies of T in K_n and $T_{[k]}$ be a fixed copy of T on vertices $[k]$ of K_n. Then, arguing as in Theorem 2.3,

$$\mathbb{E}(\hat{X}_k^2) = \sum_{T_1, T_2 \in \mathscr{T}} \mathbb{P}(T_2 \subseteq_i \mathbb{G}_{n,p} \mid T_1 \subseteq_i \mathbb{G}_{n,p}) \mathbb{P}(T_1 \subseteq_i \mathbb{G}_{n,p})$$

$$= \mathbb{E}\hat{X}_k \left(1 + \sum_{\substack{T_2 \in \mathscr{T} \\ V(T_2) \cap [k] = \emptyset}} \mathbb{P}(T_2 \subseteq_i \mathbb{G}_{n,p} \mid T_{[k]} \subseteq_i \mathbb{G}_{n,p}) \right)$$

$$\leq \mathbb{E}\hat{X}_k \left(1 + (1-p)^{-k^2} \mathbb{E}X_k \right).$$

Notice that the $(1-p)^{-k^2}$ factor comes from conditioning on the event $T_{[k]} \subseteq_i \mathbb{G}_{n,p}$, which forces the non-existence of fewer than k^2 edges.

Hence, by the Second Moment Method,

$$\mathbb{P}(\hat{X}_k > 0) \geq \frac{(\mathbb{E}\hat{X}_k)^2}{\mathbb{E}\hat{X}_k \left(1 + (1-p)^{-k^2} \mathbb{E}\hat{X}_k \right)} \to 1.$$

Then, by a similar reasoning to that in the proof of Theorem 2.3,

$$\mathbb{P}(\mathbb{G}_m \text{ contains a copy of } T) \to 1,$$

as $n \to \infty$. □

Combining the two above theorems we arrive at the following conclusion.

Corollary 2.7 *The function* $m^*(n) = n^{\frac{k-2}{k-1}}$ *is the threshold for the property that a random graph* $\mathbb{G}_m$ *contains a tree with* $k \geq 3$ *vertices, i.e.*

$$\mathbb{P}(\mathbb{G}_m \supseteq k\text{-vertex-tree}) = \begin{cases} o(1) & \text{if } m \ll n^{\frac{k-2}{k-1}} \\ 1 - o(1) & \text{if } m \gg n^{\frac{k-2}{k-1}}. \end{cases}$$

In the next theorem we show that "on the threshold" for k vertex trees, i.e. if $m = cn^{\frac{k-2}{k-1}}$, where c is a constant, $c > 0$, the number of tree components of a given order asymptotically follows the Poisson distribution. This time we formulate both the result and its proof in terms of $\mathbb{G}_m$.

Theorem 2.8 *If* $m = cn^{\frac{k-2}{k-1}}$*, where* $c > 0$*, and* T *is a fixed tree with* $k \geq 3$ *vertices, then*

$$\mathbb{P}(\mathbb{G}_m \text{ contains an isolated copy of tree } T) \to 1 - e^{-\lambda},$$

as $n \to \infty$*, where* $\lambda = \frac{(2c)^{k-1}}{aut(T)}$.

More precisely, the number of copies of T *is asymptotically distributed as the Poisson distribution with expectation* λ.

Proof Let $T_1, T_2, \ldots, T_M$ be an enumeration of the copies of some k vertex tree T in K_n.

Let

$$A_i = \{T_i \text{ occurs as a component in } \mathbb{G}_m\}.$$

Suppose $J \subseteq [M] = \{1, 2, \ldots, M\}$ with $|J| = t$, where t is fixed. Let $A_J = \bigcap_{j \in J} A_j$. We have $\mathbb{P}(A_J) = 0$ if there are $i, j \in J$ such that T_i, T_j share a vertex. Suppose $T_i, i \in J$ are vertex disjoint, then,

$$\mathbb{P}(A_J) = \frac{\binom{\binom{n-kt}{2}}{m-(k-1)t}}{\binom{N}{m}}.$$

Note that in the numerator we count the number of ways of choosing m edges so that A_J occurs.

If, say, $t \leq \log n$, then

$$\binom{n-kt}{2} = N\left(1 - \frac{kt}{n}\right)\left(1 - \frac{kt}{n-1}\right) = N\left(1 - O\left(\frac{kt}{n}\right)\right),$$

and so

$$\frac{m^2}{\binom{n-kt}{2}} \to 0.$$

Then from Lemma 21.1(d),

$$\binom{\binom{n-kt}{2}}{m-(k-1)t} = (1+o(1))\frac{\left(N\left(1-O\left(\frac{kt}{n}\right)\right)\right)^{m-(k-1)t}}{(m-(k-1)t)!}$$
$$= (1+o(1))\frac{N^{m-(k-1)t}\left(1-O\left(\frac{mkt}{n}\right)\right)}{(m-(k-1)t)!}$$
$$= (1+o(1))\frac{N^{m-(k-1)t}}{(m-(k-1)t)!}.$$

Similarly, again by Lemma 21.1,

$$\binom{N}{m} = (1+o(1))\frac{N^m}{m!},$$

and so

$$\mathbb{P}(A_J) = (1+o(1))\frac{m!}{(m-(k-1)t)!}N^{-(k-1)t} = (1+o(1))\left(\frac{m}{N}\right)^{(k-1)t}.$$

Thus, if Z_T denotes the number of components of $\mathbb{G}_m$ that are copies of T, then,

$$\mathbb{E}\binom{Z_T}{t} \approx \frac{1}{t!}\binom{n}{k,k,k,\ldots,k}\left(\frac{k!}{aut(T)}\right)^t\left(\frac{m}{N}\right)^{(k-1)t}$$
$$\approx \frac{n^{kt}}{t!(k!)^t}\left(\frac{k!}{aut(T)}\right)^t\left(\frac{cn^{(k-2)/(k-1)}}{N}\right)^{(k-1)t}$$
$$\approx \frac{\lambda^t}{t!},$$

where

$$\lambda = \frac{(2c)^{k-1}}{aut(T)}.$$

So by Corollary 20.11, the number of copies of T is asymptotically distributed as the Poisson distribution with expectation λ given above, which combined with the statements of Theorem 2.1 and Corollary 2.7 proves the theorem. Note that Theorem 2.7 implies that w.h.p. there are no non-component copies of T. □

We complete our presentation of the basic features of a random graph in its sub-critical phase of evolution with a description of the order of its largest component.

Theorem 2.9 *If $m = \frac{1}{2}cn$, where $0 < c < 1$ is a constant, then the order of the largest component of a random graph $\mathbb{G}_m$ is $O(\log n)$.*

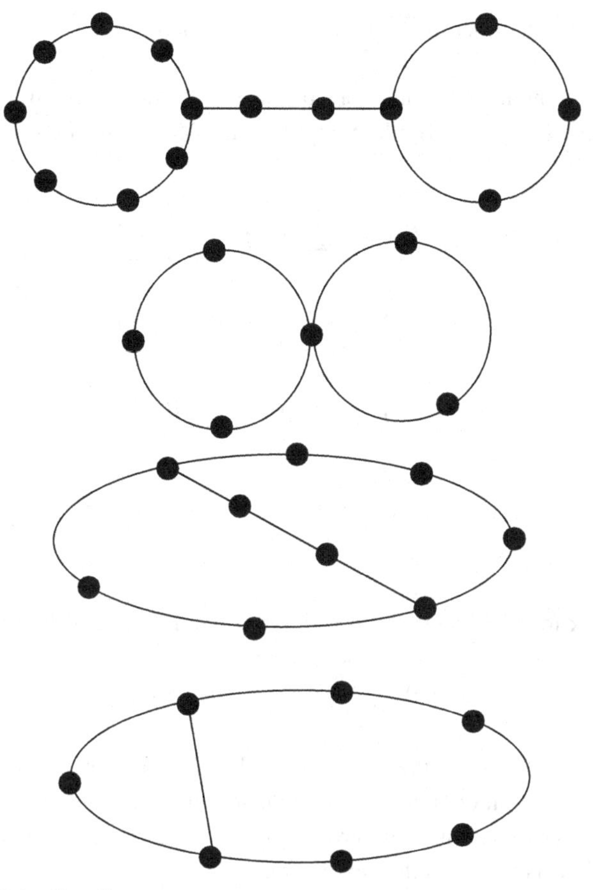

Figure 2.1. $C_1 \cup C_2$

The above theorem follows from the next three lemmas stated and proved in terms of $\mathbb{G}_{n,p}$ with $p = c/n,\ 0 < c < 1$. In fact the first of these three lemmas covers a little bit more than the case of $p = c/n,\ 0 < c < 1$.

Lemma 2.10 *If* $p \leq \frac{1}{n} - \frac{\omega}{n^{4/3}}$, *where* $\omega = \omega(n) \to \infty$, *then w.h.p. every component in* $\mathbb{G}_{n,p}$ *contains at most one cycle.*

Proof Suppose that there is a pair of cycles that are in the same component. If such a pair exists then there is *minimal* pair C_1, C_2, i.e. either C_1 and C_2 are connected by a path (or meet at a vertex) or they form a cycle with a diagonal path (see Figure 2.1). Then in either case, $C_1 \cup C_2$ consists of a path P plus another two distinct edges, one from each endpoint of P joining it to another vertex in P. The number of such graphs on k labeled vertices can be bounded by $k^2 k!$.

Let X be the number of subgraphs of the above kind (shown in Figure 2.1) in the random graph $\mathbb{G}_{n,p}$. By the First Moment Method (see Lemma 20.2),

$$\begin{aligned}
\mathbb{P}(X > 0) \leq \mathbb{E}X &\leq \sum_{k=4}^{n} \binom{n}{k} k^2 k! p^{k+1} \\
&\leq \sum_{k=4}^{n} \frac{n^k}{k!} k^2 k! \frac{1}{n^{k+1}} \left(1 - \frac{\omega}{n^{1/3}}\right)^{k+1} \\
&\leq \int_0^\infty \frac{x^2}{n} \exp\left(-\frac{\omega x}{n^{1/3}}\right) dx \\
&= \frac{2}{\omega^3} \\
&= o(1).
\end{aligned} \tag{2.3}$$

□

We remark for later use that if $p = c/n$, $0 < c < 1$ then (2.3) implies

$$\mathbb{P}(X > 0) \leq \sum_{k=4}^{n} k^2 c^{k+1} n^{-1} = O(n^{-1}). \tag{2.4}$$

Hence, in determining the order of the largest component we may concentrate our attention on unicyclic components and tree-components (isolated trees). However the number of vertices on unicyclic components tends to be rather small, as is shown in the next lemma.

Lemma 2.11 *If $p = c/n$, where $0 < c < 1$ is a constant, then in $\mathbb{G}_{n,p}$ w.h.p. the number of vertices in components with exactly one cycle, is $o(\omega)$ for any growing function ω.*

Proof Let X_k be the number of vertices on unicyclic components with k vertices. Then

$$\mathbb{E}X_k \leq \binom{n}{k} k^{k-2} \binom{k}{2} p^k (1-p)^{k(n-k)+\binom{k}{2}-k}. \tag{2.5}$$

The factor $k^{k-2}\binom{k}{2}$ in (2.5) is the number of choices for a tree plus an edge on k vertices in $[k]$. This bounds the number $C(k,k)$ of graphs on $[k]$ with k edges. This is off by a factor $O(k^{1/2})$ from the exact formula, which is given below for completeness, it is due to Rényi [643]:

$$C(k,k) = \sum_{r=3}^{k} \binom{k}{r} \frac{(r-1)!}{2} r k^{k-r-1} \approx \sqrt{\frac{\pi}{8}} k^{k-1/2}. \tag{2.6}$$

The remaining factor, other than $\binom{n}{k}$, in (2.5) is the probability that the k edges of the unicyclic component exist and that there are no other edges on $\mathbb{G}_{n,p}$ incident with the k chosen vertices.

Noting also that by Lemma 21.1(d),

$$\binom{n}{k} \leq \frac{n^k}{k!} e^{-\frac{k(k-1)}{2n}},$$

and so we get

$$\begin{aligned}\mathbb{E} X_k &\leq \frac{n^k}{k!} e^{-\frac{k(k-1)}{2n}} k^k \frac{c^k}{n^k} e^{-ck+\frac{ck(k-1)}{2n}+\frac{ck}{2n}} \\ &\leq \frac{e^k}{k^k} e^{-\frac{k(k-1)}{2n}} k^k c^k e^{-ck+\frac{k(k-1)}{2n}+\frac{c}{2}} \\ &= \left(ce^{1-c}\right)^k e^{\frac{c}{2}}.\end{aligned}$$

So,

$$\mathbb{E} \sum_{k=3}^{n} X_k \leq \sum_{k=3}^{n} \left(ce^{1-c}\right)^k e^{\frac{c}{2}} = O(1), \tag{2.7}$$

since $ce^{1-c} < 1$ for $c \neq 1$. By Markov's inequality, if $\omega = \omega(n) \to \infty$, (see Lemma 20.1)

$$\mathbb{P}\left(\sum_{k=3}^{n} X_k \geq \omega\right) = O\left(\frac{1}{\omega}\right) \to 0 \text{ as } n \to \infty,$$

and the Lemma follows. □

After proving the first two lemmas one can easily see that the only remaining candidate for the largest component of our random graph is an isolated tree.

Lemma 2.12 *Let $p = \frac{c}{n}$, where $c \neq 1$ is a constant, $\alpha = c - 1 - \log c$, and $\omega = \omega(n) \to \infty$, $\omega = o(\log\log n)$. Then*

(i) w.h.p. there exists an isolated tree of order

$$k_- = \frac{1}{\alpha}\left(\log n - \frac{5}{2}\log\log n\right) - \omega,$$

(ii) w.h.p. there is no *isolated tree of order at least*

$$k_+ = \frac{1}{\alpha}\left(\log n - \frac{5}{2}\log\log n\right) + \omega.$$

Proof Note that our assumption on c means that α is a positive constant.

Let X_k be the number of isolated trees of order k, then,

$$\mathbb{E} X_k = \binom{n}{k} k^{k-2} p^{k-1} (1-p)^{k(n-k)+\binom{k}{2}-k+1}. \tag{2.8}$$

To prove *(i)*, suppose $k = O(\log n)$, then by using Lemma 21.1(a), (b) and Stirlings approximation (1.4) for $k!$ we see that

$$\begin{aligned}
\mathbb{E}X_k &= (1+o(1))\frac{n}{c}\frac{k^{k-2}}{k!}(ce^{-c})^k \\
&= \frac{(1+o(1))}{c\sqrt{2\pi}}\frac{n}{k^{5/2}}(ce^{1-c})^k \\
&= \frac{(1+o(1))}{c\sqrt{2\pi}}\frac{n}{k^{5/2}}e^{-\alpha k}.
\end{aligned} \tag{2.9}$$

Putting $k = k_-$ we see that

$$\mathbb{E}X_k = \frac{(1+o(1))}{c\sqrt{2\pi}}\frac{n}{k^{5/2}}\frac{e^{\alpha\omega}(\log n)^{5/2}}{n} \geq Ae^{\alpha\omega}, \tag{2.10}$$

for some constant $A > 0$.

We continue via the Second Moment Method, this time using the Chebyshev inequality as we need a little extra precision for the proof of Theorem 2.14. Using essentially the same argument as for a fixed tree T of order k (see Theorem 2.6), we obtain

$$\mathbb{E}X_k^2 \leq \mathbb{E}X_k\left(1+(1-p)^{-k^2}\,\mathbb{E}X_k\right).$$

So

$$\begin{aligned}
\operatorname{Var} X_k &\leq \mathbb{E}X_k + (\mathbb{E}X_k)^2\left((1-p)^{-k^2}-1\right) \\
&\leq \mathbb{E}X_k + 2ck^2(\mathbb{E}X_k)^2/n.
\end{aligned}$$

Thus, by the Chebyshev inequality (see Lemma 20.3), we see that for any fixed $\varepsilon > 0$,

$$\mathbb{P}(|X_k - \mathbb{E}X_k| \geq \varepsilon\,\mathbb{E}X_k) \leq \frac{1}{\varepsilon^2\,\mathbb{E}X_k} + \frac{2ck^2}{\varepsilon^2 n} = o(1). \tag{2.11}$$

Thus w.h.p. $X_k \geq Ae^{\omega}/2$ and this completes the proof of *(i)*.

For *(ii)* we go back to the formula (2.8) and write, for some new constant $A > 0$,

$$\begin{aligned}
\mathbb{E}X_k &\leq \frac{A}{\sqrt{k}}\left(\frac{ne}{k}\right)^k k^{k-2}\left(1-\frac{k}{2n}\right)^{k-1}\left(\frac{c}{n}\right)^{k-1} e^{-ck+\frac{ck^2}{2n}} \\
&\leq \frac{2An}{\widehat{c}_k k^{5/2}}\left(\widehat{c}_k e^{1-\widehat{c}_k}\right)^k,
\end{aligned}$$

where $\widehat{c}_k = c\left(1-\frac{k}{2n}\right)$.

In the case $c < 1$ we have $\widehat{c}_k e^{1-\widehat{c}_k} \leq ce^{1-c}$ and $\widehat{c}_k \approx c$ and so we can write

$$\sum_{k=k_+}^{n} \mathbb{E}X_k \leq \frac{3An}{c} \sum_{k=k_+}^{n} \frac{\left(ce^{1-c}\right)^k}{k^{5/2}} \leq \frac{3An}{ck_+^{5/2}} \sum_{k=k_+}^{\infty} e^{-\alpha k}$$

$$= \frac{3Ane^{-\alpha k_+}}{ck_+^{5/2}(1-e^{-\alpha})} = \frac{(3A+o(1))\alpha^{5/2}e^{-\alpha\omega}}{c(1-e^{-\alpha})} = o(1). \tag{2.12}$$

If $c > 1$ then for $k \leq \frac{n}{\log n}$ we use $\widehat{c}_k e^{1-\widehat{c}_k} = e^{-\alpha+O(1/\log n)}$ and for $k > \frac{n}{\log n}$ we use $c_k \geq c/2$ and $\widehat{c}_k e^{1-\widehat{c}_k} \leq 1$ and replace (2.12) by

$$\sum_{k=k_+}^{n} \mathbb{E}X_k \leq \frac{3An}{ck_+^{5/2}} \sum_{k=k_+}^{n/\log n} e^{-(\alpha+O(1/\log n))k} + \frac{6An}{c} \sum_{k=n/\log n}^{n} \frac{1}{k^{5/2}} = o(1).$$

□

We remark that (2.9) gives the expected number of trees of a given size and Pittel [623] proved a central limit theorem for this number. Finally, applying Lemmas 2.11 and 2.12 we can prove the following useful identity: suppose that $x = x(c)$ is given as

$$x = x(c) = \begin{cases} c & c \leq 1 \\ \text{The solution in } (0,1) \text{ to } xe^{-x} = ce^{-c} & c > 1. \end{cases}$$

Note that xe^{-x} increases continuously as x increases from 0 to 1 and then decreases. This justifies the existence and uniqueness of x.

Lemma 2.13 *If $p = \frac{c}{n}$ and $c > 0$, $c \neq 1$ is a constant, and $x = x(c)$ is defined above, then*

$$\frac{1}{x}\sum_{k=1}^{\infty} \frac{k^{k-1}}{k!} \left(ce^{-c}\right)^k = 1.$$

Proof Assume first that $c < 1$ and let X be the total number of vertices of $\mathbb{G}_{n,p}$ that lie in non-tree components. Let X_k be the number of tree-components of order k, then,

$$n = \sum_{k=1}^{n} kX_k + X,$$

so,

$$n = \sum_{k=1}^{n} k\mathbb{E}X_k + \mathbb{E}X.$$

Now,

(i) by (2.4) and (2.7), $\mathbb{E}X = O(1)$,

(ii) by (2.9), if $k < k_+$ then

$$\mathbb{E}X_k = (1+o(1))\frac{n}{ck!}k^{k-2}\left(ce^{-c}\right)^k.$$

So, by Lemma 2.12,

$$\begin{aligned} n &= o(n) + \frac{n}{c}\sum_{k=1}^{k_+}\frac{k^{k-1}}{k!}\left(ce^{-c}\right)^k \\ &= o(n) + \frac{n}{c}\sum_{k=1}^{\infty}\frac{k^{k-1}}{k!}\left(ce^{-c}\right)^k. \end{aligned}$$

Now divide through by n and let $n \to \infty$.

This proves the identity for the case $c < 1$. Suppose now that $c > 1$. Then, since x is a solution of equation $ce^{-c} = xe^{-x}$, $0 < x < 1$, we have

$$\sum_{k=1}^{\infty}\frac{k^{k-1}}{k!}\left(ce^{-c}\right)^k = \sum_{k=1}^{\infty}\frac{k^{k-1}}{k!}\left(xe^{-x}\right)^k = x,$$

by the first part of the proof (for $c < 1$). □

We note that in fact, Lemma 2.13 is also true for $c = 1$.

2.2 Super-Critical Phase

The structure of a random graph $\mathbb{G}_m$ changes dramatically when $m = \frac{1}{2}cn$ where $c > 1$ is a constant. We give a precise characterization of this phenomenon, presenting results in terms of $\mathbb{G}_m$ and proving them for $\mathbb{G}_{n,p}$ with $p = c/n$, $c > 1$.

Theorem 2.14 *If $m = cn/2$, $c > 1$, then w.h.p. $\mathbb{G}_m$ consists of a unique giant component, with $\left(1 - \frac{x}{c} + o(1)\right)n$ vertices and $\left(1 - \frac{x^2}{c^2} + o(1)\right)\frac{cn}{2}$ edges. Here $0 < x < 1$ is the solution of the equation $xe^{-x} = ce^{-c}$. The remaining components are of order at most $O(\log n)$.*

Proof Suppose that Z_k is the number of components of order k in $\mathbb{G}_{n,p}$. Then, bounding the number of such components by the number of trees with k

vertices that span a component, we obtain

$$
\begin{aligned}
\mathbb{E} Z_k &\leq \binom{n}{k} k^{k-2} p^{k-1} (1-p)^{k(n-k)} \\
&\leq \frac{A}{\sqrt{k}} \left(\frac{ne}{k}\right)^k k^{k-2} \left(\frac{c}{n}\right)^{k-1} e^{-ck+ck^2/n} \\
&\leq \frac{An}{k^{5/2}} \left(ce^{1-c+ck/n}\right)^k .
\end{aligned}
\tag{2.13}
$$

Now let $\beta_1 = \beta_1(c)$ be small enough so that

$$
ce^{1-c+c\beta_1} < 1,
$$

and let $\beta_0 = \beta_0(c)$ be large enough so that

$$
\left(ce^{1-c+o(1)}\right)^{\beta_0 \log n} < \frac{1}{n^2}.
$$

If we choose β_1 and β_0 as above then it follows that w.h.p. there is no component of order $k \in [\beta_0 \log n, \beta_1 n]$.

Our next task is to estimate the number of vertices on small components, i.e. those of size at most $\beta_0 \log n$.

We first estimate the total number of vertices on small tree components, i.e. on isolated trees of order at most $\beta_0 \log n$.

Assume first that $1 \leq k \leq k_0$, where $k_0 = \frac{1}{2\alpha} \log n$, where α is from Lemma 2.12. It follows from (2.9) that

$$
\begin{aligned}
\mathbb{E}\left(\sum_{k=1}^{k_0} kX_k\right) &\approx \frac{n}{c} \sum_{k=1}^{k_0} \frac{k^{k-1}}{k!} \left(ce^{-c}\right)^k \\
&\approx \frac{n}{c} \sum_{k=1}^{\infty} \frac{k^{k-1}}{k!} \left(ce^{-c}\right)^k ,
\end{aligned}
$$

using $k^{k-1}/k! < e^k$, and $ce^{-c} < e^{-1}$ for $c \neq 1$ to extend the summation from k_0 to infinity.

Putting $\varepsilon = 1/\log n$ and using (2.11), we see that the probability that any X_k, $1 \leq k \leq k_0$, deviates from its mean by more than $1 \pm \varepsilon$ is at most

$$
\sum_{k=1}^{k_0} \left[\frac{(\log n)^2}{n^{1/2-o(1)}} + O\left(\frac{(\log n)^4}{n}\right) \right] = o(1),
$$

where the $n^{1/2-o(1)}$ term comes from putting $\omega \approx k_0/2$ in (2.10).

Thus, if $x = x(c)$, $0 < x < 1$ is the unique solution in $(0,1)$ of the equation $xe^{-x} = ce^{-c}$, then w.h.p.,

$$\sum_{k=1}^{k_0} kX_k \approx \frac{n}{c} \sum_{k=1}^{\infty} \frac{k^{k-1}}{k!} \left(xe^{-x}\right)^k = \frac{nx}{c},$$

by Lemma 2.13.

Now consider $k_0 < k \leq \beta_0 \log n$.

$$\mathbb{E}\left(\sum_{k=k_0+1}^{\beta_0 \log n} kX_k\right) \leq \frac{n}{c} \sum_{k=k_0+1}^{\beta_0 \log n} \left(ce^{1-c+ck/n}\right)^k = O\left(n(ce^{1-c})^{k_0}\right) = O\left(n^{1/2+o(1)}\right).$$

So, by the Markov inequality (see Lemma 20.1), w.h.p.,

$$\sum_{k=k_0+1}^{\beta_0 \log n} kX_k = o(n).$$

Now consider the number Y_k of non-tree components with k vertices, $1 \leq k \leq \beta_0 \log n$.

$$\mathbb{E}\left(\sum_{k=1}^{\beta_0 \log n} kY_k\right) \leq \sum_{k=1}^{\beta_0 \log n} \binom{n}{k} k^{k-1} \binom{k}{2} \left(\frac{c}{n}\right)^k \left(1 - \frac{c}{n}\right)^{k(n-k)} \leq \sum_{k=1}^{\beta_0 \log n} k\left(ce^{1-c+ck/n}\right)^k = O(1).$$

So, again by the Markov inequality, w.h.p.,

$$\sum_{k=1}^{\beta_0 \log n} kY_k = o(n).$$

Summarizing, we have proved so far that w.h.p. there are approximately $\frac{nx}{c}$ vertices on components of order k, where $1 \leq k \leq \beta_0 \log n$ and all the remaining *giant* components are of size at least $\beta_1 n$.

We complete the proof by showing the uniqueness of the giant component. Let

$$c_1 = c - \frac{\log n}{n} \quad \text{and} \quad p_1 = \frac{c_1}{n}.$$

Define p_2 by

$$1 - p = (1 - p_1)(1 - p_2)$$

and note that $p_2 \geq \frac{\log n}{n^2}$, then, see Section 1.2,

$$\mathbb{G}_{n,p} = \mathbb{G}_{n,p_1} \cup \mathbb{G}_{n,p_2}.$$

If $x_1 e^{-x_1} = c_1 e^{-c_1}$, then $x_1 \approx x$ and so, by our previous analysis, w.h.p., $\mathbb{G}_{n,p_1}$ has no components with number of vertices in the range $[\beta_0 \log n, \beta_1 n]$.

Suppose there are components $C_1, C_2, \ldots, C_l$ with $|C_i| > \beta_1 n$. Here $l \leq 1/\beta_1$. Now we add edges of $\mathbb{G}_{n,p_2}$ to $\mathbb{G}_{n,p_1}$, then

$$\begin{aligned}\mathbb{P}\left(\exists i,j : \text{ no } \mathbb{G}_{n,p_2} \text{ edge joins } C_i \text{ with } C_j\right) &\leq \binom{l}{2}(1-p_2)^{(\beta_1 n)^2} \\ &\leq l^2 e^{-\beta_1^2 \log n} \\ &= o(1).\end{aligned}$$

So w.h.p. $\mathbb{G}_{n,p}$ has a unique component with more than $\beta_0 \log n$ vertices and it has $\approx \left(1 - \frac{x}{c}\right) n$ vertices.

We now consider the number of edges in the giant C_0. Now we switch to $G = \mathbb{G}_{n,m}$. Suppose that the edges of G are $e_1, e_2, \ldots, e_m$ in random order. We estimate the probability that $e = e_1 = \{x, y\}$ is an edge of the giant. Let G_1 be the graph induced by $\{e_2, e_3, \ldots, e_m\}$. G_1 is distributed as $\mathbb{G}_{n,m-1}$ and so we know that w.h.p. G_1 has a unique giant C_1 and other components are of size $O(\log n)$. So the probability that e is an edge of the giant is $o(1)$ plus the probability that x or y is a vertex of C_1. Thus,

$$\begin{aligned}\mathbb{P}\left(e \notin C_0 \mid |C_1| \approx n\left(1 - \frac{x}{c}\right)\right) &= \mathbb{P}\left(e \cap C_1 = \emptyset \mid |C_1| \approx n\left(1 - \frac{x}{c}\right)\right) \\ &= \left(1 - \frac{|C_1|}{n}\right)\left(1 - \frac{|C_1| - 1}{n}\right) \approx \left(\frac{x}{c}\right)^2.\end{aligned} \tag{2.14}$$

It follows that the expected number of edges in the giant is as claimed. To prove concentration, it is simplest to use the Chebyshev inequality, see Lemma 20.3. So, now fix $i, j \leq m$ and let C_2 denote the unique giant component of $G_{n,p} - \{e_i, e_j\}$. Then, arguing as for (2.14),

$$\mathbb{P}(e_i, e_j \subseteq C_0) = o(1) + \mathbb{P}(e_j \cap C_2 \neq \emptyset \mid e_i \cap C_2 \neq \emptyset)\mathbb{P}(e_i \cap C_2 \neq \emptyset)$$
$$= (1 + o(1))\mathbb{P}(e_i \subseteq C_0)\mathbb{P}(e_j \subseteq C_0).$$

In the $o(1)$ term, we hide the probability of the event

$$\{e_i \cap C_2 \neq \emptyset, e_j \cap C_2 = \emptyset, e_i \cap e_j \neq \emptyset\}$$

which has probability $o(1)$. We should double this $o(1)$ probability here to account for switching the roles of i, j.

The Chebyshev inequality can now be used to show that the number of edges is concentrated as claimed. □

We will see later, see Lemma 2.17, that w.h.p. each of the small components have at most one cycle.

From the above theorem and the results of previous sections we see that, when $m = cn/2$ and c passes the critical value equal to 1, the typical structure of a random graph changes from a scattered collection of small trees and unicyclic components to a coagulated lump of components (the giant component) that dominates the graph. This short period when the giant component emerges is called the *phase transition*. We will look at this fascinating period of the evolution more closely in Section 2.3.

We know that w.h.p. the giant component of $\mathbb{G}_{n,m}, m = cn/2, c > 1$ has $\approx 1 - \frac{x}{c}$ vertices and $\approx \left(1 - \frac{x^2}{c^2}\right) \frac{cn}{2}$ edges. So, if we look at the graph H induced by the vertices outside the giant, then w.h.p. H has $\approx n_1 = \frac{nx}{c}$ vertices and $\approx m_1 = xn_1/2$ edges. Thus we should expect H to resemble $\mathbb{G}_{n_1.m_1}$, which is sub-critical since $x < 1$. This can be made precise, but the intuition is clear.

Now increase m further and look on the outside of the giant component. The giant component subsequently consumes the small components not yet attached to it. When m is such that $m/n \to \infty$ then unicyclic components disappear and a random graph $\mathbb{G}_m$ achieves the structure described in the next theorem.

Theorem 2.15 *Let $\omega = \omega(n) \to \infty$ as $n \to \infty$ be some slowly growing function. If $m \geq \omega n$ but $m \leq n(\log n - \omega)/2$, then $\mathbb{G}_m$ is disconnected and all components, with the exception of the giant, are trees w.h.p.*

Tree-components of order k die out in the reverse order they were born, i.e. larger trees are "swallowed" by the giant earlier than smaller ones.

Cores

Given a positive integer k, the k-core of a graph $G=(V,E)$ is the largest set $S\subseteq V$ such that the minimum degree δ_S in the vertex induced subgraph $G[S]$ is at least k. This is unique because if $\delta_S\geq k$ and $\delta_T\geq k$ then $\delta_{S\cup T}\geq k$. Cores were first discussed by Bollobás [129]. It was shown by Łuczak [539] that for $k\geq 3$ either there is no k-core in $G_{n,p}$ or one of linear size, w.h.p. The precise size and first occurrence of k-cores for $k\geq 3$ was established in Pittel, Spencer and Wormald [628]. The 2-core, C_2, which is the set of vertices that lie on at least one cycle, behaves differently to the other cores, $k\geq 3$. It grows gradually. We will need the following result in Section 16.2.

Lemma 2.16 *Suppose that $c>1$ and that $x<1$ is the solution to $xe^{-x}=ce^{-c}$. Then w.h.p. the 2-core C_2 of $G_{n,p}, p=c/n$ has $(1-x)\left(1-\frac{x}{c}+o(1)\right)n$ vertices and $\left(1-\frac{x}{c}+o(1)\right)^2\frac{cn}{2}$ edges.*

Proof Fix $v\in[n]$. We estimate $\mathbb{P}(v\in C_2)$. Let C_1 denote the unique giant component of $G_1=\mathbb{G}_{n,p}-v$. Now G_1 is distributed as $\mathbb{G}_{n-1,p}$ and so C_1 exists w.h.p. To be in C_2, either (i) v has two neighbors in C_1 or (ii) v has two neighbors in some other component. Now because all components other than C_1 have size $O(\log n)$ w.h.p., we see that

$$\mathbb{P}((ii))=o(1)+n\binom{O(\log n)}{2}\left(\frac{c}{n}\right)^2=o(1).$$

Now w.h.p. $|C_1|\approx\left(1-\frac{x}{c}\right)n$ and it is independent of the edges incident with v and so

$$\begin{aligned}\mathbb{P}((i))&=o(1)+1-\mathbb{P}(0\text{ or }1\text{ neighbors in }C_1)\\&=o(1)+(1+o(1))\mathbb{E}\left(1-\left(\left(1-\frac{c}{n}\right)^{|C_1|}+|C_1|\left(1-\frac{c}{n}\right)^{|C_1|-1}\frac{c}{n}\right)\right)\\&=o(1)+1-(e^{-c+x}+(c-x)e^{-c+x})\\&=o(1)+(1-x)\left(1-\frac{x}{c}\right),\end{aligned}\tag{2.15}$$

where the last line follows from the fact that $e^{-c+x}=\frac{x}{c}$. Also, one has to be careful when estimating something like $\mathbb{E}\left(1-\frac{c}{n}\right)^{|C_1|}$. For this we note that Jensen's inequality implies that

$$\mathbb{E}\left(1-\frac{c}{n}\right)^{|C_1|}\geq\left(1-\frac{c}{n}\right)^{\mathbb{E}|C_1|}=e^{-c+x+o(1)}.$$

On the other hand, if $n_g = \left(1-\frac{x}{c}\right)n$,

$$\mathbb{E}\left(1-\frac{c}{n}\right)^{|C_1|} \leq \mathbb{E}\left(\left(1-\frac{c}{n}\right)^{|C_1|} \bigg| |C_1| \geq (1-o(1))n_g\right)\mathbb{P}\left(|C_1| \geq (1-o(1))n_g\right)$$
$$+\mathbb{P}\left(|C_1| \leq (1-o(1))n_g\right) = e^{-c+x+o(1)}.$$

It follows from (2.15) that $\mathbb{E}(|C_2|) \approx (1-x)\left(1-\frac{x}{c}\right)n$. To prove concentration of $|C_2|$, we can use the Chebyshev inequality as we did in the proof of Theorem 2.14 to prove concentration for the number of edges in the giant.

To estimate the expected number of edges in C_2, we proceed as in Theorem 2.14 and turn to $G = \mathbb{G}_{n,m}$ and estimate the probability that $e_1 \subseteq C_2$. If $G' = G \setminus e$ and C_1' is the giant of G' then e_1 is an edge of C_2 if and only if $e_1 \subseteq C_1'$ or e_1 is contained in a small component. This latter condition is unlikely. Thus,

$$\mathbb{P}(e_1 \subseteq C_2) = o(1) + \mathbb{E}\left(\frac{|C_1'|}{n}\right)^2 = o(1) + \left(1-\frac{x}{c}\right)^2.$$

The estimate for the expectation of the number of edges in the 2-core follows immediately and one can prove concentration using the Chebyshev inequality.
□

2.3 Phase Transition

In the previous two sections we studied the asymptotic behavior of $\mathbb{G}_m$ (and $\mathbb{G}_{n,p}$) in the "sub-critical phase" when $m = cn, c < 1/2$ $(p = c/n, c < 1)$, as well as in the "super-critical phase" when $m > n/2$ $(p = c/n, c > 1)$ of its evolution.

We have learned that when $m = cn, c, 1/2$ our random graph consists w.h.p. of tree components and components with exactly one cycle (see Theorem 2.1 and Lemma 2.11). We call such components *simple*, while components which are not simple, i.e. components with at least two cycles, are called *complex*.

All components during the sub-critical phase are rather small, of order $\log n$, tree-components dominate the typical structure of $\mathbb{G}_m$, and there is no significant gap in the order of the first and the second largest component. This follows from Lemma 2.12. The proof of this lemma shows that w.h.p. there are many trees of size k. The situation changes when $m > n/2$, i.e. when we enter the super-critical phase and then w.h.p. $\mathbb{G}_m$ consists of a single giant complex component (of the order comparable to n), and some number of simple components, i.e. tree components and components with exactly one cycle (see Theorem 2.14). We can also observe a clear gap between the order of the largest component (the giant) and the second largest component, which is of the order $O(\log n)$. This phenomenon of dramatic change of the typical structure of a random graph is called its *phase transition.*

A natural question arises as to what happens when $m/n \to 1/2$, either from below or above, as $n \to \infty$. It appears that we can establish, a so-called, *scaling window* or *critical window* for the phase transition in which $\mathbb{G}_m$ is undergoing a rapid change in its typical structure. A characteristic feature of this period is that a random graph can w.h.p. consist of *more than one* complex component (recall: there are no complex components in the sub-critical phase and there is a unique complex component in the super-critical phase).

Erdős and Rényi [276] studied the size of the largest tree in the random graph $\mathbb{G}_{n,m}$ when $m = n/2$ and showed that it was likely to be around $n^{2/3}$. They called the transition from $O(\log n)$ through $\Theta(n^{2/3})$ to $\Omega(n)$ the "double jump." They did not study the regime $m = n/2 + o(n)$. Bollobás [128] opened the detailed study of this and Łuczak [537] continued this analysis. He established the precise size of the "scaling window" by removing a logarithmic factor from Bollobás's estimates. The component structure of $\mathbb{G}_{n,m}$ for $m = n/2 + o(n)$ is rather complicated and the proofs are technically challenging. We begin by stating several results that give an idea of the component structure in this range, referring the reader elsewhere for proofs: Chapter 5 of Janson, Łuczak and Ruciński [432]; Aldous [15]; Bollobás [128]; Janson [419]; Janson, Knuth, Łuczak and Pittel [436]; Łuczak [537], [538], [542]; Łuczak, Pittel and Wierman [545]. We finish with a proof by Nachmias and Peres that when $p = 1/n$ the largest component is likely to have size of order $n^{2/3}$.

The first theorem is a refinement of Lemma 2.10.

Theorem 2.17 *Let $m = \frac{n}{2} - s$, where $s = s(n) \geq 0$.*

(a) The probability that $\mathbb{G}_{n,m}$ contains a complex component is at most $n^2/4s^3$.
(b) If $s \gg n^{2/3}$ then w.h.p. the largest component is a tree of size asymptotic to $\frac{n}{2s^2} \log \frac{s^3}{n}$.

The next theorem indicates when the phase in which we may have more than one complex component "ends", i.e. when a single giant component emerges.

Theorem 2.18 *Let $m = \frac{n}{2} + s$, where $s = s(n) \geq 0$. Then the probability that $\mathbb{G}_{n,m}$ contains more than one complex component is at most $6n^{2/9}/s^{1/3}$.* □

For larger s, the next theorem gives a precise estimate of the size of the largest component for $s \gg n^{2/3}$. For $s > 0$ we let $\bar{s} > 0$ be defined by

$$\left(1 - \frac{2\bar{s}}{n}\right) \exp\left\{\frac{2\bar{s}}{n}\right\} = \left(1 + \frac{2s}{n}\right) \exp\left\{\frac{2s}{n}\right\}.$$

Theorem 2.19 *Let $m = \frac{n}{2} + s$ where $s \gg n^{2/3}$. Then with probability at least $1 - 7n^{2/9}/s^{1/3}$,*

$$\left| L_1 - \frac{2(s+\bar{s})n}{n+2s} \right| \leq \frac{n^{2/3}}{5}$$

where L_1 is the size of the largest component in $\mathbb{G}_{n,m}$. In addition, the largest component is complex and all other components are either trees or unicyclic components.

To get a feel for this estimate of L_1 we remark that

$$\bar{s} = s - \frac{4s^2}{3n} + O\left(\frac{s^3}{n^2}\right).$$

The next theorem gives some information about ℓ-components inside the scaling window $m = n/2 + O(n^{2/3})$. An ℓ-component is one that has ℓ more edges than vertices. So trees are (-1)-components.

Theorem 2.20 *Let $m = \frac{n}{2} + O(n^{2/3})$ and let r_ℓ denote the number of ℓ-components in $\mathbb{G}_{n,m}$. For every $0 < \delta < 1$ there exists C_δ such that if n is sufficiently large, then with probability at least $1 - \delta$, $\sum_{\ell \geq 3} \ell r_\ell \leq C_\delta$ and the number of vertices on complex components is at most $C_\delta n^{2/3}$.*

One of the difficulties in analyzing the phase transition stems from the need to estimate $C(k,\ell)$, which is the number of connected graphs with vertex set $[k]$ and ℓ edges. We need good estimates for use in first moment calculations. We have seen the values for $C(k,k-1)$ (Cayley's formula) and $C(k,k)$, see (2.6). For $\ell > 0$, things become more tricky. Wright [725], [726], [727] showed that $C_{k,k+\ell} \approx \gamma_\ell k^{k+(3\ell-1)/2}$ for $\ell = o(k^{1/3})$ where the Wright coefficients γ_ℓ satisfy an explicit recurrence and have been related to Brownian motion, see Aldous [15] and Spencer [687]. In a breakthrough paper, Bender, Canfield and McKay [72] gave an asymptotic formula valid for all k. Łuczak [536] in a beautiful argument simplified a large part of their argument, see Exercise (4.3.6). Bollobás [130] proved the useful simple estimate $C_{k,k+\ell} \leq c\ell^{-\ell/2} k^{k+(3\ell-1)/2}$ for some absolute constant $c > 0$. It is difficult to prove tight statements about $\mathbb{G}_{n,m}$ in the phase transition window without these estimates. Nevertheless, it is possible to see that the largest component should be of size order $n^{2/3}$, using a nice argument from Nachmias and Peres. They have published a stronger version of this argument in [599].

Theorem 2.21 *Let $p = \frac{1}{n}$ and A be a large constant. Let Z be the size of the largest component in $\mathbb{G}_{n,p}$, then*

$$(i)\quad \mathbb{P}\left(Z \leq \frac{1}{A}n^{2/3}\right) = O(A^{-1}),$$

$$(ii)\quad \mathbb{P}\left(Z \geq An^{2/3}\right) = O(A^{-1}).$$

Proof We will prove part (i) of the theorem first. This is a standard application of the First Moment Method, see for example Bollobás [130]. Let X_k be the number of tree components of order k and let $k \in \left[\frac{1}{A}n^{2/3}, An^{2/3}\right]$, then, see also (2.8),

$$\mathbb{E}X_k = \binom{n}{k} k^{k-2} p^{k-1} (1-p)^{k(n-k)+\binom{k}{2}-k+1},$$

but

$$\begin{aligned}
(1-p)^{k(n-k)+\binom{k}{2}-k+1} &\approx (1-p)^{kn-k^2/2} \\
&= \exp\{(kn - k^2/2)\log(1-p)\} \\
&\approx \exp\left\{-\frac{kn-k^2/2}{n}\right\}.
\end{aligned}$$

Hence, by the above and Lemma 21.2,

$$\mathbb{E}X_k \approx \frac{n}{\sqrt{2\pi}\, k^{5/2}} \exp\left\{-\frac{k^3}{6n^2}\right\}. \tag{2.16}$$

So if

$$X = \sum_{\frac{1}{A}n^{2/3}}^{An^{2/3}} X_k,$$

then

$$\begin{aligned}
\mathbb{E}X &\approx \frac{1}{\sqrt{2\pi}} \int_{x=\frac{1}{A}}^{A} \frac{e^{-x^3/6}}{x^{5/2}} dx \\
&= \frac{4}{3\sqrt{\pi}} A^{3/2} + O(A^{1/2}).
\end{aligned}$$

Arguing as in Lemma 2.12 we see that

$$\mathbb{E}X_k^2 \leq \mathbb{E}X_k + (1+o(1))(\mathbb{E}X_k)^2,$$

$$\mathbb{E}(X_k X_l) \leq (1+o(1))(\mathbb{E}X_k)(\mathbb{E}X_l), \quad k \neq l.$$

It follows that

$$\mathbb{E}X^2 \leq \mathbb{E}X + (1+o(1))(\mathbb{E}X)^2.$$

Applying the Second Moment Method, Lemma 20.6, we see that

$$\mathbb{P}(X>0) \geq \frac{1}{(\mathbb{E}X)^{-1}+1+o(1)}$$
$$= 1-O(A^{-1}),$$

which completes the proof of part (*i*).

To prove (*ii*) we first consider a breadth first search (BFS) starting from, say, vertex x. We construct a sequence of sets $S_1 = \{x\}, S_2, \ldots,$ where

$$S_{i+1} = \{v \notin S_i : \exists w \in S_i \text{ such that } (v,w) \in E(\mathbb{G}_{n,p})\}.$$

We have

$$\mathbb{E}(|S_{i+1}|\ |S_i) \leq (n-|S_i|)\left(1-(1-p)^{|S_i|}\right)$$
$$\leq (n-|S_i|)|S_i|p$$
$$\leq |S_i|.$$

So

$$\mathbb{E}\,|S_{i+1}| \leq \mathbb{E}\,|S_i| \leq \cdots \leq \mathbb{E}\,|S_1| = 1. \tag{2.17}$$

We prove next that

$$\pi_k = \mathbb{P}(S_k \neq \emptyset) \leq \frac{4}{k}. \tag{2.18}$$

This is clearly true for $k \leq 4$ and we obtain (2.18) by induction from

$$\pi_{k+1} \leq \sum_{i=1}^{n-1} \binom{n-1}{i} p^i (1-p)^{n-1-i} (1-(1-\pi_k)^i). \tag{2.19}$$

To explain the above inequality, note that we can couple the construction of $S_1, S_2, \ldots, S_k$ with a (branching) process where $T_1 = \{1\}$ and T_{k+1} is obtained from T_k as follows: each T_k independently spawns $\text{Bin}(n-1,p)$ individuals. Note that $|T_k|$ stochastically dominates $|S_k|$. This is because in the BFS process, each $w \in S_k$ gives rise to at most $\text{Bin}(n-1,p)$ new vertices. Inequality (2.19) follows, because $T_{k+1} \neq \emptyset$ implies that at least one of 1's children gives rise to descendants at level k. Going back to (2.19) we obtain

$$\begin{aligned}
\pi_{k+1} &\leq 1-(1-p)^{n-1}-(1-p+p(1-\pi_k))^{n-1}+(1-p)^{n-1} \\
&= 1-(1-p\pi_k)^{n-1} \\
&\leq 1-1+(n-1)p\pi_k-\binom{n-1}{2}p^2\pi_k^2+\binom{n-1}{3}\pi_k^3 \\
&\leq \pi_k-\left(\frac{1}{2}+o(1)\right)\pi_k^2+\left(\frac{1}{6}+o(1)\right)\pi_k^3 \\
&= \pi_k\left(1-\pi_k\left(\left(\frac{1}{2}+o(1)\right)-\left(\frac{1}{6}+o(1)\right)\pi_k\right)\right) \\
&\leq \pi_k\left(1-\frac{1}{4}\pi_k\right).
\end{aligned}$$

This expression increases for $0 \leq \pi_k \leq 1$ and immediately gives $\pi_5 \leq 3/4 \leq 4/5$. In general we have by induction that

$$\pi_{k+1} \leq \frac{4}{k}\left(1-\frac{1}{k}\right) \leq \frac{4}{k+1},$$

completing the inductive proof of (2.18).

Let C_x be the component containing x and let $\rho_x = \max\{k : S_k \neq \emptyset\}$ in the BFS from x. Let

$$X = \left|\left\{x : |C_x| \geq n^{2/3}\right\}\right| \leq X_1 + X_2,$$

where

$$X_1 = \left|\left\{x : |C_x| \geq n^{2/3} \text{ and } \rho_x \leq n^{1/3}\right\}\right|,$$

$$X_2 = \left|\left\{x : \rho_x > n^{1/3}\right\}\right|.$$

It follows from (2.18) that

$$\mathbb{P}(\rho_x > n^{1/3}) \leq \frac{4}{n^{1/3}}$$

and so

$$\mathbb{E}X_2 \leq 4n^{2/3}.$$

Furthermore,

$$\begin{aligned}
&\mathbb{P}\left\{|C_x| \geq n^{2/3} \text{ and } \rho_x \leq n^{1/3}\right\} \\
&\leq \mathbb{P}\left(|S_1| + \ldots + |S_{n^{1/3}}| \geq n^{2/3}\right) \\
&\leq \frac{\mathbb{E}(|S_1| + \ldots + |S_{n^{1/3}}|)}{n^{2/3}} \\
&\leq \frac{1}{n^{1/3}},
\end{aligned}$$

after using (2.17). So $\mathbb{E} X_1 \leq n^{2/3}$ and $\mathbb{E} X \leq 5n^{2/3}$.

Now let C_{max} denote the size of the largest component. Now

$$C_{max} \leq |X| + n^{2/3}$$

where the addition of $n^{2/3}$ accounts for the case where $X = 0$.

So we have

$$\mathbb{E}\, C_{max} \leq 6n^{2/3}$$

and part (ii) of the theorem follows from the Markov inequality (see Lemma 20.1).

□

2.4 Exercises

2.4.1 Prove Theorem 2.15.

2.4.2 Show that if $p = \omega/n$ where $\omega = \omega(n) \to \infty$, then w.h.p. $\mathbb{G}_{n,p}$ contains no unicyclic components. (A component is *unicyclic* if it contains exactly one cycle, i.e. is a tree plus one extra edge.)

2.4.3 Prove Theorem 2.17.

2.4.4 Suppose that $m = cn/2$ where $c > 1$ is a constant. Let C_1 denote the giant component of $G_{n,m}$, assuming that it exists. Suppose that C_1 has $n' \leq n$ vertices and $m' \leq m$ edges. Let G_1, G_2 be two connected graphs with n' vertices from $[n]$ and m' edges. Show that

$$\mathbb{P}(C_1 = G_1) = \mathbb{P}(C_1 = G_2),$$

(i.e. C_1 is a uniformly random connected graph with n' vertices and m' edges).

2.4.5 Suppose that Z is the length of the cycle in a randomly chosen connected unicyclic graph on vertex set $[n]$. Show that, where $N = \binom{n}{2}$,

$$\mathbb{E} Z = \frac{n^{n-2}(N - n + 1)}{C(n,n)}.$$

2.4.6 Suppose that $c < 1$. Show that w.h.p. the length of the longest path in $\mathbb{G}_{n,p}$, $p = \frac{c}{n}$ is $\approx \frac{\log n}{\log 1/c}$.

2.4.7 Let $\mathbb{G}_{n,n,p}$ denote the random bipartite graph derived from the complete bipartite graph $K_{n,n}$ where each edge is included independently with probability p. Show that if $p = c/n$ where $c > 1$ is a constant then w.h.p. $\mathbb{G}_{n,n,p}$ has a unique giant component of size $\approx 2G(c)n$ where $G(c)$ is as in Theorem 2.14.

2.4.8 Let $p = \frac{1+\varepsilon}{n}$. Show that if ε is a small positive constant then w.h.p. $\mathbb{G}_{n,p}$ contains a giant component of size $(2\varepsilon + O(\varepsilon^2))n$.

2.4.9 Let $m = \frac{n}{2} + s$, where $s = s(n) \geq 0$. Show that if $s \gg n^{2/3}$ then w.h.p. the random graph $\mathbb{G}_m$ contains exactly one complex component. (A component C is *complex* if it contains at least two distinct cycles. In terms of edges, C is complex if and only if it contains at last $|C| + 1$ edges.)

2.4.10 Let $m_k(n) = n(\log n + (k-1)\log\log n + \omega)/(2k)$, where $|\omega| \to \infty, |\omega| = o(\log n)$. Show that

$$\mathbb{P}(\mathbb{G}_{m_k} \not\supseteq k\text{-vertex-tree-component}) = \begin{cases} o(1) & \text{if } \omega \to -\infty \\ 1 - o(1) & \text{if } \omega \to \infty. \end{cases}$$

2.4.11 Let k be fixed and let $p = \frac{c}{n}$. Show that if c is sufficiently large, then w.h.p. the k-core of $\mathbb{G}_{n,p}$ is non-empty.

2.4.12 Let k be fixed and let $p = \frac{c}{n}$. Show that there exists $\theta = \theta(c,k)$ such that w.h.p. all vertex sets S with $|S| \leq \theta n$ contain fewer than $k|S|/2$ edges. Deduce that w.h.p. either the k-core of $\mathbb{G}_{n,p}$ is empty or it has size of at least θn.

2.4.13 Suppose that $p = \frac{c}{n}$ where $c > 1$ is a constant. Show that w.h.p. the giant component of $\mathbb{G}_{n,p}$ is non-planar. (Hint: Assume that $c = 1 + \varepsilon$ where ε is small. Remove a few vertices from the giant so that the girth is large. Now use Euler's formula.)

2.4.14 Show that if $\omega = \omega(n) \to \infty$ then w.h.p. $\mathbb{G}_{n,p}$ has at most ω complex components.

2.4.15 Suppose that $np \to \infty$ and $3 \leq k = O(1)$. Show that $\mathbb{G}_{n,p}$ contains a k-cycle w.h.p.

2.4.16 Suppose that $p = c/n$ where $c > 1$ is constant and let $\beta = \beta(c)$ be the smallest root of the equation

$$\frac{1}{2}c\beta + (1-\beta)ce^{-c\beta} = \log\left(c(1-\beta)^{(\beta-1)/\beta}\right).$$

(1) Show that if $\omega \to \infty$ and $\omega \leq k \leq \underline{n}$ then w.h.p. $\mathbb{G}_{n,p}$ contains no *maximal* induced tree of size k.

(2) Show that w.h.p. $\mathbb{G}_{n,p}$ contains an induced tree of size $(\log n)^2$.
(3) Deduce that w.h.p. $\mathbb{G}_{n,p}$ contains an induced tree of size at least βn.

2.4.17 Consider the bipartite random graph $\mathbb{G}_{n,n,p=c/n}$, with constant $c > 1$. Define $0 < x < 1$ to be the solution to $xe^{-x} = ce^{-c}$. Prove that w.h.p. $\mathbb{G}_{n,n,p=c/n}$ has a 2-core with $2(1-x)\left(1-\frac{x}{c}\right)n$ vertices and $c\left(1-\frac{x}{c}\right)^2 n$ edges.

2.4.18 Show that if $c \neq 1$ and $xe^{-x} = ce^{-c}$ where $0 < x < 1$ then

$$\frac{1}{c}\sum_{k=1}^{\infty}\frac{k^{k-2}}{k!}(ce^{-c})^k = \begin{cases} 1-\frac{c}{2} & c < 1. \\ \frac{x}{c}\left(1-\frac{x}{2}\right) & c > 1. \end{cases}$$

2.5 Notes

Phase transition

The paper by Łuczak, Pittel and Wierman [545] contains a great deal of information about the phase transition. In particular, [545] shows that if $m = n/2 + ln^{2/3}$ then the probability that $\mathbb{G}_{n,m}$ is planar tends to a limit $p(l)$, where $p(l) \to 0$ as $l \to \infty$. The landmark paper by Janson, Knuth, Łuczak and Pittel [436] gives the most detailed analysis to date of the events in the scaling window.

Outside the critical window $\frac{n}{2} \pm O(n^{2/3})$, the size of the largest component is asymptotically determined. Theorem 2.17 describes $\mathbb{G}_{n,m}$ before reaching the window. On the other hand a unique "giant" component of size $\approx 4s$ begins to emerge at around $m = \frac{n}{2} + s$, for $s \gg n^{2/3}$. Ding, Kim, Lubetzky and Peres [245] give a useful model for the structure of this giant.

Achlioptas processes

Dimitris Achlipotas proposed the following variation on the basic graph process. Suppose that instead of adding a random edge e_i to add to G_{i-1} to create G_i, one is given a choice of two random edges e_i, f_i and one chooses one of them to add. He asked whether it was possible to come up with a choice rule that would delay the occurrence of some graph property $\mathcal{P}$. As an initial challenge he asked whether it was possible to delay the production of a giant component beyond $n/2$. Bohman and Frieze [112] showed that this was possible using a simple rule. Since that time, this has grown into a large area of research. Kang, Perkins and Spencer [463] have given a more detailed analysis of the "Bohman–Frieze" process. Bohman and Kravitz [120] and, in greater generality, Spencer and Wormald [689] analyzed "bounded size algorithms" in respect of avoiding giant components. Flaxman, Gamarnik and Sorkin [306] considered how to speed up the occurrence of a giant component. Riordan and Warnke [648] discussed the speed of transition at a critical point in an Achlioptas process.

The above papers concern component structure. Krivelevich, Loh and Sudakov [505] considered rules for avoiding specific subgraphs. Krivelevich, Lubetzky and Sudakov [506] discussed rules for speeding up Hamiltonicity.

Graph Minors

Fountoulakis, Kühn and Osthus [312] showed that for every $\varepsilon > 0$ there exists C_ε such that if $np > C_\varepsilon$ and $p = o(1)$ then w.h.p. $\mathbb{G}_{n,p}$ contains a complete minor of size $(1 \pm \varepsilon)\left(\frac{n^2 p}{\log np}\right)$. This improves earlier results of Bollobás, Catlin and Erdős [134] and Krivelevich and Sudakov [511]. Ajtai, Komlós and Szemerédi [9] showed that if $np \geq 1 + \varepsilon$ and $np = o(n^{1/2})$ then w.h.p. $\mathbb{G}_{n,p}$ contains a topological clique of size almost as large as the maximum degree. If we know that $\mathbb{G}_{n,p}$ is non-planar w.h.p. then it makes sense to determine its *thickness*. This is the minimum number of planar graphs whose union is the whole graph. Cooper [197] showed that the thickness of $\mathbb{G}_{n,p}$ is strongly related to its *arboricity* and is asymptotic to $np/2$ for a large range of p.

3
Vertex Degrees

In this chapter we study some typical properties of the degree sequence of a random graph. We begin by discussing the typical degrees in a sparse random graph, i.e. one with $O(n)$ edges and prove some results on the asymptotic distribution of degrees. Next we look at the typical values of the minimum and maximum degrees in dense random graphs. We then describe a simple canonical labeling algorithm for the graph isomorphism problem on a dense random graph.

3.1 Degrees of Sparse Random Graphs

Recall that the degree of an individual vertex of $\mathbb{G}_{n,p}$ is a Binomial random variable with parameters $n-1$ and p. One should also notice that the degrees of different vertices are only mildly correlated.

We first prove some simple but often useful properties of vertex degrees when $p = o(1)$. Let $X_0 = X_{n,0}$ be the number of isolated vertices in $\mathbb{G}_{n,p}$. In Lemma 1.11, we established the sharp threshold for "disappearance" of such vertices. Now we will be more precise and determine the asymptotic distribution of X_0 "below," "on" and "above" the threshold. Obviously,

$$\mathbb{E}X_0 = n(1-p)^{n-1},$$

and an easy computation shows that, as $n \to \infty$,

$$\mathbb{E}X_0 \to \begin{cases} \infty & \text{if } np - \log n \to -\infty \\ e^{-c} & \text{if } np - \log n \to c, \; c < \infty, \\ 0 & \text{if } np - \log n \to \infty, \end{cases} \tag{3.1}$$

We denote by $Po(\lambda)$ a random variable with the Poisson distribution with parameter λ, while $N(0,1)$ denotes the random variable with the standard normal distribution. We write $X_n \xrightarrow{D} X$ to say that a random variable X_n *converges in distribution* to a random variable X, as $n \to \infty$.

The following theorem shows that the asymptotic distribution of X_0 passes through three phases: it starts in the normal phase; next when isolated vertices are close to "dying out," it moves through a Poisson phase; it finally ends up at the distribution concentrated at 0.

Theorem 3.1 *Let X_0 be the random variable counting isolated vertices in a random graph $\mathbb{G}_{n,p}$. Then, as $n \to \infty$,*

(i) $\tilde{X}_0 = (X_0 - \mathbb{E} X_0)/(\operatorname{Var} X_0)^{1/2} \xrightarrow{D} \mathrm{N}(0,1)$,
if $n^2 p \to \infty$ *and* $np - \log n \to -\infty$,
(ii) $X_0 \xrightarrow{D} \mathrm{Po}(e^{-c})$, *if* $np - \log n \to c, \ c < \infty$,
(iii) $X_0 \xrightarrow{D} 0$, *if* $np - \log n \to \infty$.

Proof For the proof of (*i*) we refer the reader to Chapter 6 of Janson, Łuczak and Ruciński [432] (or to [60] and [500]).

To prove (*ii*) one has to show that if $p = p(n)$ is such that $np - \log n \to c$, then

$$\lim_{n\to\infty} \mathbb{P}(X_0 = k) = \frac{e^{-ck}}{k!} e^{-e^{-c}}, \tag{3.2}$$

for $k = 0, 1, \dots$. Now,

$$X_0 = \sum_{v \in V} I_v,$$

where

$$I_v = \begin{cases} 1 & \text{if } v \text{ is an isolated vertex in } \mathbb{G}_{n,p} \\ 0 & \text{otherwise.} \end{cases}$$

So

$$\begin{aligned} \mathbb{E} X_0 &= \sum_{v \in V} \mathbb{E} I_v = n(1-p)^{n-1} \\ &= n \exp\{(n-1)\log(1-p)\} \\ &= n \exp\left\{-(n-1)\sum_{k=1}^{\infty} \frac{p^k}{k}\right\} \\ &= n \exp\left\{-(n-1)p + O(np^2)\right\} \\ &= n \exp\left\{-(\log n + c) + O\left(\frac{(\log n)^2}{n}\right)\right\} \\ &\approx e^{-c}. \end{aligned} \tag{3.3}$$

The easiest way to show that (3.2) holds is to apply the Method of Moments (see Chapter 20). Briefly, since the distribution of the random variable X_0 is uniquely determined by its moments, it is enough to show, that either the kth factorial moment $\mathbb{E}X_0(X_0-1)\cdots(X_0-k+1)$ of X_0, or its Binomial moment $\mathbb{E}\binom{X_0}{k}$, tend to the respective moments of the Poisson distribution, i.e. to either e^{-ck} or $e^{-ck}/k!$. We choose the Binomial moments, and so let

$$B_k^{(n)} = \mathbb{E}\binom{X_0}{k},$$

then, for every non-negative integer k,

$$\begin{aligned} B_k^{(n)} &= \sum_{1\leq i_1<i_2<\cdots<i_k\leq n} \mathbb{P}(I_{v_{i_1}}=1, I_{v_{i_2}}=1,\ldots,I_{v_{i_k}}=1), \\ &= \binom{n}{k}(1-p)^{k(n-k)+\binom{k}{2}}. \end{aligned}$$

Hence

$$\lim_{n\to\infty} B_k^{(n)} = \frac{e^{-ck}}{k!},$$

and part (ii) of the theorem follows by Corollary 20.11, with $l=e^{-c}$.

For part (iii), suppose that $np=\log n+\omega$ where $\omega\to\infty$. We repeat the calculation estimating $\mathbb{E}X_0$ and replace $\approx e^{-c}$ in (3.3) by $\leq(1+o(1))e^{-\omega}\to 0$ and apply the First Moment Method. □

From the above theorem we immediately see that if $np-\log n\to c$ then

$$\lim_{n\to\infty}\mathbb{P}(X_0=0)=e^{-e^{-c}}. \tag{3.4}$$

We next give a more general result describing the asymptotic distribution of the number $X_d=X_{n,d}$, $d\geq 1$ of vertices of any fixed degree d in a random graph.

Recall, that the degree of a vertex in $\mathbb{G}_{n,p}$ has the Binomial distribution $\mathrm{Bin}(n-1,p)$. Hence,

$$\mathbb{E}X_d = n\binom{n-1}{d}p^d(1-p)^{n-1-d}. \tag{3.5}$$

Therefore, as $n \to \infty$,

$$\mathbb{E}X_d \to \begin{cases} 0 & \text{if } p \ll n^{-(d+1)/d}, \\ \lambda_1 & \text{if } p \approx cn^{-(d+1)/d},\ c < \infty, \\ \infty & \text{if } p \gg n^{-(d+1)/d)} \text{ but} \\ & \quad pn - \log n - d\log\log n \to -\infty, \\ \lambda_2 & \text{if } pn - \log n - d\log\log n \to c,\ c < \infty, \\ 0 & \text{if } pn - \log n - d\log\log n \to \infty, \end{cases} \tag{3.6}$$

where

$$\lambda_1 = \frac{c^d}{d!} \quad \text{and} \quad \lambda_2 = \frac{e^{-c}}{d!}. \tag{3.7}$$

The asymptotic behavior of the expectation of the random variable X_d suggests possible asymptotic distributions for X_d, for a given edge probability p.

Theorem 3.2 *Let $X_d = X_{n,d}$ be the number of vertices of degree d, $d \geq 1$, in $\mathbb{G}_{n,p}$ and let λ_1, λ_2 be given by (3.7). Then, as $n \to \infty$,*

(i) $X_d \xrightarrow{D} 0$ *if* $p \ll n^{-(d+1)/d}$,
(ii) $X_d \xrightarrow{D} \mathrm{Po}(\lambda_1)$ *if* $p \approx cn^{-(d+1)/d}$, $c < \infty$,
(iii) $\tilde{X}_d := (X_d - \mathbb{E}X_d)/(\mathrm{Var}\,X_d)^{1/2} \xrightarrow{D} \mathrm{N}(0,1)$ *if* $p \gg n^{-(d+1)/d}$, *but* $pn - \log n - d\log\log n \to -\infty$
(iv) $X_d \xrightarrow{D} \mathrm{Po}(\lambda_2)$ *if* $pn - \log n - d\log\log n \to c$, $-\infty < c < \infty$,
(v) $X_d \xrightarrow{D} 0$ *if* $pn - \log n - d\log\log n \to \infty$.

Proof The proofs of statements (*i*) and (*v*) are straightforward applications of the First Moment Method, while the proofs of (*ii*) and (*iv*) can be found in Chapter 3 of Bollobás [123] (see also Karoński and Ruciński [471] for estimates of the rate of convergence). The proof of (iii) can be found in [60]. □

The next theorem shows the concentration of X_d around its expectation when in $\mathbb{G}_{n,p}$ the edge probability $p = c/n$, i.e. when the average vertex degree is c.

Theorem 3.3 *Let $p = c/n$ where c is a constant. Let X_d denote the number of vertices of degree d in $\mathbb{G}_{n,p}$. Then, for $d = O(1)$, w.h.p.*

$$X_d \approx \frac{c^d e^{-c}}{d!} n.$$

Proof Assume that vertices of $\mathbb{G}_{n,p}$ are labeled $1,2,\ldots,n$. We first compute $\mathbb{E}X_d$. Thus,

$$\begin{aligned}
\mathbb{E}X_d &= n\mathbb{P}(\deg(1)=d) \\
&= n\binom{n-1}{d}\left(\frac{c}{n}\right)^d\left(1-\frac{c}{n}\right)^{n-1-d} \\
&= n\frac{n^d}{d!}\left(1+O\left(\frac{d^2}{n}\right)\right)\left(\frac{c}{n}\right)^d \exp\left\{-(n-1-d)\left(\frac{c}{n}+O\left(\frac{1}{n^2}\right)\right)\right\} \\
&= n\frac{c^d e^{-c}}{d!}\left(1+O\left(\frac{1}{n}\right)\right).
\end{aligned}$$

We now compute the second moment. For this we need to estimate

$$\begin{aligned}
&\mathbb{P}(\deg(1)=\deg(2)=d) \\
&\quad = \frac{c}{n}\left(\binom{n-2}{d-1}\left(\frac{c}{n}\right)^{d-1}\left(1-\frac{c}{n}\right)^{n-1-d}\right)^2 \\
&\quad\quad + \left(1-\frac{c}{n}\right)\left(\binom{n-2}{d}\left(\frac{c}{n}\right)^d\left(1-\frac{c}{n}\right)^{n-2-d}\right)^2 \\
&\quad = \mathbb{P}(\deg(1)=d)\,\mathbb{P}(\deg(2)=d)\left(1+O\left(\frac{1}{n}\right)\right).
\end{aligned}$$

The first line here accounts for the case where $\{1,2\}$ is an edge and the second line deals with the case where it is not.

Thus

$$\begin{aligned}
\operatorname{Var} X_d &= \sum_{i=1}^n\sum_{j=1}^n\left[\mathbb{P}(\deg(i)=d,\deg(j)=d)-\mathbb{P}(\deg(1)=d)\,\mathbb{P}(\deg(2)=d)\right] \\
&\le \sum_{i\neq j=1}^n O\left(\frac{1}{n}\right)+\mathbb{E}X_d \le An,
\end{aligned}$$

for some constant $A=A(c)$.

Applying the Chebyshev inequality (Lemma 20.3), we obtain

$$\mathbb{P}(|X_d-\mathbb{E}X_d|\ge tn^{1/2})\le\frac{A}{t^2},$$

which completes the proof. □

We conclude this section with a look at the asymptotic behavior of the maximum vertex degree, when a random graph is sparse.

Theorem 3.4 *Let $\Delta(\mathbb{G}_{n,p})$ $(\delta(\mathbb{G}_{n,p}))$ denotes the maximum (minimum) degree of vertices of $\mathbb{G}_{n,p}$.*

(i) If $p = c/n$ for some constant $c > 0$ then w.h.p.

$$\Delta(\mathbb{G}_{n,p}) \approx \frac{\log n}{\log\log n}.$$

(ii) If $np = \omega \log n$ where $\omega \to \infty$, then w.h.p. $\delta(\mathbb{G}_{n,p}) \approx \Delta(\mathbb{G}_{n,p}) \approx np$.

Proof (*i*) Let $d_\pm = \left\lceil \frac{\log n}{\log\log n \pm 2\log\log\log n} \right\rceil$. Then, if $d = d_-$,

$$\begin{aligned}
\mathbb{P}(\exists v : deg(v) \geq d) &\leq n\binom{n-1}{d}\left(\frac{c}{n}\right)^d \\
&\leq n\left(\frac{ce}{d}\right)^d \\
&= \exp\{\log n - d\log d + O(d)\}. \qquad (3.8)
\end{aligned}$$

Let $l = \frac{\log\log\log n}{\log\log n}$, then

$$\begin{aligned}
d\log d &\geq \frac{\log n}{\log\log n} \cdot \frac{1}{1-2l} \cdot (\log\log n - \log\log\log n + o(1)) \\
&= \frac{\log n}{\log\log n}(1 + 2l + O(l^2))(\log\log n - \log\log\log n + o(1)) \\
&= \frac{\log n}{\log\log n}(\log\log n + \log\log\log n + o(1)). \qquad (3.9)
\end{aligned}$$

Plugging this into (3.8) shows that $\Delta(\mathbb{G}_{n,p}) \leq d_-$ w.h.p.

Now let $d = d_+$ and let X_d be the number of vertices of degree d in $\mathbb{G}_{n,p}$, then,

$$\begin{aligned}
\mathbb{E}(X_d) &= n\binom{n-1}{d}\left(\frac{c}{n}\right)^d\left(1 - \frac{c}{n}\right)^{n-d-1} \\
&= \exp\{\log n - d\log d + O(d)\} \\
&= \exp\left\{\log n - \frac{\log n}{\log\log n}(\log\log n - \log\log\log n + o(1)) + O(d)\right\} \\
&\to \infty. \qquad (3.10)
\end{aligned}$$

Here (3.10) is obtained by using $-l$ in place of l in the argument for (3.9). Now, for vertices v, w, by the same argument as in the proof of Theorem 3.3, we have

$$\mathbb{P}(deg(v) = deg(w) = d) = (1 + o(1))\mathbb{P}(deg(v) = d)\mathbb{P}(deg(w) = d),$$

and the Chebyshev inequality implies that $X_d > 0$ w.h.p. This completes the proof of (*i*).

Statement (*ii*) is an easy consequence of the Chernoff bounds (Corollary 21.7). Let $\varepsilon = \omega^{-1/3}$. then

$$\mathbb{P}(\exists v : |deg(v) - np| \geq \varepsilon np) \leq 2e^{-\varepsilon^2 np/3} = 2n^{-\omega^{1/3}/3} = o(n^{-1}).$$

□

3.2 Degrees of Dense Random Graphs

In this section we concentrate on the case where edge probability p is constant and see how the degree sequence can be used to solve the graph isomorphism problem w.h.p. The main result deals with the maximum vertex degree in dense random graph and is instrumental in the solution of this problem.

Theorem 3.5 *Let* $d_{\pm} = (n-1)p + (1 \pm \varepsilon)\sqrt{2(n-1)pq\log n}$, *where* $q = 1 - p$. *If* p *is constant and* $\varepsilon > 0$ *is a small constant, then w.h.p.*

(*i*) $d_- \leq \Delta(\mathbb{G}_{n,p}) \leq d_+$.
(*ii*) *There is a unique vertex of maximum degree.*

Proof We break the proof of Theorem 3.5 into two lemmas.

Lemma 3.6 *Let* $d = (n-1)p + x\sqrt{(n-1)pq}$, p *be constant,* $x \leq n^{1/3}(\log n)^2$, *where* $q = 1 - p$. *Then*

$$B_d = \binom{n-1}{d} p^d (1-p)^{n-1-d} = (1+o(1))\sqrt{\frac{1}{2\pi npq}}e^{-x^2/2}.$$

Proof Stirling's formula gives

$$B_d = (1+o(1))\sqrt{\frac{1}{2\pi npq}}\left(\left(\frac{(n-1)p}{d}\right)^{\frac{d}{n-1}}\left(\frac{(n-1)q}{n-1-d}\right)^{1-\frac{d}{n-1}}\right)^{n-1}. \tag{3.11}$$

Now

$$\begin{aligned}
\left(\frac{d}{(n-1)p}\right)^{\frac{d}{n-1}} &= \left(1 + x\sqrt{\frac{q}{(n-1)p}}\right)^{\frac{d}{n-1}} \\
&= \exp\left\{\left(x\sqrt{\frac{q}{(n-1)p}} - \frac{x^2 q}{2(n-1)p} + O\left(\frac{x^3}{n^{3/2}}\right)\right)\left(p + x\sqrt{\frac{pq}{n-1}}\right)\right\} \\
&= \exp\left\{x\sqrt{\frac{pq}{n-1}} + \frac{x^2 q}{2(n-1)} + O\left(\frac{x^3}{n^{3/2}}\right)\right\},
\end{aligned}$$

whereas

$$\left(\frac{n-1-d}{(n-1)q}\right)^{1-\frac{d}{n-1}} = \left(1 - x\sqrt{\frac{p}{(n-1)q}}\right)^{1-\frac{d}{n-1}}$$

$$= \exp\left\{-\left(x\sqrt{\frac{p}{(n-1)q}} - \frac{x^2 p}{2(n-1)q} + O\left(\frac{x^3}{n^{3/2}}\right)\right)\left(q - x\sqrt{\frac{pq}{n-1}}\right)\right\}$$

$$= \exp\left\{-x\sqrt{\frac{pq}{n-1}} + \frac{x^2 p}{2(n-1)} + O\left(\frac{x^3}{n^{3/2}}\right)\right\},$$

So

$$\left(\frac{d}{(n-1)p}\right)^{\frac{d}{n-1}} \left(\frac{n-1-d}{(n-1)q}\right)^{1-\frac{d}{n-1}} = \exp\left\{\frac{x^2}{2(n-1)} + O\left(\frac{x^3}{n^{3/2}}\right)\right\},$$

and lemma follows from (3.11). □

The next lemma proves a strengthening of Theorem 3.5.

Lemma 3.7 *Let $\varepsilon = 1/10$, and p be constant and $q = 1-p$. If*

$$d_{\pm} = (n-1)p + (1 \pm \varepsilon)\sqrt{2(n-1)pq\log n},$$

then w.h.p.

(i) $\Delta(\mathbb{G}_{n,p}) \leq d_+$,
(ii) there are $\Omega(n^{2\varepsilon(1-\varepsilon)})$ *vertices of degree at least* d_-,
(iii) $\nexists\, u \neq v$ *such that* $deg(u), deg(v) \geq d_-$ *and* $|deg(u) - deg(v)| \leq 10$.

Proof We first prove that as $x \to \infty$,

$$\frac{1}{x}e^{-x^2/2}\left(1 - \frac{1}{x^2}\right) \leq \int_x^\infty e^{-y^2/2}dy \leq \frac{1}{x}e^{-x^2/2}. \tag{3.12}$$

To see this, notice

$$\int_x^\infty e^{-y^2/2}dy = -\int_x^\infty \frac{1}{y}\left(e^{-y^2/2}\right)' dy$$

$$= -\left[\frac{1}{y}e^{-y^2/2}\right]_x^\infty - \int_x^\infty \frac{1}{y^2}e^{-y^2/2}dy$$

$$= \frac{1}{x}e^{-x^2/2} + \left[\frac{1}{y^3}e^{-y^2/2}\right]_x^\infty + 3\int_x^\infty \frac{1}{y^4}e^{-y^2/2}dy$$

$$= \frac{1}{x}e^{-x^2/2}\left(1 - \frac{1}{x^2}\right) + O\left(\frac{1}{x^4}e^{-x^2/2}\right).$$

We can now prove statement (i).

Let X_d be the number of vertices of degree d. Then $\mathbb{E}X_d = nB_d$ and so Lemma 3.6 implies that

$$\mathbb{E}X_d = (1+o(1))\sqrt{\frac{n}{2\pi pq}}\exp\left\{-\frac{1}{2}\left(\frac{d-(n-1)p}{\sqrt{(n-1)pq}}\right)^2\right\}$$

assuming that

$$d \leq d_L = (n-1)p + (\log n)^2\sqrt{(n-1)pq}.$$

Also, if $d > (n-1)p$, then

$$\frac{B_{d+1}}{B_d} = \frac{(n-d-1)p}{(d+1)q} < 1$$

and so if $d \geq d_L$,

$$\mathbb{E}X_d \leq \mathbb{E}X_{d_L} \leq n\exp\{-\Omega(\log n)^4\}.$$

It follows that

$$\Delta(\mathbb{G}_{n,p}) \leq d_L \quad w.h.p. \tag{3.13}$$

Now if $Y_d = X_d + X_{d+1} + \cdots + X_{d_L}$ for $d = d_\pm$ then

$$\begin{aligned}
\mathbb{E}\,Y_d &\approx \sum_{l=d}^{d_L}\sqrt{\frac{n}{2\pi pq}}\exp\left\{-\frac{1}{2}\left(\frac{l-(n-1)p}{\sqrt{(n-1)pq}}\right)^2\right\} \\
&\approx \sum_{l=d}^{\infty}\sqrt{\frac{n}{2\pi pq}}\exp\left\{-\frac{1}{2}\left(\frac{l-(n-1)p}{\sqrt{(n-1)pq}}\right)^2\right\} \qquad (3.14) \\
&\approx \sqrt{\frac{n}{2\pi pq}}\int_{\lambda=d}^{\infty}\exp\left\{-\frac{1}{2}\left(\frac{\lambda-(n-1)p}{\sqrt{(n-1)pq}}\right)^2\right\}d\lambda.
\end{aligned}$$

The justification for (3.14) comes from

$$\sum_{l=d_L}^{\infty}\sqrt{\frac{n}{2\pi pq}}\exp\left\{-\frac{1}{2}\left(\frac{l-(n-1)p}{\sqrt{(n-1)pq}}\right)^2\right\} =$$

$$= O(n)\sum_{x=(\log n)^2}^{\infty} e^{-x^2/2} = O(e^{-(\log n)^2/3}),$$

and

$$\sqrt{\frac{n}{2\pi pq}}\exp\left\{-\frac{1}{2}\left(\frac{d_+-(n-1)p}{\sqrt{(n-1)pq}}\right)^2\right\} = n^{-O(1)}.$$

If $d = (n-1)p + x\sqrt{(n-1)pq}$, then from (3.12) we have

$$\begin{aligned}
\mathbb{E}\,Y_d &\approx \sqrt{\frac{n}{2\pi pq}} \int_{\lambda=d}^{\infty} \exp\left\{-\frac{1}{2}\left(\frac{\lambda-(n-1)p}{\sqrt{(n-1)pq}}\right)^2\right\} d\lambda \\
&= \sqrt{\frac{n}{2\pi pq}} \sqrt{(n-1)pq} \int_{y=x}^{\infty} e^{-y^2/2} dy \\
&\approx \frac{n}{\sqrt{2\pi}} \frac{1}{x} e^{-x^2/2} \\
&\begin{cases} \leq n^{-2\varepsilon} & d = d_+ \\ \geq n^{\varepsilon} & d = d_-. \end{cases}
\end{aligned} \tag{3.15}$$

Part (i) follows from (3.15).

When $d = d_-$ we see from (3.15) that $\mathbb{E}\,Y_d \to \infty$. We use the Second Moment Method to show that $Y_{d_-} \neq 0$ w.h.p.

$$\begin{aligned}
\mathbb{E}\,Y_d(Y_{d-1} - 1) &= n(n-1) \sum_{d \leq d_1, d_2}^{d_L} \mathbb{P}(deg(1) = d_1, deg(2) = d_2) \\
&= n(n-1) \sum_{d \leq d_1, d_2}^{d_L} (p\,\mathbb{P}(\hat{d}(1) = d_1 - 1, \hat{d}(2) = d_2 - 1) \\
&\quad + (1-p)\,\mathbb{P}(\hat{d}(1) = d_1, \hat{d}(2) = d_2)),
\end{aligned}$$

where $\hat{d}(x)$ is the number of neighbors of x in $\{3, 4, \ldots, n\}$. Note that $\hat{d}(1)$ and $\hat{d}(2)$ are independent, and

$$\begin{aligned}
\frac{\mathbb{P}(\hat{d}(1) = d_1 - 1)}{\mathbb{P}(\hat{d}(1) = d_1)} = \frac{\binom{n-2}{d_1-1}(1-p)}{\binom{n-2}{d_1}p} &= \frac{d_1(1-p)}{(n-1-d_1)p} \\
&= 1 + \tilde{O}(n^{-1/2}).
\end{aligned}$$

In $\tilde{O}$ we ignore polylog factors.

Hence

$$\begin{aligned}
&\mathbb{E}(Y_d(Y_{d-1} - 1)) \\
&= n(n-1) \sum_{d \leq d_1, d_2}^{d_L} \left[\mathbb{P}(\hat{d}(1) = d_1)\,\mathbb{P}(\hat{d}(2) = d_2)(1 + \tilde{O}(n^{-1/2}))\right] \\
&= n(n-1) \sum_{d \leq d_1, d_2}^{d_L} \left[\mathbb{P}(deg(1) = d_1)\,\mathbb{P}(deg(2) = d_2)(1 + \tilde{O}(n^{-1/2}))\right] \\
&= \mathbb{E}\,Y_d(\mathbb{E}\,Y_d - 1)(1 + \tilde{O}(n^{-1/2})),
\end{aligned}$$

since

$$\frac{\mathbb{P}(\hat{d}(1)=d_1)}{\mathbb{P}(deg(1)=d_1)} = \frac{\binom{n-2}{d_1}}{\binom{n-1}{d_1}}(1-p)^{-2}$$
$$= 1+\tilde{O}(n^{-1/2}).$$

So, with $d = d_-$

$$\mathbb{P}\left(Y_d \le \frac{1}{2}\mathbb{E}\,Y_d\right)$$
$$\le \frac{\mathbb{E}(Y_d(Y_d-1)) + \mathbb{E}\,Y_d - (\mathbb{E}\,Y_d)^2}{(\mathbb{E}\,Y_d)^2/4}$$
$$= \tilde{O}\left(\frac{1}{n^{\varepsilon}}\right)$$
$$= o(1).$$

This completes the proof of statement *(ii)*. Finally,

$$\mathbb{P}(\neg(iii)) \le o(1) + \binom{n}{2}\sum_{d_1=d_-}^{d_L}\sum_{|d_2-d_1|\le 10} \mathbb{P}(deg(1)=d_1, deg(2)=d_2)$$
$$= o(1) + \binom{n}{2}\sum_{d_1=d_-}^{d_L}\sum_{|d_2-d_1|\le 10} \Big[p\,\mathbb{P}(\hat{d}(1)=d_1-1)\,\mathbb{P}(\hat{d}(2)=d_2-1)$$
$$+(1-p)\,\mathbb{P}(\hat{d}(1)=d_1)\,\mathbb{P}(\hat{d}(2)=d_2)\Big],$$

Now

$$\sum_{d_1=d_-}^{d_L}\sum_{|d_2-d_1|\le 10} \mathbb{P}(\hat{d}(1)=d_1-1)\,\mathbb{P}(\hat{d}(2)=d_2-1)$$
$$\le 21(1+\tilde{O}(n^{-1/2}))\sum_{d_1=d_-}^{d_L}\Big[\mathbb{P}(\hat{d}(1)=d_1-1)\Big]^2,$$

and by Lemma 3.6 and by (3.12) we have with

$$x = \frac{d_- - (n-1)p}{\sqrt{(n-1)pqn}} \approx (1-\varepsilon)\sqrt{2\log n},$$

$$\sum_{d_1=d_-}^{d_L}\left[\mathbb{P}(\hat{d}(1)=d_1-1)\right]^2 \approx \frac{1}{2\pi pqn}\int_{y=x}^{\infty} e^{-y^2}dy$$
$$= \frac{1}{\sqrt{8\pi pqn}}\int_{z=x\sqrt{2}}^{\infty} e^{-z^2/2}dz$$
$$\approx \frac{1}{\sqrt{8\pi pqn}}\frac{1}{x\sqrt{2}}n^{-2(1-\varepsilon)^2},$$

We get a similar bound for $\sum_{d_1=d_-}^{d_L}\sum_{|d_2-d_1|\le 10}\left[\mathbb{P}(\hat{d}(1)=d_1\right]^2$. Thus

$$\mathbb{P}(\neg(iii)) = o\left(n^{2-1-2(1-\varepsilon)^2}\right)$$
$$= o(1).$$

□

Application to graph isomorphism

In this section we describe a procedure for *canonically labeling* a graph G. It is taken from Babai, Erdős and Selkow [42]. If the procedure succeeds then it is possible to quickly tell whether $G \cong H$ for any other graph H. (Here $\cong$ stands for graph isomorphism.)

Algorithm *LABEL*

Step 0: Input graph G and parameter L.

Step 1: Re-label the vertices of G so that they satisfy

$$deg(v_1) \ge deg(v_2) \ge \cdots \ge deg(v_n).$$

If there exists $i < L$ such that $deg(v_i) = deg(v_{i+1})$, then **FAIL**.

Step 2: For $i > L$ let

$$X_i = \{j \in \{1,2,\ldots,L\} : \{v_i, v_j\} \in E(G)\}.$$

Re-label vertices $v_{L+1}, v_{L+2}, \ldots, v_n$ so that these sets satisfy

$$X_{L+1} \succ X_{L+2} \succ \cdots \succ X_n$$

where $\succ$ denotes lexicographic order.

If there exists $i < n$ such that $X_i = X_{i+1}$ then **FAIL**.

Suppose now that the above ordering/labeling procedure LABEL succeeds for G. Given an n vertex graph H, we run LABEL on H.

(i) If LABEL fails on H then $G \not\cong H$.

(ii) Suppose that the ordering generated on $V(H)$ is $w_1, w_2, \ldots, w_n$, then

$$G \cong H \Leftrightarrow v_i \to w_i \text{ is an isomorphism.}$$

It is straightforward to verify (i) and (ii).

Theorem 3.8 *Let p be a fixed constant, $q = 1-p$, and let $\rho = p^2 + q^2$ and let $L = 3\log_{1/\rho} n$. Then w.h.p. LABEL succeeds on $\mathbb{G}_{n,p}$.*

Proof Lemma 3.7 implies that Step 1 succeeds w.h.p. We must now show that w.h.p. $X_i \neq X_j$ for all $i \neq j > L$. There is a slight problem because the edges from $v_i, i > L$ to $v_j, j \leq L$ are conditioned by the fact that the latter vertices are those of highest degree.

Now fix $i, j > L$ and let $\hat{G} = \mathbb{G}_{n,p} \setminus \{v_i, v_j\}$. It follows from Lemma 3.7 that w.h.p. the L largest degree vertices of $\hat{G}$ and $\mathbb{G}_{n,p}$ coincide. So, w.h.p., we can compute X_i, X_j with respect to $\hat{G}$ and therefore avoid our conditioning problem. Denote by $N_{\hat{G}}(v)$ the set of the neighbors of vertex v in graph $\hat{G}$. Then

$$\begin{aligned}
&\mathbb{P}(\text{Step 2 fails}) \\
&\leq o(1) + \mathbb{P}(\exists v_i, v_j : N_{\hat{G}}(v_i) \cap \{v_1, \ldots, v_L\} = N_{\hat{G}}(v_j) \cap \{v_1, \ldots, v_L\}) \\
&\leq o(1) + \binom{n}{2}(p^2+q^2)^L \\
&= o(1).
\end{aligned}$$

□

Application to edge coloring

The *chromatic index* $\chi'(G)$ of a graph G is the minimum number of colors that can be used to color the edges of G so that if two edges share a vertex, then they have a different color. Vizing's theorem states that

$$\Delta(G) \leq \chi'(G) \leq \Delta(G) + 1.$$

Also, if there is a unique vertex of maximum degree, then $\chi'(G) = \Delta(G)$. So, it follows from Theorem 3.5 (*ii*) that, for constant p, w.h.p. we have $\chi'(\mathbb{G}_{n,p}) = \Delta(\mathbb{G}_{n,p})$.

3.3 Exercises

3.3.1 Suppose that $m = dn/2$ where d is constant. Prove that the number of vertices of degree d in $\mathbb{G}_{n,m}$ is asymptotically equal to $\frac{d^d e^{-d}}{d!} n$ for any fixed positive integer d.

3.3.2 Suppose that $c > 1$ and that $x < 1$ is the solution to $xe^{-x} = ce^{-c}$. Show that if $d = O(1)$ is fixed then w.h.p. the giant component of $G_{n,p}, p = \frac{c}{n}$ has $\approx \frac{y^d e^{-y}}{d!} n$ vertices of degree d, where $y = c - x$.

3.3.3 Suppose that $p \le \frac{1+\varepsilon_n}{n}$ where $n^{1/4}\varepsilon_n \to 0$. Show that if Γ is the sub-graph of $G_{n,p}$ induced by the 2-core C_2, then Γ has maximum degree at most three.

3.3.4 Let $p = \frac{\log n + d \log\log n + c}{n}$, $d \ge 1$. Using the Method of Moments, prove that the number of vertices of degree d in $\mathbb{G}_{n,p}$ is asymptotically Poisson with mean $\frac{e^{-c}}{d!}$.

3.3.5 Prove parts (i) and (v) of Theorem 3.2.

3.3.6 Show that if $0 < p < 1$ is constant then w.h.p. the minimum degree δ in $\mathbb{G}_{n,p}$ satisfies

$$|\delta - (n-1)q - \sqrt{2(n-1)pq\log n}| \le \varepsilon\sqrt{2(n-1)pq\log n},$$

where $q = 1 - p$ and $\varepsilon = 1/10$.

3.4 Notes

For the more detailed account of the properties of the degree sequence of $\mathbb{G}_{n,p}$ the reader is referred to Chapter 3 of Bollobás [130].

Erdős and Rényi [275] and [277] were the first to study the asymptotic distribution of the number X_d of vertices of degree d in relation to the connectivity of a random graph. Bollobás [127] continued those investigations and provided a detailed study of the distribution of X_d in $\mathbb{G}_{n,p}$ when $0 < \liminf np(n)/\log n \le \limsup np(n)/\log n < \infty$. Palka [610] determined a certain range of the edge probability p for which the number of vertices of a given degree of a random graph $\mathbb{G}_{n,p}$ has a normal distribution. Barbour [57] and Karoński and Ruciński [471] studied the distribution of X_d using the Stein–Chen approach. A complete answer to the asymptotic normality of X_d was given by Barbour, Karoński and Ruciński [60] (see also Kordecki [500]). Janson [425] extended those results and showed that random variables counting vertices of given degree are jointly normal, when $p \approx c/n$ in $\mathbb{G}_{n,p}$ and $m \approx cn$ in $\mathbb{G}_{n,m}$, where c is a constant.

Ivchenko [414] was the first to analyze the asymptotic behavior of the kth-largest and kth smallest element element of the degree sequence of $\mathbb{G}_{n,p}$. In particular he analyzed the span between the minimum and the maximum degree of sparse $\mathbb{G}_{n,p}$. Similar results were obtained independently by Bollobás [125] (see also Palka [611]). Bollobás [127] answered the question for what values of $p(n)$, $\mathbb{G}_{n,p}$ w.h.p. has a unique vertex of maximum degree (see Theorem 3.5).

Bollobás [122], for constant p, $0 < p < 1$, i.e. when $\mathbb{G}_{n,p}$ is dense, gave an estimate of the probability that the maximum degree does not exceed $pn + O(\sqrt{n \log n})$. A more precise result was proved by Riordan and Selby [647], who showed that for constant p, the probability that the maximum degree of $\mathbb{G}_{n,p}$ does not exceed $pn + b\sqrt{np(1-p)}$, where b is fixed, is equal to $(c+o(1))^n$, for $c = c(b)$ the root of a certain equation. Surprisingly, $c(0) = 0.6102...$ is greater than $1/2$ and $c(b)$ is independent of p.

McKay and Wormald [571] proved that for a wide range of functions $p = p(n)$, the distribution of the degree sequence of $\mathbb{G}_{n,p}$ can be approximated by $\{(X_1, \ldots, X_n) | \sum X_i \text{ is even}\}$, where $X_1, \ldots, X_n$ are independent random variables each having the Binomial distribution $\text{Bin}(n-1, p')$, where p' is itself a random variable with a particular truncated normal distribution.

4
Connectivity

We first establish, rather precisely, the threshold for connectivity. We then view this property in terms of the graph process and show that w.h.p. the random graph becomes connected at precisely the time when the last isolated vertex joins the giant component. This "hitting time" result is the pre-cursor to several similar results. After this we deal with k-connectivity.

4.1 Connectivity

The first result of this chapter is from Erdős and Rényi [275].

Theorem 4.1 *Let* $m = \frac{1}{2}n(\log n + c_n)$*, then*

$$\lim_{n\to\infty} \mathbb{P}(\mathbb{G}_m \text{ is connected}) = \begin{cases} 0 & \text{if } c_n \to -\infty, \\ e^{-e^{-c}} & \text{if } c_n \to c \text{ (constant)} \\ 1 & \text{if } c_n \to \infty. \end{cases}$$

Proof To prove the theorem we consider, as before, a random graph $\mathbb{G}_{n,p}$. It suffices to prove that, when $p = \frac{\log n + c}{n}$,

$$\mathbb{P}(\mathbb{G}_{n,p} \text{ is connected }) \to e^{-e^{-c}}.$$

and use Theorem 1.4 to translate to $\mathbb{G}_m$ and then use (1.6) and monotonicity for $c_n \to \pm\infty$.

Let $X_k = X_{k,n}$ be the number of components with k vertices in $\mathbb{G}_{n,p}$ and consider the complement of the event that $\mathbb{G}_{n,p}$ is connected, then

$$\begin{aligned}\mathbb{P}(\mathbb{G}_{n,p} \text{ is } \textbf{not} \text{ connected }) &= \mathbb{P}\left(\bigcup_{k=1}^{n/2}(\mathbb{G}_{n,p} \text{ has a component of order } k)\right)\\ &= \mathbb{P}\left(\bigcup_{k=1}^{n/2}\{X_k > 0\}\right).\end{aligned}$$

Note that here X_1 counts isolated vertices and therefore

$$\mathbb{P}(X_1 > 0) \le \mathbb{P}(\mathbb{G}_{n,p} \text{ is } \textbf{not} \text{ connected}) \le \mathbb{P}(X_1 > 0) + \sum_{k=2}^{n/2}\mathbb{P}(X_k > 0).$$

Now

$$\sum_{k=2}^{n/2}\mathbb{P}(X_k > 0) \le \sum_{k=2}^{n/2}\mathbb{E}X_k \le \sum_{k=2}^{n/2}\binom{n}{k}k^{k-2}p^{k-1}(1-p)^{k(n-k)} = \sum_{k=2}^{n/2}u_k.$$

Now, for $2 \le k \le 10$,

$$\begin{aligned}u_k &\le e^k n^k \left(\frac{\log n + c}{n}\right)^{k-1} e^{-k(n-10)\frac{\log n+c}{n}}\\ &\le (1+o(1))e^{k(1-c)}\left(\frac{\log n}{n}\right)^{k-1},\end{aligned}$$

and for $k > 10$

$$\begin{aligned}u_k &\le \left(\frac{ne}{k}\right)^k k^{k-2}\left(\frac{\log n + c}{n}\right)^{k-1} e^{-k(\log n + c)/2}\\ &\le n\left(\frac{e^{1-c/2+o(1)}\log n}{n^{1/2}}\right)^k.\end{aligned}$$

So

$$\begin{aligned}\sum_{k=2}^{n/2}u_k &\le (1+o(1))\frac{e^{-c}\log n}{n} + \sum_{k=10}^{n/2}n^{1+o(1)-k/2}\\ &= O\left(n^{o(1)-1}\right).\end{aligned}$$

It follows that

$$\mathbb{P}(\mathbb{G}_{n,p} \text{ is connected }) = \mathbb{P}(X_1 = 0) + o(1).$$

But we already know (see Theorem 3.1) that for $p = (\log n + c)/n$ the number of isolated vertices in $\mathbb{G}_{n,p}$ has an asymptotically Poisson distribution and therefore, as in (3.4)

$$\lim_{n\to\infty} \mathbb{P}(X_1) = e^{-e^{-c}},$$

and so the theorem follows. □

It is possible to tweak the proof of Theorem 4.1 to give a more precise result stating that a random graph becomes connected exactly at the moment when the last isolated vertex disappears.

Theorem 4.2 *Consider the random graph process* $\{\mathbb{G}_m\}$. *Let*

$$m_1^* = \min\{m : \delta(\mathbb{G}_m) \geq 1\},$$

$$m_c^* = \min\{m : \mathbb{G}_m \text{ is connected}\}.$$

Then, w.h.p.,

$$m_1^* = m_c^*.$$

Proof Let

$$m_\pm = \frac{1}{2} n \log n \pm \frac{1}{2} n \log\log n,$$

and

$$p_\pm = \frac{m_\pm}{N} \approx \frac{\log n \pm \log\log n}{n}.$$

We first show that w.h.p.

(i) G_{m_-} consists of a giant connected component plus a set V_1 of at most $2\log n$ isolated vertices,
(ii) G_{m_+} is connected.

Assume (i) and (ii). It follows that w.h.p.

$$m_- \leq m_1^* \leq m_c^* \leq m_+.$$

Since $\mathbb{G}_{m_-}$ consists of a connected component and a set of isolated vertices V_1, to create $\mathbb{G}_{m_+}$ we add $m_+ - m_-$ random edges. Note that $m_1^* = m_c^*$ if none of these edges is contained in V_1. Thus

$$\begin{aligned}
\mathbb{P}(m_1^* < m_c^*) &\leq o(1) + (m_+ - m_-)\frac{\frac{1}{2}|V_1|^2}{N - m_+} \\
&\leq o(1) + \frac{2n((\log n)^2)\log\log n}{\frac{1}{2}n^2 - O(n\log n)} \\
&= o(1).
\end{aligned}$$

Thus to prove the theorem, it is sufficient to verify (i) and (ii).

Let

$$p_- = \frac{m_-}{N} \approx \frac{\log n - \log\log n}{n},$$

and let X_1 be the number of isolated vertices in $\mathbb{G}_{n,p_-}$. Then

$$\begin{aligned}\mathbb{E}X_1 &= n(1-p_-)^{n-1} \\ &\approx ne^{-np_-} \\ &\approx \log n.\end{aligned}$$

Moreover

$$\begin{aligned}\mathbb{E}X_1^2 &= \mathbb{E}X_1 + n(n-1)(1-p_-)^{2n-3} \\ &\le \mathbb{E}X_1 + \mathbb{E}X_1^2(1-p_-)^{-1}.\end{aligned}$$

So,

$$\operatorname{Var} X_1 \le \mathbb{E}X_1 + 2\mathbb{E}X_1^2 p_-,$$

and

$$\begin{aligned}\mathbb{P}(X_1 \ge 2\log n) &= \mathbb{P}(|X_1 - \mathbb{E}X_1| \ge (1+o(1))\mathbb{E}X_1) \\ &\le (1+o(1))\left(\frac{1}{\mathbb{E}X_1} + 2p_-\right) \\ &= o(1).\end{aligned}$$

Having at least $2\log n$ isolated vertices is a monotone property and so w.h.p. $\mathbb{G}_{m_-}$ has less then $2\log n$ isolated vertices.

To show that the rest of $\mathbb{G}_m$ is a single connected component we let X_k, $2 \le k \le n/2$, be the number of components with k vertices in $\mathbb{G}_{p_-}$. Repeating the calculations for p_- from the proof of Theorem 4.1, we have

$$\mathbb{E}\left(\sum_{k=2}^{n/2} X_k\right) = O\left(n^{o(1)-1}\right).$$

Let

$$\mathscr{E} = \{\exists \text{ component of order } 2 \le k \le n/2\}.$$

Then

$$\begin{aligned}\mathbb{P}(\mathbb{G}_{m_-} \in \mathscr{E}) &\le O(\sqrt{n})\mathbb{P}(\mathbb{G}_{n,p_-} \in \mathscr{E}) \\ &= o(1),\end{aligned}$$

and this completes the proof of (i).

To prove (ii) (that G_{m_+} is connected w.h.p.) we note that (ii) follows from the fact that $\mathbb{G}_{n,p}$ is connected w.h.p. for $np - \log n \to \infty$ (see Theorem 4.1). By implication, $\mathbb{G}_m$ is connected w.h.p. if $n\frac{m}{N} - \log n \to \infty$. But,

$$\begin{aligned}\frac{nm_+}{N} &= \frac{n(\frac{1}{2}n\log n + \frac{1}{2}\log\log n)}{N} \\ &\approx \log n + \log\log n.\end{aligned}$$

□

4.2 k-Connectivity

In this section we show that the threshold for the existence of vertices of degree k is also the threshold for the k-connectivity of a random graph. Recall that a graph G is k-connected if the removal of at most k vertices of G does not disconnect it. In the light of the previous result it should be expected that a random graph becomes k-connected as soon as the last vertex of degree $k-1$ disappears. This is true and follows from the results of Erdős and Rényi [277]. Here is a weaker statement.

Theorem 4.3 *Let* $m = \frac{1}{2}n(\log n + (k-1)\log\log n + c_n)$, $k = 1,2,\ldots$, *then*

$$\lim_{n\to\infty} \mathbb{P}(\mathbb{G}_m \textit{ is k-connected}) = \begin{cases} 0 & \textit{if } c_n \to -\infty \\ e^{-\frac{e^{-c}}{(k-1)!}} & \textit{if } c_n \to c \\ 1 & \textit{if } c_n \to \infty. \end{cases}$$

Proof Let

$$p = \frac{\log n + (k-1)\log\log n + c}{n}.$$

We will prove that, in $\mathbb{G}_{n,p}$, with edge probability p above,

(i) the expected number of vertices of degree at most $k-2$ is $o(1)$,
(ii) the expected number of vertices of degree $k-1$ is, approximately $\frac{e^{-c}}{(k-1)!}$.

We have

$$\begin{aligned}&\mathbb{E}(\text{number of vertices of degree } t \le k-1) \\ &\quad = n\binom{n-1}{t}p^t(1-p)^{n-1-t} \approx n\frac{n^t}{t!}\frac{(\log n)^t}{n^t}\frac{e^{-c}}{n(\log n)^{k-1}}\end{aligned}$$

and (i) and (ii) follow immediately.

The distribution of the number of vertices of degree $k-1$ is asymptotically Poisson, as may be verified by the Method of Moments. (See Exercise 3.3.4).

We now show that, if

$$\mathscr{A}(S,T) = \{T \text{ is a component of } \mathbb{G}_{n,p} \setminus S\}$$

then

$$\mathbb{P}\left(\exists S,T,\ |S| < k,\ k-|S|+1 \leq |T| \leq \frac{1}{2}(n-|S|) : \mathscr{A}(S,T)\right) = o(1).$$

This implies that if $\delta(\mathbb{G}_{n,p}) \geq k$ then $\mathbb{G}_{n,p}$ is k-connected and Theorem 4.3 follows. The lower bound on $|T|$ arises from the fact that in this case each vertex of S has at least $k-|S|+1$ neighbors outside S. Also, $|T| \geq 2$ because if $T = \{v\}$ then v has degree less than k.

We can assume that S is minimal and then $N(T) = S$ and denote $s = |S|$, $t = |T|$. We note that T minimal implies that T is connected, and so it contains a tree with $t-1$ edges. Also each vertex of S is incident with an edge from S to T and so there are at least s edges between S and T. Thus, if $p = (1+o(1))\frac{\log n}{n}$ then

$$\begin{aligned}
&\mathbb{P}(\exists S,T) \leq o(1) \\
&\quad + \sum_{s=1}^{k-1} \sum_{t=\max\{2,k-s+1\}}^{(n-s)/2} \binom{n}{s}\binom{n}{t} t^{t-2} p^{t-1} \binom{st}{s} p^s (1-p)^{t(n-s-t)} \\
&\leq p^{-1} \sum_{s=1}^{k-1} \sum_{t=\max\{2,k-s+1\}}^{(n-s)/2} \left(\frac{ne}{s} \cdot (te) \cdot p \cdot \frac{e}{s} \cdot e^{tp}\right)^s \left(\frac{ne}{t} \cdot p \cdot e^{-(n-t)p}\right)^t \\
&\leq p^{-1} \sum_{s=1}^{k-1} \sum_{t=\max\{2,k-s+1\}}^{(n-s)/2} A^t B^s \qquad (4.1)
\end{aligned}$$

where

$$\begin{aligned}
A &= nepe^{-(n-t)p} = e^{1+o(1)} n^{-1+(t+o(t))/n} \log n \\
B &= ne^3 tpe^{tp} = e^{3+o(1)} tn^{(t+o(t))/n} \log n.
\end{aligned}$$

Now if $2 \leq t \leq \log n$ then $A = n^{-1+o(1)}$ and $B = O((\log n)^2)$. On the other hand, if $t > \log n$ then we can use $A \leq n^{-1/3}$ and $B \leq n^2$ to see that the sum in (4.1) is $o(1)$. □

4.3 Exercises

4.3.1 Let $m = m_1^*$ be as in Theorem 4.2 and let $e_m = (u, v)$ where u has degree one. Let $0 < c < 1$ be a positive constant. Show that w.h.p. there is no triangle containing u or v.

4.3.2 Let $m = m_1^*$ as in Theorem 4.2 and let $e_m = (u, v)$ where u has degree one. Let $0 < c < 1$ be a positive constant. Show that w.h.p. the degree of v in G_m is at least $c \log n$.

4.3.3 Suppose that $m \gg n \log n$ and let $d = m/n$. Let $S_i(v)$ be the set of vertices at distance i from vertex v. Show that w.h.p. $|S_i(v)| \geq (d/2)^i$ for all $v \in [n]$ and $1 \leq i \leq \frac{2 \log n}{3 \log d}$.

4.3.4 Suppose that $m \gg n \log n$ and let $d = m/n$. Using the previous question, show that w.h.p. there are at least $d/2$ internally vertex disjoint paths of length at most $\frac{4 \log n}{3 \log d}$ between any pair of vertices in $G_{n,m}$.

4.3.5 Suppose that $m \gg n \log n$ and let $d = m/n$. Suppose that we randomly color the edges of $G_{n,m}$ with q colors where $q \gg \frac{(\log n)^2}{(\log d)^2}$. Show that w.h.p. there is a *rainbow* path between every pair of vertices. (A path is rainbow if each of its edges has a different color.)

4.3.6 Let $C_{k,k+\ell}$ denote the number of connected graphs with vertex set $[k]$ and $k + \ell$ edges where $\ell \to \infty$ with k and $\ell = o(k)$. Use the inequality

$$\binom{n}{k} C_{k,k+\ell} p^{k+\ell} (1-p)^{\binom{k}{2} - k - \ell + k(n-k)} \leq \frac{n}{k}$$

and a careful choice of p, n to prove (see Łuczak [536]) that

$$C_{k,k+\ell} \leq \sqrt{\frac{k^3}{\ell}} \left(\frac{e + O(\sqrt{\ell/k})}{12\ell} \right)^{\ell/2} k^{k+(3\ell-1)/2}.$$

4.3.7 Let $\mathbb{G}_{n,n,p}$ be the random bipartite graph with vertex bi-partition $V = (A, B)$, $A = [1, n], B = [n+1, 2n]$ in which each of the n^2 possible edges appears independently with probability p. Let $p = \frac{\log n + \omega}{n}$, where $\omega \to \infty$. Show that w.h.p. $\mathbb{G}_{n,n,p}$ is connected.

4.4 Notes

Disjoint Paths

Being k-connected means that we can find disjoint paths between any two sets of vertices $A = \{a_1, a_2, \dots, a_k\}$ and $B = \{b_1, b_2, \dots, b_k\}$. In this statement there is no control over the endpoints of the paths, i.e. we cannot specify a path from a_i to b_i for $i = 1, 2, \dots, k$. Specifying the endpoints leads to the notion of *linkedness*. Broder, Friczc, Suen and Upfal [163] proved that when we are above the connectivity threshold, we can w.h.p. link any two k-sets by edge disjoint paths, provided some natural restrictions apply. The result is optimal up to constants. Broder, Frieze, Suen and Upfal [162] considered the case of vertex disjoint paths. Frieze and Zhao [352] considered the edge disjoint path version in random regular graphs.

Rainbow Connection

The rainbow connection $rc(G)$ of a connected graph G is the minimum number of colors needed to color the edges of G so that there is a rainbow path between every pair of vertices. Caro, Lev, Roditty, Tuza and Yuster [171] proved that $p = \sqrt{\log n/n}$ is the sharp threshold for the property $rc(G) \leq 2$. This was sharpened to a hitting time result by Heckel and Riordan [402]. He and Liang [401] further studied the rainbow connection of random graphs. Specifically, they obtain a threshold for the property $rc(G) \leq d$ where d is constant. Frieze and Tsourakakis [351] studied the rainbow connection of $G = G(n,p)$ at the connectivity threshold $p = \frac{\log n+\omega}{n}$ where $\omega \to \infty$ and $\omega = o(\log n)$. They showed that w.h.p. $rc(G)$ is asymptotically equal to $\max\{diam(G), Z_1(G)\}$, where Z_1 is the number of vertices of degree one.

5

Small Subgraphs

Graph theory is replete with theorems stating conditions for the existence of a subgraph H in a larger graph G. For example Turán's theorem [706] states that a graph with n vertices and more than $\left(1-\frac{1}{r}\right)\frac{n^2}{2}$ edges must contain a copy of K_{r+1}. In this chapter we see instead how many random edges are required to have a particular fixed size subgraph w.h.p. In addition, we consider the distribution of the number of copies.

5.1 Thresholds

In this section we look for a threshold for the appearance of any fixed graph H, with $v_H = |V(H)|$ vertices and $e_H = |E(H)|$ edges. The property that a random graph contains H as a subgraph is clearly monotone increasing. It is also transparent that "denser" graphs appear in a random graph "later" than "sparser" ones. More precisely, denote by

$$d(H) = \frac{e_H}{v_H}, \tag{5.1}$$

the *density* of a graph H. Notice that $2d(H)$ is the average vertex degree in H. We begin with the analysis of the asymptotic behavior of the expected number of copies of H in the random graph $\mathbb{G}_{n,p}$.

Lemma 5.1 *Let X_H denote the number of copies of H in $\mathbb{G}_{n,p}$.*

$$\mathbb{E}X_H = \binom{n}{v_H}\frac{v_H!}{aut(H)}p^{e_H},$$

where $aut(H)$ is the number of automorphisms of H.

Proof The complete graph on n vertices K_n contains $\binom{n}{v_H}a_H$ distinct copies of H, where a_H is the number of copies of H in K_{v_H}. Thus,

$$\mathbb{E}X_H = \binom{n}{v_H} a_H p^{e_H},$$

and all we need to show is that

$$a_H \times aut(H) = v_H!.$$

Each permutation σ of $[v_H] = \{1,2,\ldots,v_H\}$ defines a unique copy of H as follows: A copy of H corresponds to a set of e_H edges of K_{v_H}. The copy H_σ corresponding to σ has edges $\{(x_{\sigma(i)}, y_{\sigma(i)}) : 1 \le i \le e_H\}$, where $\{(x_j, y_j) : 1 \le j \le e_H\}$ is some fixed copy of H in K_{v_H}. But $H_\sigma = H_{\tau\sigma}$ if and only if for each i there is j such that $(x_{\tau\sigma(i)}, y_{\tau\sigma(i)}) = (x_{\sigma(j)}, y_{\sigma(j)})$, i.e. if τ is an automorphism of H. □

Theorem 5.2 *Let H be a fixed graph with $e_H > 0$. Suppose $p = o\left(n^{-1/d(H)}\right)$. Then w.h.p. $\mathbb{G}_{n,p}$ contains no copies of H.*

Proof Suppose that $p = \omega^{-1} n^{-1/d(H)}$ where $\omega = \omega(n) \to \infty$ as $n \to \infty$, then

$$\mathbb{E}X_H = \binom{n}{v_H} \frac{v_H!}{aut(H)} p^{e_H} \le n^{v_H} \omega^{-e_H} n^{-e_H/d(H)} = \omega^{-e_H}.$$

Thus

$$\mathbb{P}(X_H > 0) \le \mathbb{E}X_H \to 0 \text{ as } n \to \infty.$$

□

From our previous experience, we expect that when $\mathbb{E}X_H \to \infty$ as $n \to \infty$ the random graph $\mathbb{G}_{n,p}$ contains H as a subgraph w.h.p. Let us check whether such a phenomenon also holds in this case. So consider the case when $pn^{1/d(H)} \to \infty$, i.e. where $p = \omega n^{-1/d(H)}$ and $\omega = \omega(n) \to \infty$ as $n \to \infty$. Then for some constant $c_H > 0$

$$\mathbb{E}X_H \ge c_H n^{v_H} \omega^{e_H} n^{-e_H/d(H)} = c_H \omega^{e_H} \to \infty.$$

However, as we will see, this is not always enough for $\mathbb{G}_{n,p}$ to contain a copy of a given graph H w.h.p. To see this, consider the graph H given in Figure 5.1.

Here $v_H = 6$ and $e_H = 8$. Let $p = n^{-5/7}$. Now $1/d(H) = 6/8 > 5/7$ and so

$$\mathbb{E}X_H \approx c_H n^{6-8\times 5/7} \to \infty.$$

On the other hand, if $\hat{H} = K_4$ then

$$\mathbb{E}X_{\hat{H}} \le n^{4-6\times 5/7} \to 0,$$

and so w.h.p. there are no copies of $\hat{H}$ and hence no copies of H.

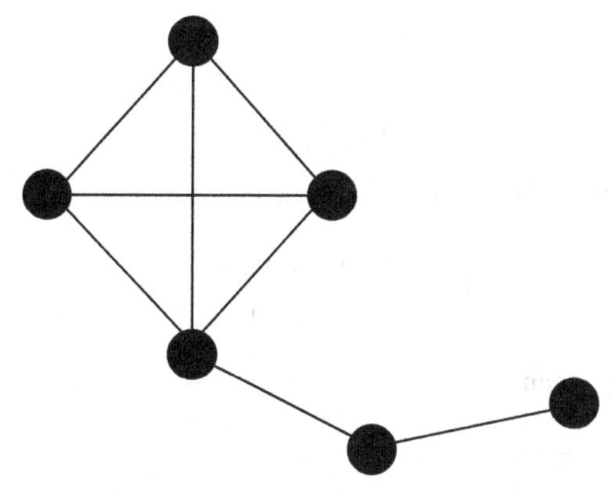

Figure 5.1. A kite

The reason for such “strange” behavior is quite simple. Our graph H is in fact *not balanced*, since its overall density is smaller than the density of one of its subgraphs, i.e. of $\hat{H} = K_4$. So we need to introduce another density characteristic of graphs, namely the maximum subgraph density defined as:

$$m(H) = \max\{d(K) : K \subseteq H\}. \tag{5.2}$$

A graph H is *balanced* if $m(H) = d(H)$. It is *strictly balanced* if $d(H) > d(K)$ for all proper subgraphs $K \subset H$.

Now we are ready to determine the threshold for the existence of a copy of H in $\mathbb{G}_{n,p}$. Erdős and Rényi [276] proved this result for balanced graphs. The threshold for any graph H was first found by Bollobás in [123] and an alternative, deterministic argument to derive the threshold was presented in [470]. A simple proof, given here, is due to Ruciński and Vince [659].

Theorem 5.3 *Let H be a fixed graph with $e_H > 0$. Then*

$$\lim_{n\to\infty} \mathbb{P}(H \subseteq \mathbb{G}_{n,p}) = \begin{cases} 0 & \text{if } pn^{1/m(H)} \to 0 \\ 1 & \text{if } pn^{1/m(H)} \to \infty. \end{cases}$$

Proof Let $\omega = \omega(n) \to \infty$ as $n \to \infty$. The first statement follows from Theorem 5.2. Notice, that if we choose $\hat{H}$ to be a subgraph of H with $d(\hat{H}) = m(H)$ (such a subgraph always exists since we do not exclude $\hat{H} = H$), then $p = \omega^{-1} n^{-1/d(\hat{H})}$ implies that $\mathbb{E} X_{\hat{H}} \to 0$. Therefore, w.h.p. $\mathbb{G}_{n,p}$ contains no copies of $\hat{H}$, and so it also does not contain H.

To prove the second statement we use the Second Moment Method. Suppose now that $p = \omega n^{-1/m(H)}$. Denote by $H_1, H_2, \ldots, H_t$ all copies of H in the complete graph on $\{1,2,\ldots,n\}$. Note that

$$t = \binom{n}{v_H} \frac{v_H!}{aut(H)}, \tag{5.3}$$

where $aut(H)$ is the number of automorphisms of H. For $i = 1,2,\ldots,t$ let

$$I_i = \begin{cases} 1 & \text{if } H_i \subseteq \mathbb{G}_{n,p}, \\ 0 & \text{otherwise.} \end{cases}$$

Let $X_H = \sum_{i=1}^t I_i$. Then

$$\begin{aligned}
\operatorname{Var} X_H &= \sum_{i=1}^t \sum_{j=1}^t \operatorname{Cov}(I_i, I_j) = \sum_{i=1}^t \sum_{j=1}^t (\mathbb{E}(I_i I_j) - (\mathbb{E} I_i)(\mathbb{E} I_j)) \\
&= \sum_{i=1}^t \sum_{j=1}^t (\mathbb{P}(I_i = 1, I_j = 1) - \mathbb{P}(I_i = 1)\mathbb{P}(I_j = 1)) \\
&= \sum_{i=1}^t \sum_{j=1}^t \left(\mathbb{P}(I_i = 1, I_j = 1) - p^{2e_H}\right).
\end{aligned}$$

Observe that random variables I_i and I_j are independent if and only if H_i and H_j are edge disjoint. In this case $\operatorname{Cov}(I_i, I_j) = 0$ and such terms vanish from the above summation. Therefore we consider only pairs (H_i, H_j) with $H_i \cap H_j = K$, for some graph K with $e_K > 0$. So,

$$\begin{aligned}
\operatorname{Var} X_H &= O\Big(\sum_{K \subseteq H, e_K > 0} n^{2v_H - v_K} \left(p^{2e_H - e_K} - p^{2e_H}\right)\Big) \\
&= O\Big(n^{2v_H} p^{2e_H} \sum_{K \subseteq H, e_K > 0} n^{-v_K} p^{-e_K}\Big).
\end{aligned}$$

On the other hand,

$$\mathbb{E} X_H = \binom{n}{v_H} \frac{v_H!}{aut(H)} p^{e_H} = \Omega\left(n^{v_H} p^{e_H}\right),$$

thus, by Lemma 20.4,

$$\begin{aligned}
\mathbb{P}(X_H = 0) \le \frac{\operatorname{Var} X_H}{(\mathbb{E} X_H)^2} &= O\left(\sum_{K \subseteq H, e_K > 0} n^{-v_K} p^{-e_K}\right) \\
&= O\left(\sum_{K \subseteq H, e_K > 0} \left(\frac{1}{\omega n^{1/d(K) - 1/m(H)}}\right)^{e_K}\right) \\
&= o(1).
\end{aligned}$$

Hence w.h.p., the random graph $\mathbb{G}_{n,p}$ contains a copy of the subgraph H when $pn^{1/m(H)} \to \infty$. $\square$

5.2 Asymptotic Distributions

We now study the asymptotic distribution of the number X_H of copies of a fixed graph H in $\mathbb{G}_{n,p}$. We start at the threshold, so we assume that $np^{m(H)} \to c$, $c > 0$, where $m(H)$ denotes as before, the maximum subgraph density of H. Now, if H is not balanced, i.e. its maximum subgraph density exceeds the density of H, then $\mathbb{E}X_H \to \infty$ as $n \to \infty$, and we can show that there is a sequence of numbers a_n, increasing with n, such that the asymptotic distribution of X_H/a_n coincides with the distribution of a random variable counting the number of copies of a subgraph K of H for which $m(H) = d(K)$. Note that K is itself a balanced graph. However, the asymptotic distribution of balanced graphs on the threshold, although computable, cannot be given in a closed form. The situation changes dramatically if we assume that the graph H whose copies in $\mathbb{G}_{n,p}$ we want to count is *strictly balanced*, i.e. when for every proper subgraph K of H, $d(K) < d(H) = m(H)$.

The following result is due to Bollobás [123], and Karoński and Ruciński [469].

Theorem 5.4 *If H is a strictly balanced graph and $np^{m(H)} \to c, c > 0$, then $X_H \xrightarrow{D} \text{Po}(\lambda)$, as $n \to \infty$, where $\lambda = c^{v_H}/aut(H)$.*

Proof Denote, as before, by $H_1, H_2, \ldots, H_t$ all copies of H in the complete graph on $\{1,2,\ldots,n\}$. For $i = 1,2,\ldots,t$, let

$$I_{H_i} = \begin{cases} 1 & \text{if } H_i \subseteq \mathbb{G}_{n,p} \\ 0 & \text{otherwise.} \end{cases}$$

Then $X_H = \sum_{i=1}^{t} I_{H_i}$ and the kth factorial moment of X_H, $k = 1,2\ldots$,

$$\mathbb{E}(X_H)_k = \mathbb{E}[X_H(X_H - 1)\cdots(X_H - k + 1)],$$

can be written as

$$\begin{aligned} \mathbb{E}(X_H)_k &= \sum_{i_1,i_2,\ldots,i_k} \mathbb{P}(I_{H_{i_1}} = 1, I_{H_{i_2}} = 1, \ldots, I_{H_{i_k}} = 1) \\ &= D_k + \overline{D}_k, \end{aligned}$$

where the summation is taken over all k-element sequences of distinct indices i_j from $\{1,2,\ldots,t\}$, while D_k and $\overline{D}_k$ denote the partial sums taken over all (ordered) k tuples of copies of H, which are, pairwise vertex disjoint (D_k) and not all pairwise vertex disjoint ($\overline{D}_k$), respectively. Now, observe that

$$\begin{aligned}D_k &= \sum_{i_1,i_2,\dots,i_k} \mathbb{P}(I_{H_{i_1}} = 1)\,\mathbb{P}(I_{H_{i_2}} = 1)\cdots\mathbb{P}(I_{H_{i_k}} = 1)\\ &= \binom{n}{v_H, v_H, \dots, v_H} (a_H p^{e_H})^k\\ &\approx (\mathbb{E}X_H)^k.\end{aligned}$$

So assuming that $np^{d(H)} = np^{m(H)} \to c$ as $n \to \infty$,

$$D_k \approx \left(\frac{c^{v_H}}{aut(H)}\right)^k. \tag{5.4}$$

However, we will show that

$$\overline{D}_k \to 0 \text{ as } n \to \infty. \tag{5.5}$$

Consider the family $\mathscr{F}_k$ of all (mutually non-isomorphic) graphs obtained by taking unions of k not all pairwise vertex disjoint copies of the graph H. Suppose $F \in \mathscr{F}_k$ has v_F vertices ($k \le v_F \le kv_H - 1$) and e_F edges, and let $d(F) = e_F/v_F$ be its density. To prove that (5.5) holds we need the following Lemma.

Lemma 5.5 *If $F \in \mathscr{F}_k$ then $d(F) > m(H)$.*

Proof Define

$$f_F = m(H)v_F - e_F. \tag{5.6}$$

We will show (by induction on $k \ge 2$) that $f_F < 0$ for all $F \in \mathscr{F}_k$. First note that $f_H = 0$ and that $f_K > 0$ for every proper subgraph K of H, since H is strictly balanced. Notice also that the function f is modular, i.e. for any two graphs F_1 and F_2,

$$f_{F_1 \cup F_2} = f_{F_1} + f_{F_2} - f_{F_1 \cap F_2}. \tag{5.7}$$

Assume that the copies of H composing F are numbered in such a way that $H_{i_1} \cap H_{i_2} \ne \emptyset$. If $F = H_{i_1} \cup H_{i_2}$ then (5.6) and $f_{H_1} = f_{H_2} = 0$ implies

$$f_{H_{i_1} \cup H_{i_2}} = -f_{H_{i_1} \cap H_{i_2}} < 0.$$

For arbitrary $k \ge 3$, let $F' = \bigcup_{j=1}^{k-1} H_{i_j}$ and $K = F' \cap H_{i_k}$. Then by the inductive assumption we have $f_{F'} < 0$ while $f_K \ge 0$ since K is a subgraph of H (in extreme cases K can be H itself or an empty graph). Therefore,

$$f_F = f_{F'} + f_{H_{i_k}} - f_K = f_{F'} - f_K < 0,$$

which completes the induction and implies that $d(F) > m(H)$. □

Let C_F be the number of sequences $H_{i_1}, H_{i_2}, \ldots, H_{i_k}$ of k distinct copies of H, such that

$$V\Big(\bigcup_{j=1}^{k} H_{i_j}\Big) = \{1, 2, \ldots, v_F\} \text{ and } \bigcup_{j=1}^{k} H_{i_j} \cong F.$$

Then, by Lemma 5.5,

$$\begin{aligned}\overline{D}_k &= \sum_{F \in \mathscr{F}_k} \binom{n}{v_F} C_F \, p^{e_F} = O(n^{v_F} p^{e_F}) \\ &= O\left(\left(np^{d(F)} \right)^{v(F)} \right) = o(1),\end{aligned}$$

and so (5.5) holds.

Summarizing,

$$\mathbb{E}(X_H)_k \approx \left(\frac{c^{v_H}}{aut(H)} \right)^k,$$

and the theorem follows by the Method of Moments (see Corollary 20.11). □

The following theorem describes the asymptotic behavior of the number of copies of a graph H in $\mathbb{G}_{n,p}$ past the threshold for the existence of a copy of H. It holds regardless of whether or not H is balanced or strictly balanced. We state the theorem but we do not supply a proof (see Ruciński [657]).

Theorem 5.6 *Let H be a fixed (not-empty) graph. If $np^{m(H)} \to \infty$ and $n^2(1 - p) \to \infty$, then $(X_H - \mathbb{E} X_H)/(\mathrm{Var} X_H)^{1/2} \xrightarrow{D} \mathrm{N}(0,1)$, as $n \to \infty$.*

5.3 Exercises

5.3.1 Draw a graph that is: (i) balanced but not strictly balanced, (ii) unbalanced.

5.3.2 Are the small graphs listed below, balanced or unbalanced: (i) a tree, (ii) a cycle, (iii) a complete graph, (iv) a regular graph, (v) the Petersen graph, (vi) a graph composed of a complete graph on 4 vertices and a triangle, sharing exactly one vertex.

5.3.3 Determine (directly, not from the statement of Theorem 5.3) thresholds $\hat{p}$ for $\mathbb{G}_{n,p} \supseteq G$, for graphs listed in exercise 5.3.2. Do the same for the thresholds of G in $\mathbb{G}_{n,m}$.

5.3.4 For a graph G a *balanced extension* of G is a graph F, such that $G \subseteq F$ and $m(F) = d(F) = m(G)$. Applying the result of Győri, Rothschild and Ruciński [390] that every graph has a balanced extension, deduce Bollobás's result (Theorem 5.3) from that of Erdős and Rényi (threshold for balanced graphs).

5.3.5 Let F be a graph obtained by taking a union of triangles such that not every pair of them is vertex-disjoint, Show (by induction) that $e_F > v_F$.

5.3.6 Let f_F be a graph function defined as

$$f_F = a\, v_F + b\, e_F,$$

where a, b are constants, while v_F and e_F denote, the number of vertices and edges of a graph F, respectively. Show that the function f_F is modular.

5.3.7 Determine (directly, using exercise 5.3.5) when the random variable counting the number of copies of a triangle in $\mathbb{G}_{n,p}$ has asymptotically the Poisson distribution.

5.3.8 Let X_e be the number of isolated edges (edge-components) in $\mathbb{G}_{n,p}$ and let

$$\omega(n) = 2pn - \log n - \log\log n.$$

Prove that

$$\mathbb{P}(X_e > 0) \to \begin{cases} 0 & \text{if } p \ll n^{-2} \quad \text{or} \quad \omega(n) \to \infty \\ 1 & \text{if } p \gg n^{-2} \quad \text{and} \quad \omega(n) \to \infty. \end{cases}$$

5.3.9 Determine when the random variable X_e defined in exercise 5.3.7 has asymptotically the Poisson distribution.

5.3.10 Use Janson's inequality, Theorem 21.12, to prove (5.8) below.

5.4 Notes

Distributional Questions

In 1982, Barbour [57] adapted the Stein–Chen technique for obtaining estimates of the rate of convergence to the Poisson and the normal distribution (see Section 20.3 or [58]) to random graphs. The method was next applied by Karoński and Ruciński [471] to prove the convergence results for semi-induced graph properties of random graphs.

Barbour, Karoński and Ruciński [60] used the original Stein's method for normal approximation to prove a general central limit theorem for the wide class of decomposable random variables. Their result is illustrated by a variety of applications to random graphs. For example, one can deduce from it the asymptotic distribution of the number of k-vertex tree-components in $\mathbb{G}_{n,p}$, as well as the number of vertices of fixed degree d in $\mathbb{G}_{n,p}$ (in fact, Theorem 3.2 is a direct consequence of the last result).

Barbour, Janson, Karoński and Ruciński [59] studied the number X_k of maximal complete subgraphs (cliques) of a given fixed size $k \geq 2$ in the random graph $\mathbb{G}_{n,p}$. They showed that if the edge probability $p = p(n)$ is such that the $\mathbb{E}X_k$ tends to a finite constant λ as $n \to \infty$, then X_k tends in distribution to the Poisson random variable with the expectation λ. When its expectation tends to infinity, X_k converges in distribution to a random variable that is normally distributed. Poisson convergence was proved using the Stein–Chen method, while for the proof of the normal part, different methods for different ranges of p were used such as the first projection method or martingale limit theorem (for details of these methods see Chapter 6 of Janson, Łuczak and Ruciński [432]).

Svante Janson in a sequence of papers [415],[416], [417], [420] (see also [433]) developed or accommodated various methods to establish the asymptotic normality of various numerical random graph characteristics. In particular, in [416] he established the normal convergence by higher semi-invariants of sums of dependent random variables with direct applications to random graphs. In [417] he proved a functional limit theorem for subgraph count statistics in random graphs (see also [433]).

In 1997, Janson [415] answered the question posed by Paul Erdős: what is the length Y_n of the first cycle appearing in the random graph process $\mathbb{G}_m$? He proved that

$$\lim_{n\to\infty} \mathbb{P}(Y_n = j) = \frac{1}{2}\int_0^1 t^{j-1} e^{t/2+t^2/4}\sqrt{1-t}\, dt, \text{ for every } j \geq 3.$$

Tails of Subgraph Counts in $\mathbb{G}_{n,p}$

Often we need exponentially small bounds for the probability that X_H deviates from its expectation. In 1990, Janson [418] showed that for fixed $\varepsilon \in (0,1]$,

$$\mathbb{P}(X_H \leq (1-\varepsilon)\mathbb{E}X_H) = \exp\{-\Theta(\Phi_H)\}, \tag{5.8}$$

where $\Phi_H = \min_{K \subseteq H: e_K > 0} n^{v_K} p^{e_K}$.

The upper tail $\mathbb{P}(X_H \geq (1+\varepsilon)\mathbb{E}X_H)$ proved to be much more elusive. To simplify the results, let us assume that ε is fixed and p is above the existence threshold, i.e. $p \gg n^{1/m(H)}$, but small enough to make sure that $(1+\varepsilon)\mathbb{E}X_H$ is at most the number of copies of H in K_n.

Given a graph G, let Δ_G be the maximum degree of G and α^*_G the *fractional independence number* of G, defined as the maximum of $\sum_{v\in V(G)} w(v)$ over all functions $w : V(G) \to [0,1]$ satisfying $w(u)+w(v) \le 1$ for every $uv \in E(G)$.

In 2004, Janson, Oleszkiewicz and Ruciński [430] proved that

$$\exp\{-O(M_H \log(1/p))\} \le \mathbb{P}(X_H \ge (1+\varepsilon)\mathbb{E}X_H) \le \exp\{-\Omega(M_H)\}, \quad (5.9)$$

where

$$M_H = \begin{cases} \min_{K\subseteq H}(n^{v_K} p^{e_K})^{1/\alpha^*_K}, & \text{if } n^{-1/m(H)} \le p \le n^{-1/\Delta_H}, \\ n^2 p^{\Delta_H}, & \text{if } p \ge n^{-1/\Delta_H}. \end{cases}$$

For example, if H is k-regular, then $M_H = n^2 p^k$ for every p.

The logarithms of the upper and lower bounds in (5.9) differ by a multiplicative factor of $\log(1/p)$. Recent results suggest that the following conjecture (stated in [232] for $\varepsilon = 1$) is likely to be true.

Conjecture: For any H and $\varepsilon > 0$,

$$\mathbb{P}(X_H \ge (1+\varepsilon)\mathbb{E}X_H) = \exp\left(-\Theta\left(\min\{\Phi_H, M_H \log(1/p)\}\right)\right). \quad (5.10)$$

A careful look reveals that the minimum in (5.10) is Φ_H just in the tiny range above the threshold of existence, i.e. when $p \le n^{-1/m(H)}(\log n)^{a_H}$ for some $a_H > 0$. So, only in this range does the upper tail behave similarly to the lower tail.

DeMarco and Kahn [232] proved (5.10) for cliques $H = K_k$, $k = 3,4,\ldots$. Adamczak and Wolff [6] proved a polynomial concentration inequality that confirms (5.10) for any cycle $H = C_k$, $k = 3,4,\ldots$ and $p \ge n^{-\frac{k-2}{2(k-1)}}$. Moreover, Lubetzky and Zhao [534], via a large deviations framework of Chatterjee and Dembo [174], showed that (5.10) holds for any H and $p \ge n^{-\alpha}$ for a sufficiently small constant $\alpha > 0$.

6
Spanning Subgraphs

The previous chapter dealt with the existence of small subgraphs of a fixed size. In this chapter we concern ourselves with the existence of large subgraphs, most notably perfect matchings and Hamilton cycles. The celebrated theorems of Hall and Tutte give necessary and sufficient conditions for a bipartite and arbitrary graph, respectively, to contain a perfect matching. Hall's theorem in particular can be used to establish that the threshold for having a perfect matching in a random bipartite graph can be identified with that of having no isolated vertices.

For general graphs, we view a perfect matching as half a Hamilton cycle and prove thresholds for the existence of perfect matchings and Hamilton cycles in a similar way.

Having dealt with perfect matchings and Hamilton cycles, we turn our attention to long paths in sparse random graphs, i.e. in those where we expect a linear number of edges. We then analyze a simple GREEDY matching algorithm using differential equations.

We then consider random subgraphs of some fixed graph G, as opposed to random subgraphs of K_n. We give sufficient conditions for the existence of long paths and cycles.

We finally consider the existence of arbitrary spanning subgraphs H where we bound the maximum degree $\Delta(H)$.

6.1 Perfect Matchings

Before we move to the problem of the existence of a perfect matching, i.e. a collection of independent edges covering all of the vertices of a graph, in our main object of study, the random graph $\mathbb{G}_{n,p}$, we analyze the same problem in a random bipartite graph. This problem is much simpler than the respective one for $\mathbb{G}_{n,p}$, but provides a general approach to finding a perfect matching in a random graph.

Bipartite Graphs

Let $\mathbb{G}_{n,n,p}$ be the random bipartite graph with vertex bi-partition $V = (A, B)$, $A = [1, n], B = [n+1, 2n]$ in which each of the n^2 possible edges appears independently with probability p. The following theorem was first proved by Erdős and Rényi [278].

Theorem 6.1 *Let $\omega = \omega(n)$, $c > 0$ be a constant, and $p = \frac{\log n + \omega}{n}$, then*

$$\lim_{n\to\infty} \mathbb{P}(\mathbb{G}_{n,n,p} \text{ has a perfect matching}) = \begin{cases} 0 & \text{if } \omega \to -\infty \\ e^{-2e^{-c}} & \text{if } \omega \to c \\ 1 & \text{if } \omega \to \infty. \end{cases}$$

Moreover,

$$\lim_{n\to\infty} \mathbb{P}(\mathbb{G}_{n,n,p} \text{ has a perfect matching}) = \lim_{n\to\infty} \mathbb{P}(\delta(\mathbb{G}_{n,n,p}) \geq 1).$$

Proof We will use Hall's condition for the existence of a perfect matching in a bipartite graph. It states that a bipartite graph contains a perfect matching if and only if the following condition is satisfied:

$$\forall S \subseteq A,\ |N(S)| \geq |S|, \tag{6.1}$$

where for a set of vertices S, $N(S)$ denotes the set of neighbors of S.

It is convenient to replace (6.1) by

$$\forall S \subseteq A,\ |S| \leq \frac{n}{2},\ \ |N(S)| \geq |S|, \tag{6.2}$$

$$\forall T \subseteq B,\ |T| \leq \frac{n}{2},\ \ |N(T)| \geq |T|. \tag{6.3}$$

This is because if $|S| > n/2$ and $|N(S)| < |S|$ then $T = B \setminus N(S)$ will violate (6.3).

Now

$$\begin{aligned} &\mathbb{P}(\exists v : v \text{ is isolated}) \leq \mathbb{P}(\nexists \text{ a perfect matching}) \\ &\leq \mathbb{P}(\exists v : v \text{ is isolated}) + 2\mathbb{P}(S \subseteq A, T \subseteq B, 2 \leq k = |S| \leq n/2, \\ &|T| = k-1, N(S) \subseteq T \quad \text{and} \quad e(S:T) \geq 2k-2). \end{aligned}$$

Here $e(S : T)$ denotes the number of edges between S and T, and to see why $e(S : T)$ must be at least $2k - 2$, take a pair S, T with $|S| + |T|$ as small as possible. Next

(i) if $|S| > |T| + 1$, remove $|S| - |T| - 1$ vertices from $|S|$.

(ii) Suppose $\exists w \in T$ such that w has less than 2 neighbors in S. Remove w and its (unique) neighbor in $|S|$.

Repeat until (i) and (ii) do not hold. Note that $|S|$ will remain at least 2 if the minimum degree $\delta \geq 1$.

Suppose now that $p = \frac{\log n + c}{n}$ for some constant c. Then let Y denote the number of sets S and T satisfying the conditions (6.2) and (6.3). Then

$$\begin{aligned}
\mathbb{E}\,Y &\leq 2\sum_{k=2}^{n/2} \binom{n}{k}\binom{n}{k-1}\binom{k(k-1)}{2k-2} p^{2k-2}(1-p)^{k(n-k)} \\
&\leq 2\sum_{k=2}^{n/2} \left(\frac{ne}{k}\right)^k \left(\frac{ne}{k-1}\right)^{k-1} \left(\frac{ke(\log n + c)}{2n}\right)^{2k-2} e^{-npk(1-k/n)} \\
&\leq \sum_{k=2}^{n/2} n\left(\frac{e^{O(1)}n^{k/n}(\log n)^2}{n}\right)^k \\
&= \sum_{k=2}^{n/2} u_k.
\end{aligned}$$

Case 1: $2 \leq k \leq n^{3/4}$.

$$u_k = n((e^{O(1)}n^{-1}\log n)^2)^k.$$

So

$$\sum_{k=2}^{n^{3/4}} u_k = O\left(\frac{1}{n^{1-o(1)}}\right).$$

Case 2: $n^{3/4} < k \leq n/2$.

$$u_k \leq n^{1-k(1/2-o(1))}.$$

So

$$\sum_{n^{3/4}}^{n/2} u_k = O\left(n^{-n^{3/4}/3}\right).$$

So

$$\mathbb{P}(\nexists \text{ a perfect matching}) = \mathbb{P}(\exists \text{ isolated vertex}) + o(1).$$

Let X_0 denote the number of isolated vertices in $\mathbb{G}_{n,n,p}$, then

$$\mathbb{E}X_0 = 2n(1-p)^n \approx 2e^{-c}.$$

By previously used techniques we have

$$\mathbb{P}(X_0 = 0) \approx e^{-2e^{-c}}.$$

To prove the case for $|\omega| \to \infty$ we can use monotonicity and (1.6) and the fact that $e^{-e^{-2c}} \to 0$ if $c \to -\infty$ and $e^{-e^{-2c}} \to 1$ if $c \to \infty$. □

Non-Bipartite Graphs

We now consider $\mathbb{G}_{n,p}$. We could try to replace Hall's theorem by Tutte's theorem. A proof along these lines was given by Erdős and Rényi [279]. We can, however, get away with a simpler approach based on simple expansion properties of $\mathbb{G}_{n,p}$. The proof here can be traced back to Bollobás and Frieze [140]. By a perfect matching, we now mean a matching of size $\lfloor n/2 \rfloor$.

Theorem 6.2 *Let $\omega = \omega(n)$, $c > 0$ be a constant, and let $p = \frac{\log n + c_n}{n}$, then*

$$\lim_{n\to\infty} \mathbb{P}(\mathbb{G}_{n,p} \textit{ has a perfect matching}) = \begin{cases} 0 & \textit{if } c_n \to -\infty \\ e^{-e^{-c}} & \textit{if } c_n \to c \\ 1 & \textit{if } c_n \to \infty. \end{cases}$$

Moreover,

$$\lim_{n\to\infty} \mathbb{P}(\mathbb{G}_{n,p} \textit{ has a perfect matching}) = \lim_{n\to\infty} \mathbb{P}(\delta(\mathbb{G}_{n,p}) \geq 1).$$

Proof We will, for convenience, only consider the case where $c_n = \omega \to \infty$ and $\omega = o(\log n)$. If $c_n \to -\infty$ then there are isolated vertices, w.h.p. and our proof can easily be modified to handle the case $c_n \to c$.

Our combinatorial tool that replaces Tutte's theorem is the following: we say that a matching M *isolates* a vertex v if no edge of M contains v. Next let $G = (V,E)$ be a graph without a matching of size $\lfloor |V(G)|/2 \rfloor$. For $v \in V$ such that v is isolated by some maximum matching, let

$$A(v) = \{w \in V : \exists \text{ a maximum matching } M \text{ that isolates } v \text{ and } w\}.$$

We have $v \in A(v)$ here, and once again for a graph $G = (V,E)$ and set $S \subseteq V$ we denote $N_G(S) = \{w \notin S : \exists v \in S \text{ such that } \{v,w\} \in E\}$.

Lemma 6.3 *Let G be a graph without a perfect matching. Let M be a maximum matching of G. If $v \in V$ and $A(v) \neq \{v\}$ then $|N_G(A(v))| < |A(v)|$.*

Proof Let $v \in V$, let M be a maximum matching that isolates v, and let $S_0 \neq \emptyset$ be the set of vertices, other than v, that are isolated by M. Let $S_1 \supseteq S_0$ be the set of vertices reachable from S_0 by a non-empty even length alternating path with respect to M. Let $x \in N_G(S_1)$ and let $y \in S_1$ be a neighbor of x. Then x is covered by M, as otherwise we can get a larger matching by using an alternating path from v to y, and then the edge $\{y,x\}$.

Let y_1 satisfy $\{x,y_1\} \in M$. We show that $y_1 \in S_1$ and this implies that $|N_G(S_1)| \leq |S_1|$ as M defines a mapping $x \to y_1$ of $N_G(S_1)$ into S_1. Let P be an even length alternating path from v terminating at y. If P contains $\{x,y_1\}$ we can truncate it to terminate with $\{x,y_1\}$, otherwise we can extend it using edges $\{y,x\}$ and $\{x,y_1\}$.

Finally, observe that $A(v) = S_1 \cup \{v\}$. □

Now let

$$p = \frac{\log n + \theta \log\log n + \omega}{n},$$

where $\theta \geq 0$ is a fixed integer and $\omega \to \infty$ and $\omega = o(\log\log n)$.

We have introduced θ so that we can use some of the following results for the Hamilton cycle problem.

We write

$$\mathbb{G}_{n,p} = \mathbb{G}_{n,p_1} \cup \mathbb{G}_{n,p_2},$$

where

$$p_1 = \frac{\log n + \theta \log\log n + \omega/2}{n}$$

and

$$1 - p = (1-p_1)(1-p_2) \text{ so that } p_2 \approx \frac{\omega}{2n}.$$

Note that Theorem 4.3 implies:

$$\text{the minimum degree in } \mathbb{G}_{n,p_1} \text{ is at least } \theta + 1 \textit{ w.h.p.} \tag{6.4}$$

We consider a process where we add the edges of $\mathbb{G}_{n,p_2}$ one at a time to $\mathbb{G}_{n,p_1}$. We want to argue that if the current graph does not have a perfect matching then there is a good chance that adding such an edge $\{x,y\}$ will increase the size of a largest matching. This will happen if $y \in A(x)$. If we know that w.h.p. every set S for which $|N_{\mathbb{G}_{n,p_1}}(S)| < |S|$ satisfies $|S| \geq \alpha n$ for some constant $\alpha > 0$, then

$$\mathbb{P}(y \in A(x)) \geq \frac{\binom{\alpha n}{2} - i}{\binom{n}{2}} \geq \frac{\alpha^2}{2}, \tag{6.5}$$

provided $i = O(n)$.

This is because the edges we add will be uniformly random and there will be at least $\binom{\alpha n}{2}$ edges $\{x,y\}$, where $y \in A(x)$. Here given an initial x we can include edges $\{x',y'\}$ where $x' \in A(x)$ and $y' \in A(x')$. We have subtracted i to account for not re-using edges in $f_1, f_2, \ldots, f_{i-1}$.

In the light of this we now argue that sets S, with $|N_{\mathbb{G}_{n,p_1}}(S)| < (1+\theta)|S|$, are w.h.p. of size $\Omega(n)$.

Lemma 6.4 *W.h.p.* $S \subseteq [n], |S| \leq \frac{n}{2000(\theta+4)}$ *implies* $|N(S)| \geq (\theta+1)|S|$, *where* $N(S) = N_{\mathbb{G}_{n,p_1}}(S)$.

Proof Let a vertex of graph $G_1 = \mathbb{G}_{n,p_1}$ be large if its degree is at least $\lambda = \frac{\log n}{100}$, and small otherwise. Denote by *LARGE* and *SMALL*, the set of large and small vertices in G_1, respectively.

***Claim* 1** *W.h.p. if* $v, w \in SMALL$ *then* $dist(v, w) \geq 5$.

Proof If v, w are small and connected by a short path P, then v, w will have few neighbors outside P and conditional on P existing, v having few neighbors outside P is independent of w having few neighbors outside P. Hence,

$$
\begin{aligned}
&\mathbb{P}(\exists v, w \in SMALL \text{ in } \mathbb{G}_{n,p_1} \text{ such that } dist(v,w) < 5)\\
&\leq \binom{n}{2}\left(\sum_{l=0}^{3}\binom{n}{l}p_1^{l+1}\right)\left(\sum_{k=0}^{l}\binom{n}{k}p_1^k(1-p_1)^{n-k-5}\right)^2\\
&\leq n(\log n)^4\left(\sum_{k=0}^{l}\frac{(\log n)^k}{k!}\cdot\frac{(\log n)^{(\theta+1)/100}\cdot e^{-\omega/2}}{n\log n}\right)^2\\
&\leq 2n(\log n)^4\left(\frac{(\log n)^l}{l!}\cdot\frac{(\log n)^{(\theta+1)/100}\cdot e^{-\omega/2}}{n\log n}\right)^2\\
&= O\left(\frac{(\log n)^{O(1)}}{n}(100e)^{\frac{2\log n}{100}}\right)\\
&= O(n^{-3/4})\\
&= o(1).
\end{aligned}
\tag{6.6}
$$

The bound in (6.6) holds since $l! \geq \left(\frac{l}{e}\right)^l$ and $\frac{u_{k+1}}{u_k} > 100$ for $k \leq l$, where

$$
u_k = \frac{(\log n)^k}{k!}\cdot\frac{(\log n)^{(\theta+1)/100}\cdot e^{-\omega/2}}{n\log n}.
$$

□

***Claim* 2** *W.h.p.* $\mathbb{G}_{n,p_1}$ *does not have a 4-cycle containing a small vertex.*

Proof

$$
\begin{aligned}
&\mathbb{P}(\exists \text{ a 4-cycle containing a small vertex })\\
&\leq 4n^4p_1^4\sum_{k=0}^{(\log n)/100}\binom{n-4}{k}p_1^k(1-p_1)^{n-4-k}\\
&\leq n^{-3/4}(\log n)^4\\
&= o(1).
\end{aligned}
$$

□

***Claim* 3** *W.h.p. in* $\mathbb{G}_{n,p_1}$ *for every* $S \subseteq [n], |S| \leq \frac{n}{(\log n)^3}$, $e(S) \leq 2|S|$.

Proof

$$
\begin{aligned}
&\mathbb{P}\left(\exists |S| \leq \frac{n}{(\log n)^3} \text{ and } e(S) > 2|S|\right) \\
&\leq \sum_{s=4}^{n/(\log n)^3} \binom{n}{s}\binom{\binom{s}{2}}{2s} p_1^{2s} \\
&\leq \sum_{s=4}^{n/(\log n)^3} \left(\frac{ne}{s}\left(\frac{se^{1+o(1)}\log n}{2n}\right)^2\right)^s \\
&\leq \sum_{s=4}^{n/(\log n)^3} \left(\frac{s}{n}\cdot\frac{e^{3+o(1)}(\log n)^2}{4}\right)^s \\
&= o(n^{-3}).
\end{aligned}
$$

□

***Claim* 4** *W.h.p. in $\mathbb{G}_{n,p_1}$, if $S \subseteq LARGE$, $|S| \leq \frac{n}{2000(\theta+4)}$ then $|N(S)| \geq (\theta+4)|S|$.*

Proof Let $T = N(S), s = |S|, t = |T|$.

Case 1: $1 \leq s \leq \frac{n}{(\log n)^4}$.

Note that $2e(S) + e(S,T) \geq \frac{\log n}{100}s$ and so from Claim 3 we can assume that $e(S,T) \geq \frac{\log n}{101}s$. If $t \leq \frac{\log n}{1000}s$ then $|S \cup T| \leq \frac{\log n}{999}s \leq \frac{n}{(\log n)^3}$ and $e(S \cup T) \geq e(S,T) \geq \frac{\log n}{101}s$. But then $\frac{e(S \cup T)}{|S \cup T|} \geq \frac{s(\log n)/101}{s(\log n)/999}$ and this contradicts Claim 3.

Case 2: $\frac{n}{(\log n)^4} \leq |S| \leq \frac{n}{\log n}$.

$$
\begin{aligned}
&\mathbb{P}\left(\exists S \subseteq LARGE: \frac{n}{(\log n)^4} \leq |S| \leq \frac{n}{\log n} \text{ and } |N(S)| < \frac{\log n}{1000}|S|\right) \\
&\leq \sum_{s=n/(\log n)^4}^{n/\log n} \sum_{t=0}^{\frac{\log n}{1000}s} \binom{n}{s}\binom{n}{t}\binom{st}{t} p^t (1-p)^{s(n-s-t)} \\
&\leq \sum_{s=n/(\log n)^4}^{n/\log n} \sum_{t=0}^{\frac{\log n}{1000}s} \left(\frac{ne}{s}\right)^s \left(\frac{ne}{t}\right)^t (sep)^t n^{-0.99s} \\
&= \sum_{s=n/(\log n)^4}^{n/\log n} \sum_{t=0}^{\frac{\log n}{1000}s} u_{s,t}.
\end{aligned}
$$

Note that

$$
\frac{u_{s,t+1}}{u_{s,t}} = \frac{ne}{t+1}\left(\frac{t}{t+1}\right)^t (sep) \geq 1000,
$$

so

$$\mathbb{P}\left(\exists S \subseteq LARGE : \frac{n}{(\log n)^4} \leq |S| \leq \frac{n}{\log n} \text{ and } |N(S)| < \frac{\log n}{1000}|S|\right)$$
$$\leq 2 \sum_{s=n/(\log n)^4}^{n/\log n} \left(e(\log n)^4 \left(10^3 e^{2+o(1)}\right)^{\frac{\log n}{1000}} n^{-0.99}\right)^s$$
$$= O\left(n^{-0.5n/(\log n)^4}\right).$$

Case 3: $\frac{n}{\log n} \leq |S| \leq \frac{n}{2000(\theta+4)}$.

Choose some set $S_1 \subseteq S$ of size $n/\log n$. Then

$$|N(S)| \geq |N(S_1)| - |S| \geq \frac{n}{\log n} \cdot \frac{\log n}{1000} - \frac{n}{2000(\theta+4)}$$
$$= \frac{n}{1000}\left(1 - \frac{1}{2(\theta+4)}\right) > (\theta+4)|S|.$$

□

We can now complete the proof of Lemma 6.4. Let $|S| \leq \frac{n}{2000(\theta+4)}$ and assume that $\mathbb{G}_{n,p_1}$ has minimum degree of at least $\theta+1$.

Let $S_1 = S \cap SMALL$ and $S_2 = S \setminus S_1$, then

$$|N(S)|$$
$$\geq |N(S_1)| + |N(S_2)| - |N(S_1) \cap S_2| - |N(S_2) \cap S_1| - |N(S_1) \cap N(S_2)|$$
$$\geq |N(S_1)| + |N(S_2)| - |S_2| - |N(S_2) \cap S_1| - |N(S_1) \cap N(S_2)|.$$

But Claim 1 and Claim 2 and a minimum degree of at least $\theta+1$ imply that

$$|N(S_1)| \geq (\theta+1)|S_1|, \; |N(S_2) \cap S_1| \leq \min\{|S_1|, |S_2|\},$$
$$|N(S_1) \cap N(S_2)| \leq |S_2|.$$

So, from this and Claim 4 we obtain

$$|N(S)| \geq (\theta+1)|S_1| + (\theta+4)|S_2| - 3|S_2| = (\theta+1)|S|.$$

□

We now go back to the proof of Theorem 6.2 for the case $c = \omega \to \infty$. Let the edges of $\mathbb{G}_{n,p_2}$ be $\{f_1, f_2, \ldots, f_s\}$ in random order, where $s \approx \omega n/4$. Let $\mathbb{G}_0 = \mathbb{G}_{n,p_2}$ and $\mathbb{G}_i = \mathbb{G}_{n,p_2} + \{f_1, f_2, \ldots, f_i\}$ for $i \geq 1$. It follows from Lemmas 6.3 and 6.4 that if $\mu(G)$ denotes the size of the largest matching in G, and if $\mu(\mathbb{G}_i) <$

$n/2$ then, assuming $\mathbb{G}_{n,p_1}$ has the expansion claimed in Lemma 6.4, with $\theta = 0$ and $\alpha = 1/8000$,

$$\mathbb{P}(\mu(\mathbb{G}_{i+1}) \geq \mu(\mathbb{G}_i) + 1 \mid f_1, f_2, \ldots, f_i) \geq \frac{\alpha^2}{2}, \tag{6.7}$$

see (6.5).

It follows that

$$\begin{aligned}&\mathbb{P}(\mathbb{G}_{n,p} \text{ does not have a perfect matching})\\ &\leq o(1) + \mathbb{P}(\text{Bin}(s, \alpha^2/2) < n/2) = o(1).\end{aligned}$$

We have used the notion of dominance, see Section 21.9, in order to use the Binomial distribution in the above inequality. □

6.2 Hamilton Cycles

This was a difficult question left open in [276]. A breakthrough came with the result of Pósa [631]. The precise theorem given below can be credited to Komlós and Szemerédi [498], Bollobás [128] and Ajtai, Komlós and Szemerédi [11].

Theorem 6.5 *Let* $p = \frac{\log n + \log\log n + c_n}{n}$, *then,*

$$\lim_{n\to\infty} \mathbb{P}(\mathbb{G}_{n,p} \text{ has a Hamilton cycle}) = \begin{cases} 0 & \text{if } c_n \to -\infty \\ e^{-e^{-c}} & \text{if } c_n \to c \\ 1 & \text{if } c_n \to \infty. \end{cases}$$

Moreover,

$$\lim_{n\to\infty} \mathbb{P}(\mathbb{G}_{n,p} \text{ has a Hamilton cycle}) = \lim_{n\to\infty} \mathbb{P}(\delta(\mathbb{G}_{n,p}) \geq 2).$$

Proof We first give a proof of the first statement under the assumption that $c_n = \omega \to \infty$ where $\omega = o(\log\log n)$. The proof of the second statement is postponed to Section 6.3. Under this assumption, we have $\delta(G_{n,p}) \geq 2$ w.h.p., see Theorem 4.3. The result for larger p follows by monotonicity.

We now set up the main tool, viz. Pósa's lemma. Let P be a path with end points a, b, as in Figure 6.1. Suppose that b does not have a neighbor outside of P.

Notice that the P' below in Figure 6.2 is a path of the same length as P, obtained by a *rotation* with vertex a as the *fixed endpoint*. To be precise, suppose that $P = (a, \ldots, x, y, y', \ldots, b', b)$ and $\{b, x\}$ is an edge where x is an

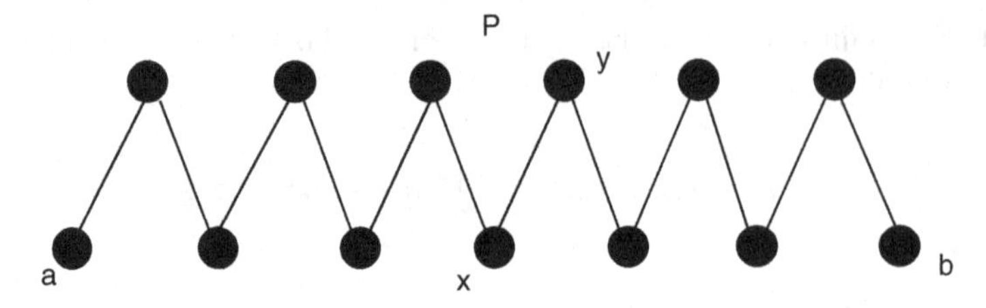

Figure 6.1. The path P

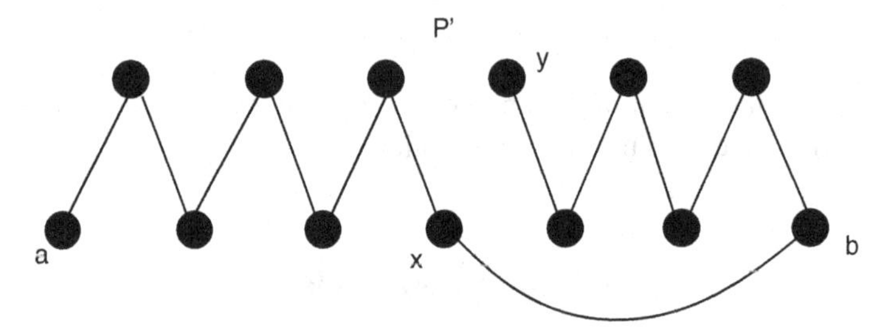

Figure 6.2. The path P' obtained after a single rotation

interior vertex of P. The path $P' = (a, \dots, x, b, b', \dots, y', y)$ is said to be obtained from P by a rotation.

Now let $END = END(P)$ denote the set of vertices v such that there exists a path P_v from a to v such that P_v is obtained from P by a sequence of rotations with vertex a fixed as in Figure 6.3.

Here the set END consists of all the white vertices on the path drawn below in Figure 6.4.

Lemma 6.6 *If $v \in P \setminus END$ and v is adjacent to $w \in END$ then there exists $x \in END$ such that the edge $\{v, x\} \in P$.*

Proof Suppose to the contrary that x, y are the neighbors of v on P and that $v, x, y \notin END$ and that v is adjacent to $w \in END$. Consider the path P_w. Let $\{r, t\}$ be the neighbors of v on P_w. Now $\{r, t\} = \{x, y\}$ because if a rotation deleted $\{v, y\}$ say then v or y becomes an endpoint. But then after a further rotation from P_w we see that $x \in END$ or $y \in END$.

□

Corollary 6.7

$$|N(END)| < 2|END|.$$

□

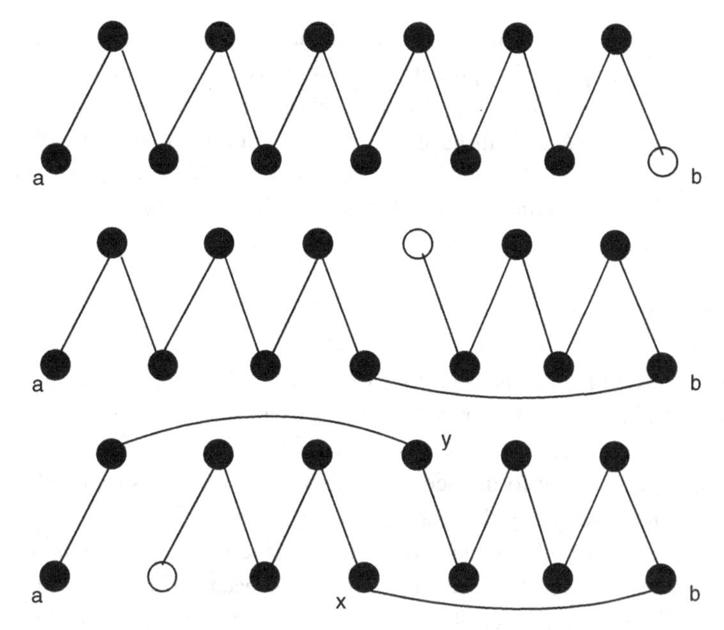

Figure 6.3. A sequence of rotations

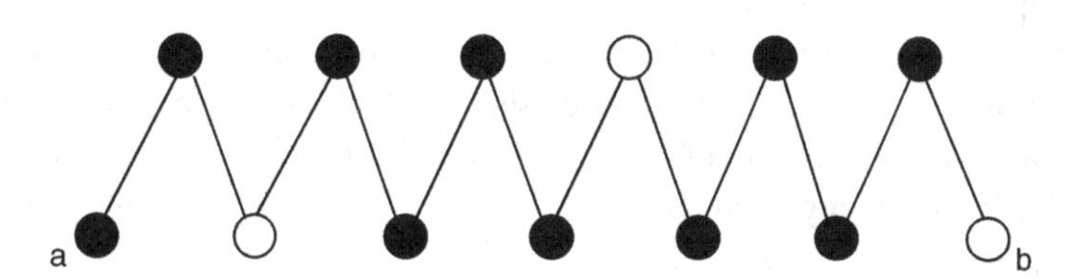

Figure 6.4. The set *END*

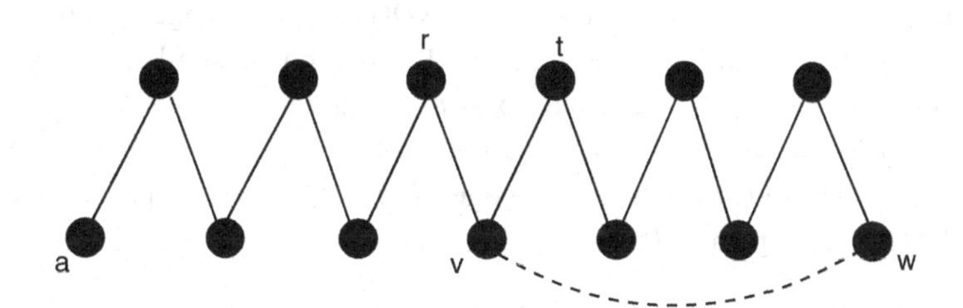

Figure 6.5. One of r, t will become an endpoint after a rotation

It follows from Lemma 6.4 with $\theta = 1$ that w.h.p. we have

$$|END| \geq \alpha n \text{ where } \alpha = \frac{1}{10000}. \tag{6.8}$$

We now consider the following algorithm that searches for a Hamilton cycle in a connected graph G. The probability p_1 is above the connectivity threshold and so $\mathbb{G}_{n,p_1}$ is connected w.h.p. Our algorithm will proceed in stages. At the beginning of Stage k we will have a path of length k in G and we will try to grow it by one vertex in order to reach Stage $k+1$. In Stage $n-1$, our aim is simply to create a Hamilton cycle, given a Hamilton path. We start the whole procedure with an arbitrary path of G.

Algorithm Pósa

(i) Let P be our path at the beginning of Stage k. Let its endpoints be x_0, y_0. If x_0 or y_0 have neighbors outside P then we can simply extend P to include one of these neighbors and move to Stage $k+1$.

(ii) Failing this, we perform a sequence of rotations with x_0 as the fixed vertex until one of two things happens: (i) We produce a path Q with an endpoint y that has a neighbor outside of Q. In this case we extend Q and proceed to Stage $k+1$. (ii) No sequence of rotations leads to Case (i). In this case let END denote the set of endpoints of the paths produced. If $y \in END$ then P_y denotes a path with endpoints x_0, y that is obtained from P by a sequence of rotations.

(iii) If we are in Case (ii) then for each $y \in END$ we let $END(y)$ denote the set of vertices z such that there exists a longest path Q_z from y to z such that Q_z is obtained from P_y by a sequence of rotations with vertex y fixed. Repeating the argument above in (b) for each $y \in END$, we either extend a path and begin Stage $k+1$ or we go to (d).

(iv) Suppose now that we do not reach Stage $k+1$ by an extension and that we have constructed the sets END and $END(y)$ for all $y \in END$. Suppose that G contains an edge (y,z) where $z \in END(y)$. Such an edge would imply the existence of a cycle $C = (z, Q_y, z)$. If this is not a Hamilton cycle then connectivity implies that there exist $u \in C$ and $v \notin C$ such that u, v are joined by an edge. Let w be a neighbor of u on C and let P' be the path obtained from C by deleting the edge (u,w). This creates a path of length $k+1$ viz. the path w, P', v, and we can move to Stage $k+1$.

A pair z, y where $z \in END(y)$ is called a *booster* in the sense that if we added this edge to $\mathbb{G}_{n,p_1}$ then it would either (i) make the graph Hamiltonian or (ii) make the current path longer. We argue now that $\mathbb{G}_{n,p_2}$ can be used to "boost" P to a Hamilton cycle, if necessary.

We observe now that when $G = \mathbb{G}_{n,p_1}$, $|END| \geq \alpha n$ w.h.p., see (6.8). Also, $|END(y)| \geq \alpha n$ for all $y \in END$. So we will have $\Omega(n^2)$ boosters.

For a graph G let $\lambda(G)$ denote the length of a longest path in G, when G is not Hamiltonian and let $\lambda(G) = n$ when G is Hamiltonian. Let the edges of $\mathbb{G}_{n,p_2}$ be $\{f_1,f_2,\ldots,f_s\}$ in random order, where $s \approx \omega n/4$. Let $\mathbb{G}_0 = \mathbb{G}_{n,p_2}$ and $\mathbb{G}_i = \mathbb{G}_{n,p_2} + \{f_1,f_2,\ldots,f_i\}$ for $i \geq 1$. It follows from Lemmas 6.3 and 6.4 that if $\lambda(\mathbb{G}_i) < n$ then, assuming $\mathbb{G}_{n,p_1}$ has the expansion claimed in Lemma 6.4,

$$\mathbb{P}(\lambda(\mathbb{G}_{i+1}) \geq \lambda(\mathbb{G}_i) + 1 \mid f_1,f_2,\ldots,f_i) \geq \frac{\alpha^2}{2}, \tag{6.9}$$

see (6.5), replacing $A(v)$ by $END(v)$.

It follows that

$$\mathbb{P}(\mathbb{G}_{n,p} \text{ is not Hamiltonian}) \leq o(1) + \mathbb{P}(\text{Bin}(s,\alpha^2/2) < n) = o(1). \tag{6.10}$$

6.3 Long Paths and Cycles in Sparse Random Graphs

In this section we study the length of the longest path and cycle in $\mathbb{G}_{n,p}$ when $p = c/n$ where $c = O(\log n)$, most importantly for the case where c is a large constant. We saw in Chapter 1 that under these conditions, $\mathbb{G}_{n,p}$ will w.h.p. have isolated vertices and so it will not be Hamiltonian. We can, however, show that it contains a cycle of length $\Omega(n)$ w.h.p.

The question of the existence of a long path/cycle was posed by Erdős and Rényi in [276]. The first positive answer to this question was given by Ajtai, Komlós and Szemerédi [10] and by de la Vega [712]. The proof we give here is due to Krivelevich, Lee and Sudakov. It is subsumed by the more general results of [504].

Theorem 6.8 *Let $p = c/n$ where c is sufficiently large but $c = O(\log n)$. Then w.h.p.*

(i) $\mathbb{G}_{n,p}$ has a path of length at least $\left(1 - \frac{6\log c}{c}\right)n$.
(ii) $\mathbb{G}_{n,p}$ has a cycle of length at least $\left(1 - \frac{12\log c}{c}\right)n$.

Proof We prove this theorem by analyzing the simple properties of Depth First Search (DFS). This is a well-known algorithm for exploring the vertices of a component of a graph. We can describe the progress of this algorithm using three sets: U is the set of *unexplored* vertices that have not yet been reached by the search; D is the set of *dead* vertices that have been fully explored and no longer take part in the process; and $A = \{a_1, a_2, \ldots, a_r\}$ is the set of active vertices and they form a path from a_1 to a_r. We start the algorithm by choosing a vertex v from which to start the process, then we let

$$A = \{v\} \text{ and } D = \emptyset \text{ and } U = [n] \setminus \{v\} \text{ and } r = 1.$$

We now describe how these sets change during one step of the algorithm.

Step (1) If there is an edge $\{a_r, w\}$ for some $w \in U$ then we choose one such w and extend the path defined by A to include w.

$$a_{r+1} \leftarrow w; A \leftarrow A \cup \{w\}; U \leftarrow U \setminus \{w\}; r \leftarrow r+1.$$

We now repeat Step (1).
If there is no such w then we do Step (2):

Step (2) We have now completely explored a_r.

$$D \leftarrow D \cup \{a_r\}; A \leftarrow A \setminus \{a_r\}; r \leftarrow r - 1.$$

If $r \geq 1$ we go to Step (1). Otherwise, if $U = \emptyset$ at this point then we terminate the algorithm. If $U \neq \emptyset$ then we choose some $v \in U$ to re-start the process with $r = 1$. We then go to Step (1).

We make the following simple observations:

- A step of the algorithm increases $|D|$ by one or decreases $|U|$ by one and so at some stage we must have $|D| = |U| = s$ for some positive integer s.
- There are no edges between D and U because we only add a_r to D when there are no a_r, U edges and U never increases from this point on.

Thus at some stage we have two disjoint sets D, U of size s with no edges between them and a path of length $|A| - 1 = n - 2s - 1$. This plus the following claim implies that $\mathbb{G}_{n,p}$ has a path P of length at least $\left(1 - \frac{6\log c}{c}\right) n$ w.h.p. Note that if c is large then

$$\alpha > \frac{3\log c}{c} \text{ implies } c > \frac{2}{\alpha} \log\left(\frac{e}{\alpha}\right).$$

Claim 5 *Let $0 < \alpha < 1$ be a positive constant. If $p = c/n$ and $c > \frac{2}{\alpha}\log\left(\frac{e}{\alpha}\right)$ then w.h.p. in $\mathbb{G}_{n,p}$, every pair of disjoint sets S_1, S_2 of size at least $\alpha n - 1$ are joined by at least one edge.*

Proof The probability that there exist sets S_1, S_2 of size (at least) $\alpha n - 1$ with no joining edge is at most

$$\binom{n}{\alpha n - 1}^2 (1-p)^{(\alpha n - 1)^2} \leq \left(\frac{e^{2+o(1)}}{\alpha^2} e^{-c\alpha}\right)^{\alpha n - 1} = o(1).$$

□

To complete the proof of the theorem, we apply the above lemma to the vertices S_1, S_2 on the two sub-paths P_1, P_2 of length $\frac{3\log c}{c}n$ at each end of P. There will w.h.p. be an edge joining S_1, S_2, creating the cycle of the claimed length. □

Krivelevich and Sudakov [512] used DFS to give simple proofs of good bounds on the size of the largest component in $\mathbb{G}_{n,p}$ for $p = \frac{1+\varepsilon}{n}$ where ε is a small constant. Exercises 6.7.19, 6.7.20 and 6.7.21 elaborate on their results.

Completing the proof of Theorem 6.5

We need to prove the last statement. So we let $1-p=(1-p_1)(1-p_2)$ where $p_2 = \frac{1}{n\log\log n}$. Then we apply Theorem 6.8(i) to argue that w.h.p. G_{n,p_1} has a path of length $n\left(1-O\left(\frac{\log\log n}{\log n}\right)\right)$.

Now, conditional on G_{n,p_1} having minimum degree at least two, the proof of the statement of Lemma 6.4 goes through without change for $\theta = 1$, i.e. $S \subseteq [n], |S| \le \frac{n}{10000}$ implies $|N(S)| \ge 2|S|$. We can then use the extension–rotation argument that we used to prove Theorem 6.5. This time we only need to close $O\left(\frac{n\log\log n}{\log n}\right)$ cycles and we have $\Omega\left(\frac{n}{\log\log n}\right)$ edges. Thus (6.10) is replaced by

$$\mathbb{P}(\mathbb{G}_{n,p} \text{ is not Hamiltonian} \mid \delta(G_{n,p_1}) \ge 2)$$
$$\le o(1) + \mathbb{P}\left(\text{Bin}\left(\frac{c_1 n}{\log\log n}, 10^{-8}\right) < \frac{c_2 n \log\log n}{\log n}\right) = o(1),$$

for some hidden constants c_1, c_2.

6.4 GREEDY Matching Algorithm

In this section we see how we can use differential equations to analyze the performance of a GREEDY algorithm for finding a large matching in a random graph. Finding a large matching is a standard problem in Combinatorial Optimization. The first polynomial time algorithm to solve this problem was devised by Edmonds in 1965 and runs in time $O(|V|^4)$ [271]. Over the years, many improvements have been made. Currently, the fastest such algorithm is that of Micali and Vazirani which dates back to 1980. Its running time is $O(|E|\sqrt{|V|})$ [578]. These algorithms are rather complicated and there is a natural interest in the performance of simpler heuristic algorithms which should find large, but not necessarily maximum matchings. One well-studied class of heuristics goes under the general title of the GREEDY heuristic.

The following simple GREEDY algorithm proceeds as follows: Beginning with graph $G = (V, E)$ we choose a random edge $e = \{u, v\} \in E$ and place it in a set M. We then delete u, v and their incident edges from G and repeat. In the following, we analyze the size of the matching M produced by this algorithm.

Algorithm *GREEDY*

begin
  $M \leftarrow \emptyset$;
  while $E(G) \neq \emptyset$ **do**
  begin
    A: Randomly choose $e = \{x, y\} \in E$
      $G \leftarrow G \setminus \{x, y\}$;
      $M \leftarrow M \cup \{e\}$
  end;
Output M
end

($G \setminus \{x, y\}$ is the graph obtained from G by deleting the vertices x, y and all incident edges.)

We will study this algorithm in the context of the pseudo-graph model $\mathbb{G}_{n,m}^{(B)}$ of Section 1.3 and apply (1.16) to bring the results back to $\mathbb{G}_{n,m}$. We argue next that if at some stage G has ν vertices and μ edges then G is equally likely to be any pseudo-graph with these parameters.

We will use the method of *deferred decisions*, a term coined in Knuth, Motwani and Pittel [489]. In this scenario, we do not expose the edges of the pseudo-graph until we actually need to. So, as a thought experiment, think that initially there are m boxes, each containing a uniformly random pair of distinct integers from $[n]$. Until the box is opened, the contents are unknown except for their distribution. Observe that opening box A and observing its contents tells us nothing more about the contents of box B. This would not be the case if, as in $\mathbb{G}_{n,m}$, we insisted that no two boxes had the same contents.

Remark 6.9 Rather than choose the edge $\{x, y\}$ at random, we let $\{x, y\}$ be the first edge. This makes (6.11) below clearer and because the edges are in random order anyway, it does not change the algorithm.

Remark 6.10 A step of GREEDY involves choosing the first unopened box at random to expose its contents x, y.

After this, the contents of the remaining boxes remain uniformly random over $\binom{V(G)}{2}$. The algorithm then asks for each box with x or y to be opened.

Other boxes remain unopened and all we learn is that their contents do not contain x or y and so they are still uniform over the remaining possible edges.

We need the following

Lemma 6.11 *Suppose that* $m = cn$ *for some constant* $c > 0$*, then w.h.p. the maximum degree in* $\mathbb{G}_{n,m}^{(B)}$ *is at most* $\log n$.

Proof The degree of a vertex is distributed as $\mathrm{Bin}(m, 2/n)$. So, if Δ denotes the maximum degree in $\mathbb{G}_{n,m}^{(B)}$, then with $\ell = \log n$,

$$\mathbb{P}(\Delta \geq \ell) \leq n\binom{m}{\ell}\left(\frac{2}{n}\right)^{\ell} \leq n\left(\frac{2ce}{\ell}\right)^{\ell} = o(1).$$

□

Now let $X(t) = (\nu(t), \mu(t)), t = 1, 2, \ldots,$ denote the number of vertices and edges in the graph at the start of the tth iterations of GREEDY. Also, let $G_t = (V_t, E_t) = G$ at this point and let $G_t' = (V_t, E_t \setminus e_1)$, where e_1 is the first edge of E_t. Thus $\nu(1) = n, \mu(1) = m$ and $G_1 = \mathbb{G}_{n,m}^{(B)}$. Now $\nu(t+1) = \nu(t) - 2$ and so $\nu(t) = n - 2t$. For $\mu(t)$ let $d_t(\cdot)$ denote degree in G_t' and let $\theta_t(x,y)$ denote the number of copies of the edge $\{x,y\}$ in G_t, excluding e_1. Then we have

$$\mathbb{E}(\mu(t+1) \mid G_t) = \mu(t) - (d_t(x) + d_t(y) + 1 + \theta_t(x,y)).$$

Taking expectations over G_t we have

$$\mathbb{E}(\mu(t+1)) = \mathbb{E}(\mu(t)) - \frac{4\mu(t)}{n-2t} - 1 + O\left(\frac{1}{n-2t}\right). \tag{6.11}$$

Here we use Remark 6.9 to argue that $\mathbb{E}\,d_t(x), \mathbb{E}\,d_t(y) = 2\frac{\mu(t)-1}{\nu(t)}$ and Remark 6.10 to argue that $\mathbb{E}\theta_t(x,y)) = O(1/(n-2t))$.

This *suggests* that w.h.p. $\mu(t) \approx nz(t/n)$, where $z(0) = c$ and $z(\tau)$ is the solution to the differential equation

$$\frac{dz}{d\tau} = -\frac{4z(\tau)}{1-2\tau} - 1.$$

This is easy to solve and gives

$$z(\tau) = \left(c + \frac{1}{2}\right)(1-2\tau)^2 - \frac{1-2\tau}{2}.$$

The smallest root of $z(\tau) = 0$ is $\tau = \frac{c}{2c+1}$. This suggests the following theorem.

Theorem 6.12 *W.h.p., running GREEDY on* $\mathbb{G}_{n,cn}$ *finds a matching of size* $\frac{c+o(1)}{2c+1}n$.

Proof We replace $\mathbb{G}_{n,m}$ by $\mathbb{G}_{n,m}^{(B)}$ and consider the random sequence $\mu(t)$, $t = 1,2,\ldots$. The number of edges in the matching found by GREEDY equals one less than the first value of t for which $\mu(t) = 0$. We show that w.h.p. $\mu(t) > 0$ if and only if $t \le \frac{c+o(1)}{2c+1}n$. We will use Theorem 22.1 of Chapter 22.

In our set up for the theorem we let

$$f(\tau, x) = -\frac{4x}{1-2\tau} - 1.$$

$$D = \left\{ (\tau, x) : 0 \le \tau \le \mathcal{T}_D = \frac{c}{2c+1}, 0 \le x \le \frac{1}{2} \right\}.$$

We let $X(t) = \mu(t)$ for the statement of the theorem. Then we have to check the conditions:

(P1) $|\mu(t)| \le cn, \ \forall t < T_D = \mathcal{T}_D n.$

(P2) $|\mu(t+1) - \mu(t)| \le 2\log n \quad \forall t < T_D.$

(P3) $|\mathbb{E}(m(t+1) - \mu(t)|H_t, \mathcal{E}) - f(t/n, X(t)/n)| \le \frac{A}{n}, \forall t < T_D.$
Here $\mathcal{E} = \{\Delta \le \log n\}$ and this is needed for (P2).

(P4) $f(t,x)$ is continuous and satisfies a Lipschitz condition
$|f(t,x) - f(t',x')| \le (4c+2)\|(t,x) - (t',x)'\|_\infty$
for $(t,x),(t',x') \in D \cap \{(t,x) : t \ge 0\}$

Now let $\beta = n^{1/5}$, $\lambda = n^{-1/20}$ and $\sigma = T_D - 10\lambda$ and apply the theorem. This shows that w.h.p. $\mu(t) = nz(t/n) + O(n^{19/20})$ for $t \le \sigma n$. □

The result in Theorem 6.12 is taken from Dyer, Frieze and Pittel [269], where a central limit theorem is proven for the size of the matching produced by GREEDY.

The use of differential equations to approximate the trajectory of a stochastic process is quite natural and is often very useful. However, it is not always best practise to try to use an "off the shelf" theorem like Theorem 22.1 to obtain a best result. It is hard to design a general theorem that can deal optimally with terms that are $o(n)$.

6.5 Random Subgraphs of Graphs with Large Minimum Degree

Here we prove an extension of Theorem 6.8. The setting is this. We have a sequence of graphs G_k with minimum degree at least k, where $k \to \infty$. We construct a random subgraph G_p of $G = G_k$ by including each edge of G, independently with probability p. Thus if $G = K_n$, G_p is $G_{n,p}$. The theorem we prove was first proved by Krivelevich, Lee and Sudakov [504]. The argument we present here is due to Riordan [646].

In the following we abbreviate $(G_k)_p$ to G_p, where the parameter k is to be understood.

Theorem 6.13 *Let G_k be a sequence of graphs with minimum degree at least k where $k \to \infty$. Let p be such that $pk \to \infty$ as $k \to \infty$. Then w.h.p. G_p contains a cycle of length at least $(1 - o(1))k$.*

Proof We assume that G has n vertices. We let T denote the *forest* produced by DFS. We also let D, U, A be as in the proof of Theorem 6.8. Let v be a vertex of the rooted forest T. There is a unique *vertical* path from v to the root of its component. We write $\mathscr{A}(v)$ for the set of ancestors of v, i.e. vertices (excluding v) on this path. We write $\mathscr{D}(v)$ for the set of descendants of v, again excluding v. Thus $w \in \mathscr{D}(v)$ if and only if $v \in \mathscr{A}(w)$. The distance $d(u,v)$ between two vertices u and v on a common vertical path is just their graph distance along this path. We write $\mathscr{A}_i(v)$ and $\mathscr{D}_i(v)$ for the set of ancestors/descendants of v at distance exactly i, and $\mathscr{A}_{\leq i}(v), \mathscr{D}_{\leq i}(v)$ for those at distance at most i. By the depth of a vertex we mean its distance from the root. The height of a vertex v is $\max\{i : \mathscr{D}_i(v) \neq \emptyset\}$. Let R denote the set of edges of G that are not tested for inclusion in G_p during the exploration.

Lemma 6.14 *Every edge e of R joins two vertices on some vertical path in T.*

Proof Let $e = (u,v)$ and suppose that u is placed in D before v. When u is placed in D, v cannot be in U, else $\{u,v\}$ would have been tested. Also, v cannot be in D by our choice of u. Therefore, at this time $v \in A$ and there is a vertical path from v to u. □

Lemma 6.15 *With high probability, at most $2n/p = o(kn)$ edges are tested during the DFS exploration.*

Proof Each time an edge is tested, the test succeeds (the edge is found to be present) with probability p. The Chernoff bound implies that the probability that more than $2n/p$ tests are made but fewer than n succeed is $o(1)$. But every successful test contributes an edge to the forest, T, so w.h.p. at most n tests are successful. □

From now on let us fix an arbitrary (small) constant $0 < \varepsilon < 1/10$. We call a vertex v *full* if it is incident with at least $(1 - \varepsilon)k$ edges in R.

Lemma 6.16 *With high probability, all but $o(n)$ vertices of T_k are full.*

Proof Since G has minimum degree at least k, each $v \in V(G) = V(T)$ that is not full is incident with at least εk tested edges. If for some constant $c > 0$

there are at least cn such vertices, then there are at least $c\varepsilon kn/2$ tested edges. But the probability of this is $o(1)$ by Lemma 6.15. □

Let us call a vertex v *rich* if $|\mathscr{D}(v)| \geq \varepsilon k$, and *poor* otherwise. In the next two lemmas, (T_k) is a sequence of rooted forests with n vertices. We suppress the dependence on k in notation.

Lemma 6.17 *Suppose that $T = T_k$ contains $o(n)$ poor vertices. Then, for any constant C, all but $o(n)$ vertices of T are at height at least Ck.*

Proof For each rich vertex v, let $P(v)$ be a set of $\lceil \varepsilon k \rceil$ descendants of v, obtained by choosing vertices of $\mathscr{D}(v)$ one by one starting with those furthest from v. For every $w \in P(v)$ we have $\mathscr{D}(w) \subseteq P(v)$, so $|\mathscr{D}(w)| < \varepsilon k$, i.e. w is poor. Consider the set S_1 of ordered pairs (v, w) with v rich and $w \in P(v)$. Each of the $n - o(n)$ rich vertices appears in at least εk pairs, so $|S_1| \geq (1 - o(1))\varepsilon kn$.

For any vertex w we have $|\mathscr{A}_{\leq i}(w)| \leq i$, since there is only one ancestor at each distance, until we hit the root. Since $(v, w) \in S_1$ implies that w is poor and $v \in \mathscr{A}(w)$, and there are only $o(n)$ poor vertices, at most $o(Ckn) = o(kn)$ pairs $(v, w) \in S_1$ satisfy $d(v, w) \leq Ck$. Thus $S_1' = \{(v, w) \in S_1 : d(v, w) > Ck\}$ satisfies $|S_1'| \geq (1 - o(1))\varepsilon kn$. Since each vertex v is the first vertex of at most $\lceil \varepsilon k \rceil \approx \varepsilon k$ pairs in $S_1 \supseteq S_1'$, it follows that $n - o(n)$ vertices v appear in pairs $(v, w) \in S_1'$. Since any such v has height at least Ck, the proof is complete. □

Let us call a vertex v *light* if $|D_{\leq (1-5\varepsilon)k}(v)| \leq (1 - 4\varepsilon)k$, and *heavy* otherwise. Let H denote the set of heavy vertices in T.

Lemma 6.18 *Suppose that $T = T_k$ contains $o(n)$ poor vertices, and let $X \subseteq V(T)$ with $|X| = o(n)$. Then, for k large enough, T contains a vertical path P of length at least $\varepsilon^{-2}k$ containing at most $\varepsilon^2 k$ vertices in $X \cup H$.*

Proof Let S_2 be the set of pairs (u, v) where u is an ancestor of v and $0 < d(u, v) \leq (1 - 5\varepsilon)k$. Since a vertex has at most one ancestor at any given distance, we have $|S_2| \leq (1 - 5\varepsilon)kn$. On the other hand, by Lemma 6.17 all but $o(n)$ vertices u are at height at least k and so appear in at least $(1 - 5\varepsilon)k$ pairs $(u, v) \in S_2$. It follows that only $o(n)$ vertices u are in more than $(1 - 4\varepsilon)k$ such pairs, i.e. $|H| = o(n)$.

Let S_3 denote the set of pairs (u, v) where $v \in X \cup H$, u is an ancestor of v, and $d(u, v) \leq \varepsilon^{-2}k$. Since a given v can only appear in $\varepsilon^{-2}k$ pairs $(u, v) \in S_3$, we see that $|S_3| \leq \varepsilon^{-2}k|X \cup H| = o(kn)$. Hence, only $o(n)$ vertices u appear in more than $\varepsilon^2 k$ pairs $(u, v) \in S_3$.

By Lemma 6.17, all but $o(n)$ vertices are at height at least $\varepsilon^{-2}k$. Let u be such a vertex appearing in at most $\varepsilon^2 k$ pairs $(u, v) \in S_3$, and let P be the vertical path from u to some $v \in D_{\varepsilon^{-2}k}(u)$. Then P has the required properties. □

Proof of Theorem 6.13

Fix $\varepsilon > 0$. It suffices to show that w.h.p. G_p contains a cycle of length at least $(1-5\varepsilon)k$, say. Explore G_p by DFS as described above. We condition on the result of the exploration, noting that the edges of R are still present independently with probability p. By Lemma 6.14, $\{u,v\} \in R$ implies that u is either an ancestor or a descendant of v. By Lemma 6.16, we may assume that all but $o(n)$ vertices are full.

Suppose that

$$|\{u : \{u,v\} \in R, d\{u,v\} \geq (1-5\varepsilon)k\}| \geq \varepsilon k, \tag{6.12}$$

for some vertex v. Then, since $\varepsilon kp \to \infty$, testing the relevant edges $\{u,v\}$ one by one, w.h.p we find one present in G_p, forming, together with T, the required long cycle. On the other hand, suppose that (6.12) fails for every v. Suppose that some vertex v is full but poor. Since v has at most εk descendants, there are at least $(1-2\varepsilon)k$ pairs $\{u,v\} \in R$ with $u \in \mathscr{A}(v)$. Since v has only one ancestor at each distance, it follows that (6.12) holds for v, a contradiction.

We have shown that we can assume that no poor vertex is full. Hence there are $o(n)$ poor vertices, and we may apply Lemma 6.18, with X the set of vertices that are not full. Let P be the path whose existence is guaranteed by the lemma, and let Z be the set of vertices on P that are full and light, so $|V(P) \setminus Z| \leq \varepsilon^2 k$. For any $v \in Z$, since v is full, there are at least $(1-\varepsilon)k$ vertices $u \in \mathscr{A}(v) \cup \mathscr{D}(v)$ with $\{u,v\} \in R$. Since (6.12) does not hold, at least $(1-2\varepsilon)k$ of these vertices satisfy $d(u,v) \leq (1-5\varepsilon)k$. Since v is light, in turn at least $2\varepsilon k$ of these u must be in $\mathscr{A}(v)$. Recalling that a vertex has at most one ancestor at each distance, we find a set $R(v)$ of at least εk vertices $u \in \mathscr{A}(v)$ with $\{u,v\} \in R$ and $\varepsilon k \leq d(u,v) \leq (1-5\varepsilon)k \leq k$.

It is now easy to find a (very) long cycle w.h.p. Recall that $Z \subseteq V(P)$ with $|V(P) \setminus Z| \leq \varepsilon^2 k$. Thinking of P as oriented upwards towards the root, let v_0 be the lowest vertex in Z. Since $|R(v_0)| \geq \varepsilon k$ and $kp \to \infty$, w.h.p. there is an edge $\{u_0, v_0\}$ in G_p with $u_0 \in R(v_0)$. Let v_1 be the first vertex below u_0 along P with $v_1 \in Z$. Note that we go up at least εk steps from v_0 to u_0 and down at most $1 + |V(P) \setminus Z| \leq 2\varepsilon^2 k$ from u_0 to v_1, so v_1 is above v_0. Again w.h.p. there is an edge $\{u_1, v_1\}$ in G_p with $u_1 \in R(v_1)$, and so at least εk steps above v_1. Continue downwards from u_1 to the first $v_2 \in Z$, and so on. Since $\varepsilon^{-1} = O(1)$, w.h.p. we may continue in this way to find overlapping chords $\{u_i, v_i\}$ for $0 \leq i \leq \lfloor 2\varepsilon^{-1} \rfloor$, say. (Note that we remain within P as each upwards step has length at most k.) These chords combine with P to give a cycle of length at least $(1 - 2\varepsilon^{-1} \times 2\varepsilon^2)k = (1-4\varepsilon)k$, as shown in Figure 6.6.

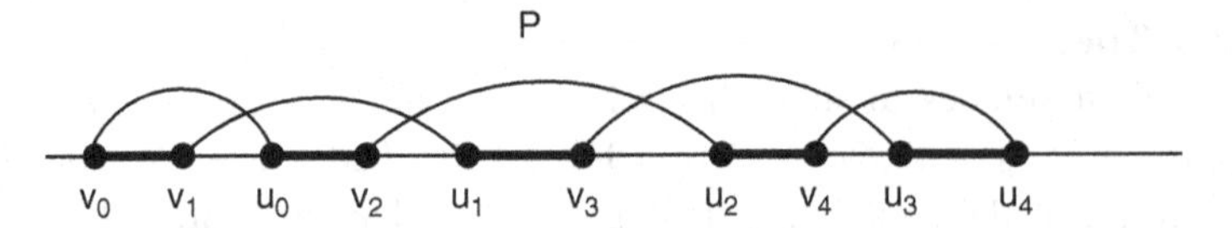

Figure 6.6. The path P, with the root off to the right. Each chord $\{v_i, u_i\}$ has length at least εk (and at most k); from u_i to v_{i+1} is at most $2\varepsilon^2 k$ steps back along P. The chords and the thick part of P form a cycle

6.6 Spanning Subgraphs

Consider a fixed sequence $H^{(d)}$ of graphs where $n = |V(H^{(d)})| \to \infty$. In particular, we consider a sequence Q_d of d-dimensional cubes where $n = 2^d$ and a sequence of 2-dimensional lattices L_d of order $n = d^2$. We ask when $\mathbb{G}_{n,p}$ or $\mathbb{G}_{n,m}$ contains a copy of $H = H^{(d)}$ w.h.p.

We give a condition that can be proved in quite an elegant and easy way. This proof is from Alon and Füredi [24].

Theorem 6.19 *Let H be fixed sequence of graphs with $n = |V(H)| \to \infty$ and maximum degree Δ, where $(\Delta^2+1)^2 < n$. If*

$$p^\Delta > \frac{10\log\lfloor n/(\Delta^2+1)\rfloor}{\lfloor n/(\Delta^2+1)\rfloor}, \tag{6.13}$$

then $\mathbb{G}_{n,p}$ contains an isomorphic copy of H w.h.p.

Proof To prove this we first apply the Hajnal–Szemerédi theorem to the square H^2 of our graph H.

Recall that we square a graph if we add an edge between any two vertices of our original graph which are at distance two. The Hajnal–Szemerédi theorem states that every graph with n vertices and maximum vertex degree at most d is $d+1$-colorable with all color classes of size $\lfloor n/(d+1)\rfloor$ or $\lceil n/(d+1)\rceil$, i.e the $(d+1)$-coloring is equitable.

Since the maximum degree of H^2 is at most Δ^2, there exists an equitable Δ^2+1-coloring of H^2 that induces a partition of the vertex set of H, say $U = U(H)$, into Δ^2+1 pairwise disjoint subsets $U_1, U_2, \dots, U_{\Delta^2+1}$, so that each U_k is an independent set in H^2 and the cardinality of each subset is either $\lfloor n/(\Delta^2+1)\rfloor$ or $\lceil n/(\Delta^2+1)\rceil$.

Next, partition the set V of vertices of the random graph $\mathbb{G}_{n,p}$ into pairwise disjoint sets $V_1, V_2, \dots, V_{\Delta^2+1}$, so that $|U_k| = |V_k|$ for $k = 1, 2, \dots, \Delta^2+1$.

We define a one-to-one function $f : U \mapsto V$, which maps each U_k onto V_k resulting in a mapping of H into an isomorphic copy of H in $\mathbb{G}_{n,p}$. In the first step, choose an arbitrary mapping of U_1 onto V_1. Now U_1 is an independent subset of H and so $\mathbb{G}_{n,p}[V_1]$ trivially contains a copy of $H[U_1]$. Assume, by induction, that we have already defined

$$f : U_1 \cup U_2 \cup \ldots \cup U_k \mapsto V_1 \cup V_2 \cup \ldots \cup V_k,$$

and that f maps the induced subgraph of H on $U_1 \cup U_2 \cup \ldots \cup U_k$ into a copy of it in $V_1 \cup V_2 \cup \ldots \cup V_k$. Now, define f on U_{k+1}, using the following construction. Suppose first that $U_{k+1} = \{u_1, u_2, \ldots, u_m\}$ and $V_{k+1} = \{v_1, v_2, \ldots, v_m\}$, where $m \in \{\lfloor n/(\Delta^2+1) \rfloor, \lceil n/(\Delta^2+1) \rceil\}$.

Next, construct a random bipartite graph $G^{(k)}_{m,m,p^*}$ with a vertex set $V = (X, Y)$, where $X = \{x_1, x_2, \ldots, x_m\}$ and $Y = \{y_1, y_2, \ldots, y_m\}$ and connect x_i and y_j with an edge if and only if in $\mathbb{G}_{n,p}$ the vertex v_j is joined by an edge to all vertices $f(u)$, where u is a neighbor of u_i in H which belongs to $U_1 \cup U_2 \cup \ldots \cup U_k$. Hence, we join x_i with y_j if and only if we can define $f(u_i) = v_j$.

Note that for each i and j, the edge probability $p^* \geq p^\Delta$ and that edges of $G^{(k)}_{m,m,p^*}$ are independent of each other, since they depend on pairwise disjoint sets of edges of $\mathbb{G}_{n,p}$. This follows from the fact that U_{k+1} is independent in H^2. Assuming that the condition (6.13) holds and that $(\Delta^2+1)^2 < n$, then by Theorem 6.1, the random graph $G^{(k)}_{m,m,p^*}$ has a perfect matching w.h.p. Moreover, we can conclude that the probability that there is no perfect matching in $G^{(k)}_{m,m,p^*}$ is at most $\frac{1}{(\Delta^2+1)n}$. It is here that we have used the extra factor 10 in the RHS of (6.13). We use a perfect matching in $G^{(k)}(m,m,p^*)$ to define f, assuming that if x_i and y_j are matched then $f(u_i) = v_j$. To define our mapping $f : U \mapsto V$ we have to find perfect matchings in all $G^{(k)}(m,m,p^*), k = 1,2,\ldots,\Delta^2+1$. The probability that we can succeed in this is at least $1 - 1/n$. This implies that $\mathbb{G}_{n,p}$ contains an isomorphic copy of H w.h.p. □

Corollary 6.20 *Let $n = 2^d$ and suppose that $d \to \infty$ and $p \geq \frac{1}{2} + o_d(1)$, where $o_d(1)$ is a function that tends to zero as $d \to \infty$. Then w.h.p. $\mathbb{G}_{n,p}$ contains a copy of a d-dimensional cube Q_d.*

Corollary 6.21 *Let $n = d^2$ and $p = \frac{\omega(n)\log n}{n^{1/4}}$, where $\omega(n), d \to \infty$. Then w.h.p. $\mathbb{G}_{n,p}$ contains a copy of the 2-dimensional lattice L_d.*

6.7 Exercises

6.7.1 Consider the bipartite graph process $\Gamma_m, m = 0,1,2,\ldots,n^2$ where we add the n^2 edges in $A \times B$ in random order, one by one. Show that w.h.p. the hitting time for Γ_m to have a perfect matching is identical with the hitting time for minimum degree at least one.

6.7.2 Show that if $p = \frac{\log n+(k-1)\log\log n+\omega}{n}$ where $k = O(1)$ and $\omega \to \infty$ then w.h.p. $G_{n,n,p}$ contains a k-regular spanning subgraph.

6.7.3 Consider the random bipartite graph G with bi-partition A, B where $|A| = |B| = n$. Each vertex $a \in A$ independently chooses $\lceil 2\log n \rceil$ random neighbors in B. Show that w.h.p. G contains a perfect matching.

6.7.4 Show that if $p = \frac{\log n+(k-1)\log\log n+\omega}{n}$ where $k = O(1)$ and $\omega \to \infty$ then w.h.p. $G_{n,p}$ contains $\lfloor k/2 \rfloor$ edge disjoint Hamilton cycles. If k is odd, show that in addition there is an edge disjoint matching of size $\lfloor n/2 \rfloor$. (Hint: Use Lemma 6.4 to argue that after "peeling off" a few Hamilton cycles, we can still use the arguments of Sections 6.1, 6.2.)

6.7.5 Let m_k^* denote the first time that G_m has minimum degree at least k. Show that w.h.p. in the graph process (i) $G_{m_1^*}$ contains a perfect matching and (ii) $G_{m_2^*}$ contains a Hamilton cycle.

6.7.6 Show that if $p = \frac{\log n+\log\log n+\omega}{n}$ where $\omega \to \infty$ then w.h.p. $G_{n,n,p}$ contains a Hamilton cycle. (Hint: Start with a 2-regular spanning subgraph from (6.7.2). Delete an edge from a cycle. Argue that rotations will always produce paths beginning and ending at different sides of the partition. Proceed more or less as in Section 6.2.)

6.7.7 Show that if $p = \frac{\log n+\log\log n+\omega}{n}$ where n is even and $\omega \to \infty$ then w.h.p. $G_{n,p}$ contains a pair of vertex disjoint $n/2$-cycles. (Hint: Randomly partition $[n]$ into two sets of size $n/2$. Then move some vertices between parts to make the minimum degree at least two in both parts.)

6.7.8 Show that if three divides n and $np^2 \gg \log n$ then w.h.p. $G_{n,p}$ contains $n/3$ vertex disjoint triangles. (Hint: Randomly partition $[n]$ into three sets A, B, C of size $n/3$. Choose a perfect matching M between A and B and then match C into M.)

6.7.9 Let $G = (X, Y, E)$ be an arbitrary bipartite graph where the bi-partition X, Y satisfies $|X| = |Y| = n$. Suppose that G has minimum degree at least $3n/4$. Let $p = \frac{K\log n}{n}$ where K is a large constant. Show that w.h.p. G_p contains a perfect matching.

6.7.10 Let $p = (1+\varepsilon)\frac{\log n}{n}$ for some fixed $\varepsilon > 0$. Prove that w.h.p. $G_{n,p}$ is Hamilton connected, i.e. every pair of vertices are the endpoints of a Hamilton path.

6.7.11 Show that for any fixed $\varepsilon > 0$ there exists c_ε such that if $c \geq c_\varepsilon$ then $\mathbb{G}_{n,p}$ contains a cycle of length $(1-\varepsilon)n$ with probability $1 - e^{-c\varepsilon^2 n/10}$.

6.7.12 Let $p = (1+\varepsilon)\frac{\log n}{n}$ for some fixed $\varepsilon > 0$. Prove that w.h.p. $G_{n,p}$ is pancyclic, i.e. it contains a cycle of length k for every $3 \leq k \leq n$. (See Cooper and Frieze [202] and Cooper [196], [198]).

6.7.13 Show that if p is constant then

$$\mathbb{P}(\mathbb{G}_{n,p} \text{ is not Hamiltonian}) = O(e^{-\Omega(np)}).$$

6.7.14 Let T be a tree on n vertices and maximum degree less than $c_1 \log n$. Suppose that T has at least $c_2 n$ leaves. Show that there exists $K = K(c_1, c_2)$ such that if $p \geq \frac{K \log n}{n}$ then $G_{n,p}$ contains a copy of T w.h.p.

6.7.15 Let $p = \frac{10}{n}$ and $G = G_{10n,p}$. Show that w.h.p. any red–blue coloring of the edges of G contains a mono-chromatic path of length n. (Hint: Apply the argument of Section 6.3 to both the red and blue sub-graphs of G.) This question is taken from Dudek and Pralat [259].

6.7.16 Suppose that $p = n^{-\alpha}$ for some constant $\alpha > 0$. Show that if $\alpha > \frac{1}{3}$ then w.h.p. $\mathbb{G}_{n,p}$ does not contain a maximal spanning planar subgraph, i.e. a planar subgraph with $3n-6$ edges. Show that if $\alpha < \frac{1}{3}$ then it contains one w.h.p. (see Bollobás and Frieze [141].)

6.7.17 Show that the hitting time for the existence of k edge-disjoint spanning trees coincides w.h.p. with the hitting time for minimum degree k, for $k = O(1)$ (see Palmer and Spencer [612]).

6.7.18 Consider the modified GREEDY matching algorithm where you first choose a random vertex x and then choose a random edge $\{x, y\}$ incident with x. Show that applied to $\mathbb{G}_{n,m}$, with $m = cn$, that w.h.p. it produces a matching of size $\left(\frac{1}{2} + o(1) - \frac{\log(2-e^{-2c})}{4c}\right) n$.

6.7.19 Let $X_1, X_2, \ldots, N = \binom{n}{2}$ be a sequence of independent Bernouilli random variables with common probability p. Let $\varepsilon > 0$ be sufficiently small (see [512]).

(i) Let $p = \frac{1-\varepsilon}{n}$ and let $k = \frac{7 \log n}{\varepsilon^2}$. Show that w.h.p. there is no interval I of length kn in $[N]$ in which at least k of the variables take the value 1.

(ii) Let $p = \frac{1+\varepsilon}{n}$ and let $N_0 = \frac{\varepsilon n^2}{2}$. Show that w.h.p.

$$\left| \sum_{i=1}^{N_0} X_i - \frac{\varepsilon(1+\varepsilon)n}{2} \right| \leq n^{2/3}.$$

6.7.20 Use the result of Exercise 6.7.19(a) to show that if $p = \frac{1-\varepsilon}{n}$ then w.h.p. the maximum component size in $\mathbb{G}_{n,p}$ is at most $\frac{7 \log n}{\varepsilon^2}$.

6.7.21 Use the result of Exercise 6.7.19(b) to show that if $p = \frac{1+\varepsilon}{n}$ then w.h.p $\mathbb{G}_{n,p}$ contains a path of length at least $\frac{\varepsilon^2 n}{5}$.

6.8 Notes

Hamilton Cycles

Multiple Hamilton Cycles

There are several results pertaining to the number of distinct Hamilton cycles in $G_{n,m}$. Cooper and Frieze [201] showed that in the graph process $G_{m_2^*}$ contains $(\log n)^{n-o(n)}$ distinct Hamilton cycles w.h.p. This number was improved by Glebov and Krivelevich [369] to $n!p^n e^{o(n)}$ for $G_{n,p}$ and $\left(\frac{\log n}{e}\right)^n e^{o(n)}$ at time m_2^*. McDiarmid [564] showed that for Hamilton cycles, perfect matchings, spanning trees the *expected* number was much higher. This comes from the fact that although there is a small probability that m_2^* is of order n^2, most of the expectation comes from here. (m_k^* is defined in Exercise 6.7.5.)

Bollobás and Frieze [140] (see Exercise 6.7.4) showed that in the graph process, $G_{m_k^*}$ contains $\lfloor k/2 \rfloor$ edge disjoint Hamilton cycles plus another edge disjoint matching of size $\lfloor n/2 \rfloor$ if n is odd. We call this property $\mathscr{A}_k$. This was the case $k = O(1)$. The more difficult case of the occurrence of $\mathscr{A}_k$ at m_k^*, where $k \to \infty$ was verified in two papers, Krivelevich and Samotij [509] and Knox, Kühn and Osthus [490].

Conditioning on Minimum Degree

Suppose that instead of taking enough edges to make the minimum degree in $\mathbb{G}_{n,m}$ two very likely, we instead condition on having a minimum degree of at least two. Let $G_{n,m}^{\delta \geq k}$ denote $G_{n,m}$ conditioned on having a minimum degree of at least $k = O(1)$. Bollobás, Fenner and Frieze [138] proved that if

$$m = \frac{n}{2}\left(\frac{\log n}{k+1} + k\log\log n + \omega(n)\right)$$

then $G_{n,m}^{\delta \geq k}$ has $\mathscr{A}_k$ w.h.p.

Bollobás, Cooper, Fenner and Frieze [135] proved that w.h.p. $G_{n,cn}^{\delta \geq k}$ has property $\mathscr{A}_{k-1}$ w.h.p. provided $3 \leq k = O(1)$ and $c \geq (k+1)^3$. For $k = 3$, Frieze [327] showed that $G_{n,cn}^{\delta \geq 3}$ is Hamiltonian w.h.p. for $c \geq 10$.

The k-core of a random graphs is distributed like $G_{\nu,\mu}^{\delta \geq k}$ for some (random) ν, μ. Krivelevich, Lubetzky and Sudakov [508] proved that when a k-core first appears, $k \geq 15$, w.h.p. it has $\lfloor (k-3)/2 \rfloor$ edge disjoint Hamilton cycles.

Algorithms for Finding Hamilton Cycles

Gurevich and Shelah [389] and Thomason [705] gave linear expected time algorithms for finding a Hamilton cycle in a sufficiently dense random graph, i.e. $G_{n,m}$ with $m \gg n^{5/3}$ in the Thomason paper. Bollobás, Fenner and Frieze [137] gave an $O(n^{3+o(1)})$ time algorithm that w.h.p. finds a Hamilton cycle in the graph $G_{m_2^*}$. Frieze and Haber [329] gave an $O(n^{1+o(1)})$ time algorithm for finding a Hamilton cycle in $G_{n,cn}^{\delta\geq 3}$ for c sufficiently large.

Resilience

Sudakov and Vu [697] introduced the notion of the *resilience* of a graph property $\mathscr{P}$. Given $G = G_{n,m} \in \mathscr{P}$ w.h.p. we define the *global resilience* of the property to be the maximum number of edges in a subgraph H of G such that $G \setminus H \in \mathscr{P}$ w.h.p. The *local resilience* is defined to be the maximum r such that if H is a subgraph H of G and $\Delta(H) \leq r$ then $G \setminus H \in \mathscr{P}$ w.h.p. In the context of Hamilton cycles, after a sequence of partial results in Frieze and Krivelevich [334], Ben-Shimon, Krivelevich and Sudakov [76], [77], Lee and Sudakov [521] proved that if $p \gg \log n/n$ then w.h.p. $G \setminus H$ is Hamiltonian for all subgraphs H for which $\Delta(H) \leq (\frac{1}{2} - o(1))np$, for a suitable $o(1)$ term. It is not difficult to see that this bound on Δ is best possible.

Long Cycles

A sequence of improvements, Bollobás [126] and Bollobás, Fenner and Frieze [139] to Theorem 6.8 in the sense of replacing $O(\log c/c)$ by something smaller, led finally to Frieze [321]. He showed that w.h.p. there is a cycle of length $n(1 - ce^{-c}(1+\varepsilon_c))$ where $\varepsilon_c \to 0$ with c. Up to the value of ε_c this is best possible.

Glebov, Naves and Sudakov [370] proved the following generalization of (part of) Theorem 6.5. They prove that if a graph G has a minimum degree of at least k and $p \geq \frac{\log k + \log\log k + \omega_k(1)}{k}$ then w.h.p. G_p has a cycle of length at least $k+1$.

Spanning Subgraphs

Riordan [644] used a second moment calculation to prove the existence of a certain (sequence of) spanning subgraphs $H = H^{(i)}$ in $\mathbb{G}_{n,p}$. Suppose that we denote the number of vertices in a graph H by $|H|$ and the number of edges by $e(H)$. Suppose that $|H| = n$. For $k \in [n]$ we let $e_H(k) = \max\{e(F) : F \subseteq H, |F| = k\}$ and $\gamma = \max_{3 \leq k \leq n} \frac{e_H(k)}{k-2}$. Riordan proved that if the following conditions hold, then $\mathbb{G}_{n,p}$ contains a copy of H w.h.p.: (i) $e(H) \geq n$, (ii) $Np, (1-p)n^{1/2} \to \infty$, (iii) $np^{\gamma} / \Delta(H)^4 \to \infty$.

This for example replaces the $\frac{1}{2}$ in Corollary 6.20 by $\frac{1}{4}$.

Spanning Trees

Gao, Pérez-Giménez and Sato [358] considered the existence of k edge disjoint spanning trees in $\mathbb{G}_{n,p}$. Using a characterization of Nash-Williams [602] they were able to show that w.h.p. one can find $\min\left\{\delta, \frac{m}{n-1}\right\}$ edge disjoint spanning trees. Here δ denotes the minimum degree and m denotes the number of edges.

When it comes to spanning trees of a fixed structure, Kahn conjectured that the threshold for the existence of any fixed bounded degree tree T, in terms of number of edges, is $O(n \log n)$. For example, a *comb* consists of a path P of length $n^{1/2}$ with each $v \in P$ being one endpoint of a path P_v of the same length. The paths P_v, P_w being vertex disjoint for $v \neq w$. Hefetz, Krivelevich and Szabó [404] proved this for a restricted class of trees, i.e. those with a linear number of leaves or with an induced path of length $\Omega(n)$. Kahn, Lubetzky and Wormald [456], [457] verified the conjecture for combs. Montgomery [585], [586] sharpened the result for combs, replacing $m = Cn \log n$ by $m = (1 + \varepsilon)n \log n$ and proved that any tree can be found w.h.p. when $m = O(\Delta n (\log n)^5)$, where Δ is the maximum degree of T.

Large Matchings

Karp and Sipser [476] analyzed a GREEDY algorithm for finding a large matching in the random graph $\mathbb{G}_{n,p}, p = c/n$ where $c > 0$ is a constant. It has a much better performance than the algorithm described in Section 6.4. It follows from their work that if $\mu(G)$ denotes the size of the largest matching in G then w.h.p.

$$\frac{\mu(\mathbb{G}_{n,p})}{n} \approx 1 - \frac{\gamma^* + \gamma_* + \gamma^* \gamma_*}{2c}$$

where γ_* is the smallest root of $x = c\exp\left\{-ce^{-x}\right\}$ and $\gamma^* = ce^{-\gamma_*}$.

Later, Aronson, Frieze and Pittel [38] tightened their analysis. This led to the consideration of the size of the largest matching in $G^{\delta \geq 2}_{n,m=cn}$. Frieze and Pittel [347] showed that w.h.p. this graph contains a matching of size $n/2 - Z$ where Z is a random variable with bounded expectation. Frieze [325] proved that in the bipartite analogue of this problem, a perfect matching exists w.h.p. Building on this work, Chebolu, Frieze and Melsted [176] showed how to find an exact maximum sized matching in $\mathbb{G}_{n,m}, m = cn$ in $O(n)$ expected time.

H-factors

By an H-factor of a graph G, we mean a collection of vertex disjoint copies of a fixed graph H that together cover all the vertices of G. Some early results on the existence of H-factors in random graphs are given in Alon and Yuster [32] and Ruciński [658]. For the case when H is a tree, Łuczak and Ruciński [547] found the precise threshold. For general H, there is a recent breakthrough paper of Johansson, Kahn and Vu [451] that gives the threshold for strictly balanced H and good estimates in general. See Gerke and McDowell [357] for some further results.

7
Extreme Characteristics

This chapter is devoted to the extremes of certain graph parameters. We look first at the diameter of random graphs, i.e. the extreme value of the shortest distance between a pair of vertices. Then we look at the size of the largest independent set and the the related value of the chromatic number. We decribe an important recent result on "interpolation" that proves certain limits exist. We end the chapter with the likely values of the first and second eigenvalues of a random graph.

7.1 Diameter

In this section we first discuss the threshold for $\mathbb{G}_{n,p}$ to have diameter d, when $d \geq 2$ is a constant. The diameter of a connected graph G is the maximum over distinct vertices v, w of $dist(v,w)$, where $dist(v,w)$ is the minimum number of edges in a path from v to w. The theorem below was proved independently by Burtin [165], [166] and by Bollobás [124]. The proof we give is due to Spencer [686].

Theorem 7.1 *Let $d \geq 2$ be a fixed positive integer. Suppose that $c > 0$ and*

$$p^d n^{d-1} = \log(n^2/c),$$

then

$$\lim_{n\to\infty} \mathbb{P}(diam(\mathbb{G}_{n,p}) = k) = \begin{cases} e^{-c/2} & \text{if } k = d \\ 1 - e^{-c/2} & \text{if } k = d+1. \end{cases}$$

Proof (a) w.h.p. $\text{diam}(G) \geq d$.

Fix $v \in V$ and let

$$N_k(v) = \{w : dist(v,w) = k\}. \tag{7.1}$$

It follows from Theorem 3.4 that w.h.p. for $0 \leq k < d$,

$$|N_k(v)| \leq \Delta^k \approx (np)^k \approx (n\log n)^{k/d} = o(n). \tag{7.2}$$

(b) w.h.p. $\mathrm{diam}(G) \leq d+1$.
Fix $v, w \in [n]$. Then for $1 \leq k < d$, define the event

$$\mathscr{F}_k = \left\{ |N_k(v)| \geq \left(\frac{np}{2}\right)^k \right\}.$$

Then for $k < \lceil d/2 \rceil$ we have

$$\begin{aligned}
&\mathbb{P}(\bar{\mathscr{F}}_k \mid \mathscr{F}_1, \dots, \mathscr{F}_{k-1}) \\
&= \mathbb{P}\left(\mathrm{Bin}\left(n - \sum_{i=0}^{k-1} |N_i(v)|, 1-(1-p)^{|N_{k-1}(v)|} \right) \leq \left(\frac{np}{2}\right)^k \right) \\
&\leq \mathbb{P}\left(\mathrm{Bin}\left(n - o(n), \frac{3}{4}\left(\frac{np}{2}\right)^{k-1} p \right) \leq \left(\frac{np}{2}\right)^k \right) \\
&\leq \exp\left\{ -\Omega\left(\left(\frac{np}{2}\right)^k \right) \right\} \\
&= O(n^{-3}).
\end{aligned}$$

So with probability $1 - O(n^{-3})$,

$$|N_{\lfloor d/2 \rfloor}(v)| \geq \left(\frac{np}{2}\right)^{\lfloor d/2 \rfloor} \quad \text{and} \quad |N_{\lceil d/2 \rceil}(w)| \geq \left(\frac{np}{2}\right)^{\lceil d/2 \rceil}.$$

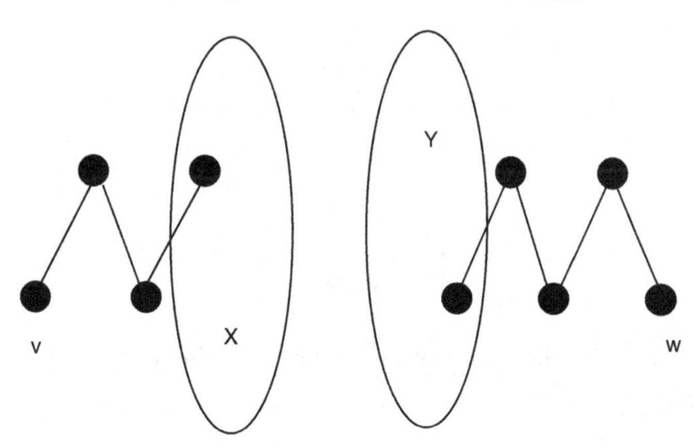

If $X = N_{\lfloor d/2 \rfloor}(v)$ and $Y = N_{\lceil d/2 \rceil}(w)$ then, either

$$X \cap Y \neq \emptyset \quad \text{and} \quad dist(v,w) \leq \lfloor d/2 \rfloor + \lceil d/2 \rceil = d,$$

or since the edges between X and Y are unconditioned by our construction,

$$\mathbb{P}(\nexists \text{ an } X:Y \text{ edge }) \le (1-p)^{(\frac{np}{2})^d} \le \exp\left\{-\left(\frac{np}{2}\right)^d p\right\}$$
$$\le \exp\{-(2-o(1))np\log n\} = o(n^{-3}).$$

So

$$\mathbb{P}(\exists v,w : dist(v,w) > d+1) = o(n^{-1}).$$

We now consider the probability that d or $d+1$ is the diameter. We will use Janson's inequality, see Section 21.6. More precisely, we will use the earlier inequality, Corollary 21.13, from Janson, Łuczak and Ruciński [431].

We first use this to estimate the probability of the following event: let $v \neq w \in [n]$ and let

$$\mathscr{A}_{v,w} = \{v,w \text{ are not joined by a path of length } d\}.$$

For $\mathbf{x} = x_1, x_2, \ldots, x_{d-1}$ let

$$\mathscr{B}_{v,\mathbf{x},w} = \{(v, x_1, x_2, \ldots, x_{d-1}, w) \text{ is a path in } \mathbb{G}_{n,p}\}.$$

Let

$$Z = \sum_{\mathbf{x}} Z_{\mathbf{x}},$$

where

$$Z_{\mathbf{x}} = \begin{cases} 1 & \text{if } \mathscr{B}_{v,\mathbf{x},w} \text{ occurs} \\ 0 & \text{otherwise.} \end{cases}$$

Janson's inequality allows us to estimate the probability that $Z = 0$, which is precisely the probability of $\mathscr{A}_{v,w}$.

Now

$$\mu = \mathbb{E}Z = (n-2)(n-3)\cdots(n-d)p^d = \log\left(\frac{n^2}{c}\right)\left(1+O\left(\frac{1}{n}\right)\right).$$

Let $\mathbf{x} = x_1, x_2, \cdots, x_{d-1}$, $\mathbf{y} = y_1, y_2, \ldots, y_{d-1}$ and

$$\Delta = \sum_{\substack{\mathbf{x},\mathbf{y}:\mathbf{x}\neq\mathbf{y}\\ v,\mathbf{x},w \text{ and } v,\mathbf{y},w\\ \text{share an edge}}} \mathbb{P}(\mathscr{B}_{\mathbf{x}} \cap \mathscr{B}_{\mathbf{y}})$$

$$\leq \sum_{t=1}^{d-1} \binom{d}{t} n^{2(d-1)-t} p^{2d-t}, \qquad t \text{ is the number of shared edges,}$$

$$= O\left(\sum_{t=1}^{d-1} n^{2(d-1)-t-\frac{d-1}{d}(2d-t)} (\log n)^{\frac{2d-t}{d}} \right)$$

$$= O\left(\sum_{t=1}^{d-1} n^{-t/d+o(1)} \right)$$

$$= o(1).$$

Applying Corollary 21.13, $\mathbb{P}(Z=0) \leq e^{-\mu+\Delta}$, we get

$$\mathbb{P}(Z=0) \leq \frac{c+o(1)}{n^2}.$$

On the other hand the Fortuin–Kasteleyn–Ginibre (FKG) inequality (see Section 21.3) implies that

$$\mathbb{P}(Z=0) \geq \left(1-p^d\right)^{(n-2)(n-3)\cdots(n-d)} = \frac{c+o(1)}{n^2}.$$

So

$$\mathbb{P}(\mathscr{A}_{v,w}) = \mathbb{P}(Z=0) = \frac{c+o(1)}{n^2}.$$

So

$$\mathbb{E}(\#v,w : \mathscr{A}_{v,w} \text{ occurs}) = \frac{c+o(1)}{2}$$

and we should expect that

$$\mathbb{P}(\nexists\, v,w : \mathscr{A}_{v,w} \text{ occurs}) \approx e^{-c/2}. \tag{7.3}$$

Indeed, if we choose $v_1, w_1, v_2, w_2, \ldots, v_k, w_k$, k constant, we find that

$$\mathbb{P}\left(\mathscr{A}_{v_1,w_1}, \mathscr{A}_{v_2,w_2}, \ldots, \mathscr{A}_{v_k,w_k}\right) \approx \left(\frac{c}{n^2}\right)^k \tag{7.4}$$

and (7.3) follows from the Method of Moments.

The proof of (7.3) is just a more involved version of the proof of the special case $k=1$ that we have just completed. We now let

$$\mathscr{B}_{\mathbf{x}} = \bigcup_{i=1}^{k} \mathscr{B}_{v_i,\mathbf{x},w_i}$$

and re-define

$$Z = \sum_{\mathbf{x}} Z_{\mathbf{x}},$$

where now

$$Z_{\mathbf{x}} = \begin{cases} 1 & \text{if } \mathscr{B}_{\mathbf{x}} \text{ occurs} \\ 0 & \text{otherwise.} \end{cases}$$

Then we have $\{Z=0\}$ is equivalent to $\bigcap_{i=1}^{k} \mathscr{A}_{v_i,w_i}$.

Now,

$$\mathbb{E}Z \le k(n-2)(n-3)\cdots(n-d)p^d = k\log\left(\frac{n^2}{c}\right)\left(1+O\left(\frac{1}{n}\right)\right).$$

We need to show that the corresponding $\Delta = o(1)$. But,

$$\begin{aligned}\Delta &\le \sum_{r,s} \sum_{\substack{\mathbf{x},\mathbf{y}:\mathbf{x}\neq\mathbf{y} \\ v_r,\mathbf{x},w_r \text{ and } v_s,\mathbf{y},w_s \\ \text{share an edge}}} \mathbb{P}(\mathscr{B}_{v_r,\mathbf{x},w_r} \cap \mathscr{B}_{v_s,\mathbf{y},w_s}) \\ &\le d^2 \sum_{t=1}^{d-1} \binom{d}{t} n^{2(d-1)-t} p^{2d-t} \\ &= o(1).\end{aligned}$$

This shows that

$$\mathbb{P}(Z=0) \le e^{-k\log(n^2/c+o(1)} = \left(\frac{c+o(1)}{n^2}\right)^k.$$

On the other hand, the FKG inequality (see Section 21.3) shows that

$$\mathbb{P}\left(\mathscr{A}_{v_1,w_1}, \mathscr{A}_{v_2,w_2}, \ldots, \mathscr{A}_{v_k,w_k}\right) \ge \prod_{i=1}^{k} \mathbb{P}\left(\mathscr{A}_{v_i,w_i}\right).$$

This verifies (7.4) and completes the proof of Theorem 7.1. □

We turn next to a sparser case and prove a somewhat weaker result.

Theorem 7.2 *Suppose that $p = \frac{\omega \log n}{n}$ where $\omega \to \infty$. Then*

$$diam(\mathbb{G}_{n,p}) \approx \frac{\log n}{\log np} \qquad w.h.p.$$

Proof Fix $v \in [n]$ and let $N_i = N_i(v)$ be as in (7.1). Let $N_{\leq k} = \bigcup_{i \leq k} N_i$. Using the proof of Theorem 3.4(b) we see that we can assume that

$$(1-\omega^{-1/3})np \leq \deg(x) \leq (1+\omega^{-1/3})np \qquad \text{for all } x \in [n]. \tag{7.5}$$

It follows that if $\gamma = \omega^{-1/3}$ and

$$k_0 = \frac{\log n - \log 3}{\log np + \gamma} \approx \frac{\log n}{\log np}$$

then w.h.p.

$$|N_{\leq k_0}| \leq \sum_{k \leq k_0} ((1+\gamma)np)^k \leq 2((1+\gamma)np)^{k_0} = \frac{2n}{3+o(1)}$$

and so the diameter of $\mathbb{G}_{n,p}$ is at least $(1-o(1))\frac{\log n}{\log np}$.

We can assume that $np = n^{o(1)}$ as larger p are dealt with in Theorem 7.1. Now fix $v, w \in [n]$ and let N_i be as in the previous paragraph. Now consider a Breadth First Search (BFS) that constructs $N_1, N_2, \ldots, N_{k_1}$ where

$$k_1 = \frac{3 \log n}{5 \log np}.$$

It follows that if (7.5) holds then for $k \leq k_1$ we have

$$|N_{i \leq k}| \leq n^{3/4} \text{ and } |N_k| p \leq n^{-1/5}. \tag{7.6}$$

Observe now that the edges from N_i to $[n] \setminus N_{\leq i}$ are unconditioned by the BFS up to layer k and so for $x \in [n] \setminus N_{\leq k}$,

$$\mathbb{P}(x \in N_{k+1} \mid N_{\leq k}) = 1-(1-p)^{|N_k|} \geq |N_k|p(1-|N_k|p) \geq \rho_k = |N_k|p(1-n^{-1/5}).$$

The events $x \in N_{k+1}$ are independent and so $|N_{k+1}|$ stochastically dominates the binomial $\text{Bin}(n-n^{3/4}, \rho_k)$. Assume inductively that $|N_k| \geq (1-\gamma)^k (np)^k$ for some $k \geq 1$. This is true w.h.p. for $k=1$ by (7.5). Let $\mathscr{A}_k$ be the event that (7.6) holds. It follows that

$$\mathbb{E}(|N_{k+1}| \mid \mathscr{A}_k) \geq np|N_k|(1-O(n^{-1/5})).$$

It then follows from the Chernoff bounds (Theorem 21.6) that

$$\mathbb{P}(|N_{k+1}| \leq ((1-\gamma)np)^{k+1} \leq \exp\left\{-\frac{\gamma^2}{4}|N_k|np\right\} = o(n^{-anyconstant}).$$

There is a small point to be made about conditioning here. We can condition on (7.5) holding and then argue that this only multiplies small probabilities by $1+o(1)$ if we use $\mathbb{P}(A \mid B) \leq \mathbb{P}(A)/\mathbb{P}(B)$.

It follows that if

$$k_2 = \frac{\log n}{2(\log np + \log(1-\gamma))} \approx \frac{\log n}{2\log np}$$

then w.h.p. we have

$$|N_{k_2}| \geq n^{1/2}.$$

Analogously, if we do BFS from w to create $N'_k, i = 1,2,\ldots,k_2$ then $|N'_{k_2}| \geq n^{1/2}$. If $N_{\leq k_2} \cap N'_{\leq k_2} \neq \emptyset$ then $dist(v,w) \leq 2k_2$ and we are done. Otherwise, we observe that the edges $N_{k_2} : N'_{k_2}$ between N_{k_2} and N'_{k_2} are unconditioned (except for (7.5)) and so

$$\mathbb{P}(N_{k_2} : N'_{k_2} = \emptyset) \leq (1-p)^{n^{1/2} \times n^{1/2}} \leq n^{-\omega}.$$

If $N_{k_2} : N'_{k_2} \neq \emptyset$ then $dist(v,w) \leq 2k_2 + 1$ and we are done. Note that given (7.5), all other unlikely events have probability $O(n^{-anyconstant})$ of occurring and so we can inflate these latter probabilities by n^2 to account for all choices of v, w. This completes the proof of Theorem 7.2. □

7.2 Largest Independent Sets

Let $\alpha(G)$ denote the size of the largest independent set in a graph G.

Dense Case

The following theorem was first proved by Matula [556].

Theorem 7.3 *Suppose* $0 < p < 1$ *is a constant and* $b = \frac{1}{1-p}$. *Then w.h.p.*

$$\alpha(\mathbb{G}_{n,p}) \approx 2\log_b n.$$

Proof Let X_k be the number of independent sets of order k.
(i) Let

$$k = \lceil 2\log_b n \rceil,$$

then,

$$\begin{aligned}
\mathbb{E}X_k &= \binom{n}{k}(1-p)^{\binom{k}{2}} \\
&\leq \left(\frac{ne}{k(1-p)^{1/2}}(1-p)^{k/2}\right)^k \\
&\leq \left(\frac{e}{k(1-p)^{1/2}}\right)^k \\
&= o(1).
\end{aligned}$$

(ii) Now let

$$k = \lfloor 2\log_b n - 4\log_b \log_b n \rfloor.$$

Let

$$\overline{\Delta} = \sum_{\substack{i,j \\ S_i \sim S_j}} \mathbb{P}(S_i, S_j \text{ are independent in } \mathbb{G}_{n,p}),$$

where $S_1, S_2, \ldots, S_{\binom{n}{k}}$ are all the k-subsets of $[n]$ and $S_i \sim S_j$ if and only if $|S_i \cap S_j| \geq 2$. By Janson's inequality, see Theorem 21.12,

$$\mathbb{P}(X_k = 0) \leq \exp\left\{-\frac{(\mathbb{E}X_k)^2}{2\overline{\Delta}}\right\}.$$

Now

$$\begin{aligned}
\frac{\overline{\Delta}}{(\mathbb{E}X_k)^2} &= \frac{\binom{n}{k}(1-p)^{\binom{k}{2}} \sum_{j=2}^{k} \binom{n-k}{k-j}\binom{k}{j}(1-p)^{\binom{k}{2}-\binom{j}{2}}}{\left(\binom{n}{k}(1-p)^{\binom{k}{2}}\right)^2} \\
&= \sum_{j=2}^{k} \frac{\binom{n-k}{k-j}\binom{k}{j}}{\binom{n}{k}}(1-p)^{-\binom{j}{2}} \\
&= \sum_{j=2}^{k} u_j.
\end{aligned}$$

Notice that for $j \geq 2$,

$$\begin{aligned}
\frac{u_{j+1}}{u_j} &= \frac{k-j}{n-2k+j+1}\frac{k-j}{j+1}(1-p)^{-j} \\
&\leq \left(1 + O\left(\frac{\log_b n}{n}\right)\right)\frac{k^2(1-p)^{-j}}{n(j+1)}.
\end{aligned}$$

Therefore,

$$\begin{aligned}
\frac{u_j}{u_2} &\leq (1+o(1))\left(\frac{k^2}{n}\right)^{j-2} \frac{2(1-p)^{-(j-1)(j+2)/2}}{j!} \\
&\leq (1+o(1))\left(\frac{2k^2 e}{nj}(1-p)^{-\frac{j+1}{2}}\right)^{j-2} \leq 1.
\end{aligned}$$

So

$$\frac{(\mathbb{E}X_k)^2}{\overline{\Delta}} \geq \frac{1}{ku_2} \geq \frac{n^2(1-p)}{k^5}.$$

Therefore,

$$\mathbb{P}(X_k = 0) \leq e^{-\Omega(n^2/(\log n)^5)}. \tag{7.7}$$

□

Matula used the Chebyshev inequality and so he was not able to prove an exponential bound like (7.7). This is important when we come to discuss the chromatic number.

Sparse Case

We now consider the case where $p = d/n$ and d is a large constant. Frieze [324] proved

Theorem 7.4 *Let $\varepsilon > 0$ be a fixed constant. Then for $d \geq d(\varepsilon)$ we have that w.h.p.*

$$\left|\alpha(\mathbb{G}_{n,p})) - \frac{2n}{d}(\log d - \log\log d - \log 2 + 1)\right| \leq \frac{\varepsilon n}{d}.$$

Dani and Moore [228] have recently given an even sharper result.

In this section we prove that if $p = d/n$ and d is sufficiently large then w.h.p.

$$\left|\alpha(\mathbb{G}_{n,p}) - \frac{2\log d}{d}n\right| \leq \frac{\varepsilon \log d}{d}n. \tag{7.8}$$

This follows from the following. Let X_k be as defined in the previous section. Let

$$k_0 = \frac{(2-\varepsilon/2)\log d}{d}n.$$

Then,

$$\mathbb{P}\left(\left|\alpha(\mathbb{G}_{n,p}) - \mathbb{E}(\alpha(\mathbb{G}_{n,p}))\right| \geq \frac{\varepsilon \log d}{2d}n\right) \leq \exp\left\{-\Omega\left(\frac{(\log d)^2}{d^2}\right)n\right\}. \tag{7.9}$$

$$\mathbb{P}(X_{k_0} > 0) \geq \exp\left\{-O\left(\frac{(\log d)^{3/2}}{d^2}\right)n\right\}. \tag{7.10}$$

Let us see how (7.8) follows from these two. Indeed, together they imply that

$$|k_0 - \mathbb{E}(\alpha(\mathbb{G}_{n,p}))| \leq \frac{\varepsilon \log d}{2d}n.$$

We obtain (7.8) by applying (7.9) once more.

Proof of (7.9): This follows directly from the Azuma–Hoeffding inequality – see Section 21.7, in particular Lemma 21.16. If $Z = \alpha(\mathbb{G}_{n,p})$ then we write $Z = Z(Y_2, Y_3, \ldots, Y_n)$, where Y_i is the set of edges between vertex i and vertices $[i-1]$ for $i \geq 2$. $Y_2, Y_3, \ldots, Y_n$ are independent and changing a single Y_i can change Z by at most one. Therefore, for any $t > 0$ we have

$$\mathbb{P}(|Z - \mathbb{E}(Z)| \geq t) \leq \exp\left\{-\frac{t^2}{2n-2}\right\}.$$

Setting $t = \frac{\varepsilon \log d}{2d}n$ yields (7.9).

Proof of (7.10): Now, after using Lemma 21.1(g),

$$\frac{1}{\mathbb{P}(X_{k_0}>0)} \leq \frac{\mathbb{E}(X_{k_0}^2)}{\mathbb{E}(X_{k_0})^2} = \sum_{j=0}^{k_0} \frac{\binom{n-k_0}{k_0-j}\binom{k_0}{j}}{\binom{n}{k_0}}(1-p)^{-\binom{j}{2}}$$

$$\leq \sum_{j=0}^{k_0} \left(\frac{k_0 e}{j} \cdot \exp\left\{\frac{jd}{2n} + O\left(\frac{jd^2}{n^2}\right)\right\}\right)^j$$

$$\times \left(\frac{k_0}{n}\right)^j \left(\frac{n-k_0}{n-j}\right)^{k_0-j} \tag{7.11}$$

$$\leq \sum_{j=0}^{k_0} \left(\frac{k_0 e}{j} \cdot \frac{k_0}{n} \cdot \exp\left\{\frac{jd}{2n} + O\left(\frac{jd^2}{n^2}\right)\right\}\right)^j$$

$$\times \exp\left\{-\frac{(k_0-j)^2}{n-j}\right\}$$

$$\leq_b \sum_{j=0}^{k_0} \left(\frac{k_0 e}{j} \cdot \frac{k_0}{n} \cdot \exp\left\{\frac{jd}{2n} + \frac{2k_0}{n}\right\}\right)^j \times \exp\left\{-\frac{k_0^2}{n}\right\}$$

$$= \sum_{j=0}^{k_0} v_j. \tag{7.12}$$

(The notation $A \leq_b B$ is shorthand for $A = O(B)$ when the latter is considered to look ugly.)

We observe first that $(A/x)^x \leq e^{A/e}$ for $A > 0$ implies that

$$\left(\frac{k_0 e}{j} \cdot \frac{k_0}{n}\right)^j \times \exp\left\{-\frac{k_0^2}{n}\right\} \leq 1.$$

So,

$$j \leq j_0 = \frac{(\log d)^{3/4}}{d^{3/2}} n \Longrightarrow v_j \leq \exp\left\{\frac{j^2 d}{2n} + \frac{2jk_0}{n}\right\}$$

$$= \exp\left\{O\left(\frac{(\log d)^{3/2}}{d^2}\right) n\right\}. \tag{7.13}$$

Now put

$$j = \frac{\alpha \log d}{d} n \text{ where } d^{-1/2} < \alpha < 2 - \varepsilon/2.$$

Then

$$\frac{k_0 e}{j} \cdot \frac{k_0}{n} \cdot \exp\left\{\frac{jd}{2n} + \frac{2k_0}{n}\right\} \le \frac{4e\log d}{\alpha d} \cdot \exp\left\{\frac{\alpha \log d}{2} + \frac{4\log d}{d}\right\}$$
$$= \frac{4e}{\alpha d^{1-\alpha/2}} \exp\left\{\frac{4\log d}{d}\right\}$$
$$< 1.$$

To see this, note that if $f(\alpha) = \alpha d^{1-\alpha/2}$ then f increases between $d^{-1/2}$ and $2/\log d$ after which it decreases. Then note that

$$\min\left\{f(d^{-1/2}), f(2-\varepsilon)\right\} > 4e\exp\left\{\frac{4\log d}{d}\right\}.$$

Thus $v_j < 1$ for $j \ge j_0$ and (7.10) follows from (7.13). □

7.3 Interpolation

The following theorem is taken from Bayati, Gamarnik and Tetali [63]. Note that it is not implied by Theorem 7.4. This paper proves a number of other results of a similar flavor for other parameters. It is an important paper in that it verifies some very natural conjectures about some graph parameters, that have not been susceptible to proof until now.

Theorem 7.5 *There exists a function $H(d)$ such that*

$$\lim_{n\to\infty} \frac{\mathbb{E}(\alpha(G_{n,\lfloor dn \rfloor}))}{n} = H(d).$$

Proof For this proof we use the model $\mathbb{G}^{(A)}_{n,m}$ of Section 1.3. This is proper since we we know that w.h.p.

$$|\alpha(\mathbb{G}^{(A)}_{n,m}) - \alpha(\mathbb{G}_{n,m})| \le ||E(\mathbb{G}^{(A)}_{n,m})| - m| \le \log n.$$

We will prove that for every $1 \le n_1, n_2 \le n-1$ such that $n_1 + n_2 = n$,

$$\mathbb{E}(\alpha(\mathbb{G}^{(A)}_{n,\lfloor dn \rfloor})) \ge \mathbb{E}(\alpha(\mathbb{G}^{(A)}_{n_1,m_1})) + \mathbb{E}(\alpha(\mathbb{G}^{(A)}_{n_2,m_2})) \tag{7.14}$$

where $m_i = \mathrm{Bin}(\lfloor dn \rfloor, n_i/n), i = 1,2$.

Assume (7.14). We have $\mathbb{E}(|m_j - \lfloor dn_j \rfloor|) = O(n^{1/2})$. This and (7.14) and the fact that adding/deleting one edge changes α by at most one implies that

$$\mathbb{E}(\alpha(\mathbb{G}^{(A)}_{n,\lfloor dn \rfloor})) \ge \mathbb{E}(\alpha(\mathbb{G}^{(A)}_{n_1,\lfloor dn_1 \rfloor})) + \mathbb{E}(\alpha(\mathbb{G}^{(A)}_{n_2,\lfloor dn_2 \rfloor})) - O(n^{1/2}). \tag{7.15}$$

Thus, the sequence $u_n = \mathbb{E}(\alpha(\mathbb{G}^{(A)}_{n,\lfloor dn \rfloor}))$ satisfies the conditions of Lemma 7.6 below and the proof of Theorem 7.5 follows.

Proof of (7.14): We begin by constructing a sequence of graphs interpolating between $\mathbb{G}^{(A)}_{n,\lfloor dn \rfloor}$ and a disjoint union of $\mathbb{G}^{(A)}_{n_1,m_1}$ and $\mathbb{G}^{(A)}_{n_2,m_2}$. Given n, n_1, n_2 such that $n_1 + n_2 = n$ and any $0 \le r \le m = \lfloor dn \rfloor$, let $\mathbb{G}(n,m,r)$ be the random (pseudo-)graph on vertex set $[n]$ obtained as follows. It contains precisely m edges. The first r edges $e_1, e_2, \ldots, e_r$ are selected randomly from $[n]^2$. The remaining $m - r$ edges $e_{r+1}, \ldots, e_m$ are generated as follows. For each $j = r+1, \ldots, m$, with probability n_j/n, e_j is selected randomly from $M_1 = [n_1]^2$ and with probability n_2/n, e_j is selected randomly from $M_2[n_1+1, n]^2$. Observe that when $r = m$ we have $\mathbb{G}(n,m,r) = \mathbb{G}^{(A)}(n,m)$ and when $r = 0$ it is the disjoint union of $\mathbb{G}^{(A)}_{n_1,m_1}$ and $\mathbb{G}^{(A)}_{n_2,m_2}$, where $m_j = \text{Bin}(m, n_j/n)$ for $j = 1,2$. We show next that

$$\mathbb{E}(\alpha(\mathbb{G}(n,m,r))) \ge \mathbb{E}(\alpha(\mathbb{G}(n,m,r-1))) \text{ for } r = 1, \ldots, m. \tag{7.16}$$

It follows immediately that

$$\begin{aligned}\mathbb{E}(\alpha(\mathbb{G}^{(A)}_{n,m})) &= \mathbb{E}(\alpha(\mathbb{G}(n,m,m))) \ge \mathbb{E}(\alpha(\mathbb{G}(n,m,0))) \\ &= \mathbb{E}(\alpha(\mathbb{G}^{(A)}_{n_1,m_1})) + \mathbb{E}(\alpha(\mathbb{G}^{(A)}_{n_2,m_2}))\end{aligned}$$

which is (7.14).

Proof of (7.16): Observe that $\mathbb{G}(n,m,r-1)$ is obtained from $\mathbb{G}(n,m,r)$ by deleting the random edge e_r and then adding an edge from M_1 or M_2. Let $\mathbb{G}_0$ be the graph obtained after deleting e_r, but before adding its replacement. Remember that

$$\mathbb{G}(n,m,r) = \mathbb{G}_0 + e_r.$$

We will show something stronger than (7.16) viz. that

$$\mathbb{E}(\alpha(\mathbb{G}(n,m,r)) \mid \mathbb{G}_0) \ge \mathbb{E}(\alpha(\mathbb{G}(n,m,r-1)) \mid \mathbb{G}_0) \text{ for } r = 1, \ldots, m. \tag{7.17}$$

Now let $O^* \subseteq [n]$ be the set of vertices that belong to every largest independent set in $\mathbb{G}_0$. Then for $e_r = (x,y)$, $\alpha(\mathbb{G}_0 + e) = \alpha(\mathbb{G}_0) - 1$ if $x, y \in O^*$ and $\alpha(\mathbb{G}_0 + e) = \alpha(\mathbb{G}_0)$ if $x \notin O^*$ or $y \notin O^*$. Because e_r is randomly chosen, we have

$$\mathbb{E}(\alpha(\mathbb{G}_0 + e_r) \mid \mathbb{G}_0) - \mathbb{E}(\alpha(\mathbb{G}_0)) = -\left(\frac{|O^*|}{n}\right)^2.$$

By a similar argument

$$\begin{aligned}
&\mathbb{E}(\alpha(\mathbb{G}(n,m,r-1) \mid \mathbb{G}_0) - \alpha(\mathbb{G}_0)\\
&= -\frac{n_1}{n}\left(\frac{|O^* \cap M_1|}{n_1}\right)^2 - \frac{n_2}{n}\left(\frac{|O^* \cap M_2|}{n_2}\right)^2\\
&\le -\left(\frac{n_1}{n}\frac{|O^* \cap M_1|}{n_1} + \frac{n_2}{n}\frac{|O^* \cap M_2|}{n_2}\right)^2\\
&= -\left(\frac{|O^*|}{n}\right)^2\\
&= \mathbb{E}(\alpha(\mathbb{G}_0 + e_r) \mid \mathbb{G}_0) - \mathbb{E}(\alpha(\mathbb{G}_0)),
\end{aligned}$$

completing the proof of (7.17). □

The proof of the following lemma is left as an exercise.

Lemma 7.6 *Given $\gamma \in (0,1)$, suppose that the non-negative sequence $u_n, n \ge 1$ satisfies*

$$u_n \ge u_{n_1} + u_{n_2} - O(n^\gamma)$$

for every n_1, n_2 such that $n_1 + n_2 = n$. Then $\lim_{n\to\infty} \frac{u_n}{n}$ exists.

7.4 Chromatic Number

Let $\chi(G)$ denote the chromatic number of a graph G, i.e. the smallest number of colors with which one can properly color the vertices of G. A coloring is proper if no two adjacent vertices have the same color.

Dense Graphs

We first describe the asymptotic behavior of the chromatic number of dense random graphs. The following theorem is a major result, due to Bollobás [131]. The upper bound without the 2 in the denominator follows directly from Theorem 7.3. An intermediate result giving 3/2 instead of 2 was already proved by Matula [557].

Theorem 7.7 *Suppose $0 < p < 1$ is a constant and $b = \frac{1}{1-p}$. Then w.h.p.*

$$\chi(\mathbb{G}_{n,p}) \approx \frac{n}{2\log_b n}.$$

Proof
(i) By Theorem 7.3

$$\chi(\mathbb{G}_{n,p}) \geq \frac{n}{\alpha(\mathbb{G}_{n,p})} \approx \frac{n}{2\log_b n}.$$

(ii) Let $\nu = \frac{n}{(\log_b n)^2}$ and $k_0 = 2\log_b n - 4\log_b \log_b n$. It follows from (7.7) that

$$\begin{aligned}
&\mathbb{P}(\exists S : |S| \geq \nu, S \text{ does not contain an independent set of order } \geq k_0) \\
&\leq \binom{n}{\nu} \exp\left\{-\Omega\left(\frac{\nu^2}{(\log n)^5}\right)\right\} \\
&= o(1).
\end{aligned}$$

So assume that every set of order at least ν contains an independent set of order at least k_0. We repeatedly choose an independent set of order k_0 among the set of uncolored vertices. Give each vertex in this set the same new color. Repeat until the number of uncolored vertices is at most ν. Give each remaining uncolored vertex its own color. The number of colors used is at most

$$\frac{n}{k_0} + \nu \approx \frac{n}{2\log_b n}.$$

□

It should be noted that Bollobás did not have the Janson inequality available to him and he had to make a clever choice of random variable for use with the Azuma–Hoeffding inequality. His choice was the maximum size of a family of edge independent independent sets. Łuczak [540] proved the corresponding result to Theorem 7.7 in the case where np→0.

Concentration

Theorem 7.8 *Suppose* $0 < p < 1$ *is a constant, then,*

$$\mathbb{P}(|\chi(\mathbb{G}_{n,p}) - \mathbb{E}\,\chi(\mathbb{G}_{n,p})| \geq t) \leq 2\exp\left\{-\frac{t^2}{2n}\right\}$$

Proof Write

$$\chi = Z(Y_1, Y_2, \ldots, Y_n), \tag{7.18}$$

where

$$Y_j = \{(i,j) \in E(\mathbb{G}_{n,p}) : i < j\}.$$

Then

$$|Z(Y_1, Y_2, \ldots, Y_n) - Z(Y_1, Y_2, \ldots, \hat{Y}_i, \ldots, Y_n)| \leq 1$$

and the theorem follows from the Azuma–Hoeffding inequality, see Section 21.7, in particular Lemma 21.16. □

GREEDY Coloring Algorithm

We show below that a simple GREEDY algorithm performs very efficiently. It uses twice as many colors as it "should" in the light of Theorem 7.7. This algorithm is discussed in Bollobás and Erdős [136] and by Grimmett and McDiarmid [385]. It starts by greedily choosing an independent set C_1 and at the same time giving its vertices color 1. C_1 is removed and then we greedily choose an independent set C_2 and give its vertices color 2 and so on, until all the vertices have been colored.

Algorithm *GREEDY*

- k is the current color.
- A is the current set of vertices that might get color k in the current round.
- U is the current set of uncolored vertices.

begin
  $k \longleftarrow 0$, $A \longleftarrow [n]$, $U \longleftarrow [n]$, $C_k \longleftarrow \emptyset$.
  while $U \neq \emptyset$ **do**
  $k \longleftarrow k+1$ $A \longleftarrow U$
  while $A \neq \emptyset$
    begin
    Choose $v \in A$ and put it into C_k
    $U \longleftarrow U \setminus \{v\}$
    $A \longleftarrow A \setminus (\{v\} \cup N(v))$
    end
end

Theorem 7.9 *Suppose* $0 < p < 1$ *is a constant and* $b = \frac{1}{1-p}$. *Then w.h.p. algorithm GREEDY uses approximately* $n/\log_b n$ *colors to color the vertices of* $\mathbb{G}_{n,p}$.

Proof At the start of an iteration the edges inside U are un-examined. Suppose that

$$|U| \geq \nu = \frac{n}{(\log_b n)^2}.$$

We show that approximately $\log_b n$ vertices get color k, i.e. at the end of round k, $|C_k| \approx \log_b n$.

Each iteration chooses a *maximal* independent set from the remaining uncolored vertices. Let $k_0 = \log_b n - 3\log_b \log_b n$. Then

$$
\begin{aligned}
&\mathbb{P}(\exists\, T : |T| \le k_0, T \text{ is maximally independent}) \\
&\le \sum_{t=1}^{k_0} \binom{n}{t} (1-p)^{\binom{t}{2}} \left(1-(1-p)^t\right)^{n-t} \\
&\le \sum_{t=1}^{k_0} \left(\frac{ne}{t} (1-p)^{\frac{t-1}{2}} \right)^t e^{-(n-t)(1-p)^t} \\
&\le \sum_{t=1}^{k_0} \left(ne^{1+(1-p)^t} \right)^t e^{-n(1-p)^t} \\
&\le k_0 \left(ne^2\right)^{k_0} e^{-(\log_b n)^3} \le e^{-\frac{1}{2}(\log_b n)^3}.
\end{aligned}
$$

So the probability that we fail to use at least k_0 colors while $|U| \ge \nu$ is at most

$$
ne^{-\frac{1}{2}(\log_b \nu)^3} = o(1).
$$

So w.h.p. GREEDY uses at most

$$
\frac{n}{k_0} + \nu \approx \frac{n}{\log_b n} \text{ colors.}
$$

We now put a lower bound on the number of colors used by GREEDY. Let

$$
k_1 = \log_b n + 2\log_b \log_b n.
$$

Consider one round. Let $U_0 = U$ and suppose $u_1, u_2, \ldots \in C_k$ and $U_{i+1} = U_i \setminus (\{u_i\}) \cup N(u_i))$.

Then

$$
\mathbb{E}(\, |U_{i+1}| \; |U_i \,) \le |U_i \,|(1-p),
$$

and so, for $i = 1, 2, \ldots$

$$
\mathbb{E}\,|U_i| \le n(1-p)^i.
$$

So

$$
\mathbb{P}(k_1 \text{ vertices colored in one round}) \le \frac{1}{(\log_b n)^2},
$$

and

$$
\mathbb{P}(2k_1 \text{ vertices colored in one round}) \le \frac{1}{n}.
$$

So let

$$
\delta_i = \begin{cases} 1 & \text{if at most } k_1 \text{ vertices are colored in round } i \\ 0 & \text{otherwise.} \end{cases}
$$

We see that

$$\mathbb{P}(\delta_i = 1|\delta_1,\delta_2,\ldots,\delta_{i-1}) = 1 - \frac{1}{(\log_b n)^2}.$$

So the number of rounds that color more than k_1 vertices is stochastically dominated by a Binomial with mean $n/(\log_b n)^2$. The Chernoff bounds imply that w.h.p. the number of rounds that color more than k_1 vertices is less than $2n/(\log_b n)^2$. Strictly speaking we need to use Lemma 21.22 to justify the use of the Chernoff bounds. Because no round colors more than $2k_1$ vertices we see that w.h.p. GREEDY uses at least

$$\frac{n - 4k_1 n/(\log_b n)^2}{k_1} \approx \frac{n}{\log_b n} \text{ colors.}$$

□

Sparse Graphs

We now consider the case of sparse random graphs. We first state an important conjecture about the chromatic number.

Conjecture: Let $k \geq 3$ be a fixed positive integer. Then there exists $d_k > 0$ such that if ε is an arbitrary positive constant and $p = \frac{d}{n}$ then w.h.p. (i) $\chi(\mathbb{G}_{n,p}) \leq k$ for $d \leq d_k - \varepsilon$ and (ii) $\chi(\mathbb{G}_{n,p}) \geq k+1$ for $d \geq d_k + \varepsilon$.

In the absence of a proof of this conjecture, we present the following result due to Łuczak [541]. It should be noted that Shamir and Spencer [679] had already proved six-point concentration.

Theorem 7.10 *If $p < n^{-5/6-\delta}, \delta > 0$, then the chromatic number of $\mathbb{G}_{n,p}$ is w.h.p. two point concentrated.*

Proof To prove this theorem we need three lemmas.

Lemma 7.11

(a) Let $0 < \delta < 1/10, 0 \leq p < 1$ and $d = np$. Then w.h.p. each subgraph H of $\mathbb{G}_{n,p}$ on less than $nd^{-3(1+2\delta)}$ vertices has less than $(3/2 - \delta)|H|$ edges.

(b) Let $0 < \delta < 1/10$ and let $0 \leq p \leq \delta/n$. Then w.h.p. each subgraph H of $\mathbb{G}_{n,p}$ has less than $3|H|/2$ edges.

The above lemma can be proved easily by the First Moment Method, see Exercise 7.6.6. Note also that Lemma 7.11 implies that each subgraph H satisfying the conditions of the lemma has minimum degree less than three, and thus is 3-colorable, due to the following simple observation (see Bollobás [132] Theorem V.1)

Lemma 7.12 *Let $k = \max_{H \subseteq G} \delta(H)$, where the maximum is taken over all induced subgraphs of G. Then $\chi(G) \le k+1$.*

Proof This is an easy exercise in graph theory. We proceed by induction on $|V(G)|$. We choose a vertex of minimum degree v, color $G - v$ inductively and then color v. □

The next lemma is an immediate consequence of the Azuma–Hoeffding inequality, see Section 21.7, in particular Lemma 21.16.

Lemma 7.13 *Let $k = k(n)$ be such that*

$$\mathbb{P}\left(\chi(\mathbb{G}_{n,p}) \ge k\right) > \frac{1}{\log\log n}. \tag{7.19}$$

Then w.h.p. all but at most $n^{1/2}\log n$ vertices of $\mathbb{G}_{n,p}$ can be properly colored using k colors.

Proof Let Z be the maximum number of vertices in $\mathbb{G}_{n,p}$ that can be properly colored with k colors. Write $Z = Z(Y_1, Y_2, \dots, Y_n)$ as in (7.18). Then we have

$$\mathbb{P}(Z = n) > \frac{1}{\log\log n} \text{ and } \mathbb{P}(|Z - \mathbb{E}Z)| \ge t) \le 2\exp\left\{-\frac{t^2}{2n}\right\}. \tag{7.20}$$

Putting $t = \frac{1}{2}n^{1/2}\log n$ into (7.20) shows that $\mathbb{E}Z \ge n - t$ and the lemma follows after applying the concentration inequality in (7.20) once again. □

Now we are ready to present Łuczak's ingenious argument to prove Theorem 7.10. Note first that when p is such that $np \to 0$ as $n \to \infty$, then by Theorem 2.1 $\mathbb{G}_{n,p}$ is a forest w.h.p. and so its chromatic number is either 1 or 2. Furthermore, for $1/\log n < p < 1.01$ the random graph $\mathbb{G}_{n,p}$ w.h.p. contains at least one edge and no subgraph with minimal degree larger than two (see Lemma 7.11), which implies that $\chi(\mathbb{G}_{n,p})$ is equal to 2 or 3 (see Lemma 7.12). Now let us assume that the edge probability p is such that $1.01 < d = np < n^{1/6-\delta}$. Observe that in this range of p the random graph $\mathbb{G}_{n,p}$ w.h.p. contains an odd cycle, so $\chi(\mathbb{G}_{n,p}) \ge 3$.

Let k be as in Lemma 7.13 and let U_0 be a set of size at most $u_0 = n^{1/2}\log n$ such that $[n] \setminus U_0$ can be properly colored with k colors. Let us construct a nested sequence of subsets of vertices $U_0 \subseteq U_1 \subseteq \ldots \subseteq U_m$ of $\mathbb{G}_{n,p}$, where we define $U_{i+1} = U_i \cup \{v,w\}$, where $v, w \notin U_i$ are connected by an edge and both v and w have a neighbor in U_i. The construction stops at $i = m$ if such a pair $\{v,w\}$ does not exist.

Notice that m can not exceed $m_0 = n^{1/2}\log n$, since if $m > m_0$ then a subgraph of $\mathbb{G}_{n,p}$ induced by vertices of U_{m_0} would have

$$|U_{m_0}| = u_0 + 2m_0 \le 3n^{1/2}\log n < nd^{-3(1+2\delta)}$$

vertices and at least $3m_0 \geq (3/2-\delta)|U_{m_0}|$ edges, contradicting the statement of Lemma 7.11.

As a result, the construction produces a set U_m in $\mathbb{G}_{n,p}$, such that its size is smaller than $nd^{-3(1+2\delta)}$ and, moreover, all neighbors $N(U_m)$ of U_m form an independent set, thus "isolating" U_m from the "outside world."

Now, the coloring of the vertices of $\mathbb{G}_{n,p}$ is an easy task. Namely, by Lemma 7.13, we can color the vertices of $\mathbb{G}_{n,p}$ outside the set $U_m \cup N(U_m)$ with k colors. Then we can color the vertices from $N(U_m)$ with color $k+1$, and finally, due to Lemmas 7.11 and 7.12, the subgraph induced by U_m is 3-colorable and we can color U_m with any three of the first k colors. □

7.5 Eigenvalues

Separation of first and remaining eigenvalues

The following theorem is a weaker version of a theorem of Füredi and Komlós [355], which was itself a strengthening of a result of Juhász [454]. See also Coja-Oghlan [190] and Vu [717]. In their papers, $2\omega\log n$ is replaced by $2+o(1)$ and this is best possible.

Theorem 7.14 *Suppose that $\omega \to \infty, \omega = o(\log n)$ and $\omega^3(\log n)^2 \leq np \leq n - \omega^3(\log n)^2$. Let A denote the adjacency matrix of $\mathbb{G}_{n,p}$. Let the eigenvalues of A be $\lambda_1 \geq \lambda_2 \geq \cdots \geq \lambda_n$, then w.h.p.*

(i) $\lambda_1 \approx np$
(ii) $|\lambda_i| \leq 2\omega\log n\sqrt{np(1-p)}$ *for* $2 \leq i \leq n$.

The proof of the above theorem is based on the following lemma.

In the following $|x|$ denotes the Euclidean norm of $x \in \mathbf{R}$.

Lemma 7.15 *Let J be the all 1s matrix and $M = pJ - A$ then w.h.p.*

$$\|M\| \leq 2\omega\log n\sqrt{np(1-p)},$$

where

$$\|M\| = \max_{|x|=1}|Mx| = \max\{|\lambda_1(M)|, |\lambda_n(M)|\}.$$

We first show that the lemma implies the theorem. Let $\mathbf{e}$ denote the all 1s vector.

Suppose that $|\xi| = 1$ and $\xi \perp \mathbf{e}$. Then $J\xi = 0$ and

$$|A\xi| = |M\xi| \leq \|M\| \leq 2\omega\log n\sqrt{np(1-p)}.$$

Now let $|x| = 1$ and let $x = \alpha u + \beta y$ where $u = \frac{1}{\sqrt{n}}\mathbf{e}$ and $y \perp \mathbf{e}$ and $|y| = 1$, then

$$|Ax| \leq |\alpha||Au| + |\beta||Ay|.$$

We have, writing $A = pJ + M$, that

$$|Au| = \frac{1}{\sqrt{n}}|A\mathbf{e}| \leq \frac{1}{\sqrt{n}}(np|\mathbf{e}| + \|M\|\,|\mathbf{e}|)$$
$$\leq np + 2\omega \log n\sqrt{np(1-p)}$$
$$|Ay| \leq 2\omega \log n\sqrt{np(1-p)}.$$

Thus

$$|Ax| \leq |\alpha|np + (|\alpha| + |\beta|)2\omega \log n\sqrt{np(1-p)}$$
$$\leq np + 3\omega \log n\sqrt{np(1-p)}.$$

This implies that $\lambda_1 \leq (1+o(1))np$.

But

$$|Au| \geq |(A+M)u| - |Mu|$$
$$= |pJu| - |Mu|$$
$$\geq np - 2\omega \log n\sqrt{np(1-p)},$$

implying $\lambda_1 \geq (1+o(1))np$, which completes the proof of (*i*).

Now

$$\lambda_2 = \min_{\eta} \max_{0 \neq \xi \perp \eta} \frac{|A\xi|}{|\xi|}$$
$$\leq \max_{0 \neq \xi \perp u} \frac{|A\xi|}{|\xi|}$$
$$\leq \max_{0 \neq \xi \perp u} \frac{|M\xi|}{|\xi|}$$
$$\leq 2\omega \log n\sqrt{np(1-p)}$$
$$\lambda_n = \min_{|\xi|=1} \xi^T A\xi \geq \min_{|\xi|=1} \xi^T A\xi - p\xi^T J\xi$$
$$= \min_{|\xi|=1} -\xi^T M\xi \geq -\|M\| \geq -2\omega \log n\sqrt{np(1-p)}.$$

This completes the proof of (*ii*).

Proof of Lemma 7.15: As in previously mentioned papers, we use the trace method of Wigner [723]. Putting $\hat{M} = M - pI_n$, we see that

$$\|M\| \leq \|\hat{M}\| + \|pI_n\| = \|\hat{M}\| + p$$

and so we bound $\|\hat{M}\|$.

Letting m_{ij} denote the (i,j)th entry of $\hat{M}$ we have

(i) $\mathbb{E}\, m_{ij} = 0$

(ii) $\text{Var}\, m_{ij} \le p(1-p) = \sigma^2$

(iii) $m_{ij}, m_{i'j'}$ are independent, unless $(i',j') = (j,i)$, in which case they are identical.

Now let $k \ge 2$ be an even integer.

$$\begin{aligned} Trace(\hat{M}^k) &= \sum_{i=1}^{n} \lambda_i(\hat{M}^k) \\ &\ge \max\left\{\lambda_1(\hat{M}^k), \lambda_n(\hat{M}^k)\right\} \\ &= \|\hat{M}^k\|. \end{aligned}$$

We estimate

$$\|\hat{M}\| \le Trace(\hat{M}^k)^{1/k},$$

where $k = \omega \log n$.

Now,

$$\mathbb{E}(Trace(\hat{M}^k)) = \sum_{i_0, i_1, \ldots, i_{k-1} \in [n]} \mathbb{E}(m_{i_0 i_1} m_{i_1 i_2} \cdots m_{i_{k-2} i_{k-1}} m_{i_{k-1} i_0}).$$

Recall that the i,jth entry of $\hat{M}^k$ is the sum over all products $m_{i,i_1} m_{i_1,i_2} \cdots m_{i_{k-1} j}$.

Continuing, we therefore have

$$\mathbb{E}\, \|\hat{M}\|^k \le \sum_{\rho=2}^{k} \mathbb{E}_{n,k,\rho},$$

where

$$\mathbb{E}_{n,k,\rho} = \sum_{\substack{i_0, i_1, \ldots, i_{k-1} \in [n] \\ |\{i_0, i_1, i_2, \ldots, i_{k-1}\}| = \rho}} \left| \mathbb{E}\left(\prod_{j=0}^{k-1} m_{i_j i_{j+1}} \right) \right|.$$

Note that as $m_{ii} = 0$ by construction of $\hat{M}$ we have that $\mathbb{E}_{n,k,1} = 0$.

Each sequence $\underline{i} = i_0, i_1, \ldots, i_{k-1}, i_0$ corresponds to a walk $W(\underline{i})$ on the graph K_n with n loops added. Note that

$$\mathbb{E}\left(\prod_{j=0}^{k-1} m_{i_j i_{j+1}} \right) = 0 \tag{7.21}$$

if the walk $W(\underline{i})$ contains an edge that is crossed exactly once, by condition (*i*). On the other hand, $|m_{ij}| \leq 1$ and so by conditions (*ii*), (*iii*),

$$\left| \mathbb{E}\left(\prod_{j=0}^{k-1} m_{i_j i_{j+1}} \right) \right| \leq \sigma^{2(\rho-1)}$$

if each edge of $W(\underline{i})$ is crossed at least twice and if $|\{i_0, i_1, \ldots, i_{k-1}\}| = \rho$.

Let $R_{k,\rho}$ denote the number of (k,ρ) walks, i.e closed walks of length k that visit ρ distinct vertices and do not cross any edge exactly once. We use the following trivial estimates:

(i) $\rho > \frac{k}{2}+1$ implies $R_{k,\rho} = 0$. (ρ this large will invoke (7.21)).
(ii) $\rho \leq \frac{k}{2}+1$ implies $R_{k,\rho} \leq n^\rho k^k$,

where n^ρ bounds from above the number of choices of ρ distinct vertices, while k^k bounds the number of walks of length k.

We have

$$\mathbb{E}\,\|\hat{M}\|^k \leq \sum_{\rho=2}^{\frac{1}{2}k+1} R_{k,\rho}\sigma^{2(\rho-1)} \leq \sum_{\rho=2}^{\frac{1}{2}k+1} n^\rho k^k \sigma^{2(\rho-1)} \leq 2n^{\frac{1}{2}k+1}k^k\sigma^k.$$

Therefore,

$$\mathbb{P}\left(\|\hat{M}\| \geq 2k\sigma n^{\frac{1}{2}}\right) = \mathbb{P}\left(\|\hat{M}\|^k \geq \left(2k\sigma n^{\frac{1}{2}}\right)^k\right) \leq \frac{\mathbb{E}\,\|\hat{M}\|^k}{\left(2k\sigma n^{\frac{1}{2}}\right)^k}$$

$$\leq \frac{2n^{\frac{1}{2}k+1}k^k\sigma^k}{\left(2k\sigma n^{\frac{1}{2}}\right)^k} = \left(\frac{(2n)^{1/k}}{2}\right)^k = \left(\frac{1}{2}+o(1)\right)^k = o(1).$$

It follows that w.h.p. $\|\hat{M}\| \leq 2\sigma\omega(\log n)^2 n^{1/2} \leq 2\omega(\log n)^2\sqrt{np(1-p)}$ and completes the proof of Theorem 7.14. □

Concentration of eigenvalues

We show here how one can use Talagrand's inequality, Theorem 21.17, to show that the eigenvalues of random matrices are highly concentrated around their median values. The result is from Alon, Krivelevich and Vu [28].

Theorem 7.16 *Let A be an $n \times n$ random symmetric matrix with independent entries $a_{i,j} = a_{j,i}$, $1 \leq i \leq j \leq n$ with absolute value at most one. Let its eigenvalues be $\lambda_1(A) \geq \lambda_2(A) \geq \cdots \geq \lambda_n(A)$. Suppose that $1 \leq s \leq n$. Let μ_s*

denote the median value of $\lambda_s(A)$ *i.e.* $\mu_s = \inf_\mu \{\mathbb{P}(\lambda_s(A) \leq \mu) \geq 1/2\}$, *then for any* $t \geq 0$ *we have*

$$\mathbb{P}(|\lambda_s(A) - \mu_s| \geq t) \leq 4e^{-t^2/32s^2}.$$

The same estimate holds for the probability that $\lambda_{n-s+1}(A)$ *deviates from its median by more than t.*

Proof We will use Talagrand's inequality, Theorem 21.17. We let $m = \binom{n+1}{2}$ and let $\Omega = \Omega_1 \times \Omega_2 \times \cdots \times \Omega_m$, where for each $1 \leq k \leq m$ we have $\Omega_k = \{a_{i,j}\}$ for some $i \leq j$. Fix a positive integer s and let M, t be real numbers. Let $\mathscr{A}$ be the set of matrices A for which $\lambda_s(A) \leq M$ and let $\mathscr{B}$ be the set of matrices for which $\lambda_s(B) \geq M + t$. When applying Theorem 21.17 it is convenient to view A as an m-vector.

Fix $B \in \mathscr{B}$ and let $\mathbf{v}^{(1)}, \mathbf{v}^{(2)}, \ldots, \mathbf{v}^{(s)}$ be an orthonormal set of eigenvectors for the s largest eigenvalues of B. Let $\mathbf{v}^{(k)} = (v_1^{(k)}, v_2^{(k)}, \ldots, v_n^{(k)})$,

$$\alpha_{i,i} = \sum_{k=1}^{s} (v_i^{(k)})^2 \quad \text{for } 1 \leq i \leq n$$

and

$$\alpha_{i,j} = 2\sqrt{\sum_{k=1}^{s} (v_i^{(k)})^2}\sqrt{\sum_{k=1}^{s} (v_j^{(k)})^2} \quad \text{for } 1 \leq i < j \leq n.$$

Lemma 7.17

$$\sum_{1 \leq i \leq j \leq n} \alpha_{i,j}^2 \leq 2s^2.$$

Proof

$$\begin{aligned}\sum_{1 \leq i \leq j \leq n} \alpha_{i,j}^2 &= \sum_{i=1}^{n} \left(\sum_{k=1}^{s} (v_i^{(k)})^2 \right)^2 + 4 \sum_{1 \leq i < j \leq n} \left(\sum_{k=1}^{s} (v_i^{(k)})^2 \sum_{k=1}^{s} (v_j^{(k)})^2 \right) \\ &\leq 2 \left(\sum_{i=1}^{n} \sum_{k=1}^{s} (v_i^{(k)})^2 \right)^2 = 2 \left(\sum_{k=1}^{s} \sum_{i=1}^{n} (v_i^{(k)})^2 \right)^2 = 2s^2,\end{aligned}$$

where we have used the fact that each $\mathbf{v}^{(k)}$ is a unit vector.

□

Lemma 7.18 *For every* $A = (a_{i,j}) \in \mathscr{A}$ *and* $B = (b_{i,j}) \in \mathscr{B}$,

$$\sum_{1 \leq i \leq j \leq n : a_{i,j} \neq b_{i,j}} \alpha_{i,j} \geq t/2.$$

Proof Fix $A \in \mathscr{A}$. Let $\mathbf{u} = \sum_{k=1}^{s} c_k \mathbf{v}^{(k)}$ be a unit vector in the span S of the vectors $\mathbf{v}^{(k)}$, $k = 1,2,\dots,s$,which is orthogonal to the eigenvectors of the $(s-1)$ largest eigenvalues of A. Recall that $\mathbf{v}^{(k)}$, $k = 1,2,\dots,s$ are eigenvectors of B. Then $\sum_{k=1}^{s} c_k^2 = 1$ and $\mathbf{u}^t A\mathbf{u} \le \lambda_s(A) \le M$, whereas $\mathbf{u}^t B\mathbf{u} \ge \min_{\mathbf{v}\in S} \mathbf{v}^t B\mathbf{v} = \lambda_s(B) \ge M + t$. Recall that all entries of A and B are bounded in absolute value by 1, implying that $|b_{i,j} - a_{i,j}| \le 2$ for all $1 \le i,j \le n$. It follows that if X is the set of ordered pairs (i,j) for which $a_{i,j} \ne b_{i,j}$ then

$$
\begin{aligned}
t \le \mathbf{u}^t(B-A)\mathbf{u} &= \sum_{(i,j)\in X} (b_{i,j} - a_{i,j}) \left(\sum_{k=1}^{s} c_k v_i^{(k)}\right)^t \sum_{k=1}^{s} c_k v_j^{(k)} \\
&\le 2 \sum_{(i,j)\in X} \left|\sum_{k=1}^{s} c_k v_i^{(k)}\right| \left|\sum_{k=1}^{s} c_k v_j^{(k)}\right| \\
&\le 2 \sum_{(i,j)\in X} \left(\sqrt{\sum_{k=1}^{s} c_k^2} \sqrt{\sum_{k=1}^{s} \left(v_i^{(k)}\right)^2}\right) \left(\sqrt{\sum_{k=1}^{s} c_k^2} \sqrt{\sum_{k=1}^{s} \left(v_j^{(k)}\right)^2}\right) = 2 \sum_{(i,j)\in X} a_{i,j}
\end{aligned}
$$

as claimed. (We obtained the third inequality by use of the Cauchy–Schwarz inequality.)

□

By the above two lemmas, and by Theorem 21.17 for every M and every $t > 0$

$$
\mathbb{P}(\lambda_s(A) \le M)\,\mathbb{P}(\lambda_s(B) \ge M + t) \le e^{-t^2/(32s^2)}. \tag{7.22}
$$

If M is the median of $\lambda_s(A)$ then $\mathbb{P}(\lambda_s(A) \le M) \ge 1/2$, by definition, implying that

$$
\mathbb{P}(\lambda_s(A) \ge M + t) \le 2e^{-t^2/(32s^2)}.
$$

Similarly, by applying (7.22) with $M+t$ being the median of $\lambda_s(A)$ we conclude that

$$
\mathbb{P}(\lambda_s(A) \le M - t) \le 2e^{-t^2/(32s^2)}.
$$

This completes the proof of Theorem 7.16 for $\lambda_s(A)$. The proof for λ_{n-s+1} follows by applying the theorem to s and $-A$. □

7.6 Exercises

7.6.1 Let $p = d/n$ where d is a positive constant. Let S be the set of vertices of degree at least $\frac{\log n}{2\log\log n}$. Show that w.h.p., S is an independent set.

7.6.2 Let $p = d/n$ where d is a large positive constant. Use the First Moment Method to show that w.h.p.

$$\alpha(G_{n,p}) \leq \frac{2n}{d}(\log d - \log\log d - \log 2 + 1 + \varepsilon)$$

for any positive constant ε.

7.6.3 Complete the proof of Theorem 7.4.
Let $m = d/(\log d)^2$ and partition $[n]$ into $n_0 = \frac{n}{m}$ sets $S_1, S_2, \ldots, S_{n_0}$ of size m. Let $\underline{(G)}$ be the maximum size of an independent set S that satisfies $|S \cap S_i| \leq 1$ for $i = 1, 2, \ldots, n_0$. Use the proof idea of Theorem 7.4 to show that w.h.p.

$$\beta(\mathbb{G}_{n,p}) \geq k_{-\varepsilon} = \frac{2n}{d}(\log d - \log\log d - \log 2 + 1 - \varepsilon).$$

7.6.4 Prove Theorem 7.4 using Talagrand's inequality, Theorem 21.21. (Hint: Let $A = \{\alpha(\mathbb{G}_{n,p}) \leq k_{-\varepsilon} - 1\}$.)

7.6.5 Prove Lemma 7.6.

7.6.6 Prove Lemma 7.11.

7.6.7 Prove that if $\omega = \omega(n) \to \infty$ then there exists an interval I of length $\omega n^{1/2}/\log n$ such that w.h.p. $\chi(G_{n,1/2}) \in I$. (See Scott [677].)

7.6.8 A *topological clique* of size s is a graph obtained from the complete graph K_s by subdividing edges. Let $tc(G)$ denote the size of the largest topological clique contained in a graph G. Prove that w.h.p. $tc(G_{n,1/2}) = \Theta(n^{1/2})$.

7.6.9 Suppose that H is obtained from $G_{n,1/2}$ by planting a clique C of size $m = n^{1/2}\log n$ inside it. Describe a polynomial time algorithm that w.h.p. finds C. (Think that an adversary adds the clique without telling you where it is.)

7.6.10 Show that if $d > k\log k$ for a positive integer $k \geq 2$ then w.h.p. $G(n, d/n)$ is not k-colorable. (Hint: Consider the expected number of proper k-colorings.)

7.6.11 Let $p = K\log n/n$ for some large constant $K > 0$. Show that w.h.p. the diameter of $\mathbb{G}_{n,p}$ is $\Theta(\log n/\log\log n)$.

7.6.12 Suppose that $1 + \varepsilon \leq np = o(\log n)$, where $\varepsilon > 0$ is constant. Show that given $A > 0$, there exists $B = B(A)$ such that

$$\mathbb{P}\left(diam(K) \geq B\frac{\log n}{\log np}\right) \leq n^{-A},$$

where K is the giant component of $\mathbb{G}_{n,p}$.

7.6.13 Let $p = d/n$ for some constant $d > 0$. Let A be the adjacency matrix of $G_{n,p}$. Show that w.h.p. $\lambda_1(A) \approx \Delta$ where Δ is the maximum degree in

$\mathbb{G}_{n,p}$. (Hint: the maximum eigenvalue of the adjacency matrix of $K_{1,m}$ is $m^{1/2}$.)

7.6.14 A proper *2-tone* k-coloring of a graph $G = (V,E)$ is an assignment of pairs of colors $C_v \subseteq [k], |C_v| = 2$ such that (i) $|C_v \cap C_w| < d(v,w)$, where $d(v,w)$ is the graph distance from v to w. If $\chi_2(G)$ denotes the minimum k for which there exists a 2-tone coloring of G, show that w.h.p. $\chi_2(\mathbb{G}_{n,p}) \approx 2\chi(\mathbb{G}_{n,p})$. (This question is taken from [45].)

7.6.15 The *set chromatic number* $\chi_s(G)$ of a graph $G = (V,E)$ is defined as follows: let C denote a set of colors. Color each $v \in V$ with a color $f(v) \in C$. Let $C_v = \{f(w) : \{v,w\} \in G\}$. The coloring is proper if $C_v \neq C_w$ whenever $\{v,w\} \in E$. χ_s is the minimum size of C in a proper coloring of G. Prove that if $0 < p < 1$ is constant then w.h.p. $\chi_s(\mathbb{G}_{n,p}) \approx r\log_2 n$ where $r = \frac{2}{\log_2 1/s}$ and $s = \min\left\{q^{2\ell} + (1-q^\ell)^2 : \ell = 1,2,\ldots\right\}$ where $q = 1-p$. (This question is taken from Dudek, Mitsche and Pralat [258].)

7.7 Notes

Chromatic number

There has been a lot of progress in determining the chromatic number of sparse random graphs. Alon and Krivelevich [25] extended the result in [541] to the range $p \leq n^{-1/2-\delta}$. A breakthrough came when Achlioptas and Naor [5] identified the two possible values for $np = d$ where $d = O(1)$: let k_d be the smallest integer k such that $d < 2k\log k$, then w.h.p. $\chi(\mathbb{G}_{n,p}) \in \{k_d, k_d + 1\}$. This implies that d_k, the (conjectured) threshold for a random graph to have chromatic number at most k, satisfies $d_k \geq 2k\log k - 2\log_k - 2 + o_k(1)$, where $o_k(1) \to 0$ as $k \to \infty$. Coja-Oghlan, Panagiotou and Steger [192] extended the result of [5] to $np \leq n^{1/4-\varepsilon}$, although here the guaranteed range is three values. More recently, Coja-Oghlan and Vilenchik [193] proved the following. Let $d_{k,cond} = 2k\log k - \log k - 2\log 2$, then w.h.p. $d_k \geq d_{k,cond} - o_k(1)$. On the other hand Coja-Oghlan [191] proved that $d_k \leq d_{k,cond} + (2\log 2 - 1) + o_k(1)$.

It follows from Chapter 2 that the chromatic number of $\mathbb{G}_{n,p}, p \leq 1/n$ is w.h.p. at most 3. Achlioptas and Moore [3] proved that in fact $\chi(\mathbb{G}_{n,p}) \leq 3$ w.h.p. for $p \leq 4.03/n$. Now a graph G is s-colorable if and only if it has a homomorphism $\varphi : G \to K_s$. (A *homomorphism* from G to H is a mapping $\varphi : V(G) \to V(H)$ such that if $\{u,v\} \in E(G)$ then $(\varphi(u), \varphi(v)) \in E(H)$). It is therefore of interest in the context of coloring to consider homomorphisms from $\mathbb{G}_{n,p}$ to other graphs. Frieze and Pegden [345] showed that for any $\ell > 1$ there is an $\varepsilon > 0$ such that with high probability, $G_{n,\frac{1+\varepsilon}{n}}$ either has odd-girth

$< 2\ell + 1$ or has a homomorphism to the odd cycle $C_{2\ell+1}$. They also showed that w.h.p. there is no homomorphism from $\mathbb{G}_{n,p}, p = 4/n$ to C_5. Previously, Hatami [397] has shown that w.h.p. there is no homomorphism from a random cubic graph to C_7.

Alon and Sudakov [31] considered how many edges one must add to $\mathbb{G}_{n,p}$ to significantly increase the chromatic number. They showed that if $n^{-1/3+\delta} \leq p \leq 1/2$ for some fixed $\delta > 0$, then w.h.p. for every set E of $\frac{2^{-12}\varepsilon^2 n^2}{(\log_b(np))^2}$ edges, the chromatic number of $\mathbb{G}_{n,p} \cup E$ is still at most $\frac{(1+\varepsilon)n}{2\log_b(np)}$.

Let L_k be an arbitrary function that assigns to each vertex of G a list of k colors. We say that G is *L_k-list-colorable* if there exists a proper coloring of the vertices such that every vertex is colored with a color from its own list. A graph is *k-choosable*, if for every such function L_k, G is L_k-list-colorable. The minimum k for which a graph is k-choosable is called the *list chromatic number*, or the *choice number*, and is denoted by $\chi_L(G)$. The study of the choice number of $\mathbb{G}_{n,p}$ was initiated in [20], where Alon proved that w.h.p., the choice number of $\mathbb{G}_{n,1/2}$ is $o(n)$. Kahn then showed (see [21]) that w.h.p. the choice number of $\mathbb{G}_{n,1/2}$ equals $(1 + o(1))\chi(\mathbb{G}_{n,1/2})$. In [503], Krivelevich showed that this holds for $p \gg n^{-1/4}$, and Krivelevich, Sudakov, Vu, and Wormald [514] improved this to $p \gg n^{-1/3}$. On the other hand, Alon, Krivelevich, Sudakov [27] and Vu [716] showed that for any value of p satisfying $2 < np \leq n/2$, the choice number is $\Theta(np/\log(np))$. Krivelevich and Vu [515] generalized this to hypergraphs; they also improved the leading constants and showed that the choice number for $C/n \leq p \leq 0.9$ (where C is a sufficiently large constant) is at most a multiplicative factor of $2 + o(1)$ away from the chromatic number, the best known factor for $p \leq n^{-1/3}$.

Algorithmic questions

We have seen that the Greedy algorithm applied to $\mathbb{G}_{n,p}$ generally produces a coloring that uses roughly twice the minimum number of colors needed. Note also that the analysis of Theorem 7.9, when $k = 1$, implies that a simple GREEDY algorithm for finding a large independent set produces one of roughly half the maximum size. In spite of much effort neither of these two results have been significantly improved. We mention some negative results. Jerrum [448] showed that the *Metropolis* algorithm was unlikely to do very well in finding an independent set that was significantly larger than GREEDY. Other earlier negative results include: Chvátal [186], who showed that for a significant set of densities, a large class of algorithms will w.h.p. take exponential time to find the size of the largest independent set and McDiarmid [561] who carried out a similar analysis for the chromatic number.

Frieze, Mitsche, Pérez-Giménez and Pralat [344] studied list coloring in an on-line setting and show that for a wide range of p, one can asymptotically match the best known constants of the off-line case. Moreover, if $pn \geq \log^{\omega} n$, then they get the same multiplicative factor of $2+o(1)$.

Randomly Coloring random graphs

A substantial amount of research in Theoretical Computer Science has been associated with the question of random sampling from complex distributions. Of relevance here is the following: let G be a graph and k be a positive integer. Then let $\Omega_k(G)$ be the set of proper k-colorings of the vertices of G. There has been a good deal of work on the problem of efficiently choosing a (near) random member of $\Omega_k(G)$. For example, Vigoda [714] has described an algorithm that produces a (near) random sample in polynomial time provided $k > 11\Delta(G)/6$. When it comes to $\mathbb{G}_{n,p}$, Dyer, Flaxman, Frieze and Vigoda [265] showed that if $p = d/n, d = O(1)$ then w.h.p. one can sample a random coloring if $k = O(\log\log n) = o(\Delta)$. The bound on k was reduced to $k = O(d^{O(1)})$ by Mossell and Sly [591] and then to $k = O(d)$ by Efthymiou [272].

Diameter of sparse random graphs

The diameter of the giant component of $\mathbb{G}_{n,p}$, $p = \lambda/n, \lambda > 1$ was considered by Fernholz and Ramachandran [299] and by Riordan and Wormald [649]. In particular, [649] proves that w.h.p. the diameter is $\frac{\log n}{\log \lambda} + 2\frac{\log n}{\log 1/\lambda^*} + W$, where $\lambda^* < 1$ and $\lambda^* e^{-\lambda^*} = \lambda e^{-\lambda}$ and $W = O_p(1)$, i.e. is bounded in probability for $\lambda = O(1)$ and $O(1)$ for $\lambda \to \infty$. In addition, when $\lambda = 1+\varepsilon$, where $\varepsilon^3 n \to \infty$, i.e. the case of the emerging giant, [649] shows that w.h.p. the diameter is $\frac{\log \varepsilon^3 n}{\log \lambda} + 2\frac{\log \varepsilon^3 n}{\log 1/\lambda^*} + W$, where $W = O_p(1/\varepsilon)$. If $\lambda = 1-\varepsilon$, where $\varepsilon^3 n \to \infty$, i.e. the sub-critical case, then Łuczak [543] showed that w.h.p. the diameter is $\frac{\log(2\varepsilon^3 n)+O_p(1)}{-\log \lambda}$.

8

Extremal Properties

A typical question in extremal combinatorics can be viewed as "how many edges of the complete graph (or hypergraph) on n vertices can a graph have without having some property $\mathscr{P}$." In recent years research has been carried out where the complete graph is replaced by a random graph.

8.1 Containers

Ramsey theory and the Turán problem constitute two of the most important areas in extremal graph theory. For a fixed graph H we can ask how large should n be so that in any r-coloring of the edges of K_n can we be sure of finding a monochromatic copy of H – a basic question in Ramsey theory. Or we can ask for the maximum $\alpha > 0$ such that we take an α proportion of the edges of K_n without including a copy of H – a basic question related to the Turán problem. Both of these questions have analogues where we replace K_n by $\mathbb{G}_{n,p}$.

There have been recent breakthroughs in transferring extremal results to the context of random graphs and hypergraphs. Conlon and Gowers [195], Schacht [673], Balogh, Morris and Samotij [54] and Saxton and Thomason [672] have proved general theorems enabling such transfers. One of the key ideas being to bound the number of independent sets in carefully chosen hypergraphs. Our presentation uses the framework of [672] where it could just as easily have used [54].

In this section, we present a special case of Theorem 2.3 of [672] that enables us to deal with Ramsey and Turán properties of random graphs. For a graph H with $e(H) \geq 2$ we let

$$m_2(H) = \max_{H' \subseteq H, e(H') > 1} \frac{e(H') - 1}{v(H') - 2}. \tag{8.1}$$

Next let

$$\pi(H) = \lim_{n\to\infty} \frac{ex(n,H)}{\binom{n}{2}} \tag{8.2}$$

where as usual, $ex(n,H)$ is the maximum number of edges in an H-free subgraph of K_n.

Theorem 8.1 *Let H be a graph with $e(H) \geq 2$ and let ε be a positive constant. For some constant $h = h(H,\varepsilon) > 0$ and n sufficiently large, there exists a collection $\mathscr{C}$ of graphs on vertex set $[n]$ and a function $C : 2^{[n]} \to 2^{[n]}$ such that the following holds. The graphs $\mathscr{C}$ are the* containers*:*

(a) For every H-free graph Γ there exists $T \subseteq \Gamma \subseteq C(T) \in \mathscr{C}$ such that $e(T) \leq hn^{2-1/m_2(H)}$.
(b) $e(C) \leq (\pi(H)+\varepsilon)\binom{n}{2}$ for every $C \in \mathscr{C}$.

We do not have the room to prove this theorem, but see Exercises 8.4.7 and 8.4.8. It is a corollary of a general theorem that is sure to be an essential tool in the development of the subject. This general theorem is concerned with independent sets in hypergraphs and is stated in the next section. To make the connection with Theorem 8.1, we must consider the $e(H)$-uniform hypergraph $\mathscr{H}$ with vertex set n and an edge for every isomorphic copy of H. In this way, an H-free graph corresponds to an independent set in $\mathscr{H}$. There is no difficulty in allowing H itself to be a hypergraph. In another vein, we obtain results on arithmetic progressions by considering the k-uniform hypergraph on vertex set $[n]$ with an edge for every k-term arithmetic progression.

Statement of the Container Theorem

Let H be an r-uniform hypergraph with vertex set $[n]$. For $S \subseteq [n]$ we define

$$d(S) = |\{e \in E(H) : \ S \subseteq e\}|. \tag{8.3}$$

Next let $d = \frac{re(H)}{n}$ be the average vertex degree. For $v \in [n]$ and $2 \leq j \leq r$ let

$$d^{(j)}(v) = \max\{d(S) : \ v \in S \subseteq [n],\ |S| = j\}.$$

For $\tau > 0$ we define δ_j by the equation

$$\delta_j \tau^{j-1} nd = \sum_v d^{(j)}(v).$$

The *co-degree function* $\delta(H,\tau)$ is defined by

$$\delta(H,\tau)=2^{\binom{r}{2}-1}\sum_{j=2}^{r}2^{-\binom{j-1}{2}}\delta_j.$$

If $d=0$ we define $\delta(H,\tau)=0$.

Next define the *degree measure* $\mu_2(S), S\subseteq[n]$ by

$$\mu_2(S)=\frac{1}{dn}\sum_{u\in S}d(u),$$

where $d(u)=d(\{u\})$ as defined in (8.3).

We can finally state the main theorem of [672], Theorem 3.4: We will use the notation $2^{A\times b}$ in place of $(2^A)^b$ for set A and positive integer b.

Theorem 8.2 *Let H be an r-uniform hypergraph with vertex set $[n]$. Let $\tau,\zeta>0$ satisfy $\delta(H,\tau)\le\zeta$. Then there exists $C:2^{[n]}\to 2^{[n]}$ such that for every independent set $I\subseteq[n]$ there exists $T\subseteq I$ such that*

(a) $I\subseteq C(T)$.
(b) $\mu(T)\le 2r\tau/\zeta$.
(c) $|T|\le 2r\tau n/\zeta^2$.
(d) $\mu(C(T))\le 1-\frac{1}{r!}+4\zeta+\frac{2r\tau}{\zeta}$.

8.2 Ramsey Properties

The investigation of the Ramsey properties of $\mathbb{G}_{n,p}$ was initiated by Łuczak, Ruciński and Voigt [548]. Later, Rödl and Ruciński [652], [653] proved that the following holds w.h.p. for some constants $0<c<C$. Here H is some fixed graph containing at least one cycle. Suppose that the edges of $\mathbb{G}_{n,m}$ are colored with r colors. If $m<cn^{2-1/m_2(H)}$ then w.h.p. there exists an r-coloring without a monochromatic copy of H, while if $m>Cn^{2-1/m_2(H)}$ then w.h.p. in every r-coloring there is a monochromatic copy of H.

We will give a proof of the 1-statement based on Theorem 8.1. We closely follow the argument in a recent paper of Nenadov and Steger [604]. The notation $G\to(H)_r^e$ means that in every r-coloring of the edges of G there is a copy of H with all edges the same color. Rödl and Ruciński [653] proved the following

Theorem 8.3 *For any graph H with $e(H) \geq v(H)$ and $r \geq 2$, there exist $c, C > 0$ such that*

$$\mathbb{P}(\mathbb{G}_{n,p} \to (H)_r^e) = \begin{cases} o(1) & p \leq cn^{-1/m_2(H)} \\ 1 - o(1) & p \geq Cn^{-1/m_2(H)}. \end{cases}$$

The density $p_0 = n^{-1/m_2(H)}$ is the threshold for every edge of $\mathbb{G}_{n,p}$ to be contained in a copy of H. When $p \leq cp_0$ for small c, the copies of H in $\mathbb{G}_{n,p}$ will be spread out and the associated 0-statement is not so surprising. We use Theorem 8.1 to prove the 1-statement for $p \geq Cp_0$. The proof of the 0-statement follows [604] and is given in Exercises 8.4.1 to 8.4.6.

We begin with a couple of lemmas:

Lemma 8.4 *For every graph H and $r \geq 2$ there exist constants $\alpha > 0$ and n_0 such that for all $n \geq n_0$ every r-coloring of the edges of K_n contains at least $\alpha n^{v(H)}$ copies of H.*

Proof From Ramsey's theorem we know that there exists $N = N(H, r)$ such that every r-coloring of the edges of K_N contains a monochromatic copy of H. Thus, in any r-coloring of K_n, every N-subset of the vertices of K_n contains at least one monochromatic copy of H. As every copy of H is contained in at most $\binom{n-v(H)}{N-v(H)}$ N-subsets, the theorem follows with $\alpha = 1/N^{v(H)}$. □

From this we get

Corollary 8.5 *For every graph H and every positive integer r there exist constants n_0 and $\delta > 0$ such that the following is true: if $n \geq n_0$, then for any $E_1, E_2, \ldots, E_r \subseteq E(K_n)$ such that for all $1 \leq i \leq r$ the set E_i contains no copy of H, we have*

$$|E(K_n) \setminus (E_1 \cup E_2 \cup \cdots \cup E_r)| \geq \delta n^2.$$

Proof Let α and n_0 be as given in Lemma 8.4 for H and $r+1$. Further, let $E_{r+1} = E(K_n) \setminus (E_1 \cup E_2 \cup \cdots \cup E_r)$, and consider the coloring $f : E(K_n) \to [r+1]$ given by $f(e) = \min_{i \in [r+1]} \{e \in E_i\}$. By Lemma 8.4 there exist at least $\alpha n^{v(H)}$ monochromatic copies of H under coloring f, and so by our assumption on the sets $E_i, 1 \leq i \leq r$, E_{r+1} must contain at least $\alpha n^{v(H)}$ copies. As every edge is contained in at most $e(H) n^{v(H)-2}$ copies, the lemma follows with $\delta = \frac{\alpha}{e(H)}$. □

We can now proceed to the proof of the 1-statement of Theorem 8.3. If $\mathbb{G}_{n,p} \not\to (H)_r^e$ then there must exist a coloring $f : E(\mathbb{G}_{n,p}) \to [r]$ such that for all $1 \leq i \leq r$ the set $E_i = f^{-1}(i)$ does not contain a copy of H. By Theorem 8.1 we have that for every such E_i there exist T_i and a container C_i such that $T_i \subseteq E_i \subseteq C_i$. The crucial observation is that $\mathbb{G}_{n,p}$ completely avoids

$E_0 = E(K_n) \setminus (C_1 \cup C_2 \cup \cdots \cup C_r)$, which by Corollary 8.5 has size at least δn^2.

Therefore, we can bound $\mathbb{P}(\mathbb{G}_{n,p} \not\to (H)_r^e)$ by the probability that there exist $\mathcal{T} = \{T_1, \ldots, T_r\}$ and $\mathcal{C} = \{C_i = C(T_i) : i = 1, 2, \ldots, r\}$ such that E_0 is edge-disjoint from $\mathbb{G}_{n,p}$. Thus,

$$\mathbb{P}((\mathbb{G}_{n,p} \not\to (H)_r^e) \\ \leq \sum_{T_i, 1 \leq i \leq r} \mathbb{P}(T_i \subseteq \mathbb{G}_{n,p}, 1 \leq i \leq r \wedge E(\mathbb{G}_{n,p}) \cap E_0 = \emptyset).$$

Note that the two events in the above probability are independent and can thus be bounded by $p^a(1-p)^b$ where $a = \left|\bigcup_i T_i\right|$ and $b = \delta n^2$. The sum can be bounded by first deciding on $a \leq rhn^{2-1/m_2(H)}$ (h from Theorem 8.1) and then choosing a edges ($\binom{\binom{n}{2}}{a}$ choices) and then deciding for every edge in which T_i it appears (r^a choices). Thus,

$$\mathbb{P}((\mathbb{G}_{n,p} \not\to (H)_r^e) \leq e^{-\delta n^2 p} \sum_{a=0}^{rhn^{2-1/m_2(H)}} \binom{\binom{n}{2}}{a} (rp)^a$$
$$\leq e^{-\delta n^2 p} \sum_{a=0}^{rhn^{2-1/m_2(H)}} \left(\frac{en^2 rp}{2a}\right)^a.$$

Recall that $p = Cn^{-1/m_2(H)}$. By choosing C sufficiently large with respect to c we get

$$\sum_{a=0}^{rhn^{2-1/m_2(H)}} \left(\frac{e2^{rh}n^2 p}{2a}\right)^a \leq n^2 \left(\frac{e2^{rh}C}{2rh}\right)^{(rh/C)n^2 p} \leq e^{\delta n^2 p/2},$$

and thus $\mathbb{P}((\mathbb{G}_{n,p} \not\to (H)_r^e) = o(1)$ as desired. (Recall that $(eA/x)^x$ is unimodal with a maximum at $x = A$ and then that if C is large, $rhn^{2-1/m_2(H)} \leq 2^{rh} n^2 p/2$.)

8.3 Turán Properties

Early success on the Turán problem for random graphs was achieved by Haxell, Kohayakawa and Łuczak [399], [400], Kohayakawa, Kreuter and Steger [492], Kohayakawa, Łuczak and Rödl [493], Gerke, Prömel, Schickinger and Steger [362], Gerke, Schickinger and Steger [363] and Łuczak [544]. It is only recently that Turán's theorem in its full generality has been transferred to $\mathbb{G}_{n,p}$.

From its definition, every H-free graph with n vertices will have $(\pi(H)+o(1))\binom{n}{2}$ edges. In this section we prove a corresponding result for random graphs. Our proof is taken from [672], although Conlon and Gowers [195] gave a proof for 2-balanced H and Schacht [673] gave a proof for general H.

Theorem 8.6 *Suppose that $0<\gamma<1$, then there exists $A>0$ such that if $p \geq An^{-1/m_2(H)}$ and n is sufficiently large then the following event occurs with probability at least $1-e^{-\gamma^3\binom{n}{2}p/384}$:*

$$\textit{Every } H\textit{-free subgraph of } \mathbb{G}_{n,p} \textit{ has at most } (\pi(H)+\gamma)\binom{n}{2}p \textit{ edges.}$$

To prove the theorem, we first prove the following lemma:

Lemma 8.7 *Given $0<\eta<1$ and $h \geq 1$, there is a constant $\varphi=\varphi(\eta,h)$ such that the following holds: Let M be a set, $|M|=N$. Let $t \geq 1$, $\varphi t/N \leq p \leq 1$ and let $\eta N/2 \leq d \leq N$. Suppose there exists $C: 2^M \to 2^M$ and $\mathscr{T} \subseteq \binom{M}{\leq t}$ such that for each $I \in \mathscr{I}$ there exists $T_I \in \mathscr{T}$ such that $T_I \subseteq I$ and $C_I = C(T_I) \subseteq M$, where $|C_I| \leq d$. Let $X \subseteq M$ be a random subset where each element is chosen independently with probability p. Then*

$$\mathbb{P}(\exists I \in \mathscr{I} : |C_I \cap X| > (1+\eta)pd \textit{ and } I \subseteq X) \leq e^{-\eta^2 dp/24}. \tag{8.4}$$

Proof For $T \in \mathscr{T}$ let E_T be the event that

$$T \subseteq X \text{ and } |C(T) \cap X| \geq (1+\eta)pd.$$

The event E_T is contained in $F_T \cap G_T$, where F_T is the event that $T \subseteq X$ and G_T is the event that $|(C(T) \setminus T) \cap X| \geq (1+\eta)dp - |T|$. Since F_T and G_T are independent, $\mathbb{P}(E_T) \leq \mathbb{P}(F_T)\mathbb{P}(G_T)$. Now $|T| \leq t \leq Np/\varphi \leq 2dp/\varphi\eta \leq \eta dp/2$ if φ is large. So by the Chernoff bound, see Lemma 21.6,

$$\mathbb{P}(G_T) \leq \mathbb{P}(\mathrm{Bin}(d,p) \geq (1+\eta/2)dp) \leq e^{-\eta^2 dp/12}.$$

Note that $\mathbb{P}(F_T) = p^{|T|}$. Let $x = Np/t \geq \varphi$, so that $t \leq Np/x \leq 2dp/\eta x$. If φ is large we may assume that $p(N-t) > t$. So

$$\sum_T \mathbb{P}(F_T) \leq \sum_{i=0}^{t} \binom{N}{i} p^i \leq 2\left(\frac{eNp}{t}\right)^t$$
$$= 2(xe)^t \leq (xe)^{2dp/\eta x} \leq e^{\eta^2 dp/24},$$

if φ, and therefore x, is large. If there exists $I \subseteq X, I \in \mathscr{I}$ with $|C(T_I) \cap X| \geq (1+\eta)dp$ then the event E_T holds. Hence the probability in (8.4) is bounded by

$$\sum_T \mathbb{P}(F_T)\mathbb{P}(G_T) \le e^{\eta^2 dp/24} e^{-\eta^2 dp/12} = e^{-\eta^2 dp/24}.$$

$\square$

With this lemma in hand, we can complete the proof of Theorem 8.6.

Let $\mathscr{I}$ be the set of H-free graphs on vertex set $[n]$. We take $M = \binom{[n]}{2}$ and $X = E(\mathbb{G}_{n,p})$ and $N = \binom{n}{2}$. For $I \in \mathscr{I}$, let T_I and $h = h(H,\varepsilon)$ be given by Theorem 8.1. Each H-free graph $I \in \mathscr{I}$ is contained in C_I and so if $\mathbb{G}_{n,p}$ contains an H-free subgraph with $(\pi(H)+\gamma)Np$ edges then there exists I such that $|X \cap C_I| \ge (\pi(H)+\gamma)Np$. Our aim is to apply Lemma 8.7 with

$$\eta = \frac{\gamma}{2},\ d = \left(\pi(H) + \frac{\gamma}{4}\right)N,\ t = hn^{2-1/m_2(H)}.$$

The conditions of Lemma 8.7 then hold after noting that $d \ge \eta N/2$ and that $p \ge An^{-1/m_2(H)} \ge \varphi t/N$ if A is large enough. Note also that $|C_I| \le d$. Now $(1+\eta)dp \le (\pi(H)+\gamma)Np$, and so the probability in the statement of the theorem is bounded by

$$e^{-\eta^2 dp/24} \le \exp\left\{-\frac{\gamma^3 Np}{384}\right\}$$

completing the proof. $\square$

8.4 Exercises

8.4.1 An edge e of G is H-open if it is contained in at most one copy of H and H-closed otherwise. The H-core $\hat{G}_H$ of G is obtained by repeatedly deleting H-open edges. Show that $G \to (H)_2^e$ implies that $\hat{G}_{H'} \to (H')_2^e$ for every $H' \subseteq H$. (Thus one only needs to prove the 0-statement of Theorem 8.3 for strictly 2-balanced H. A graph H is strictly 2-balanced if $H' = H$ is the unique maximizer in (8.1).)

8.4.2 A subgraph G' of the H-core is H-closed if it contains at least one copy of H and every copy of H in $\hat{G}_H$ is contained in G' or is edge disjoint from G'. Show that the edges of $\hat{G}_H$ can be partitioned into inclusion minimal H-closed subgraphs.

8.4.3 Show that there exists a sufficiently small $c > 0$ and a constant $L = L(H,c)$ such that if H is 2-balanced and $p \le cn^{-1/m_2(H)}$ then w.h.p. every inclusion minimal H-closed subgraph of $\mathbb{G}_{n,p}$ has size at most L. (Try $c = o(1)$ first here.)

8.4.4 Show that if $e(G)/v(G) \le m_2(H)$ and $m_2(H) > 1$ then $G \not\to (H)_2^e$.

8.4.5 Show that if H is 2-balanced and $p = cn^{-1/m_2(H)}$ then w.h.p. every subgraph G of $\mathbb{G}_{n,p}$ with $v(G) \leq L = O(1)$ satisfies $e(G)/v(G) \leq m_2(H)$.

8.4.6 Prove the 0-statement of Theorem 8.3 for $m_2(H) > 1$.

8.4.7 Let G be an ℓ-uniform hypergraph and let $r = e(G) \geq 2$. Define the r-uniform hypergraph $H(n,G)$ as follows: It has vertex set $V = \binom{[n]}{\ell}$ and an edge for each subset of $\binom{V}{r}$ that induces a copy of G. Suppose that $\gamma \leq 1$ and that n is sufficiently large. Prove that

$$\delta(H(n,G), \gamma^{-1} n^{-1/m_2(G)}) \leq r 2^{r^2} v(G)!^2 \gamma .$$

Here

$$m_2(G) = \max_{G' \subseteq G, e(G') > 1} \frac{e(G') - 1}{v(G') - \ell} .$$

8.4.8 Use Theorem 8.2 and the supersaturation theorem [280] to prove Theorem 8.1. Recall that the supersaturation theorem states: let H be an ℓ-uniform hypergraph and let $\varepsilon > 0$. There exists N_0 and $\eta > 0$ such that if C is an ℓ-uniform hypergraph on $N \geq N_0$ vertices containing at most $\eta N^{v(H)}$ copies of H then $e(C) \leq (\pi(H) + \varepsilon)\binom{N}{\ell}$. (We define $\pi(H)$ as we did in (8.2) but with $\binom{n}{2}$ replaced by $\binom{N}{\ell}$).

8.5 Notes

The Largest Triangle-Free Subgraph of a Random Graph

Babai, Simonovits and Spencer [43] proved that if $p \geq 1/2$ then w.h.p. the largest triangle-free subgraph of $\mathbb{G}_{n,p}$ is bipartite. They used Szemerédi's regularity lemma in the proof. Using the sparse version of this lemma, Brightwell, Panagiotou and Steger [158] improved the lower bound on p to n^{-c} for some (unspecified) positive constant c. DeMarco and Kahn [233] improved the lower bound to $p \geq Cn^{-1/2}(\log n)^{1/2}$, which is best possible up to the value of the constant C. And in [234] they extended their result to K_r-free graphs.

Anti-Ramsey Property

Let H be a fixed graph. A copy of H in an edge-colored graph G is said to be rainbow colored if all of its edges have a different color. The study of rainbow copies of H was initiated by Erdős, Simonovits and Sós [281]. An edge-coloring of a graph G is said to be b-bounded if no color is used more than b times. A graph is G said to have property $\mathscr{A}(b,H)$ if there is a rainbow copy of H in every b-bounded coloring. Bohman, Frieze, Pikhurko and Smyth [117] studied the threshold for $\mathbb{G}_{n,p}$ to have property $\mathscr{A}(b,H)$. For graphs H

containing at least one cycle they proved that there exists b_0 such that if $b \geq b_0$ then there exist $c_1, c_2 > 0$ such that

$$\lim_{n\to\infty} \mathbb{P}(\mathbb{G}_{n,p} \in \mathscr{A}(b,H)) = \begin{cases} 0 & p \leq c_1 n^{-1/m_2(H)} \\ 1 & p \geq c_2 n^{-1/m_2(H)} \end{cases}. \tag{8.5}$$

A reviewer of this paper pointed out a simple proof of the 1-statement. Given a b-bounded coloring of G, let the edges colored i be denoted $e_{i,1}, e_{i,2}, \ldots, e_{i,b_i}$ where $b_i \leq b$ for all i. Now consider the auxiliary coloring in which edge $e_{i,j}$ is colored with j. At most b colors are used and so in the auxiliary coloring there will be a monochromatic copy of H. The definition of the auxiliary coloring implies that this copy of H is rainbow in the original coloring. So the 1-statement follows directly from the results of Rödl and Ruciński [653], i.e. Theorem 8.3.

Nenadov, Person, Škorić and Steger [603] gave further threshold results on both the Ramsey and anti-Ramsey theories of random graphs. In particular they proved that in many cases $b_0 = 2$ in (8.5).

PART II

Basic Model Extensions

9

Inhomogeneous Graphs

Thus far we have concentrated on the properties of the random graphs $\mathbb{G}_{n,m}$ and $\mathbb{G}_{n,p}$. We first consider a generalization of $\mathbb{G}_{n,p}$ where the probability of edge (i,j) is p_{ij} is not the same for all pairs i,j. We call this the *generalized binomial graph*. Our main result on this model concerns the probability that it is connected. For this model we concentrate on its degree sequence and the existence of a giant component. After this we move onto a special case of this model, viz. the *expected degree model*. Here p_{ij} is proportional to $w_i w_j$ for weights w_i. In this model, we prove results about the size of the largest components. We finally consider another special case of the generalized binomial graph, viz. the *Kronecker random graph*.

9.1 Generalized Binomial Graph

Consider the following natural generalization of the binomial random graph $\mathbb{G}_{n,p}$, first considered by Kovalenko [501].

Let $V = \{1,2,\ldots,n\}$ be the vertex set. The random graph $\mathbb{G}_{n,\mathbf{P}}$ has vertex set V and two vertices i and j from V, $i \neq j$, are joined by an edge with probability $p_{ij} = p_{ij}(n)$, independently of all other edges. Denote by

$$\mathbf{P} = \left[p_{ij}\right]$$

the symmetric $n \times n$ matrix of edge probabilities, where $p_{ii} = 0$. Put $q_{ij} = 1 - p_{ij}$ and for $i,k \in \{1,2,\ldots,n\}$ define

$$Q_i = \prod_{j=1}^{n} q_{ij}, \quad \lambda_n = \sum_{i=1}^{n} Q_i.$$

Note that Q_i is the probability that vertex i is isolated and λ_n is the expected number of isolated vertices. Next let

$$R_{ik} = \min_{1 \le j_1 < j_2 < \cdots < j_k \le n} q_{ij_1} \cdots q_{ij_k}.$$

Suppose that the edge probabilities p_{ij} are chosen in such a way that the following conditions are simultaneously satisfied as $n \to \infty$:

$$\max_{1 \le i \le n} Q_i \to 0, \tag{9.1}$$

$$\lim_{n\to\infty} \lambda_n = \lambda = constant, \tag{9.2}$$

and

$$\lim_{n\to\infty} \sum_{k=1}^{n/2} \frac{1}{k!} \left(\sum_{i=1}^{n} \frac{Q_i}{R_{ik}} \right)^k = e^{\lambda} - 1. \tag{9.3}$$

The next two theorems are due to Kovalenko [501].

We will first give the asymptotic distribution of the number of isolated vertices in $\mathbb{G}_{n,\mathbf{P}}$, assuming that the above three conditions are satisfied. The next theorem is a generalization of the corresponding result for the classical model $\mathbb{G}_{n,p}$ (see Theorem 3.1(ii)).

Theorem 9.1 *Let X_0 denote the number of isolated vertices in the random graph $\mathbb{G}_{n,\mathbf{P}}$. If conditions (9.1), (9.2) and (9.3) hold, then*

$$\lim_{n\to\infty} \mathbb{P}(X_0 = k) = \frac{\lambda^k}{k!} e^{-\lambda}$$

for $k = 0, 1, \ldots$, i.e. the number of isolated vertices is asymptotically Poisson distributed with mean λ.

Proof Let

$$X_{ij} = \begin{cases} 1 & \text{with prob. } p_{ij} \\ 0 & \text{with prob. } q_{ij} = 1 - p_{ij}. \end{cases}$$

Denote by X_i, for $i = 1, 2, \ldots n$, the indicator of the event that vertex i is isolated in $\mathbb{G}_{n,\mathbf{P}}$. To show that X_0 converges in distribution to the Poisson random variable with mean λ one has to show (see Corollary 20.11) that for any natural number k

$$\mathbb{E}\left(\sum_{1 \le i_1 < i_2 < \ldots < i_k \le n} X_{i_1} X_{i_2} \cdots X_{i_k} \right) \to \frac{\lambda^k}{k!} \tag{9.4}$$

as $n \to \infty$. But

$$\mathbb{E}\left(X_{i_1}X_{i_2}\cdots X_{i_k}\right) = \prod_{r=1}^{k} \mathbb{P}\left(X_{i_r} = 1 | X_{i_1} = \cdots = X_{i_{r-1}} = 1\right), \tag{9.5}$$

where in the case of $r = 1$ we condition on the sure event.

Since the LHS of (9.4) is the sum of $\mathbb{E}\left(X_{i_1}X_{i_2}\cdots X_{i_k}\right)$ over all $i_1 < \cdots < i_k$, we need to find matching upper and lower bounds for this expectation. Now $\mathbb{P}\left(X_{i_r} = 1 | X_{i_1} = \cdots = X_{i_{r-1}} = 1\right)$ is the unconditional probability that i_r is not adjacent to any vertex $j \neq i_1, \ldots, i_{r-1}$ and so

$$\mathbb{P}\left(X_{i_r} = 1 | X_{i_1} = \cdots = X_{i_{r-1}} = 1\right) = \frac{\prod_{j=1}^{n} q_{i_r j}}{\prod_{s=1}^{r-1} q_{i_r i_s}}.$$

Hence

$$Q_{i_r} \leq P\left(X_{i_r} = 1 | X_{i_1} = \cdots = X_{i_{r-1}} = 1\right) \leq \frac{Q_{i_r}}{R_{i_r, r-1}} \leq \frac{Q_{i_r}}{R_{i_r k}}.$$

It follows from (9.5) that

$$Q_{i_1} \cdots Q_{i_k} \leq \mathbb{E}\left(X_{i_1} \cdots X_{i_k}\right) \leq \frac{Q_{i_1}}{R_{i_1 k}} \cdots \frac{Q_{i_k}}{R_{i_k k}}. \tag{9.6}$$

Applying conditions (9.1) and (9.2) we get that

$$\begin{aligned}
\sum_{1 \leq i_1 < \cdots < i_k \leq n} Q_{i_1} \cdots Q_{i_k} &= \frac{1}{k!} \sum_{1 \leq i_1 \neq \cdots \neq i_r \leq n} Q_{i_1} \cdots Q_{i_k} \\
&\geq \frac{1}{k!} \sum_{1 \leq i_1, \ldots, i_k \leq n} Q_{i_1} \cdots Q_{i_k} - \frac{k}{k!} \sum_{i=1}^{n} Q_i^2 \left(\sum_{1 \leq i_1, \ldots, i_{k-2} \leq n} Q_{i_1} \cdots Q_{i_{k-2}} \right) \\
&\geq \frac{\lambda_n^k}{k!} - (\max_i Q_i) \lambda_n^{k-1} = \to \frac{\lambda^k}{k!},
\end{aligned} \tag{9.7}$$

as $n \to \infty$.

Now,

$$\sum_{i=1}^{n} \frac{Q_i}{R_{ik}} \geq \lambda_n = \sum_{i=1}^{n} Q_i,$$

and if $\limsup_{n\to\infty} \sum_{i=1}^{n} \frac{Q_i}{R_{ik}} > \lambda$ then $\limsup_{n\to\infty} \sum_{k=1}^{n/2} \frac{1}{k!} \left(\sum_{i=1}^{n} \frac{Q_i}{R_{ik}} \right)^k > e^{\lambda} - 1$, which contradicts (9.3). It follows that

$$\lim_{n\to\infty} \sum_{i=1}^{n} \frac{Q_i}{R_{ik}} = \lambda.$$

Therefore

$$\sum_{1\le i_1<\ldots<i_k\le n} Q_{i_1}\cdots Q_{i_k} \le \frac{1}{k!}\left(\sum_{i=1}^{n}\frac{Q_i}{R_{ik}}\right)^k \to \frac{\lambda^k}{k!},$$

as $n\to\infty$.

Combining this with (9.7) gives us (9.4) and completes the proof of Theorem 9.1. □

We can check that the conditions of the theorem are satisfied when

$$p_{ij} = \frac{\log n + x_{ij}}{n},$$

where x_{ij}s are uniformly bounded by a constant.

The next theorem shows that under certain circumstances, the random graph $\mathbb{G}_{n,\mathbf{P}}$ behaves in a similar way to $\mathbb{G}_{n,p}$ at the connectivity threshold.

Theorem 9.2 *If the conditions (9.1), (9.2) and (9.3) hold, then*

$$\lim_{n\to\infty}\mathbb{P}(\mathbb{G}_{n,\mathbf{P}} \textit{ is connected}) = e^{-\lambda}.$$

Proof To prove the this we will show that if (9.1), (9.2) and (9.3) are satisfied then w.h.p. $\mathbb{G}_{n,\mathbf{P}}$ consists of X_0+1 connected components, i.e. $\mathbb{G}_{n,\mathbf{P}}$ consists of a single giant component plus components that are isolated vertices only. This, together with Theorem 9.1, implies the conclusion of Theorem 9.2.

Let $U\subseteq V$ be a subset of the vertex set V. We say that U is *closed* if $X_{ij}=0$ for every i and j, where $i\in U$ and $j\in V\setminus U$. Furthermore, a closed set U is called *simple* if either U or $V\setminus U$ consists of isolated vertices only. Denote the number of non-empty closed sets in $\mathbb{G}_{n,\mathbf{P}}$ by Y_1 and the number of non-empty simple sets by Y. Clearly $Y_1\ge Y$.

We will prove first that

$$\liminf_{n\to\infty}\mathbb{E}\,Y \ge 2e^{\lambda}-1. \tag{9.8}$$

Denote the set of isolated vertices in $\mathbb{G}_{n,\mathbf{P}}$ by J. If $V\setminus J$ is not empty then $Y=2^{X_0+1}-1$ (the number of non-empty subsets of J plus the number of their complements, plus V itself). If $V\setminus J=\emptyset$ then $Y=2^n-1$. Now, by Theorem 9.1, for every fixed $k=0,1,\ldots,$

$$\lim_{n\to\infty}\mathbb{P}(Y=2^{k+1}-1) = e^{-\lambda}\frac{\lambda^k}{k!}.$$

Observe that for any $\ell \geq 0$,

$$\mathbb{E}\, Y \geq \sum_{k=0}^{\ell} (2^{k+1}-1)\,\mathbb{P}(Y = 2^{k+1}-1)$$

and hence

$$\liminf_{n\to\infty} \mathbb{E}\, Y \geq \sum_{k=0}^{\ell} (2^{k+1}-1)\frac{\lambda^k e^{-\lambda}}{k!}.$$

So,

$$\liminf_{n\to\infty} \mathbb{E}\, Y \geq \lim_{\ell\to\infty} \sum_{k=0}^{\ell} (2^{k+1}-1)\frac{\lambda^k e^{-\lambda}}{k!} = 2e^{\lambda}-1$$

which completes the proof of (9.8).

We will show next that

$$\limsup_{n\to\infty} \mathbb{E}\, Y_1 \leq 2e^{\lambda}-1. \tag{9.9}$$

To prove (9.9) denote by Z_k the number of closed sets of order k in $\mathbb{G}_{n,\mathbf{P}}$, so that $Y_1 = \sum_{k=1}^{n} Z_k$. Note that

$$Z_k = \sum_{i_1<\ldots<i_k} Z_{i_1\ldots i_k},$$

where $Z_{i_1,\ldots i_k}$ indicates whether set $I_k = \{i_1 \ldots i_k\}$ is closed. Then

$$\mathbb{E}\, Z_{i_1,\ldots i_k} = \mathbb{P}(X_{ij} = 0, i \in I_k, j \notin I_k) = \prod_{i\in I_k, j\notin I_k} q_{ij}.$$

Consider first the case when $k \leq n/2$, then

$$\prod_{i\in I_k, j\notin I_k} q_{ij} = \frac{\prod_{i\in I_k, 1\leq j\leq n} q_{ij}}{\prod_{i\in I_k, j\in I_k} q_{ij}} = \prod_{i\in I_k} \frac{Q_i}{\prod_{j\in I_k} q_{ij}} \leq \prod_{i\in I_k} \frac{Q_i}{R_{ik}}.$$

Hence

$$\mathbb{E}\, Z_k \leq \sum_{i_1<\ldots<i_k} \prod_{i\in I_k} \frac{Q_i}{R_{ik}} \leq \frac{1}{k!}\left(\sum_{i=1}^{n} \frac{Q_i}{R_{ik}}\right)^k.$$

Now, (9.3) implies that

$$\limsup_{n\to\infty} \sum_{k=1}^{n/2} \mathbb{E}\, Z_k \leq e^{\lambda}-1.$$

To complete the estimation of $\mathbb{E}\, Z_k$ (and thus for $\mathbb{E}\, Y_1$), consider the case when $k > n/2$. For convenience let us switch k with $n-k$, i.e. consider $\mathbb{E}\, Z_{n-k}$, when

$0 \le k < n/2$. Notice that $\mathbb{E}Z_n = 1$ since V is closed. So for $1 \le k < n/2$

$$\mathbb{E}Z_{n-k} = \sum_{i_1<\ldots<i_k} \prod_{i\in I_k, j\notin I_k} q_{ij}.$$

But $q_{ij} = q_{ji}$ so, for such k, $\mathbb{E}Z_{n-k} = \mathbb{E}Z_k$. This gives

$$\limsup_{n\to\infty} \mathbb{E}\,Y_1 \le 2(e^{\lambda} - 1) + 1,$$

where the +1 comes from $Z_n = 1$. This completes the proof of (9.9).

Now,

$$\mathbb{P}(Y_1 > Y) = \mathbb{P}(Y_1 - Y \ge 1) \le \mathbb{E}(Y_1 - Y).$$

Estimates (9.8) and (9.9) imply that

$$\limsup_{n\to\infty} \mathbb{E}(Y_1 - Y) \le 0,$$

which in turn leads to the conclusion that

$$\lim_{n\to\infty} \mathbb{P}(Y_1 > Y) = 0,$$

i.e. asymptotically, the probability that there is a closed set that is not simple, tends to zero as $n \to \infty$. It is easy to check that $X_0 < n$ w.h.p. and therefore $Y = 2^{X_0+1} - 1$ w.h.p. and so w.h.p. $Y_1 = 2^{X_0+1} - 1$. If $\mathbb{G}_{n,\mathbf{P}}$ has more than $X_0 + 1$ connected components then the graph, after removal of all isolated vertices, would contain at least one closed set, i.e. the number of closed sets would be at least 2^{X_0+1}. But the probability of such an event tends to zero and the theorem follows. □

We finish this section by presenting a sufficient condition for $\mathbb{G}_{n,\mathbf{P}}$ to be connected w.h.p. as proven by Alon [22].

Theorem 9.3 *For every positive constant b there exists a constant $c = c(b) > 0$ so that if, for every non-trivial $S \subset V$,*

$$\sum_{i\in S, j\in V\setminus S} p_{ij} \ge c\log n,$$

then the probability that $\mathbb{G}_{n,\mathbf{P}}$ is connected is at least $1 - n^{-b}$.

Proof In fact Alon's result is much stronger. He considers a random subgraph $\mathbb{G}_{p_e}$ of a multi-graph G on n vertices, obtained by deleting each edge e independently with probability $1 - p_e$. The random graph $\mathbb{G}_{n,\mathbf{P}}$ is a special case of $\mathbb{G}_{p_e}$ when G is the complete graph K_n. Therefore, following in his footsteps, we will prove that Theorem 9.3 holds for $\mathbb{G}_{p_e}$ and thus for $\mathbb{G}_{n,\mathbf{P}}$.

So, let $G = (V,E)$ be a loopless, undirected multigraph on n vertices, with probability p_e, $0 \leq p_e \leq 1$ assigned to every edge $e \in E$ and suppose that for any non-trivial $S \subset V$ the expectation of the number E_S of edges in a cut $(S, V \setminus S)$ of $\mathbb{G}_{p_e}$ satisfies

$$\mathbb{E} E_S = \sum_{e \in (S, V \setminus S)} p_e \geq c \log n. \tag{9.10}$$

Create a new graph $G' = (V, E')$ from G by replacing each edge e by $k = c \log n$ parallel copies with the same endpoints and giving each copy e' of e a probability $p'_{e'} = p_e/k$.

Observe that for $S \subset V$

$$\mathbb{E} E'_S = \sum_{e' \in (S, V \setminus S)} p'_{e'} = \mathbb{E} E_S.$$

Moreover, for every edge e of G, the probability that no copy e' of e survives in a random subgraph $\mathbb{G}'_{p'}$ is $(1 - p_e/k)^k \geq 1 - p_e$ and hence the probability that $\mathbb{G}_{p_e}$ is connected exceeds the probability of $\mathbb{G}'_{p'_e}$ being connected, and so in order to prove the theorem it suffices to prove that

$$\mathbb{P}(\mathbb{G}'_{p'_e} \text{ is connected}) \geq 1 - n^{-b}. \tag{9.11}$$

To prove this, let $E'_1 \cup E'_2 \cup \ldots \cup E'_k$ be a partition of the set E' of the edges of G', such that each E'_i consists of a single copy of each edge of G. For $i = 0, 1, \ldots, k$ define $\mathbb{G}'_i$ as follows. $\mathbb{G}'_0$ is the subgraph of G' that has no edges, and for all $i \geq 1$, $\mathbb{G}'_i$ is the random subgraph of G' obtained from $\mathbb{G}'_{i-1}$ by adding to it each edge $e' \in E'_i$ independently, with probability $p'_{e'}$.

Let C_i be the number of connected components of $\mathbb{G}'_i$. Then we have $C_0 = n$ and we have $\mathbb{G}'_k \equiv \mathbb{G}'_{p'_e}$. Let us call the stage i, $1 \leq i \leq k$, *successful* if either $C_{i-1} = 1$ (i.e. $\mathbb{G}'_{i-1}$ is connected) or if $C_i < 0.9 C_{i-1}$. We will prove that

$$\mathbb{P}(C_{i-1} = 1 \text{ or } C_i < 0.9 C_{i-1} | \mathbb{G}'_{i-1}) \geq \frac{1}{2}. \tag{9.12}$$

To see that (9.12) holds, note first that if $\mathbb{G}'_{i-1}$ is connected then there is nothing to prove. Otherwise let $\mathbb{H}_i = (U, F)$ be the graph obtained from $\mathbb{G}'_{i-1}$ by (i) contracting every connected component of $\mathbb{G}'_{i-1}$ to a single vertex and (ii) adding to it each edge $e' \in E'_i$ independently, with probability $p'_{e'}$ and throwing away loops. Note that since for every nontrivial S, $\mathbb{E} E'_S \geq k$, we have that for every vertex $u \in U$ (connected component of $\mathbb{G}'_{i-1}$),

$$\sum_{u \in e' \in F} p'_{e'} = \sum_{e \in U : U^c} \frac{p_e}{k} \geq 1.$$

Moreover, the probability that a fixed vertex $u \in U$ is isolated in $\mathbb{H}_i$ is

$$\prod_{u \in e' \in F} (1 - p'_{e'}) \leq \exp\left\{ - \sum_{u \in e' \in F} p'_{e'} \right\} \leq e^{-1}.$$

Hence, the expected number of isolated vertices of $\mathbb{H}_i$ does not exceed $|U|e^{-1}$. Therefore, by the Markov inequality, it is at most $2|U|e^{-1}$ with probability at least 1/2. But in this case the number of connected components of $\mathbb{H}_i$ is at most

$$2|U|e^{-1} + \frac{1}{2}(|U| - 2|U|e^{-1}) = \left(\frac{1}{2} + e^{-1}\right)|U| < 0.9|U|,$$

and so (9.12) follows. Observe that if $C_k > 1$, then the total number of successful stages is strictly less than $\log n / \log 0.9 < 10 \log n$. However, by (9.12), the probability of this event is at most the probability that a Binomial random variable with parameters k and $1/2$ will attain a value of at most $10 \log n$. It follows from (21.22) that if $k = c \log n = (20 + t) \log n$ then the probability that $C_k > 1$ (i.e. that $\mathbb{G}'_{p'_e}$ is disconnected) is at most $n^{-t^2/4c}$. This completes the proof of (9.11) and the theorem follows. □

9.2 Expected Degree Model

In this section we consider a special case of Kovalenko's generalized binomial model, introduced by Chung and Lu in [181], where edge probabilities p_{ij} depend on weights assigned to vertices. This was also meant as a model for real world networks, see Chapter 17.

Let $V = \{1, 2, \ldots, n\}$ and let w_i be the *weight* of vertex i. Now insert edges between vertices $i, j \in V$ independently with probability p_{ij} defined as

$$p_{ij} = \frac{w_i w_j}{W} \quad \text{where} \quad W = \sum_{k=1}^{n} w_k.$$

We assume that $\max_i w_i^2 < W$ so that $p_{ij} \leq 1$. The resulting graph is denoted as $\mathbb{G}_{n,\mathbf{P^w}}$. Note that putting $w_i = np$ for $i \in [n]$ yields the random graph $\mathbb{G}_{n,p}$.

Notice that loops are allowed here but we will ignore them in what follows. Moreover, the expected degree for vertex $i \in V$ is

$$\sum_j \frac{w_i w_j}{W} = w_i.$$

Denote the average vertex weight by $\overline{w}$ (average expected vertex degree), i.e.

$$\overline{w} = \frac{W}{n},$$

while, for any subset U of a vertex set V define the *volume* of U as

$$\mathrm{w}(U) = \sum_{k \in U} w_k.$$

Chung and Lu in [181] and [183] proved the following results summarized in the next theorem.

Theorem 9.4 *The random graph $\mathbb{G}_{n,\mathbf{P^w}}$ with a given expected degree sequence has a unique giant component w.h.p. if the average expected degree is strictly greater than one (i.e. $\overline{w} > 1$). Moreover, if $\overline{w} > 1$ then w.h.p. the giant component has volume*

$$\lambda_0 W + O\left(\sqrt{n}(\log n)^{3.5}\right),$$

where λ_0 is the unique nonzero root of the following equation

$$\sum_{i=1}^{n} w_i e^{-w_i \lambda} = (1-\lambda) \sum_{i=1}^{n} w_i,$$

Furthermore w.h.p., the second-largest component has size at most

$$(1+o(1))\mu(\overline{w}) \log n,$$

where

$$\mu(\overline{w}) = \begin{cases} 1/(\overline{w} - 1 - \log \overline{w}) & \text{if } 1 < \overline{w} < 2, \\ 1/(1 + \log \overline{w} - \log 4) & \text{if } \overline{w} > 4/e. \end{cases}$$

Here we will prove a weaker and restricted version of the above theorem. In the current context, a giant component is one with volume $\Omega(W)$.

Theorem 9.5 *If the average expected degree $\overline{w} > 4$, then a random graph $\mathbb{G}_{n,\mathbf{P^w}}$ w.h.p. has a unique giant component and its volume is at least*

$$\left(1 - \frac{2}{\sqrt{e\overline{w}}}\right) W,$$

while the second-largest component w.h.p. has the size at most

$$(1+o(1)) \frac{\log n}{1 + \log \overline{w} - \log 4}.$$

The proof is based on a key lemma given below, proved under stronger conditions on $\overline{w}$ than in fact Theorem 9.5 requires.

Lemma 9.6 *For any positive $\varepsilon < 1$ and $\overline{w} > \frac{4}{e(1-\varepsilon)^2}$ w.h.p. every connected component in the random graph $\mathbb{G}_{n,\mathbf{P^w}}$ either has volume at least εW or has at most $\frac{\log n}{1+\log \overline{w} - \log 4 + 2\log(1-\varepsilon)}$ vertices.*

Proof We first estimate the probability of the existence of a connected component with k vertices (component of *size* k) in the random graph $\mathbb{G}_{n,\mathbf{P^w}}$. Let $S \subseteq V$ and suppose that vertices from $S = \{v_{i_1}, v_{i_2}, \ldots, v_{i_k}\}$ have respective weights $w_{i_1}, w_{i_2}, \ldots, w_{i_k}$. If the set S induces a connected subgraph of $\mathbb{G}_{n,\mathbf{P^w}}$ then it contains at least one spanning tree T. The probability of such event equals

$$\mathbb{P}(T) = \prod_{\{v_{i_j}, v_{i_l}\} \in E(T)} w_{i_j} w_{i_l} \rho,$$

where

$$\rho := \frac{1}{W} = \frac{1}{n\overline{w}}.$$

So, the probability that S induces a connected subgraph of our random graph can be bounded from above by

$$\sum_T \mathbb{P}(T) = \sum_T \prod_{\{v_{i_j}, v_{i_l}\} \in E(T)} w_{i_j} w_{i_l} \rho,$$

where T ranges over all spanning trees on S.

By the matrix-tree theorem (see West [722]) the above sum equals the determinant of any $k-1$ by $k-1$ principal sub-matrix of $(D-A)\rho$, where A is defined as

$$A = \begin{pmatrix} 0 & w_{i_1} w_{i_2} & \cdots & w_{i_1} w_{i_k} \\ w_{i_2} w_{i_1} & 0 & \cdots & w_{i_2} w_{i_k} \\ \vdots & \vdots & \ddots & \vdots \\ w_{i_k} w_{i_1} & w_{i_k} w_{i_2} & \cdots & 0 \end{pmatrix},$$

while D is the diagonal matrix

$$D = \operatorname{diag}\left(w_{i_1}(W - w_{i_1}), \ldots, w_{i_k}(W - w_{i_k})\right).$$

(To evaluate the determinant of the first principal co-factor of $D-A$, delete row and column k of $D-A$; take out a factor $w_{i_1} w_{i_2} \cdots w_{i_{k-1}}$; add the last $k-2$ rows to row 1; row 1 is now $(w_{i_k}, w_{i_k}, \ldots, w_{i_k})$, so we can take out a factor w_{i_k}; now subtract column 1 from the remaining columns to get a $(k-1) \times (k-1)$ upper triangular matrix with diagonal equal to $\operatorname{diag}(1, \mathrm{w}(S), \mathrm{w}(S), \ldots, \mathrm{w}(S))$.)

It follows that

$$\sum_T \mathbb{P}(T) = w_{i_1} w_{i_2} \cdots w_{i_k} \mathrm{w}(S)^{k-2} \rho^{k-1}. \tag{9.13}$$

To show that this subgraph is in fact a component, we have to multiply by the probability that there is no edge leaving S in $\mathbb{G}_{n,\mathbf{P^w}}$. Obviously, this probability equals $\prod_{v_i \in S, v_j \notin S}(1 - w_i w_j \rho)$ and can be bounded from above

$$\prod_{v_i \in S, v_j \in V \setminus S} (1 - w_i w_j \rho) \le e^{-\rho \mathrm{w}(S)(W - \mathrm{w}(S))}. \tag{9.14}$$

Let X_k be the number of components of size k in $\mathbb{G}_{n,\mathbf{P^w}}$, then, using bounds from (9.13) and (9.14) we obtain

$$\mathbb{E} X_k \le \sum_S \mathrm{w}(S)^{k-2} \rho^{k-1} e^{-\rho \mathrm{w}(S)(W - \mathrm{w}(S))} \prod_{i \in S} w_i,$$

where the sum ranges over all $S \subseteq V$, $|S| = k$. Now, we focus our attention on k-vertex components whose volume is at most εW. We call such components *small* or *ε-small*. So, if Y_k is the number of small components of size k in $\mathbb{G}_{n,\mathbf{P^w}}$ then

$$\mathbb{E} Y_k \le \sum_{\text{small } S} \mathrm{w}(S)^{k-2} \rho^{k-1} e^{-\mathrm{w}(S)(1-\varepsilon)} \prod_{i \in S} w_i = f(k). \tag{9.15}$$

Now, using the arithmetic-geometric mean inequality, we have

$$f(k) \le \sum_{\text{small } S} \left(\frac{\mathrm{w}(S)}{k} \right)^k \mathrm{w}(S)^{k-2} \rho^{k-1} e^{-\mathrm{w}(S)(1-\varepsilon)}.$$

The function $x^{2k-2} e^{-x(1-\varepsilon)}$ achieves its maximum at $x = (2k-2)/(1-\varepsilon)$.
Therefore,

$$\begin{aligned}
f(k) &\le \binom{n}{k} \frac{\rho^{k-1}}{k^k} \left(\frac{2k-2}{1-\varepsilon} \right)^{2k-2} e^{-(2k-2)} \\
&\le \left(\frac{ne}{k} \right)^k \frac{\rho^{k-1}}{k^k} \left(\frac{2k-2}{1-\varepsilon} \right)^{2k-2} e^{-(2k-2)} \\
&\le \frac{(n\rho)^k}{4\rho(k-1)^2} \left(\frac{2}{1-\varepsilon} \right)^{2k} e^{-k} \\
&= \frac{1}{4\rho(k-1)^2} \left(\frac{4}{e\overline{w}(1-\varepsilon)^2} \right)^k \\
&= \frac{e^{-ak}}{4\rho(k-1)^2},
\end{aligned}$$

where

$$a = 1 + \log \overline{w} - \log 4 + 2 \log(1-\varepsilon) > 0$$

under the assumption of Lemma 9.6.

Let $k_0 = \frac{\log n}{a}$. When k satisfies $k_0 < k < 2k_0$, we have

$$f(k) \leq \frac{1}{4n\rho(k-1)^2} = o\left(\frac{1}{\log n}\right),$$

while, when $\frac{2\log n}{a} \leq k \leq n$, we have

$$f(k) \leq \frac{1}{4n^2\rho(k-1)^2} = o\left(\frac{1}{n\log n}\right).$$

So, the probability that there exists an ε-small component of size exceeding k_0 is at most

$$\sum_{k>k_0} f(k) \leq \frac{\log n}{a} \times o\left(\frac{1}{\log n}\right) + n \times o\left(\frac{1}{n\log n}\right) = o(1).$$

This completes the proof of Lemma 9.6. □

To prove Theorem 9.5 assume that for some fixed $\delta > 0$ we have

$$\overline{w} = 4 + \delta = \frac{4}{e(1-\varepsilon)^2} \text{ where } \varepsilon = 1 - \frac{2}{(e\overline{w})^{1/2}} \tag{9.16}$$

and suppose that $w_1 \geq w_2 \geq \cdots \geq w_n$. We show next that there exists $i_0 \geq n^{1/3}$ such that

$$w_{i_0} \geq \sqrt{\frac{\left(1+\frac{\delta}{8}\right)W}{i_0}}. \tag{9.17}$$

Suppose the contrary, i.e. for all $i \geq n^{1/3}$,

$$w_i < \sqrt{\frac{\left(1+\frac{\delta}{8}\right)W}{i}}.$$

Then

$$\begin{aligned} W &\leq n^{1/3}W^{1/2} + \sum_{i=n^{1/3}}^{n} \sqrt{\frac{\left(1+\frac{\delta}{8}\right)W}{i}} \\ &\leq n^{1/3}W^{1/2} + 2\sqrt{\left(1+\frac{\delta}{8}\right)Wn}. \end{aligned}$$

Hence

$$W^{1/2} \leq n^{1/3} + 2\left(1+\frac{\delta}{8}\right)n^{1/2}.$$

This is a contradiction since for our choice of $\overline{w}$

$$W = n\overline{w} \geq 4(1+\delta)n.$$

We have therefore verified the existence of i_0 satisfying (9.17).

Now consider the subgraph G of $\mathbb{G}_{n,\mathbf{P^w}}$ on the first i_0 vertices. The probability that there is an edge between vertices v_i and v_j, for any $i,j \le i_0$, is at least

$$w_i w_j \rho \ge w_{i_0}^2 \rho \ge \frac{1+\frac{\delta}{8}}{i_0}.$$

So the asymptotic behavior of G can be approximated by a random graph $\mathbb{G}_{n,p}$ with $n = i_0$ and $p > 1/i_0$. So, w.h.p. G has a component of size $\Theta(i_0) = \Omega(n^{1/3})$. Applying Lemma 9.6 with ε as in (9.16) we see that any component with size $\gg \log n$ has volume at least εW.

Finally, consider the volume of a giant component. Suppose first that there exists a giant component of volume cW which is ε-small, i.e. $c \le \varepsilon$. By Lemma 9.6, the size of the giant component is then at most $\frac{\log n}{2\log 2}$. Hence, there must be at least one vertex with weight w greater than or equal to the average

$$w \ge \frac{2cW \log 2}{\log n}.$$

But it implies that $w^2 \gg W$, which contradicts the general assumption that all $p_{ij} < 1$.

We now prove uniqueness in the same way that we proved the uniqueness of the giant component in $G_{n,p}$. Choose $\eta > 0$ such that $\overline{w}(1-\eta) > 4$. Then define $w_i' = (1-\eta)w_i$ and decompose

$$\mathbb{G}_{n,\mathbf{P^w}} = G_1 \cup G_2,$$

where the edge probability in G_1 is $p_{ij}' = \frac{w_i' w_j'}{(1-\eta)W}$ and the edge probability in G_2 is p_{ij}'', where $1 - \frac{w_i w_j}{W} = (1-p_{i,j}')(1-p_{ij}'')$. Simple algebra gives $p_{ij}'' \ge \frac{\eta w_i w_j}{W}$. It follows from the previous analysis that G_1 contains between one and $1/\varepsilon$ giant components. Let C_1, C_2 be two such components. The probability that there is no G_2 edge between them is at most

$$\prod_{\substack{i \in C_1 \\ j \in C_2}} \left(1 - \frac{\eta w_i w_j}{W}\right) \le \exp\left\{-\frac{\eta w(C_1) w(C_2)}{W}\right\} \le e^{-\eta W} = o(1).$$

As $1/\varepsilon < 4$, this completes the proof of Theorem 9.5. □

To add to the picture of the asymptotic behavior of the random graph $\mathbb{G}_{n,\mathbf{P^w}}$ we present one more result from [181]. Denote by $\overline{w^2}$ the expected second-order average degree, i.e.

$$\overline{w^2} = \sum_j \frac{w_j^2}{W}.$$

Notice that

$$\overline{w^2} = \frac{\sum_j w_j^2}{W} \geq \frac{W}{n} = \overline{w}.$$

Chung and Lu [181] proved the following.

Theorem 9.7 *If the average expected square degree* $\overline{w^2} < 1$ *then, with probability at least* $1 - \frac{\overline{w}\left(\overline{w^2}\right)^2}{C^2\left(1-\overline{w^2}\right)}$, *all components of* $\mathbb{G}_{n,\mathbf{P^w}}$ *have volume at most* $C\sqrt{n}$.

Proof Let

$$x = \mathbb{P}(\exists S : w(S) \geq Cn^{1/2} \text{and } S \text{ is a component}).$$

Randomly choose two vertices u and v from V, each with probability proportional to its weight. Then, for each vertex, the probability that it is in a set S with $w(S) \geq C\sqrt{n}$ is at least $C\sqrt{n}\rho$. Hence, the probability that both vertices are in the same component is at least

$$x(C\sqrt{n}\rho)^2 = C^2 xn\rho^2. \tag{9.18}$$

On the other hand, for any two fixed vertices, say u and v, the probability $P_k(u,v)$ of u and v being connected via a path of length $k+1$ can be bounded from above as follows

$$P_k(u,v) \leq \sum_{i_1,i_2,\ldots,i_k} (w_u w_{i_1}\rho)(w_{i_1} w_{i_2}\rho)\cdots(w_{i_k} w_v \rho) \leq w_u w_v \rho (\overline{w^2})^k.$$

So the probability that u and v belong to the same component is at most

$$\sum_{k=0}^{n} P_k(u,v) \leq \sum_{k=0}^{\infty} w_u w_v \rho (\overline{w^2})^k = \frac{w_u w_v \rho}{1-\overline{w^2}}.$$

Recall that the probabilities of u and v being chosen from V are $w_u\rho$ and $w_v\rho$, respectively, so the probability that a random pair of vertices are in the same component is at most

$$\sum_{u,v} w_u\rho \; w_v\rho \; \frac{w_u w_v \rho}{1-\overline{w^2}} = \frac{\left(\overline{w^2}\right)^2 \rho}{1-\overline{w^2}}.$$

Combining this with (9.18) we have

$$C^2 xn\rho^2 \leq \frac{\left(\overline{w^2}\right)^2 \rho}{1-\overline{w^2}},$$

which implies

$$x \leq \frac{\overline{w}\left(\overline{w^2}\right)^2}{C^2\left(1-\overline{w^2}\right)},$$

and Theorem 9.7 follows. □

9.3 Kronecker Graphs

Kronecker random graphs were introduced by Leskovec, Chakrabarti, Kleinberg and Faloutsos [525] (see also [524]). It is meant as a model of real world networks, see Chapter 17. Here we consider a special case of this model of an inhomogeneous random graph. To construct it we begin with a *seed* matrix

$$\mathbf{P} = \begin{bmatrix} \alpha & \beta \\ \beta & \gamma \end{bmatrix},$$

where $0 < \alpha, \beta, \gamma < 1$, and let $\mathbf{P}^{[k]}$ be the kth Kronecker power of $\mathbf{P}$. Here $\mathbf{P}^{[k]}$ is obtained from $\mathbf{P}^{[k-1]}$ as in the diagram below:

$$\mathbf{P}^{[k]} = \begin{bmatrix} \alpha\mathbf{P}^{[k-1]} & \beta\mathbf{P}^{[k-1]} \\ \beta\mathbf{P}^{[k-1]} & \gamma\,\mathbf{P}^{[k-1]} \end{bmatrix}$$

and so, for example,

$$\mathbf{P}^{[2]} = \begin{bmatrix} \alpha^2 & \alpha\beta & \beta\alpha & \beta^2 \\ \alpha\beta & \alpha\gamma & \beta^2 & \beta\gamma \\ \beta\alpha & \beta^2 & \gamma\alpha & \gamma\beta \\ \beta^2 & \beta\gamma & \gamma\beta & \gamma^2 \end{bmatrix}.$$

Note that $\mathbf{P}^{[k]}$ is symmetric and has size $2^k \times 2^k$.

We define a *Kronecker random graph* as a copy of $\mathbb{G}_{n,\mathbf{P}^{[k]}}$ for some $k \geq 1$ and $n = 2^k$. Thus each vertex is a binary string of length k, and between any two such vertices (strings) $\mathbf{u}, \mathbf{v}$ we put an edge independently with probability

$$p_{\mathbf{u},\mathbf{v}} = \alpha^{\mathbf{uv}}\gamma^{(1-\mathbf{u})(1-\mathbf{v})}\beta^{k-\mathbf{uv}-(1-\mathbf{u})(1-\mathbf{v})},$$

or equivalently

$$p_{\mathbf{uv}} = \alpha^i\beta^j\gamma^{k-i-j},$$

where i is the number of positions t such that $u_t = v_t = 1$, j is the number of t where $u_t \neq v_t$ and hence $k-i-j$ is the number of t that $u_t = v_t = 0$. We observe that when $\alpha = \beta = \gamma$, then $\mathbb{G}_{n,\mathbf{P}^{[\mathbf{k}]}}$ becomes $\mathbb{G}_{n,p}$ with $n = 2^k$ and $p = \alpha^k$.

Connectivity

We first examine, following Mahdian and Xu [550], conditions under which $\mathbb{G}_{n,\mathbf{P[k]}}$ is connected w.h.p.

Theorem 9.8 *Suppose that $\alpha \geq \beta \geq \gamma$. The random graph $\mathbb{G}_{n,\mathbf{P[k]}}$ is connected w.h.p. (for $k \to \infty$) if and only if either (i) $\beta + \gamma > 1$ or (ii) $\alpha = \beta = 1, \gamma = 0$.*

Proof We show first that $\beta + \gamma \geq 1$ is a necessary condition. Denote by $\mathbf{0}$ the vertex with all 0s. Then the expected degree of vertex $\mathbf{0}$ is

$$\sum_{\mathbf{v}} p_{\mathbf{0v}} = \sum_{j=0}^{k} \binom{k}{j} \beta^j \gamma^{k-j} = (\beta + \gamma)^k = o(1), \quad \text{when } \beta + \gamma < 1.$$

Thus in this case vertex $\mathbf{0}$ is isolated w.h.p.

Moreover, if $\beta + \gamma = 1$ and $0 < \beta < 1$ then $\mathbb{G}_{n,\mathbf{P[k]}}$ cannot be connected w.h.p. since the probability that vertex $\mathbf{0}$ is isolated is bounded away from 0. Indeed, $0 < \beta < 1$ implies that $\beta^j \gamma^{k-j} \leq \zeta < 1, 0 \leq j \leq k$ for some absolute constant ζ. Thus, using Lemma 21.1(b),

$$\mathbb{P}(\mathbf{0} \text{ is isolated}) = \prod_{\mathbf{v}} (1 - p_{\mathbf{0v}}) \geq \prod_{j=0}^{k} \left(1 - \beta^j \gamma^{k-j}\right)^{\binom{k}{j}}$$

$$\geq \exp\left\{ - \sum_{j=0}^{k} \frac{\binom{k}{j} \beta^j \gamma^{k-j}}{1 - \zeta} \right\} = e^{-1/\zeta}.$$

Now when $\alpha = \beta = 1, \gamma = 0$, the vertex with all 1s has degree $n - 1$ with probability one and so $\mathbb{G}_{n,\mathbf{P[k]}}$ will be connected w.h.p. in this case.

It remains to show that the condition $\beta + \gamma > 1$ is also sufficient. To show that $\beta + \gamma > 1$ implies connectivity we will apply Theorem 9.3. Notice that the expected degree of vertex $\mathbf{0}$, excluding its self-loop, given that β and γ are constants independent of k and $\beta + \gamma > 1$, is

$$(\beta + \gamma)^k - \gamma^k \geq 2c \log n,$$

for some constant $c > 0$, which can be as large as needed.

Therefore, the cut $(\mathbf{0}, V \setminus \{\mathbf{0}\})$ has weight at least $2c \log n$. Remove vertex $\mathbf{0}$ and consider any cut $(S, V \setminus S)$. Then at least one side of the cut gets at least half of the weight of vertex $\mathbf{0}$. Without loss of generality assume that it is S, i.e.

$$\sum_{\mathbf{u} \in S} p_{\mathbf{0u}} \geq c \log n.$$

Take any vertices $\mathbf{u},\mathbf{v}$ and note that $p_{\mathbf{uv}} \geq p_{\mathbf{u0}}$ because we have assumed that $\alpha \geq \beta \geq \gamma$. Therefore,

$$\sum_{\mathbf{u}\in S}\sum_{\mathbf{v}\in V\setminus S} p_{\mathbf{uv}} \geq \sum_{\mathbf{u}\in S} p_{\mathbf{u0}} > c\log n,$$

and so the claim follows by Theorem 9.3. □

To add to the picture of the structure of $\mathbb{G}_{n,\mathbf{P[k]}}$ when $\beta+\gamma > 1$ we state (without proof) the following result on the diameter of $\mathbb{G}_{n,\mathbf{P[k]}}$.

Theorem 9.9 *If $\beta+\gamma > 1$ then w.h.p. $\mathbb{G}_{n,\mathbf{P[k]}}$ has constant diameter.*

Giant Component

We now consider when $\mathbb{G}_{n,\mathbf{P[k]}}$ has a giant component (see Horn and Radcliffe [412]).

Theorem 9.10 *$\mathbb{G}_{n,\mathbf{P[k]}}$ has a giant component of order $\Theta(n)$ w.h.p., if and only if $(\alpha+\beta)(\beta+\gamma) > 1$.*

Proof We prove a weaker version of the Theorem 9.10, assuming that for $\alpha \geq \beta \geq \gamma$ as in [550]. For the proof of the more general case, see [412].

We show first that the above condition is necessary. We prove that if

$$(\alpha+\beta)(\beta+\gamma) \leq 1,$$

then w.h.p. $\mathbb{G}_{n,\mathbf{P[k]}}$ has $n-o(n)$ isolated vertices. First let

$$(\alpha+\beta)(\beta+\gamma) = 1-\varepsilon,\ \varepsilon > 0.$$

First consider those vertices with weight (counted as the number of 1s in their label) less than $k/2+k^{2/3}$ and let $X_{\mathbf{u}}$ be the degree of a vertex $\mathbf{u}$ with weight l where $l = 0,\ldots,k$. It is easily observed that

$$\mathbb{E}X_{\mathbf{u}} = (\alpha+\beta)^l(\beta+\gamma)^{k-l}. \tag{9.19}$$

Indeed, if for vertex $\mathbf{v}$, $i = i(\mathbf{v})$ is the number of bits that $u_r = v_r = 1$, $r = 1,\ldots,k$ and $j = j(\mathbf{v})$ is the number of bits where $u_r = 0$ and $v_r = 1$, then the probability of an edge between $\mathbf{u}$ and $\mathbf{v}$ equals

$$p_{\mathbf{uv}} = \alpha^i\beta^{j+l-i}\gamma^{k-l-j}.$$

Hence,

$$\mathbb{E}X_{\mathbf{u}} = \sum_{\mathbf{v} \in V} p_{\mathbf{uv}} = \sum_{i=0}^{l} \sum_{j=0}^{k-l} \binom{l}{i} \binom{k-l}{j} \alpha^i \beta^{j+l-i} \gamma^{k-l-j}$$
$$= \sum_{i=0}^{l} \binom{l}{i} \alpha^i \beta^{l-i} \sum_{j=0}^{k-l} \beta^j \gamma^{k-l-l}$$

and (9.19) follows. So, if $l < k/2 + k^{2/3}$, then assuming that $\alpha \geq \beta \geq \gamma$,

$$\begin{aligned}
\mathbb{E}X_{\mathbf{u}} \leq & (\alpha+\beta)^{k/2+k^{2/3}} (\beta+\gamma)^{k-(k/2+k^{2/3})} \\
= & ((\alpha+\beta)(\beta+\gamma))^{k/2} \left(\frac{\alpha+\beta}{\beta+\gamma}\right)^{k^{2/3}} \\
= & (1-\varepsilon)^{k/2} \left(\frac{\alpha+\beta}{\beta+\gamma}\right)^{k^{2/3}} \\
= & o(1).
\end{aligned} \tag{9.20}$$

Suppose now that $l \geq k/2 + k^{2/3}$ and let Y be the number of 1s in the label of a randomly chosen vertex of $\mathbb{G}_{n,\mathbf{P[k]}}$. Since $\mathbb{E}\, Y = k/2$, the Chernoff bound (see (21.26)) implies that

$$\mathbb{P}\left(Y \geq \frac{k}{2} + k^{2/3}\right) \leq e^{-k^{4/3}/(3k/2)} \leq e^{-k^{1/3}/2} = o(1).$$

Therefore, there are $o(n)$ vertices with $l \geq k/2 + k^{2/3}$. It then follows from (9.20) that the expected number of non-isolated vertices in $\mathbb{G}_{n,\mathbf{P[k]}}$ is $o(n)$ and the Markov inequality then implies that this number is $o(n)$ w.h.p.

Next, when $\alpha + \beta = \beta + \gamma = 1$, which implies that $\alpha = \beta = \gamma = 1/2$, then random graph $\mathbb{G}_{n,\mathbf{P[k]}}$ is equivalent to $\mathbb{G}_{n,p}$ with $p = 1/n$ and so by Theorem 2.21 it does not have a component of order n, w.h.p.

To prove that the condition $(\alpha+\beta)(\beta+\gamma) > 1$ is sufficient, we show that the subgraph of $\mathbb{G}_{n,\mathbf{P[k]}}$ induced by the vertices of H of weight $l \geq k/2$ is connected w.h.p. This will suffice as there are at least $n/2$ such vertices. Notice that for any vertex $\mathbf{u} \in H$ its expected degree, by (9.19), is at least

$$((\alpha+\beta)(\beta+\gamma))^{k/2} \gg \log n. \tag{9.21}$$

We first show that for $\mathbf{u} \in V$,

$$\sum_{\mathbf{v} \in H} p_{\mathbf{uv}} \geq \frac{1}{4} \sum_{\mathbf{v} \in V} p_{\mathbf{uv}}. \tag{9.22}$$

For the given vertex $\mathbf{u}$ let l be the weight of $\mathbf{u}$. For a vertex $\mathbf{v}$ let $i(\mathbf{v})$ be the number of bits where $u_r = v_r = 1$, $r = 1,\dots,k$, while $j(\mathbf{v})$ stands for the number of bits where $u_r = 0$ and $v_r = 1$. Consider the partition

$$V \setminus H = S_1 \cup S_2 \cup S_3,$$

where

$$S_1 = \{\mathbf{v} : i(\mathbf{v}) \geq l/2,\ j(\mathbf{v}) < (k-l)/2\},$$

$$S_2 = \{\mathbf{v} : i(\mathbf{v}) < l/2,\ j(\mathbf{v}) \geq (k-l)/2\},$$

$$S_3 = \{\mathbf{v} : i(\mathbf{v}) < l/2,\ j(\mathbf{v}) < (k-l)/2\}.$$

Next, take a vertex $\mathbf{v} \in S_1$ and turn it into $\mathbf{v}'$ by flipping the bits of $\mathbf{v}$ which correspond to 0s of $\mathbf{u}$. Surely, $i(\mathbf{v}') = i(\mathbf{v})$ and

$$j(\mathbf{v}') \geq (k-l)/2 > j(\mathbf{v}).$$

Notice that the weight of $\mathbf{v}'$ is at least $k/2$ and so $\mathbf{v}' \in H$. Notice also that $\alpha \geq \beta \geq \gamma$ implies that $p_{\mathbf{uv}'} \geq p_{\mathbf{uv}}$. Different vertices $\mathbf{v} \in S_1$ map to different $\mathbf{v}'$. Hence

$$\sum_{\mathbf{v} \in H} p_{\mathbf{uv}} \geq \sum_{\mathbf{v} \in S_1} p_{\mathbf{uv}}. \tag{9.23}$$

The same bound (9.23) holds for S_2 and S_3 in place of S_1. To prove the same relationship for S_2 one has to flip the bits of $\mathbf{v}$ corresponding to 1s in $\mathbf{u}$, while for S_3 one has to flip all the bits of $\mathbf{v}$. Adding up these bounds over the partition of $V \setminus H$ we obtain

$$\sum_{\mathbf{v} \in V \setminus H} p_{\mathbf{uv}} \leq 3 \sum_{\mathbf{v} \in H} p_{\mathbf{uv}}$$

and so the bound (9.22) follows.

Notice that combining (9.22) with the bound given in (9.21) we get that for $\mathbf{u} \in H$ we have

$$\sum_{\mathbf{v} \in H} p_{\mathbf{uv}} > 2c \log n, \tag{9.24}$$

where c can be a large as needed.

We finish the proof by showing that a subgraph of $\mathbb{G}_{n,\mathbf{P}[\mathbf{k}]}$ induced by vertex set H is connected w.h.p. For that we make use of Theorem 9.3. So, we will show that for any cut $(S, H \setminus S)$

$$\sum_{\mathbf{u} \in S} \sum_{\mathbf{v} \in H \setminus S} p_{\mathbf{uv}} \geq c \log n.$$

Without loss of generality assume that vertex $\mathbf{1} \in S$. Equation (9.24) implies that for any vertex $\mathbf{u} \in H$ either

$$\sum_{\mathbf{v} \in S} p_{\mathbf{uv}} \geq c \log n, \tag{9.25}$$

or

$$\sum_{\mathbf{v} \in H \setminus S} p_{\mathbf{uv}} \geq c \log n. \tag{9.26}$$

If there is a vertex $\mathbf{u}$ such that (9.26) holds then since $\mathbf{u} \leq \mathbf{1}$ and $\alpha \geq \beta \geq \gamma$,

$$\sum_{\mathbf{u} \in S} \sum_{\mathbf{v} \in H \setminus S} p_{\mathbf{uv}} \geq \sum_{\mathbf{v} \in H \setminus S} p_{\mathbf{1v}} \geq \sum_{\mathbf{v} \in H \setminus S} p_{\mathbf{uv}} > c \log n.$$

Otherwise, (9.25) is true for every vertex $\mathbf{u} \in H$. Since at least one such vertex is in $H \setminus S$, we have

$$\sum_{\mathbf{u} \in S} \sum_{\mathbf{v} \in H \setminus S} p_{\mathbf{uv}} \geq c \log n,$$

and the Theorem follows. □

9.4 Exercises

9.4.1 Prove Theorem 9.3 (with $c = 10$) using the result of Karger and Stein [464] that in any weighted graph on n vertices the number of r-minimal cuts is $O\left((2n)^{2r}\right)$. (A cut $(S, V \setminus S), S \subseteq V$, in a weighted graph G is called *r-minimal* if its weight, i.e. the sum of weights of the edges connecting S with $V \setminus S$, is at most r times the weight of minimal weighted cut of G.)

9.4.2 Suppose that the entries of an $n \times n$ symmetric matrix A are all non-negative. Show that for any positive constants $c_1, c_2, \ldots, c_n$, the largest eigenvalue $\lambda(A)$ satisfies

$$\lambda(A) \leq \max_{1 \leq i \leq n} \left(\frac{1}{c_i} \sum_{j=1}^{n} c_j a_{i,j} \right).$$

9.4.3 Let A be the adjacency matrix of $\mathbb{G}_{n,\mathbf{P}^w}$ and for a fixed value of x let

$$c_i = \begin{cases} w_i & w_i > x \\ x & w_i \leq x \end{cases}.$$

Let $m = \max\{w_i : i \in [n]\}$. Let $X_i = \frac{1}{c_i} \sum_{j=1}^{n} c_j a_{i,j}$. Show that

$$\mathbb{E} X_i \leq \overline{w^2} + x \text{ and } \operatorname{Var} X_i \leq \frac{m}{x} \overline{w^2} + x.$$

9.4.4 Apply Theorem 21.10 with a suitable value of x to show that w.h.p.

$$\lambda(A) \le \overline{w^2} + (6(m\log n)^{1/2}(\overline{w^2} + \log n))^{1/2} + 3(m\log n)^{1/2}.$$

9.4.5 Show that if $\overline{w^2} > m^{1/2}\log n$ then w.h.p. $\lambda(A) = (1+o(1))\overline{w^2}$.

9.4.6 Suppose that $1 \le w_i \ll W^{1/2}$ for $1 \le i \le n$ and that $w_i w_j \overline{w^2} \gg W \log n$. Show that w.h.p. $diameter(\mathbb{G}_{n,\mathbf{P^w}}) \le 2$.

9.4.7 Prove, by the Second Moment Method, that if $\alpha + \beta = \beta + \gamma = 1$ then w.h.p. the number Z_d of the vertices of degree d in the random graph $\mathbb{G}_{n,\mathbf{P[k]}}$, is concentrated around its mean, i.e. $Z_d = (1+o(1))\mathbb{E}Z_d$.

9.4.8 Fix $d \in \mathbb{N}$ and let Z_d denote the number of vertices of degree d in the Kronecker random graph $\mathbb{G}_{n,\mathbf{P[k]}}$. Show that

$$\mathbb{E}Z_d = (1+o(1))\sum_{w=0}^{k}\binom{k}{w}\frac{(\alpha+\beta)^{dw}(\beta+\gamma)^{d(k-w)}}{d!}\times$$
$$\times \exp\left(-(\alpha+\beta)^w(\beta+\gamma)^{k-w}\right) + o(1).$$

9.4.9 Depending on the configuration of the parameters $0 < \alpha, \beta, \gamma < 1$, show that we have either

$$\mathbb{E}Z_d = \Theta\left(\left((\alpha+\beta)^d + (\beta+\gamma)^d\right)^k\right),$$

or

$$\mathbb{E}Z_d = o(2^k).$$

9.5 Notes

General model of inhomogeneous random graph

The most general model of inhomogeneous random graph was introduced by Bollobás, Janson and Riordan in their seminal paper [143]. They concentrate on the study of the phase transition phenomenon of their random graphs, which includes as special cases the models presented in this chapter as well as, among others, Dubins' model (see Kalikow and Weiss [459] and Durrett [261]), the mean-field scale-free model (see Riordan [645]), the CHKNS model (see Callaway, Hopcroft, Kleinberg, Newman and Strogatz [169]) and Turova's model (see [707], [708] and [709]).

The model of Bollobás, Janson and Riordan is an extension of one defined by Söderberg [685]. The formal description of their model is as follows. Consider a *ground space* being a pair $(\mathscr{S}, \mu)$, where $\mathscr{S}$ is a separable metric space and μ is a Borel probability measure on $\mathscr{S}$. Let $\mathscr{V} = (\mathscr{S}, \mu, (\mathbf{x}_n)_{n\ge1})$ be the *vertex space*, where $(\mathscr{S}, \mu)$ is a ground space and $(\mathbf{x}_n)_{n\ge1})$ is a random sequence $(x_1, x_2, \ldots, x_n)$ of n points of $\mathscr{S}$ satisfying the condition that for every μ-continuity set A, $A \subseteq \mathscr{S}$, $|\{i : x_i \in A\}|/n$ converges in probability to $\mu(A)$. Finally, let κ be a kernel on the vertex space $\mathscr{V}$ (understood here as a kernel on a ground space $(\mathscr{S}, \mu)$), i.e. a symmetric non-negative (Borel) measurable function on $S \times S$. Given the (random) sequence $(\mathbf{x}_1, \mathbf{x}_2, \ldots, \mathbf{x}_n)$ we let $G_{\mathscr{V}}(n, \kappa)$ be the random graph $G_{\mathscr{V}}(n, (p_{ij}))$ with $p_{ij} := \min\{\kappa(\mathbf{x}_i, \mathbf{x}_j)/n, 1\}$. In other words, $G_{\mathscr{V}}(n, \kappa)$ has n vertices and, given $\mathbf{x}_1, \mathbf{x}_2, \ldots, \mathbf{x}_n$, an edge $\{i,j\}$ (with $i \neq j$) exists with probability p_{ij}, independently for all other unordered pairs $\{i,j\}$.

Bollobás, Janson and Riordan presented in [143] a wide range of results describing various properties of the random graph $G_{\mathscr{V}}(n, \kappa)$. They gave a necessary and sufficient condition for the existence of a giant component, showed its uniqueness and determined the asymptotic number of edges in the giant component. They also studied the stability of the component, i.e. they showed that its size does not change much if we add or delete a few edges. They also established bounds on the size of small components, the asymptotic distribution of the number of vertices of given degree and studied the distances between vertices (diameter). Finally, they turn their attention to the phase transition of $G_{\mathscr{V}}(n, \kappa)$ where the giant component first emerges.

Janson and Riordan [434] studied the susceptibility, i.e. the mean size of the component containing a random vertex, in a general model of inhomogeneous random graphs. They related the susceptibility to a quantity associated with a corresponding branching process, and studied both quantities in various examples.

Devroye and Fraiman [237] found conditions for the connectivity of inhomogeneous random graphs with intermediate density. They draw n independent points X_i from a general distribution on a separable metric space, and let their indices form the vertex set of a graph. An edge ij is added with probability $\min\{1, \kappa(X_i, X_j) \log n/n\}$, where $\kappa > 0$ is a fixed kernel. They showed that, under reasonably weak assumptions, the connectivity threshold of the model can be determined.

Lin and Reinert [528] showed via a multivariate normal and a Poisson process approximation that, for graphs which have independent edges, with a possibly inhomogeneous distribution, only when the degrees are large can we reasonably approximate the joint counts for the number of vertices with given degrees as independent (note that in a random graph, such counts will typically be dependent). The proofs are based on Stein's method and the Stein–Chen method (see Section 20.3) with a new size-biased coupling for such inhomogeneous random graphs.

Rank one model

An important special case of the general model of Bollobás, Janson and Riordan is the so-called *rank one model*, where the kernel κ has the form $\kappa(x,y) = \psi(x)\psi(y)$, for some function $\psi > 0$ on $\mathscr{S}$. In particular, this model includes the Chung–Lu model (*expected degree model*) discussed earlier in this chapter. Recall that in their approach we attach edges (independently) with probabilities

$$p_{ij} = \min\left\{\frac{w_i w_j}{W}, 1\right\} \text{ where } W = \sum_{k=1}^{n} w_k.$$

Similarly, Britton, Deijfen and Martin-Löf [160] defined edge probabilities as

$$p_{ij} = \frac{w_i w_j}{W + w_i w_j},$$

while Norros and Reittu [608] attached edges with probabilities

$$p_{ij} = \exp\left(-\frac{w_i w_j}{W}\right).$$

For these models several characteristics are studied, such as the size of the giant component ([182], [183] and [608]) and its volume ([182]) as well as spectral properties ([184] and [185]). It should also be mentioned here that Janson [427] established conditions under which all those models are asymptotically equivalent.

Recently, van der Hofstad [408], Bhamidi, van der Hofstad and Hooghiemstra [87], van der Hofstad, Kliem and van Leeuwaarden [410] and Bhamidi, Sen and Wang [88] undertook systematic and detailed studies of various aspects of the rank one model in its general setting.

Finally, consider *random dot product graphs* (see Young [732] and Young and Scheinerman [733]) where a vector in $\mathbb{R}^d$ is assigned to each vertex and we allow each edge to be present with probability proportional to the inner product of the vectors assigned to its endpoints. The paper [733] treats these as models of social networks.

Kronecker Random Graph

Radcliffe and Young [640] analyzed the connectivity and the size of the giant component in a generalized version of the Kronecker random graph. Their results imply that the threshold for connectivity in $\mathbb{G}_{n,\mathbf{P[k]}}$ is $\beta + \gamma = 1$. Tabor [701] proved that it is also the threshold for a k-factor. Kang, Karoński, Koch and Makai [461] studied the asymptotic distribution of small subgraphs (trees and cycles) in $\mathbb{G}_{n,\mathbf{P[k]}}$.

Leskovec, Chakrabarti, Kleinberg and Faloutsos [526] and [527] have shown empirically that Kronecker random graphs resemble several real world networks. Later, Leskovec, Chakrabarti, Kleinberg, Faloutsos and Ghahramani [527] fitted the model to several real world networks such as the Internet, citation graphs and online social networks.

The R-MAT model, introduced by Chakrabarti, Zhan and Faloutsos [173], is closely related to the Kronecker random graph. The vertex set of this model is also $\mathbb{Z}_2^n$ and one also has parameters α, β, γ. However, in this case we need the additional condition that $\alpha + 2\beta + \gamma = 1$.

The process of generating a random graph in the R-MAT model creates a multigraph with m edges and then merges the multiple edges. The advantage of the R-MAT model over the random Kronecker graph is that it can be generated significantly faster when m is small. The degree sequence of this model has been studied by Groër, Sullivan and Poole [386] and by Seshadhri, Pinar and Kolda [678] when $m = \Theta(2^n)$, i.e. the number of edges is linear in the number of vertices. They have shown, as in Kang, Karoński, Koch and Makai [461] for $\mathbb{G}_{n,\mathbf{P}^{[k]}}$, that the degree sequence of the model does not follow a power law distribution. However, no rigorous proof exists for the equivalence of the two models and in the Kronecker random graph there is no restriction on the sum of the values of α, β, γ.

Further extensions of Kronecker random graphs can be found [108] and [109].

10
Fixed Degree Sequence

The graph $\mathbb{G}_{n,m}$ is chosen uniformly at random from the set of graphs with vertex set $[n]$ and m edges. It is of great interest to refine this model so that all the graphs chosen have a fixed degree sequence $\mathbf{d} = (d_1, d_2, \ldots, d_n)$. Of particular interest is the case where $d_1 = d_2 = \cdots = d_n = r$, i.e. the graph chosen is a uniformly random r-regular graph. It is not obvious how to do this and this is the subject of the current chapter. We discuss the configuration model in the next section and show its usefulness in (i) estimating the number of graphs with a given degree sequence and (ii) showing that w.h.p. random d-regular graphs are connected w.h.p., for $3 \leq d = O(1)$.

We finish by showing in Section 10.3 how for large r, $\mathbb{G}_{n,m}$ can be embedded in a random r-regular graph. This allows one to extend some results for $\mathbb{G}_{n,m}$ to the regular case.

10.1 Configuration Model

Let $\mathbf{d} = (d_1, d_2, \ldots, d_n)$ where $d_1 + d_2 + \cdots + d_n = 2m$ is even. Let

$$\mathcal{G}_{n,\mathbf{d}} = \{\text{simple graphs with vertex set } [n] \text{ s.t. degree } d(i) = d_i,\ i \in [n]\}$$

and let $\mathbb{G}_{n,\mathbf{d}}$ be chosen randomly from $\mathcal{G}_{n,\mathbf{d}}$. We assume that

$$d_1, d_2, \ldots, d_n \geq 1 \quad \text{and} \quad \sum_{i=1}^{n} d_i(d_i - 1) = \Omega(n).$$

We describe a generative model of $\mathbb{G}_{n,\mathbf{d}}$ due to Bollobás [122]. It is referred to as the *configuration model*. Let $W_1, W_2, \ldots, W_n$ be a partition of a set of *points* W, where $|W_i| = d_i$ for $1 \leq i \leq n$ and call the W_is *cells*, see Figure 10.1. We assume some total order $<$ on W and that $x < y$ if $x \in W_i, y \in W_j$ where $i < j$. For $x \in W$ define $\varphi(x)$ by $x \in W_{\varphi(x)}$. Let F be a partition of W into m pairs (a

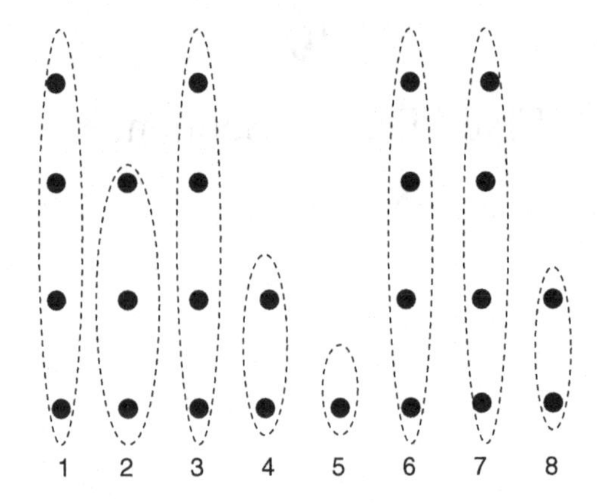

Figure 10.1. Partition of W into cells $W_1, \ldots, W_8$

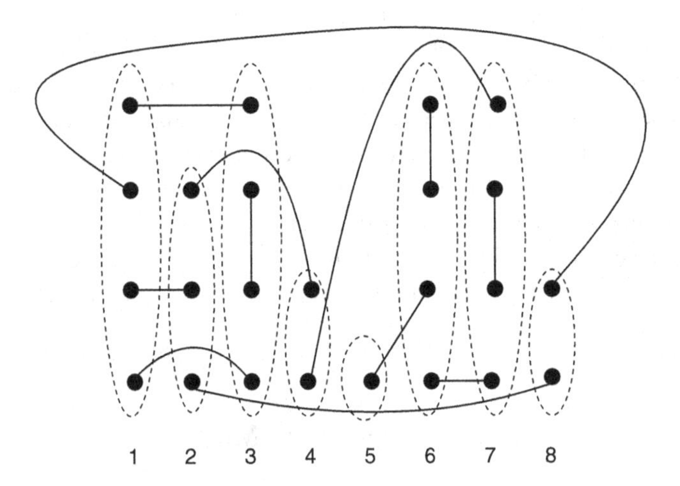

Figure 10.2. A partition F of W into $m = 12$ pairs

configuration, see Figures 10.2 and 10.3). Given F we define the (multi)graph $\gamma(F)$ as

$$\gamma(F) = ([n], \{(\varphi(x), \varphi(y)) : (x, y) \in F\}).$$

Let us consider the following example of $\gamma(F)$. Let $n = 8$ and $d_1 = 4, d_2 = 3, d_3 = 4, d_4 = 2, d_5 = 1, d_6 = 4, d_7 = 4, d_8 = 2$. The accompanying diagrams, Figures 10.1, 10.2 and 10.3 show a partition of W into $W_1, \ldots, W_8$, a configuration and its corresponding multi-graph.

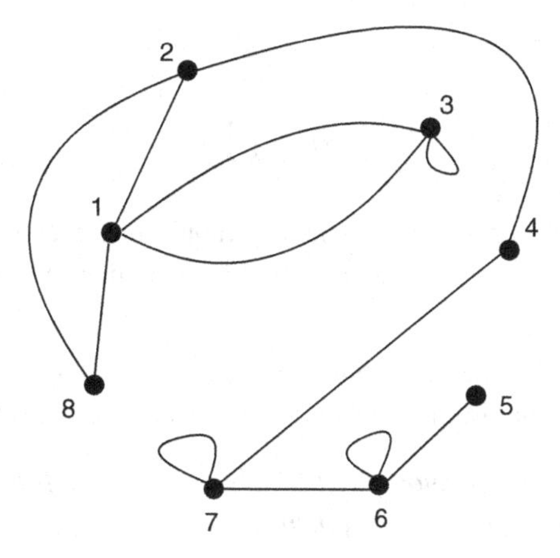

Figure 10.3. Graph $\gamma(F)$

Denote by Ω the set of all configurations defined above for $d_1+\cdots+d_n=2m$ and notice that

$$|\Omega| = \frac{(2m)!}{m!2^m} = (2m)!! \,. \qquad (10.1)$$

To see this, take d_i "distinct" copies of i for $i=1,2,\ldots,n$ and take a permutation $\sigma_1,\sigma_2,\ldots,\sigma_{2m}$ of these $2m$ symbols. Read off F, pair by pair $\{\sigma_{2i-1},\sigma_{2i}\}$ for $i=1,2,\ldots,m$. Each distinct F arises in $m!2^m$ ways.

We can also give an algorithmic construction of a random element F of the family Ω.

Algorithm *F-GENERATOR*
begin
$U \longleftarrow W$, $F \longleftarrow \emptyset$
for $i=1,2,\ldots,n$ **do**
begin
Choose x arbitrarily from U;
Choose y randomly from $U \setminus \{x\}$;
$F \longleftarrow F \cup \{(x,y)\}$;
$U \longleftarrow U \setminus \{(x,y)\}$
end
end

Note that F arises with probability $1/[(2m-1)(2m-3)\cdots 1] = |\Omega|^{-1}$.

Observe that the following relationship between a simple graph $G \in \mathcal{G}_{n,\mathbf{d}}$ and the number of configurations F for which $\gamma(F) = G$.

Lemma 10.1 *If $G \in \mathcal{G}_{n,\mathbf{d}}$, then*

$$|\gamma^{-1}(G)| = \prod_{i=1}^{n} d_i!\,.$$

Proof Arrange the edges of G in lexicographic order. Now go through the sequence of $2m$ symbols, replacing each i by a new member of W_i. We obtain all F for which $\gamma(F) = G$. □

The above lemma implies that we can use random configurations to "approximate" random graphs with a given degree sequence.

Corollary 10.2 *If F is chosen uniformly at random from the set of all configurations Ω and $G_1, G_2 \in \mathcal{G}_{n,\mathbf{d}}$ then*

$$\mathbb{P}(\gamma(F) = G_1) = \mathbb{P}(\gamma(F) = G_2).$$

So instead of sampling from the family $\mathcal{G}_{n,\mathbf{d}}$ and counting graphs with a given property, we can choose a random F and accept $\gamma(F)$ if and only if there are no loops or multiple edges, i.e. if and only if $\gamma(F)$ is a simple graph.

This is only a useful exercise if $\gamma(F)$ is simple with sufficiently high probability. We assume for the remainder of this section that

$$\Delta = \max\{d_1, d_2, \ldots, d_n\} \leq n^{1/6}.$$

We will prove later (see Lemma 10.6 and Corollary 10.7) that if F is chosen uniformly (at random) from Ω,

$$\mathbb{P}(\gamma(F) \text{ is simple}) = (1 + o(1))e^{-\lambda(\lambda+1)}, \tag{10.2}$$

where

$$\lambda = \frac{\sum d_i(d_i - 1)}{2\sum d_i}.$$

Hence, (10.1) and (10.2) tell us not only how large is $\mathcal{G}_{n,\mathbf{d}}$, (Theorem 10.4) but also lead to the following conclusion.

Theorem 10.3 *Suppose that $\Delta \leq n^{1/6}$. For any (multi)graph property $\mathcal{P}$*

$$\mathbb{P}(\mathbb{G}_{n,\mathbf{d}} \in \mathcal{P}) \leq (1 + o(1))e^{\lambda(\lambda+1)}\,\mathbb{P}(\gamma(F) \in \mathcal{P}).$$

The above statement is particularly useful if $\lambda = O(1)$, e.g. for random r-regular graphs, where r is a constant, since then $\lambda = \frac{r-1}{2}$. In the next section we apply the above result to establish the connectedness of random regular graphs.

Bender and Canfield [70] gave an asymptotic formula for $|\mathscr{G}_{n,\mathbf{d}}|$ when $\Delta = O(1)$. The paper [122] by Bollobás gives the same asymptotic formula when $\Delta < (2\log n)^{1/2}$. The following theorem allows for some more growth in Δ. Its proof uses the notion of *switching* introduced by McKay [569] and McKay and Wormald [570].

Theorem 10.4 *Suppose that* $\Delta \leq n^{1/6}$.

$$|\mathscr{G}_{n,\mathbf{d}}| \approx e^{-\lambda(\lambda+1)} \frac{(2m)!!}{\prod_{i=1}^{n} d_i!}.$$

In preparation we first prove

Lemma 10.5 *Suppose that* $\Delta \leq n^{1/6}$. *Let F be chosen uniformly (at random) from* Ω. *Then w.h.p.* $\gamma(F)$ *has*

(*a*) *no double loops,*
(*b*) *at most* $\Delta \log n$ *loops,*
(*c*) *no adjacent double edges,*
(*d*) *no triple edges.*
(*e*) *at most* $\Delta^2 \log n$ *double edges,*

Proof We will use the following inequality repeatedly.

Let $f_i = \{x_i, y_i\}, i = 1,2,\ldots,k$ be k pairwise disjoint pairs of points. Then,

$$\mathbb{P}(f_i \in F, i = 1,2,\ldots,k) \leq \frac{1}{(2m-2k)^k}. \tag{10.3}$$

This follows immediately from

$$\mathbb{P}(f_i \in F \mid f_1, f_2, \ldots, f_{i-1} \in F) = \frac{1}{2m-2i+1}.$$

This follows from considering **Algorithm** *F-GENERATOR* with $x = x_i$ and $y = y_i$ in the main loop.

(a) Using (10.3) we obtain

$$\begin{aligned}\mathbb{P}(F \text{ contains a double loop}) &\leq \sum_{i=1}^{n} 3\binom{d_i}{4}\left(\frac{1}{2m-4}\right)^2 \\ &\leq n\Delta^4 m^{-2} = o(1).\end{aligned}$$

(b) Let $k_1 = \Delta \log n$.

$$
\begin{aligned}
&\mathbb{P}(F \text{ has at least } k_1 \text{ loops}) \\
&\leq o(1) + \sum_{\substack{x_1+\cdots+x_n=k_1,\\ x_i=0,1}} \prod_{i=1}^{n} \left(\binom{d_i}{2} \cdot \frac{1}{2m-2k_1} \right)^{x_i} \\
&\leq o(1) + \left(\frac{\Delta}{2m} \right)^{k_1} \sum_{\substack{x_1+\cdots+x_n=k_1,\\ x_i=0,1}} \prod_{i=1}^{n} d_i^{x_i} \\
&\leq o(1) + \left(\frac{\Delta}{2m} \right)^{k_1} \frac{(d_1+\cdots+d_n)^{k_1}}{k_1!} \\
&\leq o(1) + \left(\frac{\Delta e}{k_1} \right)^{k_1} \\
&= o(1).
\end{aligned} \tag{10.4}
$$

The $o(1)$ term in (10.4) accounts for the probability of having a double loop.

(c)

$$
\begin{aligned}
&\mathbb{P}(F \text{ contains two adjacent double edges}) \\
&\leq \sum_{i=1}^{n} \binom{d_i}{2}^2 \left(\frac{\Delta}{2m-8} \right)^2 \\
&\leq \frac{\Delta^5}{(2m-8)^2} \sum_{i=1}^{n} d_i \\
&= o(1).
\end{aligned}
$$

(d)

$$
\begin{aligned}
&\mathbb{P}(F \text{ contains a triple edge}) \\
&\leq \sum_{1 \leq i < j \leq n} 6 \binom{d_i}{3} \binom{d_j}{3} \left(\frac{1}{2m-6} \right)^3 \\
&\leq \Delta^4 \left(\sum_{1 \leq i \leq n} d_i \right)^2 m^{-3} \\
&= o(1).
\end{aligned}
$$

(e) Let $k_2 = \Delta^2 \log n$.

$$
\begin{aligned}
&\mathbb{P}(F \text{ has at least } k_2 \text{ double edges})\\
&\leq o(1) + \sum_{\substack{x_1+\cdots+x_n=k_2,\\ x_i=0,1}} \prod_{i=1}^{n} \left(\binom{d_i}{2} \cdot \frac{\Delta}{2m-4k_2}\right)^{x_i}\\
&\leq o(1) + \left(\frac{\Delta^2}{m}\right)^{k_2} \sum_{\substack{x_1+\cdots+x_n=k_2,\\ x_i=0,1}} \prod_{i=1}^{n} d_i^{x_i}\\
&\leq o(1) + \left(\frac{\Delta^2}{m}\right)^{k_2} \frac{(d_1+\cdots+d_n)^{k_2}}{k_2!}\\
&\leq o(1) + \left(\frac{2\Delta^2 e}{k_2}\right)^{k_2}\\
&= o(1).
\end{aligned} \tag{10.5}
$$

The $o(1)$ term in (10.5) accounts for adjacent multiple edges and triple edges. The $\Delta/(2m-4k_2)$ term can be justified as follows: We have chosen two points x_1, x_2 in W_i in $\binom{d_i}{2}$ ways and this term bounds the probability that x_2 chooses a partner in the same cell as x_1. □

Let now $\Omega_{i,j}$ be the set of all $F \in \Omega$ such that F has i loops; j double edges; and no double loops or triple edges and no vertex incident with two double edges. The notation $\tilde{O}$ ignores factors of order $(\log n)^{O(1)}$.

Lemma 10.6 Switching lemma *Suppose that* $\Delta \leq n^{1/6}$. *Let* $M_1 = 2m$ *and* $M_2 = \sum_i d_i(d_i-1)$. *For* $i \leq k_1$ *and* $j \leq k_2$, *where* $k_1 = \Delta \log n$ *and* $k_2 = \Delta^2 \log n$,

$$\frac{|\Omega_{i-1,j}|}{|\Omega_{i,j}|} = \frac{2iM_1}{M_2}\left(1 + \tilde{O}\left(\frac{\Delta^3}{n}\right)\right),$$

and

$$\frac{|\Omega_{0,j-1}|}{|\Omega_{0,j}|} = \frac{4jM_1^2}{M_2^2}\left(1 + \tilde{O}\left(\frac{\Delta^3}{n}\right)\right).$$

The corollary that follows is an immediate consequence of the switching lemma. It immediately implies Theorem 10.4.

Corollary 10.7 *Suppose that* $\Delta \leq n^{1/6}$, *then,*

$$\frac{|\Omega_{0,0}|}{|\Omega|} = (1+o(1))e^{-\lambda(\lambda+1)},$$

where

$$\lambda = \frac{M_2}{2M_1}.$$

Proof It follows from the switching lemma that $i \le k_1$ and $j \le k_2$ implies

$$\frac{|\Omega_{i,j}|}{|\Omega_{0,0}|} = (1+\tilde{O}(n^{-1/2}))\frac{\lambda^{i+2j}}{i!\,j!}.$$

Therefore, because $\lambda = o(k_1)$ and $k_1 \le k_2$, we have

$$\begin{aligned}(1-o(1))|\Omega| &= (1+o(1))|\Omega_{0,0}|\sum_{i=0}^{k_1}\sum_{j=0}^{k_2}\frac{\lambda^{i+2j}}{i!\,j!}\\ &= (1+o(1))|\Omega_{0,0}|e^{\lambda(\lambda+1)}.\end{aligned}$$

□

To prove the switching lemma we need to introduce two specific operations on configurations, called an "l-switch" and a "d-switch."

Figure 10.4 illustrates the "loop removal switch" ("l-switch") operation. Here we have six points $x_1,x_2,\dots,x_6$ and three pairs $\{x_1,x_6\}$, $\{x_2,x_3\}$, $\{x_4,x_5\}$ from five different cells, where x_2 and x_3 are in the same cell. We assume that $x_2 < x_3$. The l-switch operation replaces these pairs by a new set of pairs: $\{x_1,x_2\}$, $\{x_3,x_4\}$, $\{x_5,x_6\}$ and, in this operation, no new loops or multiple edges are created.

In general, an l-switch operation takes F, a member of $\Omega_{i,j}$, to F', a member $\Omega_{i-1,j}$, see Figure 10.5. To estimate the number of choices η for a loop removal switch we note that for a forward switching operation,

$$iM_1^2 - \tilde{O}(iM_1\Delta^2) \le \eta \le iM_1^2, \tag{10.6}$$

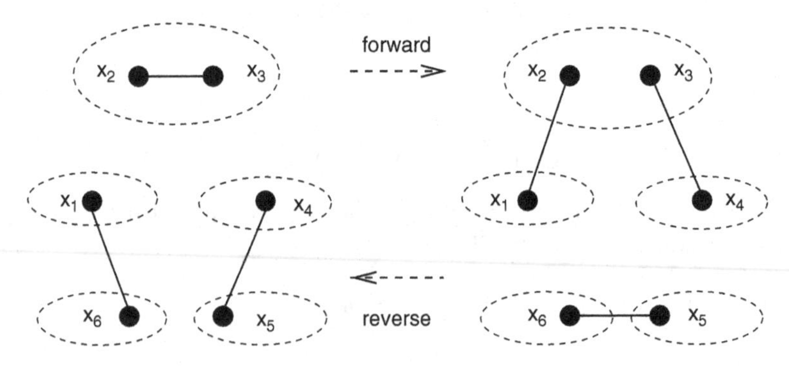

Figure 10.4. l-switch

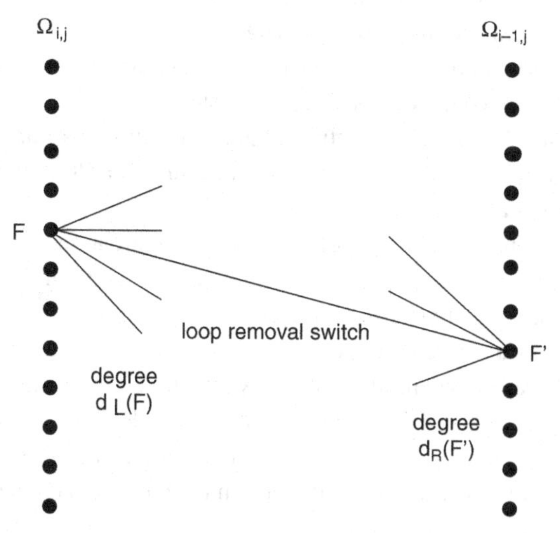

Figure 10.5. Moving between $\Omega_{i,j}$ and $\Omega_{i-1,j}$

while, for the reverse procedure, taking $F' \in \Omega_{i-1,j}$ to $F \in \Omega_{i,j}$, we have

$$M_1M_2/2 - \tilde{O}(\Delta^3 M_1) \leq \eta \leq M_1M_2/2. \tag{10.7}$$

Proof of (10.6): In the case of the forward loop removal switch, given $\Omega_{i,j}$, we can choose points x_1 and x_4 in $M_1(M_1-1)$ ways and the point x_2 in i ways, giving the upper bound in (10.6). For some choices of x_1, x_2 and x_4 the switch does not lead to a member of $\Omega_{i-1,j}$, when the switch itself cannot be performed due to the creation of or destruction of other loops or multiple pairs (edges). The number of such "bad" choices has to be subtracted from iM_1^2. We estimate those cases by $\tilde{O}(iM_1\Delta^2)$ as follows: the factor i accounts for our choice of loop to destroy. So, consider a fixed loop $\{x_2, x_3\}$. A forward switch will be good unless

(i) $\{x_1, x_6\}$ **or** $\{x_4, x_5\}$ **form a loop or multiple edge.**
There are at most k_1 choices ($2k_2$, respectively) for $\{x_1, x_6\}$ as a loop (multiple edge, respectively) and then $2M_1$ choices for x_4, x_5. This removes at most $2k_1M_1$ ($4k_2M_1$, respectively) choices. Thus, all in all, this case accounts for $\tilde{O}(M_1\Delta^2)$ choices.

(ii) $\{x_1, x_2\}$ **or** $\{x_3, x_4\}$ **form a loop.**
Having chosen x_2 there are at most Δ choices for x_1 to make a loop with it. After this there are at most M_1 choices for x_4. Thus, all in all, this case accounts for at most $2\Delta M_1$ choices.

(iii) $\{x_1,x_2\}$ **or** $\{x_3,x_4\}$ **form a multiple edge.**
Having chosen x_2 there are at most Δ^2 choices for x_1 to make a multiple edge with it. Indeed, we choose x_2' in the same cell as x_2. Then we choose x_1 in the same cell as the partner of x_2'. After this there are at most M_1 choices for x_4. Thus, all in all, this case accounts for $O(\Delta^2 M_1)$ choices.

(iv) $\{x_5,x_6\}$ **forms a loop.**
Having chosen x_1,x_6, in at most M_1 ways, there are at most Δ choices for x_5 in the same cell as x_6 that make $\{x_5,x_6\}$ a loop. Thus, all in all, this case accounts for at most $O(\Delta M_1)$ choices.

(v) $\{x_5,x_6\}$ **forms a multiple edge**.
Having chosen x_1,x_6, in at most M_1 ways, there are at most Δ^2 choices for $\{x_4,x_5\}$ that make $\{x_5,x_6\}$ a multiple edge. Indeed, we choose x_6' in the same cell as x_6. Then we choose x_5 in the same cell as the partner of x_6'. Thus, all in all, this case accounts for at most $O(\Delta^2 M_1)$ choices.

Proof of (10.7): For the reverse removal switch, to obtain an upper bound, we can choose a pair $\{x_2,x_3\}$ contained in a cell in $M_2/2$ ways and then x_5,x_6 in M_1 ways. For the lower bound, there are several things that can go wrong and not allow the move from $\Omega_{i-1,j}$ to $\Omega_{i,j}$:

(i) **The cell containing** x_2,x_3 **contains a loop of** F**.**
This creates a double loop. There are i choices for $\{x_2,x_3\}$ and given this, there are at most M_1 choices for x_5. Thus, all in all, this case accounts for $O(k_1 M_1)$ choices.

(ii) $\{x_5,x_6\}$ **is a loop or multiple edge of** F**.**
In this case there are at most k_1+k_2 choices for $\{x_5,x_6\}$ and then at most M_1 choices for x_2,x_3. Thus, all in all, this case accounts for $O((k_1+k_2)M_1)$ choices.

(iii) $\{x_1,x_6\}$ **or** $\{x_4,x_5\}$ **become loops.**
In this case there are at most M_1 choices for $\{x_5,x_6\}$ and then at most Δ choices for x_1 and then at most Δ choices for x_3, if the loop created is $\{x_1,x_6\}$. Thus, all in all, this case accounts for $O(\Delta^2 M_1)$ choices.

(iv) $\{x_1,x_6\}$ **or** $\{x_4,x_5\}$ **become multiple edges.**
In this case there are at most M_1 choices for $\{x_5,x_6\}$ and then at most Δ^2 choices for x_1. Indeed we choose x_6' in the same cell as x_6 and then x_1 in the same cell as the partner x_1' of x_5'. After this there are at most Δ choices for x_3, if the multiple edge created is $\{x_1,x_6\}$. Thus, all in all, this case accounts for $O(\Delta^3 M_1)$ choices.

This completes the proof of (10.6) and (10.7).

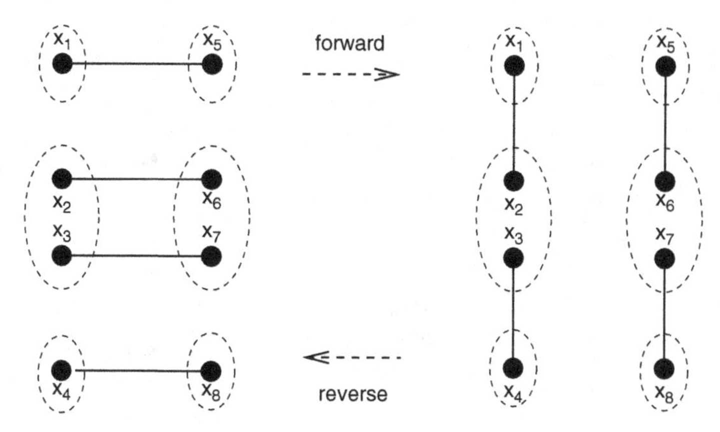

Figure 10.6. d-switch

Now for $F \in \Omega_{i,j}$ let $d_L(F)$ denote the number of $F' \in \Omega_{i-1,j}$ that can be obtained from F by an l-switch. Similarly, for $F' \in \Omega_{i-1,j}$ let $d_R(F')$ denote the number of $F \in \Omega_{i,j}$ that can be transformed into F' by an l-switch. Then,

$$\sum_{F \in \Omega_{i,j}} d_L(F) = \sum_{F' \in \Omega_{i-1,j}} d_R(F').$$

But

$$iM_1^2 |\Omega_{i,j}|(1 - \tilde{O}(\Delta^2/M_1)) \leq \sum_{F \in \Omega_{i,j}} d_L(F) \leq iM_1^2 |\Omega_{i,j}|,$$

while

$$\frac{1}{2} M_1 M_2 |\Omega_{i-1,j}| \left(1 - \tilde{O}\left(\frac{\Delta^3}{M_2}\right)\right) \leq \sum_{F' \in \Omega_{i-1,j}} d_R(F') \leq \frac{1}{2} M_1 M_2 |\Omega_{i-1,j}|.$$

So,

$$\frac{|\Omega_{i-1,j}|}{|\Omega_{i,j}|} = \frac{2iM_1}{M_2} \left(1 + \tilde{O}\left(\frac{\Delta^3}{n}\right)\right),$$

which shows that the first statement of the Switching Lemma holds.

Now consider the second operation on configurations, described as a "double edge removal switch"("d-switch"), Figure 10.6. Here we have eight points $x_1, x_2, \ldots, x_8$ from six different cells, where x_2 and x_3 are in the same cell, as are x_6 and x_7. Take pairs $\{x_1, x_5\}$, $\{x_2, x_6\}$, $\{x_3, x_7\}$ and $\{x_4, x_8\}$, where $\{x_2, x_6\}$ and $\{x_3, x_7\}$ are double pairs (edges). The d-switch operation replaces these pairs by a set of new pairs: $\{x_1, x_2\}$, $\{x_3, x_4\}$, $\{x_5, x_6\}$ and $\{x_7, x_8\}$ and, in this operation, none of the pairs created or destroyed is allowed to be a part of a multiple pair, except the double pair $\{x_2, x_6\}$ and $\{x_3, x_7\}$. Note also that those

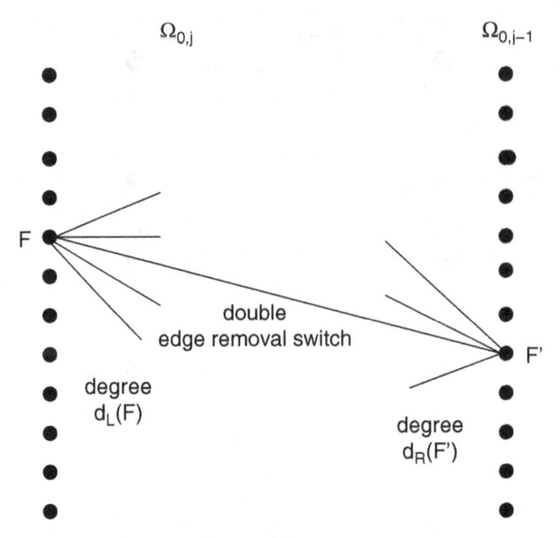

Figure 10.7. Moving between $\Omega_{0,j}$ and $\Omega_{0,j-1}$

four new pairs (edges) form two 2-paths. In general, a d-switch operation takes a member F of $\Omega_{0,j}$ to a member F' of $\Omega_{0,j-1}$, unless it creates new loops or double edges. In particular, it takes a member of $\Omega_{0,j}$ to a member of $\Omega_{0,j-1}$ (see Figure 10.7). We estimate the number of choices η during a double edge removal switch of $F \in \Omega_{0,j}$. For a forward switching operation

$$jM_1^2 - \tilde{O}(jM_1\Delta^2) \le \eta \le jM_1^2, \tag{10.8}$$

while, for the reverse procedure,

$$M_2^2/4 - \tilde{O}(M_2\Delta^3) \le \eta \le M_2^2/4. \tag{10.9}$$

Proof of (10.8): To see why the above bounds hold, note that in the case of the forward double edge removal switch, for each of j double edges we have $M_1(M_1 - 1)$ choices of two additional edges. To get the lower bound we subtract the number of "bad" choices as in the case of the forward operation of the l-switch above. We can enumerate these bad choices as follows: The factor j accounts for our choice of double edge to destroy. So we consider a fixed double edge.

(a) $\{x_1,x_5\}$ **or** $\{x_4,x_8\}$ **are part of a double pair.**
There are at most $4j$ choices for part of a pair and after this there are at most M_1 choices for the other edge. All in all, this case accounts for $O(k_2M_1)$ choices.

(b) **One of the edges created, $\{x_1,x_2\}$ etc., forms a loop.**
Given x_2 there are at most Δ choices of x_1 that will make a loop. Given this, there are at most M_1 choices for the other edge. All in all, this case accounts for $O(\Delta M_1)$ choices.

(c) **One of the edges created, $\{x_1,x_2\}$ etc., forms part of a double edge.**
Given x_2 there are at most Δ^2 choices of x_1 that will make a double edge. Indeed, we choose x_2' in the same cell as x_2 and then x_1 in the same cell as the point x_1' that is paired with x_2'. Given this, there are at most M_1 choices for the other edge. All in all, this case accounts for at most $O(\Delta^2 M_1)$ choices.

Proof of (10.9): In the reverse procedure, we choose a pair $\{x_2,x_3\}$ in $M_2/2$ ways and a pair $\{x_6,x_7\}$ also in $M_2/2$ ways, to arrive at the upper bound. For the lower bound, as in the reverse l-switch, there are several things that can go wrong and not allow the move from $\Omega_{0,j-1}$ to $\Omega_{0,j}$.

(a) $\{x_2,x_3\}$ **or** $\{x_6,x_7\}$ **are part of a double edge.**
There are $2k_2$ choices for $\{x_2,x_3\}$ as part of a double edge and then there are at most $M_2/2$ choices for the other edge. All in all, this case accounts for $O(k_2 M_2)$ choices.

(b) $\{x_1,x_5\}$ **or** $\{x_4,x_8\}$ **form loops.**
Our choice of the pair $\{x_2,x_3\}$, in M_2 ways, determines x_1 and x_4. Then there are at most Δ choices for x_5 in the same cell as x_1. Choosing x_5 determines x_6 and now there are only at most Δ choices left for x_7. All in all, this case accounts for at most $O(\Delta^2 M_2)$ choices.

(c) $\{x_1,x_5\}$ **or** $\{x_4,x_8\}$ **are part of a double edge.**
Given our choice of the pair $\{x_2,x_3\}$ there are at most Δ^2 choices for x_5 that make $\{x_1,x_5\}$ part of a double edge. Namely, choose x_1' in the same cell as x_1 and then x_5 in the same cell as the partner x_5' of x_1'. Finally, there will be at most Δ choices for x_7. All in all, this case accounts for at most $O(\Delta^3 M_2)$ choices.

Hence, the lower bound for the reverse procedure holds. Now for $F \in \Omega_{0,j}$ let $d_L(F)$ denote the number of $F' \in \Omega_{0,j-1}$ that can be obtained from F by a d-switch. Similarly, for $F' \in \Omega_{0,j-1}$ let $d_R(F')$ denote the number of $F \in \Omega_{0,j}$ that can be transformed into F' by a d-switch. Then,

$$\sum_{F \in \Omega_{0,j}} d_L(F) = \sum_{F' \in \Omega_{0,j-1}} d_R(F').$$

But

$$jM_1^2|\Omega_{0,j}|(1-\tilde{O}(\Delta^2/M_1)) \le \sum_{F\in\Omega_{0,j}} d_L(F) \le jM_1^2|\Omega_{0,j}|,$$

while

$$\frac{1}{4}M_2^2|\Omega_{0,j-1}|\left(1-\tilde{O}\left(\frac{\Delta^3}{M_2}\right)\right) \le \sum_{F'\in\Omega_{0,j-1}} d_R(F') \le \frac{1}{4}M_2^2|\Omega_{0,j-1}|.$$

So

$$\frac{|\Omega_{0,j-1}|}{|\Omega_{0,j}|} = \frac{4jM_1^2}{M_2^2}\left(1+\tilde{O}\left(\frac{\Delta^3}{n}\right)\right).$$

□

10.2 Connectivity of Regular Graphs

For an excellent survey of results on random regular graphs, see Wormald [728].

Bollobás [122] used the configuration model to prove the following: let $\mathbb{G}_{n,r}$ denote a random r-regular graph with vertex set $[n]$ and $r \ge 3$ constant.

Theorem 10.8 *$\mathbb{G}_{n,r}$ is r-connected, w.h.p.*

Since an r-regular, r-connected graph, with n even, has a perfect matching, the above theorem immediately implies the following corollary.

Corollary 10.9 *Let $\mathbb{G}_{n,r}$ be a random r-regular graph, $r \ge 3$ constant, with vertex set $[n]$ even. Then w.h.p. $\mathbb{G}_{n,r}$ has a perfect matching.*

Proof (*of Theorem 10.8*)
Partition the vertex set $V=[n]$ of $\mathbb{G}_{n,r}$ into three parts, K, L and $V\setminus(K\cup L)$, such that $L=N(K)$, i.e. such that L separates K from $V\setminus(K\cup L)$ and $|L|=l\le r-1$. We will show that w.h.p. there are no such K, L for k ranging from 2 to $n/2$. We will use the configuration model and the relationship stated in Theorem 10.3. We will divide the whole range of k into three parts.

(i) $2\le k\le 3$.

Put $S:=K\cup L$, $s=|S|=k+l\le r+2$. The set S contains at least $2r-1$ edges ($k=2$) or at least $3r-3$ edges ($k=3$). In both cases this is at least

$s+1$ edges.

$$
\begin{aligned}
&\mathbb{P}(\exists S, s=|S| \le r+2 : S \text{ contains } s+1 \text{ edges}) \\
&\le \sum_{s=4}^{r+2} \binom{n}{s} \binom{rs}{s+1} \left(\frac{rs}{rn}\right)^{s+1} \qquad (10.10)\\
&\le \sum_{s=4}^{r+2} n^s 2^{rs} s^{s+1} n^{-s-1} \\
&= o(1).
\end{aligned}
$$

Explanation for (10.10): Having chosen a set of s vertices, spanning rs points R, we choose $s+1$ of these points T, say $\frac{rs}{rn}$ bounds the probability that one of these points in T is paired with something in a cell associated with S. This bound holds conditional on other points of R being so paired.

(ii) $4 \le k \le ne^{-10}$.

The number of edges incident with the set K, $|K| = k$, is at least $(rk+l)/2$. Indeed, let a be the number of edges contained in K and b be the number of $K : L$ edges. Then $2a + b = rk$ and $b \ge l$. This gives $a + b \ge (rk+l)/2$. So,

$$
\begin{aligned}
\mathbb{P}(\exists K, L) &\le \sum_{k=4}^{ne^{-10}} \sum_{l=0}^{r-1} \binom{n}{k} \binom{n}{l} \binom{rk}{\frac{rk+l}{2}} \left(\frac{r(k+l)}{rn}\right)^{(rk+l)/2} \\
&\le \sum_{k=4}^{ne^{-10}} \sum_{l=0}^{r-1} n^{-\left(\frac{r}{2}-1\right)k+\frac{l}{2}} \frac{e^{k+l}}{k^k l^l} 2^{rk} (k+l)^{(rk+l)/2}.
\end{aligned}
$$

Now

$$
\left(\frac{k+l}{l}\right)^{l/2} \le e^{k/2} \quad \text{and} \quad \left(\frac{k+l}{k}\right)^{k/2} \le e^{l/2},
$$

and so

$$
(k+l)^{(rk+l)/2} \le l^{l/2} k^{rk/2} e^{(lr+k)/2}.
$$

Therefore, with C_r a constant,

$$\begin{aligned}\mathbb{P}(\exists K,L) &\le C_r \sum_{k=4}^{ne^{-10}} \sum_{l=0}^{r-1} n^{-(\frac{r}{2}-1)k+\frac{l}{2}} e^{3k/2} 2^{rk} k^{(r-2)k/2} \\ &= C_r \sum_{k=4}^{ne^{-10}} \sum_{l=0}^{r-1} \left(n^{-(\frac{r}{2}-1)+\frac{l}{2k}} e^{3/2} 2^r k^{\frac{r}{2}-1}\right)^k \\ &= o(1).\end{aligned}$$

(iii) $ne^{-10} < k \le n/2$

Assume that there are $rl-a$ edges between sets L and $V \setminus (K \cup L)$. Denote also

$$\varphi(2m) = \frac{(2m)!}{m!\,2^m} \approx 2^{1/2}\left(\frac{2m}{e}\right)^m.$$

Then, remembering that $r,l,a = O(1)$ we can estimate that

$$\begin{aligned}&\mathbb{P}(\exists K,L) \\ &\le \sum_{k,l,a} \binom{n}{k}\binom{n}{l}\binom{rl}{a} \frac{\varphi(rk+rl-a)\varphi(r(n-k-l)+a)}{\varphi(rn)} \qquad (10.11) \\ &\le C_r \sum_{k,l,a} \left(\frac{ne}{k}\right)^k \left(\frac{ne}{l}\right)^l \times \\ &\qquad \frac{(rk+rl-a)^{rk+rl-a}(r(n-k-l)+a)^{r(n-k-l)+a}}{(rn)^{rn}} \\ &\le C_r' \sum_{k,l,a} \left(\frac{ne}{k}\right)^k \left(\frac{ne}{l}\right)^l \frac{(rk)^{rk+rl-a}(r(n-k))^{r(n-k-l)+a}}{(rn)^{rn}} \\ &\le C_r'' \sum_{k,l,a} \left(\frac{ne}{k}\right)^k \left(\frac{ne}{l}\right)^l \left(\frac{k}{n}\right)^{rk} \left(1-\frac{k}{n}\right)^{r(n-k)} \\ &\le C_r'' \sum_{k,l,a} \left(\left(\frac{k}{n}\right)^{r-1} e^{1-r/2} n^{r/k}\right)^k \\ &= o(1).\end{aligned}$$

Explanation of (10.11): Having chosen K,L we choose a points in $W_{K\cup L} = \bigcup_{i\in K\cup L} W_i$ that will be paired outside $W_{K\cup L}$. This leaves $rk+rl-a$ points in $W_{K\cup L}$ to be paired up in $\varphi(rk+rl-a)$ ways and then the remaining points can be paired up in $\varphi(r(n-k-l)+a)$ ways. We then multiply by the probability $1/\varphi(rn)$ of the final pairing. □

10.3 $\mathbb{G}_{n,r}$ versus $\mathbb{G}_{n,p}$

The configuration model is most useful when the maximum degree is bounded. When r is large, we can learn a lot about random r-regular graphs from the following theorem of Kim and Vu [483]. They proved that if $\log n \ll r \ll n^{1/3}/(\log n)^2$ then there is a joint distribution $G_0, G = \mathbb{G}_{n,r}, G_1$ such that w.h.p. (i) $G_0 \subseteq G$, (ii) the maximum degree $\Delta(G_1 \setminus G) \leq \frac{(1+o(1))\log n}{\log(\varphi(r)/\log n)}$ where $\varphi(r)$ is any function satisfying $(r \log n)^{1/2} \leq \varphi(r) \ll r$. Here $G_i = \mathbb{G}_{n,p_i}, i = 0, 1$, where $p_0 = (1-o(1))\frac{r}{n}$ and $p_1 = (1+o(1))\frac{r}{n}$. In this way we can deduce properties of $\mathbb{G}_{n,r}$ from $\mathbb{G}_{n,r/n}$. For example, G_0 is Hamiltonian w.h.p. implies that $\mathbb{G}_{n,r}$ is Hamiltonian w.h.p. Recently, Dudek, Frieze, Ruciński and Šilekis [256] have increased the range of r for which (i) holds. The cited paper deals with random hypergraphs and here we describe the simpler case of random graphs.

Theorem 10.10 *There is a positive constant C such that if*

$$C\left(\frac{r}{n} + \frac{\log n}{r}\right)^{1/3} \leq \gamma = \gamma(n) < 1,$$

and $m = \lfloor (1-\gamma)nr/2 \rfloor$, then there is a joint distribution of $\mathbb{G}_{n,m}$ and $\mathbb{G}_{n,r}$ such that

$$\mathbb{P}(\mathbb{G}_{n,m} \subset \mathbb{G}_{n,r}) \to 1.$$

Corollary 10.11 *Let $\mathscr{Q}$ be an increasing property of graphs such that $\mathbb{G}_{n,m}$ satisfies $\mathscr{Q}$ w.h.p. for some $m = m(n)$, $n \log n \ll m \ll n^2$. Then $\mathbb{G}_{n,r}$ satisfies $\mathscr{Q}$ w.h.p. for $r = r(n) \approx \frac{2m}{n}$.*

Our approach to proving Theorem 10.10 is to represent $\mathbb{G}_{n,m}$ and $\mathbb{G}_{n,r}$ as the outcomes of two graph processes that behave similarly enough to permit a good coupling. For this let $M = nr/2$ and define

$$\mathbb{G}_M = (\varepsilon_1, \ldots, \varepsilon_M)$$

to be an ordered random uniform graph on the vertex set $[n]$, i.e. $\mathbb{G}_{n,M}$ with a random uniform ordering of edges. Similarly, let

$$\mathbb{G}_r = (\eta_1, \ldots, \eta_M)$$

be an ordered random r-regular graph on $[n]$, i.e. $\mathbb{G}_{n,r}$ with a random uniform ordering of edges. Further, write $\mathbb{G}_M(t) = (\varepsilon_1, \ldots, \varepsilon_t)$ and $\mathbb{G}_r(t) = (\eta_1, \ldots, \eta_t)$, $t = 0, \ldots, M$.

For every ordered graph G of size t and every edge $e \in K_n \setminus G$ we have

$$\Pr(\varepsilon_{t+1} = e \,|\, \mathbb{G}_M(t) = G) = \frac{1}{\binom{n}{2} - t}.$$

This is not true if we replace $\mathbb{G}_M$ by $\mathbb{G}_r$, except for the very first step $t = 0$. However, it turns out that for most of time the conditional distribution of the next edge in the process $\mathbb{G}_r(t)$ is approximately uniform, which is made precise in the lemma below. For $0 < \varepsilon < 1$, and $t = 0, \ldots, M$ consider the inequalities

$$\Pr(\eta_{t+1} = e \mid \mathbb{G}_r(t)) \geq \frac{1-\varepsilon}{\binom{n}{2} - t} \text{ for every } e \in K_n \setminus \mathbb{G}_r(t), \tag{10.12}$$

and define a stopping time

$$T_\varepsilon = \max\{u : \forall t \leq u \text{ condition (10.12) holds }\}.$$

Lemma 10.12 *There is a positive constant C' such that if*

$$C'\left(\frac{r}{n} + \frac{\log n}{r}\right)^{1/3} \leq \varepsilon = \varepsilon(n) < 1, \tag{10.13}$$

then

$$T_\varepsilon \geq (1-\varepsilon)M \quad w.h.p.$$

From Lemma 10.12, which is proved later in Section 10.3, we deduce Theorem 10.10 using a coupling.

Proof of Theorem 10.10: Let $C = 3C'$, where C' is the constant from Lemma 10.12. Let $\varepsilon = \gamma/3$. The distribution of $\mathbb{G}_r$ is uniquely determined by the conditional probabilities

$$p_{t+1}(e|G) := \Pr(\eta_{t+1} = e \mid \mathbb{G}_r(t) = G), \quad t = 0, \ldots, M-1. \tag{10.14}$$

Our aim is to couple $\mathbb{G}_M$ and $\mathbb{G}_r$ up to the time T_ε. For this we will define a graph process $\mathbb{G}'_r := (\eta'_t), t = 1, \ldots, M$ such that the conditional distribution of (η'_t) coincides with that of (η_t) and w.h.p. (η'_t) shares many edges with $\mathbb{G}_M$.

Suppose that $G_r = G'_r(t)$ and $G_M = G_M(t)$ have been exposed and for every $e \notin G_r$ the inequality

$$p_{t+1}(e|G_r) \geq \frac{1-\varepsilon}{\binom{n}{2} - t} \tag{10.15}$$

holds (we have such a situation, in particular, if $t \leq T_\varepsilon$). Generate a Bernoulli $(1-\varepsilon)$ random variable ξ_{t+1} independently of everything that has been revealed so far; expose the edge ε_{t+1}. Moreover, generate a random edge $\zeta_{t+1} \in K_n \setminus G_r$ according to the distribution

$$\mathbb{P}(\zeta_{t+1} = e | \mathbb{G}'_r(t) = G_r, \mathbb{G}_M(t) = G_M) := \frac{p_{t+1}(e|G_r) - \frac{1-\varepsilon}{\binom{n}{2}-t}}{\varepsilon} \geq 0,$$

where the inequality holds because of the assumption (10.15). Observe also that

$$\sum_{e \notin G_r} \mathbb{P}(\zeta_{t+1} = e | \mathbb{G}'_r(t) = G_r, \mathbb{G}_M(t) = G_M) = 1,$$

so ζ_{t+1} has a well-defined distribution. Finally, fix a bijection $f_{G_r,G_M} : G_r \setminus G_M \to G_M \setminus G_r$ between the sets of edges and define

$$\eta'_{t+1} = \begin{cases} \varepsilon_{t+1}, & \text{if } \xi_{t+1} = 1, \varepsilon_{t+1} \notin G_r, \\ f_{G_r,G_M}(\varepsilon_{t+1}), & \text{if } \xi_{t+1} = 1, \varepsilon_{t+1} \in G_r, \\ \zeta_{t+1}, & \text{if } \xi_{t+1} = 0. \end{cases}$$

Note that

$$\xi_{t+1} = 1 \qquad \Rightarrow \qquad \varepsilon_{t+1} \in \mathbb{G}'_r(t+1). \tag{10.16}$$

We keep generating ξ_ts even after the stopping time has passed, i.e. for $t > T_\varepsilon$, whereas η'_{t+1} is then sampled according to probabilities (10.14), without coupling. Note that ξ_t's are i.i.d. and independent of $\mathbb{G}_M$. We check that

$$\begin{aligned} &\mathbb{P}(\eta'_{t+1} = e \mid \mathbb{G}'_r(t) = G_r, \mathbb{G}_M(t) = G_M) \\ &= \mathbb{P}(\varepsilon_{t+1} = e)\,\mathbb{P}(\xi_{t+1} = 1) + \mathbb{P}(\zeta_{t+1} = e)\,\mathbb{P}(\xi_{t+1} = 0) \\ &= \frac{1-\varepsilon}{\binom{n}{2} - t} + \left(\frac{p_{t+1}(e|G_r) - \frac{1-\varepsilon}{\binom{n}{2}-t}}{\varepsilon} \right) \varepsilon \\ &= p_{t+1}(e|G_r) \end{aligned}$$

for all admissible $\mathbb{G}_r, \mathbb{G}_M$, i.e. such that $\mathbb{P}(\mathbb{G}_r(t) = G_r, \mathbb{G}_M(t) = G_M) > 0$, and for all $e \notin G_r$.

Further, define a set of edges that are potentially shared by $\mathbb{G}_M$ and $\mathbb{G}_r$:

$$S := \{\varepsilon_i : \xi_i = 1\,, 1 \le i \le (1-\varepsilon)M\}\,.$$

Note that

$$|S| = \sum_{i=1}^{\lfloor (1-\varepsilon)M \rfloor} \xi_i$$

is distributed as $\mathrm{Bin}(\lfloor (1-\varepsilon)M \rfloor, 1-\varepsilon)$.

Since (ξ_i) and (ε_i) are independent, conditioning on $|S| \ge m$, the first m edges in the set S comprise a graph that is distributed as $\mathbb{G}_{n,m}$. Moreover, if $T_\varepsilon \ge (1-\varepsilon)M$, then by (10.16) we have $S \subset \mathbb{G}_r$, therefore

$$\mathbb{P}\left(\mathbb{G}_{n,m} \subset \mathbb{G}_{n,r}\right) \ge \mathbb{P}(|S| \ge m, T_\varepsilon \ge (1-\varepsilon)M)\,.$$

We have $\mathbb{E}|S| \geq (1-2\varepsilon)M$. Recall that $\varepsilon = \gamma/3$ and therefore $m = \lfloor(1-\gamma)M\rfloor = \lfloor(1-3\varepsilon)M\rfloor$. Applying the Chernoff bounds and our assumption on ε, we get

$$\mathbb{P}(|S| < m) \leq e^{-\Omega(\gamma^2 m)} = o(1).$$

Finally, by Lemma 10.12 we have $T_\varepsilon \geq (1-\varepsilon)M$ w.h.p., which completes the proof of the theorem. □

Proof of Lemma 10.12

In all the proofs in this section we assume the condition (10.13). To prove Lemma 10.12 we start with a fact that allows one to control the degrees of the evolving graph $\mathbb{G}_r(t)$.

For a vertex $v \in [n]$ and $t = 0,\ldots,M$, let

$$\deg_t(v) = |\{i \leq t : v \in \eta_i\}|.$$

Lemma 10.13 *Let $\tau = 1 - t/M$. We have that w.h.p.*

$$\forall t \leq (1-\varepsilon)M, \quad \forall v \in [n], \quad |\deg_t(v) - tr/M| \leq 6\sqrt{\tau r \log n}. \tag{10.17}$$

In particular w.h.p.

$$\forall t \leq (1-\varepsilon)M, \quad \forall v \in [n], \quad \deg_t(v) \leq (1-\varepsilon/2)r. \tag{10.18}$$

Proof Observe that if we fix an r-regular graph H and condition $\mathbb{G}_r$ to be a permutation of the edges of H, then $X := \deg_t(v)$ is a hypergeometric random variable with expected value $tr/M = (1-\tau)r$. Using the result of Section 21.5 and Theorem 21.10, and checking that the variance of X is at most τr, we obtain

$$\mathbb{P}(|X - tr/M| \geq x) \leq 2\exp\left\{-\frac{x^2}{2(\tau r + x/3)}\right\}.$$

Let $x = 6\sqrt{\tau r \log n}$. From (10.13), assuming $C' \geq 1$, we get

$$\frac{x}{\tau r} = 6\sqrt{\frac{\log n}{\tau r}} \leq 6\sqrt{\frac{\log n}{\varepsilon r}} \leq 6\varepsilon,$$

and so $x \leq 6\tau r$. Using this, we obtain

$$\frac{1}{2}\mathbb{P}(|X - tr/M| \geq x) \leq \exp\left\{-\frac{36\tau r \log n}{2(\tau r + 2\tau r)}\right\} = n^{-6}.$$

Inequality (10.17) now follows by taking a union bound over $nM \leq n^3$ choices of t and v.

To obtain (10.18), it is enough to prove the inequality for $t = (1-\varepsilon)M$. Inequality (10.17) implies

$$\deg_{(1-\varepsilon)M}(v) \le (1-\varepsilon)r + 6\sqrt{\varepsilon r \log n}.$$

Thus it suffices to show that

$$6\sqrt{\varepsilon r \log n} \le \varepsilon r/2,$$

or, equivalently, $\varepsilon \ge 144 \log n/r$, which is implied by (10.13) with $C' \ge 144$. □

Given an ordered graph $G = (e_1, \ldots, e_t)$, we say that an ordered r-regular graph H is an *extension* of G if the first t edges of H are equal to G. Let $\mathcal{G}_G(n,r)$ be the family of extensions of G and $\mathbb{G}_G = \mathbb{G}_G(n,r)$ be a graph chosen uniformly at random from $\mathcal{G}_G(n,r)$.

Further, for a graph $H \in \mathcal{G}_G(n,r)$ and $u, v \in [n]$ let

$$\deg_{H|G}(u,v) = |\{w \in [n] : \{u,w\} \in H \setminus G, \{v,w\} \in H\}|.$$

Note that $\deg_{H|G}(u,v)$ is not in general symmetric in u and v, but for $G = \emptyset$ coincides with the usual co-degree in a graph H.

The next fact is used in the proof of Lemma 10.15 only.

Lemma 10.14 *Let graph G with $t \le (1-\varepsilon)M$ edges be such that $\mathcal{G}_G(n,r)$ is non-empty. For each $e \notin G$ we have*

$$\mathbb{P}(e \in \mathbb{G}_G) \le \frac{4r}{\varepsilon n}. \tag{10.19}$$

Moreover, if $l \ge l_0 := 4r^2/(\varepsilon n)$, then for every $u, v \in [n]$ we have

$$\mathbb{P}\left(\deg_{\mathbb{G}_G|G}(u,v) > l\right) \le 2^{-(l-l_0)}. \tag{10.20}$$

Proof To prove (10.19) define the families

$$\mathcal{G}_{e\in} = \{H \in \mathcal{G}_G(n,r) : e \in H\} \quad \text{and} \quad \mathcal{G}_{e\notin} = \left\{H' \in \mathcal{G}_G(n,r) : e \notin H'\right\}.$$

Let us define an auxiliary bipartite graph B between $\mathcal{G}_{e\in}$ and $\mathcal{G}_{e\notin}$ in which $H \in \mathcal{G}_{e\in}$ is connected to $H' \in \mathcal{G}_{e\notin}$ whenever H' can be obtained from H by the following switching operation. Fix an ordered edge $\{w,x\}$ in $H \setminus G$ that is disjoint from $e = \{u,v\}$ and such that there are no edges between $\{u,v\}$ and $\{w,x\}$ and replace the edges $\{u,v\}$ and $\{w,x\}$ by $\{u,w\}$ and $\{v,x\}$ to obtain H'. Writing $f(H)$ for the number of graphs $H' \in \mathcal{G}_{e\notin}$ that can be obtained from H by a switching, and $b(H')$ for the number of graphs $H \in \mathcal{G}_{e\in}$ such that H' can be obtained from H by a switching, we get that

$$|\mathcal{G}_{e\in}| \min_H f(H) \le |E(B)| \le |\mathcal{G}_{e\notin}| \max_{H'} b(H'). \tag{10.21}$$

We have $b(H') \le \deg_{H'}(u)\deg_{H'}(v) \le r^2$. On the other hand, recalling that $t \le (1-\varepsilon)M$, for every $H \in \mathcal{G}_{e\in}$ we obtain

$$f(H) \ge M - t - 2r^2 \ge \varepsilon M\left(1 - \frac{2r^2}{\varepsilon M}\right) \ge \frac{\varepsilon M}{2},$$

because, assuming $C' \ge 8$, we have

$$\frac{2r^2}{\varepsilon M} \le \frac{4r}{C'n}\left(\frac{n}{r}\right)^{1/3} \le \frac{4}{C'} \le \frac{1}{2}.$$

Therefore, (10.21) implies that

$$\mathbb{P}(e \in \mathbb{G}_G) \le \frac{|\mathcal{G}_{e\in}|}{|\mathcal{G}_{e\notin}|} \le \frac{2r^2}{\varepsilon M} = \frac{4r}{\varepsilon n},$$

which concludes the proof of (10.19).

To prove (10.20), fix $u, v \in [n]$ and define the families

$$\mathcal{G}(l) = \left\{H \in \mathcal{G}_G(n,r) : \deg_{H|G}(u,v) = l\right\}, \qquad l = 0, 1, \dots.$$

We compare sizes of $\mathcal{G}(l)$ and $\mathcal{G}(l-1)$ in a similar way as above. For this we define the following switching which maps a graph $H \in \mathcal{G}(l)$ to a graph $H' \in \mathcal{G}(l-1)$. Select a vertex w contributing to $\deg_{H|G}$, i.e. such that $\{u,w\} \in H \setminus G$ and $\{v,w\} \in H$; pick an ordered pair $u', w' \in [n] \setminus \{u,v,w\}$ such that $\{u',w'\} \in H \setminus G$ and there are no edges of H between $\{u,v,w\}$ and $\{u',w'\}$; replace edges $\{u,w\}$ and $\{u',w'\}$ by $\{u,u'\}$ and $\{w,w'\}$ (see Figure 10.8).

The number of ways to apply a forward switching to H is

$$f(H) \ge 2l(M - t - 3r^2) \ge 2l\varepsilon M\left(1 - \frac{3r^2}{\varepsilon M}\right) \ge l\varepsilon M,$$

since, assuming $C' \ge 12$ we have

$$\frac{3r^2}{\varepsilon M} = \frac{6r}{\varepsilon n} \le \frac{6}{C'}\left(\frac{r}{n}\right)^{2/3} \le \frac{1}{2},$$

and the number of ways to apply a backward switching is $b(H) \le r^3$. So,

$$\frac{|\mathcal{G}(l)|}{|\mathcal{G}(l-1)|} \le \frac{\max_{H\in\mathcal{G}(l-1)} b(H)}{\min_{H\in\mathcal{G}(l)} f(H)} \le \frac{2r^2}{\varepsilon l n} \le \frac{1}{2},$$

Figure 10.8. Switching between $\mathcal{G}(l)$ and $\mathcal{G}(l-1)$: Before and after

by the assumption $l \geq l_0 := 4r^2/(\varepsilon n)$. Then

$$\mathbb{P}\left(\deg_{\mathbb{G}_G|G}(u,v) > l\right) \leq \sum_{i>l} \frac{|\mathscr{G}(i)|}{|\mathscr{G}_G(n,r)|} \leq \sum_{i>l} \frac{|\mathscr{G}(i)|}{|\mathscr{G}(l_0)|}$$

$$= \sum_{i>l} \prod_{j=l_0+1}^{i} \frac{|\mathscr{G}(j)|}{|\mathscr{G}(j-1)|} \leq \sum_{i>l} 2^{-(i-l_0)} = 2^{-(l-l_0)},$$

which completes the proof of (10.20). □

For the last lemma, which will be directly used in Lemma 10.12, we need to provide a few more definitions regarding random r-regular multigraphs.

Let G be an ordered graph with t edges. Let $\mathbb{M}_G(n,r)$ be a random *multigraph extension* of G to an ordered r-regular multigraph. Namely, $\mathbb{M}_G(n,r)$ is a sequence of M edges (some of which may be loops), the first t of which comprise G, while the remaining ones are generated by taking a uniform random permutation Π of the multiset $\{1,\ldots,1,\ldots,n,\ldots,n\}$ with multiplicities $r-\deg_G(v)$, $v \in [n]$, and splitting it into consecutive pairs.

Recall that the number of such permutations is

$$N_G := \frac{(2(M-t))!}{\prod_{v\in[n]}\left(r-\deg_G(v)\right)!},$$

and note that if a multigraph extension H of G has l loops, then

$$\mathbb{P}(\mathbb{M}_G(n,r) = H) = 2^{M-t-l}/N_G. \tag{10.22}$$

Thus, $\mathbb{M}_G(n,r)$ is not uniformly distributed over all multigraph extensions of G, but it is uniform over $\mathscr{G}_G(n,r)$. Thus, $\mathbb{M}_G(n,r)$, conditioned on being simple, has the same distribution as $\mathbb{G}_G(n,r)$. Further, for every edge $e \notin G$, let us write

$$\mathbb{M}_e = \mathbb{M}_{G\cup e}(n,r) \quad \text{and} \quad \mathscr{G}_e = \mathscr{G}_{G\cup e}(n,r). \tag{10.23}$$

The next claim shows that the probabilities of simplicity $\mathbb{P}(\mathbb{M}_e \in \mathscr{G}_e)$ are asymptotically the same for all $e \notin G$.

Lemma 10.15 *Let G be a graph with $t \leq (1-\varepsilon)M$ edges such that $\mathscr{G}_G(n,r)$ is non-empty. If $\Delta_G \leq (1-\varepsilon/2)r$, then for every $e', e'' \notin G$ we have*

$$\frac{\mathbb{P}(\mathbb{M}_{e''} \in \mathscr{G}_{e''})}{\mathbb{P}(\mathbb{M}_{e'} \in \mathscr{G}_{e'})} \geq 1 - \frac{\varepsilon}{2}.$$

Proof Set

$$\mathbb{M}' = \mathbb{M}_{e'} \quad \mathbb{M}'' = \mathbb{M}_{e''}, \quad \mathscr{G}' = \mathscr{G}_{e'}, \quad \text{and} \quad \mathscr{G}'' = \mathscr{G}_{e''}, \tag{10.24}$$

for convenience. We start by constructing a coupling of $\mathbb{M}'$ and $\mathbb{M}''$ in which they differ in at most three positions (counting in the replacement of e' by e'' at the $(t+1)$st position).

Let $e' = \{u', v'\}$ and $e'' = \{u'', v''\}$. Suppose first that e' and e'' are disjoint. Let Π' be the permutation underlying the multigraph $\mathbb{M}'$. Let Π^* be obtained from Π' by replacing a uniform random copy of u'' by u' and a uniform random copy of v'' by v'. If e' and e'' share a vertex, then assume, without loss of generality, that $v' = v''$, and define Π^* by replacing only a random u'' in Π' by u'. Then define $\mathbb{M}^*$ by splitting Π^* into consecutive pairs and appending them to $G \cup e''$.

It is easy to see that Π^* is uniform over permutations of the multiset $\{1,\dots, 1,\dots,n,\dots,n\}$ with multiplicities $d - \deg_{G\cup e''}(v), v \in [n]$, and therefore $\mathbb{M}^*$ has the same distribution as $\mathbb{M}''$. Thus, we will further identify $\mathbb{M}^*$ and $\mathbb{M}''$.

Observe that if we condition $\mathbb{M}'$ on being a simple graph H, then $\mathbb{M}^* = \mathbb{M}''$ can be equivalently obtained by choosing an edge incident to u'' in $H \setminus (G \cup e')$ uniformly at random, say, $\{u'', w\}$, and replacing it by $\{u', w\}$, and then repeating this operation for v'' and v'. The crucial idea is that such a switching of edges is unlikely to create loops or multiple edges.

It is, however, possible that for certain H this is not true. For example, if $e'' \in H \setminus (G \cup e')$, then the random choice of two edges described above is unlikely to destroy this e'', but e' in the non-random part will be replaced by e'', thus creating a double edge e''. Moreover, if almost every neighbor of u'' in $H \setminus (G \cup e')$ is also a neighbor of u', then most likely the replacement of u'' by u' will create a double edge. To avoid such instances, we want to assume that

(i) $e'' \notin H$
(ii) $\max\left(\deg_{H|G\cup e'}(u', u''), \deg_{H|G\cup e'}(v', v'')\right) \le l_0 + \log_2 n$,

where $l_0 = 4r^2/\varepsilon n$ is as in Lemma 10.14. Define the following subfamily of simple extensions of $G \cup e'$:

$$\mathscr{G}'_{\text{nice}} = \left\{H \in \mathscr{G}' : H \text{ satisfies (i) and (ii)}\right\}.$$

Since $\mathbb{M}'$, conditioned on $\mathbb{M}' \in \mathscr{G}'$, is distributed as $\mathbb{G}_{G\cup e'}(n,r)$, by Lemma 10.14 and the assumption (10.13) with $C' \ge 20$,

$$\begin{aligned}\Pr\left(\mathbb{M}' \notin \mathscr{G}'_{\text{nice}} \mid \mathbb{M}' \in \mathscr{G}'\right) &= \mathbb{P}\left(\mathbb{G}_{G\cup e'}(n,r) \notin \mathscr{G}'_{\text{nice}}\right) \\ &\le \frac{4r}{\varepsilon n} + 2 \times 2^{-\log_2 n} \le \frac{\varepsilon}{4}. \end{aligned} \qquad (10.25)$$

We have

$$\begin{aligned}
&\Pr\left(\mathbb{M}'' \in \mathscr{G}'' \mid \mathbb{M}' \in \mathscr{G}'_{\text{nice}}\right) \Pr\left(\mathbb{M}' \in \mathscr{G}'_{\text{nice}} \mid \mathbb{M}' \in \mathscr{G}'\right) \\
&\quad = \frac{\mathbb{P}(\mathbb{M}'' \in \mathscr{G}'', \mathbb{M}' \in \mathscr{G}'_{\text{nice}})}{\mathbb{P}(\mathbb{M}' \in \mathscr{G}'_{\text{nice}})} \cdot \frac{\mathbb{P}(\mathbb{M}' \in \mathscr{G}'_{\text{nice}}, \mathbb{M}' \in \mathscr{G}')}{\mathbb{P}(\mathbb{M}' \in \mathscr{G}')} \\
&\quad \le \frac{\mathbb{P}\left(\mathbb{M}'' \in \mathscr{G}''\right)}{\mathbb{P}\left(\mathbb{M}' \in \mathscr{G}'\right)}. \qquad (10.26)
\end{aligned}$$

To complete the proof of the claim, it suffices to show that

$$\Pr\left(\mathbb{M}'' \in \mathscr{G}'' \mid \mathbb{M}' \in \mathscr{G}'_{\text{nice}}\right) \ge 1 - \frac{\varepsilon}{4}, \qquad (10.27)$$

since plugging (10.25) and (10.27) into (10.26) completes the proof of the statement.

To prove (10.27), fix $H \in \mathscr{G}'_{\text{nice}}$ and condition on $\mathbb{M}' = H$. A loop can only be created in $\mathbb{M}''$ when u'' is incident to u' in $H \setminus (G \cup e')$ and the randomly chosen edge is $\{u', u''\}$, or, provided $v' \neq v''$, when v'' is incident to v' in $H \setminus (G \cup e')$ and we randomly choose $\{v', v''\}$. Therefore, recalling that $\Delta_G \le (1 - \varepsilon/2)r$, we obtain

$$\begin{aligned}
\Pr\left(\mathbb{M}'' \text{ has a loop} \mid \mathbb{M}' = H\right) &\le \frac{1}{\deg_{H\setminus(G\cup e')}(u'')} + \frac{1}{\deg_{H\setminus(G\cup e')}(v'')} \\
&\le \frac{4}{\varepsilon r} \le \frac{\varepsilon}{8}, \qquad (10.28)
\end{aligned}$$

where the second term is present only if $e' \cap e'' = \emptyset$, and the last inequality is implied by (10.13).

A multiple edge can be created in three ways: (i) by choosing, among the edges incident to u'', an edge $\{u'', w\} \in H \setminus (G \cup e')$ such that $\{u', w\} \in H$; (ii) similarly for v'' (if $v' \neq v''$); and (iii) choosing both edges $\{u'', v'\}$ and $\{v'', u'\}$ (provided they exist in $H \setminus (G \cup e')$). Therefore, by (ii) and assumption $\Delta_G \le (1 - \varepsilon/2)r$,

$$\begin{aligned}
&\Pr\left(\mathbb{M}'' \text{ has a multiple edge } \mid \mathbb{M}' = H\right) \\
&\quad \le \frac{\deg_{H|G\cup e'}(u'', u')}{\deg_{H\setminus(G\cup e')}(u'')} + \frac{\deg_{H|G\cup e'}(v'', v')}{\deg_{H\setminus(G\cup e')}(v'')} \\
&\qquad + \frac{1}{\deg_{H\setminus(G\cup e')}(u'')\deg_{H\setminus(G\cup e')}(v'')} \\
&\quad \le 2\left(\frac{8r}{\varepsilon^2 n} + \frac{2\log_2 n}{\varepsilon r}\right) + \frac{4}{\varepsilon^2 r^2} \le \frac{\varepsilon}{8}, \qquad (10.29)
\end{aligned}$$

because (10.13) implies $\varepsilon > C'(r/n)^{1/3}$ and

$$\varepsilon > C'(\log n/r)^{1/3} > C'(\log n/r)^{1/2}$$

and we can choose arbitrarily large C'. (Again, in case when $|e' \cap e''| = 1$, the RHS of (10.29) reduces to only the first summand.)

Combining (10.28) and (10.29), we have shown (10.27). □

Proof of Lemma 10.12 In view of Lemma 10.13, it suffices to show that

$$\Pr(\eta_{t+1} = e \mid \mathbb{G}_r(t) = G) \geq \frac{1-\varepsilon}{\binom{n}{2} - t}, \qquad e \notin G.$$

for every $t \leq (1-\varepsilon)M$ and G such that

$$r(\tau + \delta) \geq r - \deg_G(v) \geq r(\tau - \delta) \geq \frac{\varepsilon r}{2}, \qquad v \in [n], \tag{10.30}$$

where

$$\tau = 1 - t/M, \qquad \delta = 6\sqrt{\tau \log n/r}.$$

For every $e', e'' \notin G$ we have (recall the definitions (10.23) and (10.24))

$$\frac{\Pr(\eta_{t+1} = e'' \mid \mathbb{G}_r(t) = G)}{\Pr(\eta_{t+1} = e' \mid \mathbb{G}_r(t) = G)} = \frac{|\mathscr{G}_{G \cup e''}(n,r)|}{|\mathscr{G}_{G \cup e'}(n,r)|} = \frac{|\mathscr{G}''|}{|\mathscr{G}'|}. \tag{10.31}$$

By (10.22) we have

$$\mathbb{P}(\mathbb{M}' \in \mathscr{G}') = \frac{|\mathscr{G}'| 2^{M-t}}{N_G} = \frac{|\mathscr{G}'| 2^{M-t} \prod_{v \in [n]} (r - \deg_{G \cup e'}(v))!}{(2(M-t))!},$$

and similarly for the family $\mathscr{G}''$. This yields, after a few cancellations, that

$$\frac{|\mathscr{G}''|}{|\mathscr{G}'|} = \frac{\prod_{v \in e'' \setminus e'} (r - \deg_G(v))}{\prod_{v \in e' \setminus e''} (r - \deg_G(v))} \cdot \frac{\mathbb{P}(\mathbb{M}'' \in \mathscr{G}'')}{\mathbb{P}(\mathbb{M}' \in \mathscr{G}')}. \tag{10.32}$$

By (10.30), the ratio of products in (10.32) is at least

$$\left(\frac{\tau - \delta}{\tau + \delta}\right)^2 \geq \left(1 - \frac{2\delta}{\tau}\right)^2 \geq 1 - 24\sqrt{\frac{\log n}{\tau r}} \geq 1 - 24\sqrt{\frac{\log n}{\varepsilon r}} \geq 1 - \frac{\varepsilon}{2},$$

where the last inequality holds by the assumption (10.13). Since by Lemma 10.15 the ratio of probabilities in (10.32) is

$$\frac{\mathbb{P}(\mathbb{M}'' \in \mathscr{G}'')}{\mathbb{P}(\mathbb{M}' \in \mathscr{G}')} \geq 1 - \frac{\varepsilon}{2},$$

we have obtained that

$$\frac{\Pr(\eta_{t+1} = e'' \mid \mathbb{G}_r(t) = G)}{\Pr(\eta_{t+1} = e' \mid \mathbb{G}_r(t) = G)} \geq 1 - \varepsilon.$$

Finally, noting that

$$\max_{e' \notin G} \Pr\left(\eta_{t+1} = e' \mid \mathbb{G}_r(t) = G\right)$$

is at least as large as the average over all $e' \notin G$, which is $\frac{1}{\binom{n}{2}-t}$, we conclude that for every $e \notin G$

$$\Pr\left(\eta_{t+1} = e \mid \mathbb{G}_r(t) = G\right) \geq (1-\varepsilon) \max_{e' \notin G} \Pr\left(\eta_{t+1} = e' \mid \mathbb{G}_r(t) = G\right) \geq \frac{1-\varepsilon}{\binom{n}{2}-t},$$

which finishes the proof. □

10.4 Exercises

10.4.1 Suppose that $\max\{d_i, i = 1,2,\ldots,n\} = O(1)$. Show that $G_{n,\mathbf{d}}$ has a component of size $\Omega(n)$ w.h.p. if and only if (see Molloy and Reed [583])

$$\sum_{i=1}^{n} d_i(d_i - 2) = \Omega(n).$$

10.4.2 Let H be a subgraph of $\mathbb{G}_{n,r}, r \geq 3$ obtained by independently including each vertex with probability $\frac{1+\varepsilon}{r-1}$, where $\varepsilon > 0$ is small and positive. Show that w.h.p. H contains a component of size $\Omega(n)$.

10.4.3 Let $\mathbf{x} = (x_1, x_2, \ldots, x_{2m})$ be chosen uniformly at random from $[n]^{2m}$. Let $G_{\mathbf{x}}$ be the multigraph with vertex set $[n]$ and edges $(x_{2i-1}, x_{2i}), i = 1,2,\ldots,m$. Let $d_{\mathbf{x}}(i)$ be the number of times that i appears in $\mathbf{x}$. Show that conditional on $d_{\mathbf{x}}(i) = d_i, i \in [n]$, $G_{\mathbf{x}}$ has the same distribution as the multigraph $\gamma(F)$ of Section 10.1.

10.4.4 Suppose that we condition on $d_{\mathbf{x}}(i) \geq k$ for some non-negative integer k. For $r \geq 0$, let

$$f_r(x) = e^x - 1 - x - \cdots - \frac{x^{k-1}}{(k-1)!}.$$

Let Z be a random variable taking values in $\{k, k+1, \ldots,\}$ such that

$$\mathbb{P}(Z = i) = \frac{\lambda^i e^{-\lambda}}{i! f_k(\lambda)} \qquad \text{for } i \geq k,$$

where λ is arbitrary and positive. Show that the degree sequence of $G_{\mathbf{x}}$ is distributed as independent copies $Z_1, Z_2, \ldots, Z_n$ of Z, subject to $Z_1 + Z_2 + \cdots + Z_n = 2m$.

10.4.5 Show that

$$\mathbb{E}(Z) = \frac{\lambda f_{k-1}(\lambda)}{f_k(\lambda)}.$$

Show using the Local Central Limit Theorem (see e.g. Durrett [263]) that if $\mathbb{E}(Z) = \frac{2m}{n}$ then

$$\mathbb{P}\left(\sum_{j=1}^{\upsilon} Z_j = 2m - k\right) = \frac{1}{\sigma\sqrt{2\pi n}}\left(1 + O((k^2+1)\upsilon^{-1}\sigma^{-2})\right)$$

where $\sigma^2 = \mathbb{E}(Z^2) - \mathbb{E}(Z)^2$ is the variance of Z.

10.4.6 Show that if $c = 1 + \varepsilon$ and ε is sufficiently small and $\omega \to \infty$ then w.h.p. the 2-core of $G_{n,p}, p = c/n$ does not contain a cycle C, $|C| = \omega$ in which more than 10% of the vertices are of degree three or more.

10.4.7 Let $\mathbb{G} = \mathbb{G}_{n,r}, r \geq 3$ be the random r-regular configuration multigraph of Section 10.2. Let X denote the number of Hamilton cycles in $\mathbb{G}$. Show that

$$\mathbb{E}(X) \approx \sqrt{\frac{\pi}{2n}}\left((r-1)\left(\frac{r-2}{r}\right)^{(r-2)/2}\right)^n.$$

10.4.8 Show that w.h.p. $\mathbb{G}_{n,2}$ consists of $O(\log n)$ disjoint cycles.

10.4.9 Show that if graph $G = G_1 \cup G_2$ then its rainbow connection satisfies $rc(G) \leq rc(G_1) + rc(G_2) + |E(G_1) \cap E(G_2)|$. Using the contiguity of $\mathbb{G}_{n,r}$ to the union of r independent matchings, (see Chapter 19), show that $rc(\mathbb{G}_{n,r}) = O(\log_r n)$ for $r \geq 6$.

10.4.10 Show that w.h.p. $\mathbb{G}_{n,3}$ is not planar.

10.5 Notes

Giant Components and Cores

Molloy and Reed [583] introduced the criterion for a giant component discussed in Exercise 10.4.1. They allowed the maximum degree to grow with n and found the likely size of the giant component. Hatami and Molloy [398] discussed the size of the largest component in the scaling window for a random graph with a fixed degree sequence.

Cooper [199] and Janson and Luczak [429] discussed the sizes of the cores of random graphs with a given degree sequence.

Hamilton cycles

Robinson and Wormald [650], [651] showed that random r-regular graphs are Hamiltonian for $3 \leq r = O(1)$. In doing this, they introduced the important new method of small subgraph conditioning. It is a refinement on the Chebyshev inequality. Somewhat later Cooper, Frieze and Reed [224] and Krivelevich, Sudakov, Vu and Wormald [513] removed the restriction $r = O(1)$. Frieze, Jerrum, Molloy, Robinson and Wormald [331] gave a polynomial time algorithm that w.h.p. finds a Hamilton cycle in a random regular graph. Cooper, Frieze and Krivelevich [218] considered the existence of Hamilton cycles in $G_{n,\mathbf{d}}$ for certain classes of degree sequence.

Chromatic number

Frieze and Łuczak [340] proved that w.h.p. $\chi(G_{n,r}) = (1+o_r(1))\frac{r}{2\log r}$ for $r = O(1)$. Here $o_r(1) \to 0$ as $r \to \infty$. Achlioptas and Moore [4] determined the chromatic number of a random r-regular graph to within three values, w.h.p. Kemkes, Pérez-Giménez and Wormald [480] reduced the range to two values. Shi and Wormald [683], [684] considered the chromatic number of $G_{n,r}$ for small r. In particular they showed that w.h.p. $\chi(\mathbb{G}_{n,4}) = 3$. Frieze, Krivelevich and Smyth [335] gave estimates for the chromatic number of $\mathbb{G}_{n,\mathbf{d}}$ for certain classes of degree sequence.

Eigenvalues

The largest eigenvalue of the adjacency matrix of $\mathbb{G}_{n,r}$ is always r. Kahn and Szemerédi [458] showed that w.h.p. the second eigenvalue is of order $O(r^{1/2})$. Friedman [317] proved that w.h.p. the second eigenvalue is at most $2(r-1)^{1/2}+o(1)$. Broder, Frieze, Suen and Upfal [163] considered $\mathbb{G}_{n,\mathbf{d}}$ where $C^{-1}d \le d_i \le Cd$ for some constant $C > 0$ and $d \le n^{1/10}$. They showed that w.h.p. the second eigenvalue of the adjacency matrix is $O(d^{1/2})$.

First Order Logic

Haber and Krivelevich [391] studied the first order language on random d-regular graphs. They showed that if $r = \Omega(n)$ or $r = n^{\alpha}$ where α is irrational, then $\mathbb{G}_{n,r}$ obeys a 0-1 law.

Rainbow Connection

Dudek, Frieze and Tsourakakis [257] studied the rainbow connection of random regular graphs. They showed that if $4 \le r = O(1)$ then $rc(\mathbb{G}_{n,r}) = O(\log n)$. This is best possible up to constants, since $rc(\mathbb{G}_{n,r}) \ge diam(\mathbb{G}_{n,r}) = \Omega(\log n)$. Kamčev, Krivelevich and Sudakov [460] gave a simpler proof when $r \ge 5$, with a better hidden constant.

11
Intersection Graphs

Let G be a (finite, simple) graph. We say that G is an *intersection graph* if we can assign a set S_v to each vertex $v \in V(G)$, so that $\{v,w\} \in E(G)$ exactly when $S_v \cap S_w \neq \emptyset$. In this case, we say G is the intersection graph of the family of sets $\mathscr{S} = \{S_v : v \in V(G)\}$.

Although all graphs are intersection graphs (see Marczewski [555]) some classes of intersection graphs are of special interest. Depending on the choice of family $\mathscr{S}$, often reflecting some geometric configuration, one can consider, for example, *interval graphs* defined as the intersection graphs of intervals on the real line and *unit disc graphs* defined as the intersection graphs of unit discs on the plane. In this chapter we discuss properties of *random intersection graphs*, where the family $\mathscr{S}$ is generated in a random manner.

11.1 Binomial Random Intersection Graphs

Binomial random intersection graphs were introduced by Karoński, Scheinerman and Singer-Cohen in [472] as a generalization of the classical model of the Binomial random graph $\mathbb{G}_{n,p}$.

Let n,m be positive integers and let $0 \leq p \leq 1$. Let $V = \{1,2,\dots,n\}$ be the set of vertices and for every $1 \leq k \leq n$, let S_k be a random subset of the set $M = \{1,2,\dots,m\}$ formed by selecting each element of M independently with probability p. We define a *Binomial random intersection graph* $G(n,m,p)$ as the intersection graph of sets S_k, $k = 1,2,\dots n$. Here $S_1,S_2,\dots,S_n$ are generated independently. Hence, two vertices i and j are adjacent in $G(n,m,p)$ if and only if $S_i \cap S_j \neq \emptyset$.

There are other ways to generate Binomial random intersection graphs. For example, we may start with a classical bipartite random graph $\mathbb{G}_{n,m,p}$, with vertex set bipartition

$$(V,M), V = \{1,2,\ldots,n\}, M = \{1,2,\ldots,m\},$$

where each edge between V and M is drawn independently with probability p. Next, one can generate a graph $G(n,m,p)$ with vertex set V and vertices i and j of $G(n,m,p)$ connected if and only if they share a common neighbor (in M) in the random graph $\mathbb{G}_{n,m,p}$. Here the graph $\mathbb{G}_{n,m,p}$ is treated as a *generator* of $G(n,m,p)$.

One observes that the probability that there is an edge $\{i,j\}$ in $G(n,m,p)$ equals $1-(1-p^2)^m$, since the probability that sets S_i and S_j are disjoint is $(1-p^2)^m$, however, in contrast with $\mathbb{G}_{n,p}$, the edges do not occur independently of each other.

Another simple observation leads to some natural restrictions on the choice of probability p. Note that the expected number of edges of $G(n,m,p)$ is,

$$\binom{n}{2}(1-(1-p^2)^m) \approx n^2mp^2,$$

provided $mp^2 \to 0$ as $n \to \infty$. Therefore, if we take $p = o((n\sqrt{m})^{-1})$ then the expected number of edges of $G(n,m,p)$ tends to 0 as $n \to \infty$ and therefore w.h.p. $G(n,m,p)$ is empty.

On the other hand, the expected number of non-edges in $G(n,m,p)$ is

$$\binom{n}{2}(1-p^2)^m \leq n^2e^{-mp^2}.$$

Thus, if we take $p = (2\log n + \omega(n))/m)^{1/2}$, where $\omega(n) \to \infty$ as $n \to \infty$, then the random graph $G(n,m,p)$ is complete w.h.p. One can also easily show that when $\omega(n) \to -\infty$ then $G(n,m,p)$ is w.h.p. not complete. So, when studying the evolution of $G(n,m,p)$ we may restrict ourselves to values of p in the range between $\omega(n)/(n\sqrt{m})$ and $((2\log n - \omega(n))/m)^{1/2}$, where $\omega(n) \to \infty$.

Equivalence

One of the first interesting problems to be considered is the question of when the random graphs $G(n,m,p)$ and $\mathbb{G}_{n,p}$ have asymptotically the same properties. Intuitively, it should be the case when the edges of $G(n,m,p)$ occur "almost independently," i.e. when there are no vertices of degree greater than two in M in the generator $\mathbb{G}_{n,m,p}$ of $G(n,m,p)$. Then each of its edges is induced by a vertex of degree two in M, "almost" independently of other edges. One can show that this happens w.h.p. when $p = o\left(1/(nm^{1/3})\right)$, which in turn implies that both random graphs are asymptotically equivalent for all graph properties

$\mathscr{P}$. Recall that a graph property $\mathscr{P}$ is defined as a subset of the family of all labeled graphs on vertex set $[n]$, i.e. $\mathscr{P} \subseteq 2^{\binom{n}{2}}$. The following equivalence result is due to Rybarczyk [662] and Fill, Scheinerman and Singer-Cohen [300].

Theorem 11.1 *Let* $0 \le a \le 1$, $\mathscr{P}$ *be any graph property,* $p = o\left(1/(nm^{1/3})\right)$ *and*

$$\hat{p} = 1 - \exp\left(-mp^2(1-p)^{n-2}\right). \tag{11.1}$$

Then

$$\mathbb{P}(\mathbb{G}_{n,\hat{p}} \in \mathscr{P}) \to a$$

if and only if

$$\mathbb{P}(G(n,m,p) \in \mathscr{P}) \to a$$

as $n \to \infty$.

Proof Let X and Y be random variables taking values in a common finite (or countable) set S. Consider the probability measures $\mathscr{L}(X)$ and $\mathscr{L}(Y)$ on S whose values at $A \subseteq S$ are $\mathbb{P}(X \in A)$ and $\mathbb{P}(Y \in A)$. Define the total variation distance between $\mathscr{L}(X)$ and $\mathscr{L}(Y)$ as

$$d_{TV}(\mathscr{L}(X), \mathscr{L}(Y)) = \sup_{A \subseteq S} |\mathbb{P}(X \in A) - \mathbb{P}(Y \in A)|,$$

which is equivalent to

$$d_{TV}(\mathscr{L}(X), \mathscr{L}(Y)) = \frac{1}{2}\sum_{s \in S} |\mathbb{P}(X = s) - \mathbb{P}(Y = s)|.$$

Notice (see Fact 4 of [300]) that if there exists a probability space on which random variables X' and Y' are both defined, with $\mathscr{L}(X) = \mathscr{L}(X')$ and $\mathscr{L}(Y) = \mathscr{L}(Y')$, then

$$d_{TV}(\mathscr{L}(X), \mathscr{L}(Y)) \le \mathbb{P}(X' \ne Y'). \tag{11.2}$$

Furthermore, (see Fact 3 in [300]) if there exist random variables Z and Z' such that $\mathscr{L}(X|Z = z) = \mathscr{L}(Y|Z' = z)$, for all z, then

$$d_{TV}(\mathscr{L}(X), \mathscr{L}(Y)) \le 2d_{TV}(\mathscr{L}(Z), \mathscr{L}(Z')). \tag{11.3}$$

We need one more observation. Suppose that a random variable X has distribution the $\mathrm{Bin}(n,p)$, while a random variable Y has the Poisson distribution, and $\mathbb{E}X = \mathbb{E}Y$. Then

$$d_{TV}(X, Y) = O(p). \tag{11.4}$$

We leave the proofs of (11.2), (11.3) and (11.4) as exercises.

To prove Theorem 11.1 we also need some auxiliary results on a special coupon collector scheme.

Let Z be a non-negative integer valued random variable, r a non-negative integer and γ a real, such that $r\gamma \leq 1$. Assume we have r coupons $Q_1, Q_2, \ldots, Q_r$ and one blank coupon B. We make Z independent draws (with replacement), such that in each draw,

$$\mathbb{P}(Q_i \text{ is chosen}) = \gamma, \quad \text{for } i = 1, 2, \ldots, r,$$

and

$$\mathbb{P}(B \text{ is chosen}) = 1 - r\gamma.$$

Let $N_i(Z)$, $i = 1, 2, \ldots, r$ be a random variable counting the number of times that coupon Q_i was chosen. Furthermore, let

$$X_i(Z) = \begin{cases} 1 & \text{if } N_i(Z) \geq 1, \\ 0 & \text{otherwise.} \end{cases}$$

The number of different coupons selected is given by

$$X(Z) = \sum_{i=1}^{r} X_i(Z). \tag{11.5}$$

With the above definitions we observe that the following holds.

Lemma 11.2 *If a random variable Z has the Poisson distribution with expectation λ then $N_i(Z)$, $i = 1, 2, \ldots, r$, are independent and identically Poisson distributed random variables, with expectation $\lambda\gamma$. Moreover, the random variable $X(Z)$ has the distribution $Bin(r, 1 - e^{-\lambda\gamma})$.*

Let us consider the following special case of the scheme defined above, assuming that $r = \binom{n}{2}$ and $\gamma = 1/\binom{n}{2}$. Here each coupon represents a distinct edge of K_n.

Lemma 11.3 *Suppose $p = o(1/n)$ and let a random variable Z be the $Bin\left(m, \binom{n}{2} p^2 (1-p)^{n-2}\right)$ distributed, while a random variable Y be the $Bin\left(\binom{n}{2}, 1 - e^{-mp^2(1-p)^{n-2}}\right)$ distributed, then*

$$d_{TV}(\mathscr{L}(X(Z)), \mathscr{L}(Y)) = o(1).$$

Proof Let Z' be a Poisson random variable with the same expectation as Z, i.e.

$$\mathbb{E}Z' = m\binom{n}{2} p^2 (1-p)^{n-2}.$$

By Lemma 11.2, $X(Z')$ has the Binomial distribution

$$\mathrm{Bin}\left(\binom{n}{2}, 1 - e^{-mp^2(1-p)^{n-2}}\right),$$

and so, by (11.3) and (11.4), we have

$$\begin{aligned} &d_{TV}(\mathscr{L}(Y), \mathscr{L}(X(Z))) \\ &\quad = d_{TV}(\mathscr{L}(X(Z')), \mathscr{L}(X(Z))) \leq 2d_{TV}(\mathscr{L}(Z'), \mathscr{L}(Z)) \\ &\quad \leq O\left(\binom{n}{2} p^2(1-p)^{n-2}\right) = O\left(n^2p^2\right) = o(1). \end{aligned}$$

□

Now define a random intersection graph $G_2(n,m,p)$ as follows. Its vertex set is $V = \{1,2,\ldots,n\}$, while $e = \{i,j\}$ is an edge in $G_2(n,m,p)$ if and only if in a (generator) bipartite random graph $\mathbb{G}_{n,m,p}$, there is a vertex $w \in M$ of degree two such that both i and j are connected by an edge with w.

To complete the proof of our theorem, notice that,

$$\begin{aligned} &d_{TV}(\mathscr{L}(G(n,m,p)), \mathscr{L}(\mathbb{G}_{n,\hat{p}})) \leq \\ &\quad d_{TV}(\mathscr{L}(G(n,m,p)), \mathscr{L}(G_2(n,m,p))) + d_{TV}(\mathscr{L}(G_2(n,m,p)), \mathscr{L}(\mathbb{G}_{n,\hat{p}})) \end{aligned}$$

where $\hat{p}$ is defined in (11.1). Now, by (11.2)

$$\begin{aligned} &d_{TV}(\mathscr{L}(G(n,m,p)), \mathscr{L}(G_2(n,m,p))) \\ &\quad \leq \mathbb{P}(\mathscr{L}(G(n,m,p)) \neq \mathscr{L}(G_2(n,m,p))) \\ &\quad \leq \mathbb{P}(\exists w \in M \text{ of } \mathbb{G}_{n,m,p} \text{ s.t. } deg(w) > 2) \leq m\binom{n}{3}p^3 = o(1), \end{aligned}$$

for $p = o(1/(nm^{1/3})$.

Hence it remains to show that

$$d_{TV}(\mathscr{L}(G_2(n,m,p)), \mathscr{L}(\mathbb{G}_{n,\hat{p}})) = o(1). \tag{11.6}$$

Let Z be distributed as $\mathrm{Bin}\left(m, \binom{n}{2}p^2(1-p)^{n-2}\right)$, $X(Z)$ is defined as in (11.5) and let Y be distributed as $\mathrm{Bin}\left(\binom{n}{2}, 1 - e^{-mp^2(1-p)^{n-2}}\right)$. Then the number of edges $|E(G_2(n,m,p))| = X(Z)$ and $|E(\mathbb{G}_{n,\hat{p}}))| = Y$. Moreover for any two graphs G and G' with the same number of edges

$$\mathbb{P}(G_2(n,m,p) = G) = \mathbb{P}(G_2(n,m,p) = G')$$

and

$$\mathbb{P}(\mathbb{G}_{n,\hat{p}} = G) = \mathbb{P}(\mathbb{G}_{n,\hat{p}} = G').$$

Equation (11.6) now follows from Lemma 11.3. The theorem follows immediately. □

For monotone properties (see Chapter 1) the relationship between the classical Binomial random graph and the respective intersection graph is more precise and was established by Rybarczyk [662].

Theorem 11.4 *Let $0 \leq a \leq 1$, $m = n^{\alpha}, \alpha \geq 3$, Let $\mathcal{P}$ be any monotone graph property. For $\alpha > 3$, assume*

$$\Omega(1/(nm^{1/3})) = p = O(\sqrt{\log n/m}),$$

while for $\alpha = 3$ assume $\left(1/(nm^{1/3})\right) = o(p)$. Let

$$\hat{p} = 1 - \exp\left(-mp^2(1-p)^{n-2}\right).$$

If for all $\varepsilon = \varepsilon(n) \to 0$

$$\mathbb{P}(\mathbb{G}_{n,(1+\varepsilon)\hat{p}} \in \mathcal{P}) \to a,$$

then

$$\mathbb{P}(G(n,m,p) \in \mathcal{P}) \to a$$

as $n \to \infty$.

Small subgraphs

Let H be any fixed graph. A *clique cover* $\mathscr{C}$ is a collection of subsets of vertex set $V(H)$ such that, each induces a complete subgraph (*clique*) of H, and for every edge $\{u,v\} \in E(H)$, there exists $C \in \mathscr{C}$, such that $u,v \in C$. Hence, the cliques induced by sets from $\mathscr{C}$ exactly cover the edges of H. A clique cover is allowed to have more than one copy of a given set. We say that $\mathscr{C}$ is *reducible* if for some $C \in \mathscr{C}$, the edges of H induced by C are contained in the union of the edges induced by $\mathscr{C} \setminus C$, otherwise $\mathscr{C}$ is *irreducible*. Note that if $C \in \mathscr{C}$ and $\mathscr{C}$ is irreducible, then $|C| \geq 2$.

In this section, $|\mathscr{C}|$ stands for the number of cliques in $\mathscr{C}$, while $\sum \mathscr{C}$ denotes the sum of clique sizes in $\mathscr{C}$, and we put $\sum \mathscr{C} = 0$ if $\mathscr{C} = \emptyset$.

Let $\mathscr{C} = \{C_1, C_2, \dots, C_k\}$ be a clique cover of H. For $S \subseteq V(H)$ define the following two *restricted clique covers*

$$\mathscr{C}_t[S] := \{C_i \cap S : |C_i \cap S| \geq t, \ i = 1,2,\dots,k\},$$

where $t = 1,2$. For a given S and $t = 1,2$, let

$$\tau_t = \tau_t(H,\mathscr{C},S) = \left(n^{|S|/\sum \mathscr{C}_t[S]} m^{|\mathscr{C}_t[S]|/\sum \mathscr{C}_t[S]}\right)^{-1}.$$

Finally, let

$$\tau(H) = \min_{\mathscr{C}} \max_{S \subseteq V(H)} \{\tau_1, \tau_2\},$$

where the minimum is taken over all clique covers $\mathscr{C}$ of H. In this calculation we can restrict our attention to irreducible covers.

Karoński, Scheinerman and Singer-Cohen [472] proved the following theorem.

Theorem 11.5 *Let H be a fixed graph and $mp^2 \to 0$, then*

$$\lim_{n\to\infty} \mathbb{P}(H \subseteq G(n,m,p)) = \begin{cases} 0 & \text{if } p/\tau(H) \to 0 \\ 1 & \text{if } p/\tau(H) \to \infty. \end{cases}$$

As an illustration, we will use this theorem to show the threshold for complete graphs in $G(n,m,p)$, when $m = n^{\alpha}$, for different ranges of $\alpha > 0$.

Corollary 11.6 *For a complete graph K_h with $h \geq 3$ vertices and $m = n^{\alpha}$, we have*

$$\tau(K_h) = \begin{cases} n^{-1}m^{-1/h} & \text{for } \alpha \leq 2h/(h-1) \\ n^{-1/(h-1)}m^{-1/2} & \text{for } \alpha \geq 2h/(h-1). \end{cases}$$

Proof There are many possibilities for clique covers to generate a copy of a complete graph K_h in $G(n,m,p)$. However, in the case of K_h only two play a dominating role. Indeed, we will show that for $\alpha \leq \alpha_0$, $\alpha_0 = 2h/(h-1)$ the clique cover $\mathscr{C} = \{V(K_h)\}$ composed of one set containing all h vertices of K_h only matters, while for $\alpha \geq \alpha_0$ the clique cover $\mathscr{C} = \binom{K_h}{2}$, consisting of $\binom{h}{2}$ pairs of endpoints of the edges of K_h, takes the leading role.

Let $V = V(K_h)$ and denote those two clique covers by $\{V\}$ and $\{E\}$, respectively. Observe that for the cover $\{V\}$ the following equality holds.

$$\max_{S \subseteq V}\{\tau_1(K_h,\{V\},S), \tau_2(K_h,\{V\},S)\} = \tau_1(K_h,\{V\},V). \tag{11.7}$$

To see this, check first that for $|S| = h$,

$$\tau_1(K_h,\{V\},V) = \tau_2(K_h,\{V\},V) = n^{-1}m^{-1/h}.$$

For S of size $|S| = s$, $2 \leq s \leq h-1$ restricting the clique cover $\{V\}$ to S, gives a single s-clique, so for $t = 1,2$

$$\tau_t(K_h,\{V\},S) = n^{-1}m^{-1/s} < n^{-1}m^{-1/h}.$$

Finally, when $|S| = 1$, then $\tau_1 = (nm)^{-1}$, while $\tau_2 = 0$, both smaller than $n^{-1}m^{-1/h}$, and so Equation (11.7) follows.

For the edge-clique cover $\{E\}$ we have a similar expression, viz.

$$\max_{S\subseteq V}\{\tau_1(K_h,\{E\},S),\tau_2(K_h,\{E\},S)\} = \tau_1(K_h,\{E\},V). \tag{11.8}$$

To see this, check first that for $|S| = h$,

$$\tau_1(K_h,\{E\},V) = n^{-1/(h-1)}m^{-1/2}.$$

Let $S \subset V$, with $s = |S| \leq h-1$, and consider restricted clique covers with cliques of size at most two, and exactly two.

For τ_1, the clique cover restricted to S is the edge-clique cover of K_s, plus a 1-clique for each of the $h-s$ external edges for each vertex of K_s, so

$$\begin{aligned}
&\tau_1(K_h,\{E\},S)\\
&= \left(n^{s/[s(s-1)+s(h-s)]}m^{[s(s-1)/2+s(h-s)]/[s(s-1)+s(h-s)]}\right)^{-1}\\
&= \left(n^{1/(h-1)}m^{[h-(s+1)/2]/(h-1)}\right)^{-1}\\
&\leq \left(n^{1/(h-1)}m^{h/(2(h-1))}\right)^{-1}\\
&< \left(n^{1/(h-1)}m^{1/2}\right)^{-1},
\end{aligned}$$

while for τ_2 we have

$$\tau_2(K_h,\{E\},S) = \left(n^{1/(s-1)}m^{1/2}\right)^{-1} < \left(n^{1/(h-1)}m^{1/2}\right)^{-1},$$

thus verifying equation (11.8).

Let $\mathscr{C}$ be any irreducible clique cover of K_h (hence each clique has size at least two). We will show that for any fixed α

$$\tau_1(K_h,\mathscr{C},V) \geq \begin{cases} \tau_1(K_h,\{V\},V) & \text{for } \alpha \leq 2h/(h-1) \\ \tau_1(K_h,\{E\},V) & \text{for } \alpha \geq 2h/(h-1). \end{cases}$$

Thus,

$$\tau_1(K_h,\mathscr{C},V) \geq \min\{\tau_1(K_h,\{V\},V),\tau_1(K_h,\{E\},V)\}. \tag{11.9}$$

Because $m = n^\alpha$ we see that

$$\tau_1(K_h,\mathscr{C},V) = n^{-x_{\mathscr{C}}(\alpha)},$$

where

$$x_{\mathscr{C}}(\alpha) = \frac{h}{\sum \mathscr{C}} + \frac{|\mathscr{C}|}{\sum \mathscr{C}}\alpha, \quad x_{\{V\}}(\alpha) = 1+\frac{\alpha}{h}, \quad x_{\{E\}}(\alpha) = \frac{1}{h-1}+\frac{\alpha}{2}.$$

(To simplify notation, below we have replaced $x_{\{V\}}, x_{\{E\}}$ by x_V, x_E, respectively.) Notice, that for $\alpha_0 = 2h/(h-1)$ exponents

$$x_V(\alpha_0) = x_E(\alpha_0) = 1 + \frac{2}{h-1}.$$

Moreover, for all values of $\alpha < \alpha_0$ the function $x_V(\alpha) > x_E(\alpha)$, while for $\alpha > \alpha_0$ the function $x_V(\alpha) < x_E(\alpha)$.

Now, observe that $x_{\mathscr{C}}(0) = \frac{h}{\sum \mathscr{C}} \leq 1$ since each vertex is in at least one clique of $\mathscr{C}$. Hence $x_{\mathscr{C}}(0) \leq x_V(0) = 1$. We will also show that $x_{\mathscr{C}}(\alpha) \leq x_V(\alpha)$ for $\alpha > 0$. To see this we need to bound $|\mathscr{C}|/\sum \mathscr{C}$.

Suppose that $u \in V(K_h)$ appears in the fewest number of cliques of $\mathscr{C}$, and let r be the number of cliques $C_i \in \mathscr{C}$ to which u belongs. Then

$$\sum \mathscr{C} = \sum_{i:C_i \ni u} |C_i| + \sum_{i:C_i \not\ni u} |C_i| \geq ((h-1)+r) + 2(|\mathscr{C}| - r),$$

where $h-1$ counts all other vertices aside from u since they must appear in some clique with u.
For any $v \in V(K_h)$ we have

$$\begin{aligned}
\sum \mathscr{C} + |\{i : C_i \ni v\}| - (h-1) &\geq \sum \mathscr{C} + r - (h-1) \\
&\geq (h-1) + r + 2(|\mathscr{C}| - r) + r - (h-1) \\
&= 2|\mathscr{C}|.
\end{aligned}$$

Summing the above inequality over all $v \in V(K_h)$,

$$h \sum \mathscr{C} + \sum \mathscr{C} - h(h-1) \geq 2h|\mathscr{C}|,$$

and dividing both sides by $2h \sum \mathscr{C}$, we finally obtain

$$\frac{|\mathscr{C}|}{\sum \mathscr{C}} \leq \frac{h+1}{2h} - \frac{h-1}{2\sum \mathscr{C}}.$$

Now, using the above bound,

$$\begin{aligned}
x_{\mathscr{C}}(\alpha_0) &= \frac{h}{\sum \mathscr{C}} + \frac{|\mathscr{C}|}{\sum \mathscr{C}} \left(\frac{2h}{h-1} \right) \\
&\leq \frac{h}{\sum \mathscr{C}} + \left(\frac{h+1}{2h} - \frac{h-1}{2\sum \mathscr{C}} \right) \left(\frac{2h}{h-1} \right) \\
&= 1 + \frac{2}{h-1} \\
&= x_V(\alpha_0).
\end{aligned}$$

Now, since $x_{\mathscr{C}}(\alpha) \le x_V(\alpha)$ at both $\alpha = 0$ and $\alpha = \alpha_0$, and both functions are linear, $x_{\mathscr{C}}(\alpha) \le x_V(\alpha)$ throughout the interval $(0, \alpha_0)$.

Since $x_E(\alpha_0) = x_V(\alpha_0)$ we also have $x_{\mathscr{C}}(\alpha_0) \le x_E(\alpha_0)$. The slope of $x_{\mathscr{C}}(\alpha)$ is $\frac{|\mathscr{C}|}{\sum \mathscr{C}}$, and by the assumption that $\mathscr{C}$ consists of cliques of size at least 2, this is at most 1/2. But the slope of $x_E(\alpha)$ is exactly 1/2. Thus for all $\alpha \ge \alpha_0$, $x_{\mathscr{C}}(\alpha) \le x_E(\alpha)$. Hence the bounds given by formula (11.9) hold.

One can show (see [667]) that for any irreducible clique-cover $\mathscr{C}$ that is not $\{V\}$ nor $\{E\}$,

$$\max_S \{\tau_1(K_h, \mathscr{C}, S), \tau_2(K_h, \mathscr{C}, S)\} \ge \tau_1(K_h, \mathscr{C}, V).$$

Hence, by (11.9),

$$\max_S \{\tau_1(K_h, \mathscr{C}, S), \tau_2(K_h, \mathscr{C}, S)\} \ge \min\{\tau_1(K_h, \{V\}, V), \tau_1(K_h, \{E\}, V)\}.$$

This implies that

$$\tau(K_h) = \begin{cases} n^{-1} m^{-1/h} & \text{for } \alpha \le \alpha_0 \\ n^{-1/(h-1)} m^{-1/2} & \text{for } \alpha \ge \alpha_0, \end{cases}$$

which completes the proof of Corollary 11.6. □

To add to the picture of asymptotic behavior of small cliques in $G(n,m,p)$ we quote the result of Rybarczyk and Stark [667], who with use of Stein's method (see Section 20.3) obtained an upper bound on the total variation distance between the distribution of the number of h-cliques and a respective Poisson distribution for any fixed h.

Theorem 11.7 *Let $G(n,m,p)$ be a random intersection graph, where $m = n^{\alpha}$. Let $c > 0$ be a constant and $h \ge 3$ a fixed integer, and X_n be the random variable counting the number of copies of a complete graph K_h in $G(n,m,p)$.*

(i) If $\alpha < \frac{2h}{h-1}, p \approx cn^{-1} m^{-1/h}$ then

$$\lambda_n = \mathbb{E} X_n \approx c^h / h!$$

and

$$d_{TV}(\mathscr{L}(X_n), \mathrm{Po}(\lambda_n)) = O\left(n^{-\alpha/h}\right);$$

(ii) If $\alpha = \frac{2h}{h-1}, p \approx cn^{-(h+1)/(h-1)}$ then

$$\lambda_n = \mathbb{E} X_n \approx \left(c^h + c^{h(h-1)}\right) / h!$$

and

$$d_{TV}(\mathscr{L}(X_n), \mathrm{Po}(\lambda_n)) = O\left(n^{-2/(h-1)}\right);$$

(iii) If $\alpha > \frac{2h}{h-1}, p \approx cn^{-1/(h-1)} m^{-1/2}$ *then*

$$\lambda_n = \mathbb{E}X_n \approx c^{h(h-1)}/h!$$

and

$$d_{TV}(\mathscr{L}(X_n), \mathrm{Po}(\lambda_n)) = O\left(n^{\left(h - \frac{\alpha(h-1)}{2} - \frac{2}{h-1}\right)} + n^{-1}\right).$$

11.2 Random Geometric Graphs

The graphs we consider in this section are the intersection graphs that we obtain from the intersections of balls in the d-dimensional unit cube, $D = [0,1]^d$ where $d \geq 2$. For simplicity we only consider $d = 2$ in the text.

We let $\mathscr{X} = \{X_1, X_2, \ldots, X_n\}$ be independently and uniformly chosen from $D = [0,1]^2$. For $r = r(n)$ let $G_{\mathscr{X},r}$ be the graph with vertex set $\mathscr{X}$. We join X_i, X_j by an edge if and only if X_j lies in the disk

$$B(X_i, r) = \left\{X \in [0,1]^2 : |X - X_i| \leq r\right\}.$$

Here $|\ |$ denotes Euclidean distance.

For a given set $\mathscr{X}$ we see that increasing r can only add edges and so thresholds are usually expressed in terms of upper/lower bounds on the size of r.

The book by Penrose [619] gives a detailed exposition of this model. Our aim here is to prove some simple results that are not intended to be best possible.

Connectivity

The threshold (in terms of r) for connectivity was shown to be identical with that for minimum degree one, by Gupta and Kumar [388]. This was extended to k-connectivity by Penrose [618]. We do not aim for tremendous accuracy. The simple proof of connectivity was provided to us by Tobias Müller [593].

Theorem 11.8 *Let* $\varepsilon > 0$ *be arbitrarily small and let* $r_0 = r_0(n) = \sqrt{\frac{\log n}{\pi n}}$, *then w.h.p.*

$$G_{\mathscr{X},r} \text{ contains isolated vertices if } r \leq (1-\varepsilon)r_0 \tag{11.10}$$

$$G_{\mathscr{X},r} \text{ is connected if } r \geq (1+\varepsilon)r_0 \tag{11.11}$$

Proof First consider (11.10) and the degree of X_1. Let $\mathscr{E}_1$ be the event that X_1 is within distance r of the boundary ∂D of D. Then

$$\mathbb{P}(X_1 \text{ is isolated} \mid \bar{\mathscr{E}}_1) \geq (1-\pi r^2)^{n-1}.$$

The factor $(1-\pi r^2)^{n-1}$ is the probability that none of $X_2, X_3, \ldots, X_n$ lies in $B(X_1, r)$, given that $B(X_1, r) \subseteq D$.

Now

$$(1-\pi r^2)^{n-1} \geq \left(1 - \frac{(1-\varepsilon)\log n}{n}\right)^n = n^{\varepsilon-1+o(1)}.$$

Now the area with distance r of ∂D is $4r(1-r)$ and so $\mathbb{P}(\bar{\mathscr{E}}_1) = 1 - 4r(1-r)$.

So if I is the set of isolated vertices at distance greater than r of ∂D then $\mathbb{E}(|I|) \geq n^{\varepsilon-1+o(1)}(1-4r) \to \infty$. Now

$$\mathbb{P}(X_1 \in I \mid X_2 \in I) \leq (1-4r(1-r))\left(1 - \frac{\pi r^2}{1-\pi r^2}\right)^{n-2}$$
$$\leq (1+o(1))\,\mathbb{P}(X_1 \in I).$$

The expression $\left(1 - \frac{\pi r^2}{1-\pi r^2}\right)$ is the probability that a random point lies in $B(X_1, r)$, given that it does not lie in $B(X_2, r)$, and that $|X_2 - X_1| \geq 2r$. Equation (11.10) now follows from the Chebyshev inequality Lemma 20.3.

Now consider (11.11). Let $\eta \ll \varepsilon$ be a sufficiently small constant and divide D into ℓ_0^2 sub-squares of side length ηr, where $\ell_0 = 1/\eta r$. We refer to these sub-squares as cells. We can assume that η is chosen so that ℓ_0 is an integer. We say that a cell is *good* if contains at least $i_0 = \eta^3 \log n$ members of $\mathscr{X}$ and *bad* otherwise.

We next let $K = 100/\eta^2$ and consider the number of bad cells in a $K \times K$ square block of cells.

Lemma 11.9 *Let B be a $K \times K$ square block of cells. The following hold w.h.p.:*

(i) If B is further than $100r$ from the closest boundary edge of D then B contains at most $k_0 = (1-\varepsilon/10)\pi/\eta^2$ bad cells.
(ii) If B is within distance $100r$ of exactly one boundary edge of D then B contains at most $k_0/2$ bad cells.
(iii) If B is within distance $100r$ of two boundary edges of D then B contains no bad cells.

Proof (i) There are less than $\ell_0^2 < n$ such blocks. Furthermore, the probability that a fixed block contains k_0 or more bad cells is at most

$$\binom{K^2}{k_0}\left(\sum_{i=0}^{i_0}\binom{n}{i}(\eta r^2)^i(1-\eta^2 r^2)^{n-i}\right)^{k_0}$$
$$\leq \left(\frac{K^2 e}{k_0}\right)^{k_0}\left(2\left(\frac{ne}{i_0}\right)^{i_0}(\eta r^2)^{i_0}e^{-\eta^2 r^2(n-i_0)}\right)^{k_0}. \qquad (11.12)$$

Here we have used Corollary 21.4 to obtain the LHS of (11.12).

Now

$$\left(\frac{ne}{i_0}\right)^{i_0}(\eta r^2)^{i_0}e^{-\eta^2 r^2(n-i_0)}$$
$$\leq n^{O(\eta^3\log(1/\eta))-\eta^2(1+\varepsilon-o(1))/\pi} \leq n^{-\eta^2(1+\varepsilon/2)/\pi}, \qquad (11.13)$$

for η sufficiently small. So we can bound the RHS of (11.12) by

$$\left(\frac{2K^2 e n^{-\eta^2(1+\varepsilon/2)/\pi}}{(1-\varepsilon/10)\pi/\eta^2}\right)^{(1-\varepsilon/10)\pi/\eta^2} \leq n^{-1-\varepsilon/3}. \qquad (11.14)$$

Part (i) follows after inflating the RHS of (11.14) by n to account for the number of choices of block.

(ii) Replacing k_0 by $k_0/2$ replaces the LHS of (11.14) by

$$\left(\frac{4K^2 e n^{-\eta^2(1+\varepsilon/2)/\pi}}{(1-\varepsilon/10)\pi/2\eta^2}\right)^{(1-\varepsilon/10)\pi/2\eta^2} \leq n^{-1/2-\varepsilon/6}. \qquad (11.15)$$

Observe now that the number of choices of block is $O(\ell_0) = o(n^{1/2})$ and then Part (ii) follows after inflating the RHS of (11.15) by $o(n^{1/2})$ to account for the number of choices of block.

(iii) Equation (11.13) bounds the probability that a single cell is bad. The number of cells in question in this case is $O(1)$ and (iii) follows. □

We now do a simple geometric computation to place a lower bound on the number of cells within a ball $B(X,r)$.

Lemma 11.10 *A half-disk of radius $r_1 = r(1-\eta\sqrt{2})$ with diameter part of the grid of cells contains at least $(1-2\eta^{1/2})\pi/2\eta^2$ cells.*

Proof We place the half-disk in a $2r_1 \times r_1$ rectangle. Then we partition the rectangle into $\zeta_1 = r_1/r\eta$ rows of $2\zeta_1$ cells. The circumference of the circle cuts the ith row at a point that is $r_1(1-i^2\eta^2)^{1/2}$ from the center of the row. Thus the ith row contains at least $2\lfloor r_1(1-i^2\eta^2)^{1/2}/r\eta\rfloor$ complete cells. So the half-disk contains at least

$$\frac{2r_1}{r\eta}\sum_{i=1}^{1/\eta}((1-i^2\eta^2)^{1/2}-\eta) \geq \frac{2r_1}{r\eta}\int_{x=1}^{1/\eta-1}((1-x^2\eta^2)^{1/2}-\eta)dx$$
$$= \frac{2r_1}{r\eta^2}\int_{\theta=\arcsin(\eta)}^{\arcsin(1-\eta)}(\cos^2(\theta)-\eta\cos(\theta))d\theta$$
$$\geq \frac{2r_1}{r\eta^2}\left[\frac{\theta}{2}-\frac{\sin(2\theta)}{4}-\eta\right]_{\theta=\arcsin(\eta)}^{\arcsin(1-\eta)}.$$

Now

$$\arcsin(1-\eta) \geq \frac{\pi}{2}-2\eta^{1/2} \text{ and } \arcsin(\eta) \leq 2\eta.$$

So the number of cells is at least

$$\frac{2r_1}{r\eta^2}\left(\frac{\pi}{4}-\eta^{1/2}-\eta\right).$$

This completes the proof of Lemma 11.10. □

We deduce from Lemmas 11.9 and 11.10 that w.h.p

$$X \in \mathscr{X} \text{ implies that } B(X,r_1)\cap D \text{ contains at least one good cell.} \tag{11.16}$$

Now let Γ be the graph whose vertex set consists of the good cells and where cells c_1, c_2 are adjacent if and only if their centers are within distance r_1. Note that if c_1, c_2 are adjacent in Γ, then any point in $\mathscr{X}\cap c_1$ is adjacent in $G_{\mathscr{X},r}$ to any point in $\mathscr{X}\cap c_2$. It follows from (11.16) that all we need to do now is show that Γ is connected.

It follows from Lemma 11.9 that at most π/η^2 rows of a $K\times K$ block contain a bad cell. Thus more than 95% of the rows and of the columns of such a block are free of bad cells. We call such a row or column good. The cells in a good row or column of some $K\times K$ block form part of the same component of Γ. Two neighboring blocks must have two touching good rows or columns so the cells in a good row or column of some block form part of a single component of Γ. Any other component C must be in a block bounded by good rows and columns. But the existence of such a component means that it is surrounded by bad cells and then by Lemma 11.10 that there is a block with at least $(1-2\eta^{1/2})\pi/\eta^2$ bad cells if it is far from the boundary and at least half of this if it is close to the boundary. But this contradicts Lemma 11.9. We have used the following geometric fact. If $A\subseteq\mathbb{R}^2$ then $|A\oplus B(0,r_1)|^{1/2}\geq|A|^{1/2}+|B(0,r_1)|^{1/2}$, where $|A|$ denotes area and $A \oplus B(0,r) = \{z : z = x + y \text{ where } x \in A, y \in B(0,r_1)\}$ is the set of points within distance r of a point in A. The inequality itself is the Brunn–Minkowski inequality, see for example Schneider [676]. When C is close to a boundary, we can work with D and the reflection of D in this boundary. □

Hamiltonicity

The first inroads on the Hamilton cycle problem were made by Diaz, Mitsche and Pérez-Giménez [242]. Best possible results were later given by Balogh, Bollobás, Krivelevich, Müller and Walters [53] and by Müller, Pérez and Wormald [594]. As one might expect Hamiltonicity has a threshold at r close to r_0, where r_0 is given in Theorem 11.8. We now have enough to prove the result from [242].

We start with a simple lemma, taken from [53].

Lemma 11.11 *The subgraph Γ contains a spanning tree of maximum degree at most six.*

Proof Consider a spanning tree T of γ that minimizes the sum of the lengths of the edges joining the centers of the cells. Then T does not have any vertex of degree greater than 6. This is because, if center v were to have degree at least 7, then there are two neighboring centers u, w of v such that the angle between the line segments $[v,u]$ and $[v,w]$ is strictly less than 60 degrees. We can assume without loss of generality that $[v,u]$ is shorter than $[v,w]$. Note that if we remove the edge $\{v,w\}$ and add the edge $\{u,w\}$ then we obtain another spanning tree but with strictly smaller total edge-length, a contradiction. Hence T has maximum degree at most 6. □

Theorem 11.12 *Suppose that $r \geq (1+\varepsilon)r_0$. Then w.h.p. $G_{\mathscr{X},r}$ is Hamiltonian.*

Proof We begin with the tree T promised by Lemma 11.11. Let c be a good cell. We partition the points of $\mathscr{X} \cap c$ into $2d$ roughly equal size sets $P_1, P_2, \ldots, P_{2d}$ where $d \leq 6$ is the degree of c in T. Since, the points of $\mathscr{X} \cap c$ form a clique in $G = G_{\mathscr{X},r}$ we can form $2d$ paths in G from this partition.

We next do a walk W through T, e.g. by BFS that goes through each edge of T twice and passes through each node of Γ a number of times equal to twice its degree in Γ. Each time we pass through a node we traverse the vertices of a new path described in the previous paragraph. In this way we create a cycle H that goes through all the points in $\mathscr{X}$ that lie in good cells.

Now consider the points P in a bad cell c with center x. We create a path in G through P with endpoints x, y, say. Now choose a good cell c' contained in the ball $B(x, r_1)$ and then choose an edge $\{u,v\}$ of H in the cell c'. We merge the points in P into H by deleting $\{u,v\}$ and adding $\{x,u\},\{y,v\}$. To make this work, we must be careful to ensure that we only use an edge of H at most once. But there are $\Omega(\log n)$ edges of H in each good cell and there are $O(1)$ bad cells within distance $2r$ say of any good cell and so this is easily done. □

Chromatic number

We look at the chromatic number of $G_{\mathscr{X},r}$ in a limited range. Suppose that $n\pi r^2 = \frac{\log n}{\omega_r}$ where $\omega_r \to \infty, \omega_r = O(\log n)$. We are below the threshold for connectivity here. We will show that w.h.p.

$$\chi(G_{\mathscr{X},r}) \approx \Delta(G_{\mathscr{X},r}) \approx cl(G_{\mathscr{X},r})$$

where we use cl to denote the size of the largest clique. This is a special case of a result of McDiarmid [565].

We first bound the maximum degree.

Lemma 11.13

$$\Delta(G_{\mathscr{X},r}) \approx \frac{\log n}{\log \omega_r} \quad w.h.p.$$

Proof Let Z_k denote the number of vertices of degree k and let $Z_{\geq k}$ denote the number of vertices of degree at least k. Let $k_0 = \frac{\log n}{\omega_d}$ where $\omega_d \to \infty$ and $\omega_d = o(\omega_r)$, then

$$\mathbb{E}(Z_{\geq k_0}) \leq n\binom{n}{k_0}(\pi r^2)^{k_0} \leq n\left(\frac{ne\omega_d \log n}{n\omega_r \log n}\right)^{\frac{\log n}{\omega_d}} = n\left(\frac{e\omega_d}{\omega_r}\right)^{\frac{\log n}{\omega_d}}.$$

So,

$$\log(\mathbb{E}(Z_{\geq k_0})) \leq \frac{\log n}{\omega_d}(\omega_d + 1 + \log\omega_d - \log\omega_r). \tag{11.17}$$

Now let $\varepsilon_0 = \omega_r^{-1/2}$. Then if

$$\omega_d + \log\omega_d + 1 \leq (1-\varepsilon_0)\log\omega_r$$

then (11.17) implies that $\mathbb{E}(Z_k) \to 0$. This verifies the upper bound on Δ claimed in the lemma.

Now let $k_1 = \frac{\log n}{\widehat{\omega}_d}$ where $\widehat{\omega}_d$ is the solution to

$$\widehat{\omega}_d + \log\widehat{\omega}_d + 1 = (1+\varepsilon_0)\log\omega_r.$$

Next let M denote the set of vertices that are at distance greater than r from any edge of D. Let M_k be the set of vertices of degree k in M. If $\widehat{Z}_k = |M_k|$ then

$$\mathbb{E}(\widehat{Z}_{k_1}) \geq n\mathbb{P}(X_1 \in M) \times \binom{n-1}{k_1}(\pi r^2)^{k_1}(1-\pi r^2)^{n-1-k_1}.$$

$\mathbb{P}(X_1 \in M) \geq 1 - 4r$. Using Lemma 21.1 we obtain

$$\mathbb{E}(\widehat{Z}_{k_1}) \geq (1-4r)\frac{n}{3k_1^{1/2}}\left(\frac{(n-1)e}{k_1}\right)^{k_1}(\pi r^2)^{k_1}e^{-n\pi r^2/(1-\pi r^2)}$$

$$\geq (1-o(1))\frac{n^{1-1/\omega_r}}{3k_1^{1/2}}\left(\frac{e\widehat{\omega}_d}{\omega_r}\right)^{\frac{\log n}{\widehat{\omega}_d}}.$$

So,

$$\log(\mathbb{E}(\widehat{Z}_{k_1})) \geq -o(1) - O(\log\log n) + \frac{\log n}{\widehat{\omega}_d}\left(\widehat{\omega}_d + 1 + \log\widehat{\omega}_d - \log\omega_r - \frac{\widehat{\omega}_d}{\omega_r}\right)$$

$$= \Omega\left(\frac{\varepsilon_0 \log n \log\omega_r}{\widehat{\omega}_d}\right) = \Omega\left(\frac{\log n}{\omega_r^{1/2}}\right) \to \infty.$$

An application of the Chebyshev inequality finishes the proof of the lemma. Indeed,

$$\mathbb{P}(X_1, X_2 \in M_k) \leq \mathbb{P}(X_1 \in M)\,\mathbb{P}(X_2 \in M)$$

$$\times\left(\mathbb{P}(X_2 \in B(X_1, r)) + \left(\binom{n-1}{k_1}(\pi r^2)^{k_1}(1-\pi r^2)^{n-2k_1-2}\right)^2\right)$$

$$\leq (1+o(1))\,\mathbb{P}(X_1 \in M_k)\,\mathbb{P}(X_2 \in M_k).$$

□

Now $\mathrm{cl}(G_{\mathscr{X},r}) \leq \Delta(G_{\mathscr{X},r}) + 1$ and so we now lower bound $\mathrm{cl}(G_{\mathscr{X},r})$ w.h.p. But this is easy. It follows from Lemma 11.13 that w.h.p. there is a vertex X_j with at least $(1-o(1))\frac{\log n}{\log(4\omega_r)}$ vertices in its $r/2$ ball $B(X_j, r/2)$. But such a ball provides a clique of size $(1-o(1))\frac{\log n}{\log(4\omega_r)}$. We have therefore proved the following theorem:

Theorem 11.14 *Suppose that* $n\pi r^2 = \frac{\log n}{\omega_r}$ *where* $\omega_r \to \infty, \omega_r = O(\log n)$*, then w.h.p.*

$$\chi(G_{\mathscr{X},r}) \approx \Delta(G_{\mathscr{X},r}) \approx cl(G_{\mathscr{X},r}) \approx \frac{\log n}{\log\omega_r}.$$

We now consider larger r.

Theorem 11.15 *Suppose that* $n\pi r^2 = \omega_r \log n$*, where* $\omega_r \to \infty, \omega_r = o(n/\log n)$*, then w.h.p.*

$$\chi(G_{\mathscr{X},r}) \approx \frac{\omega_r\sqrt{3}\log n}{2\pi}.$$

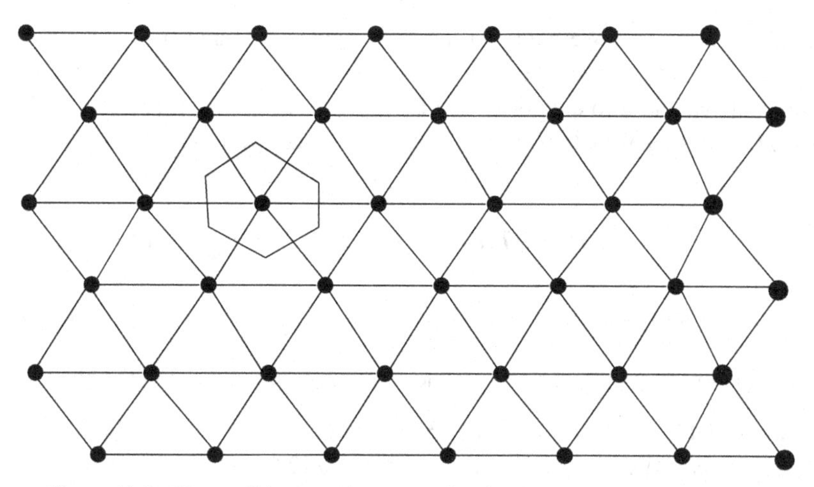

Figure 11.1. The small hexagon is an example of a C_v

Proof First consider the triangular lattice in the plane. This is the set of points $T = \{m_1 a + m_2 b : m_1, m_2 \in \mathbb{Z}\}$ where $a = (0,1), b = (1/2, \sqrt{3}/2)$, see Figure 11.1.

As in the diagram, each $v \in T$ can be placed at the center of a hexagon C_v. The C_vs intersect on a set of measure zero and each C_v has area $\sqrt{3}/2$ and is contained in $B(v, 1/\sqrt{3})$. Let $\Gamma(T,d)$ be the graph with vertex set T where two vertices $x, y \in T$ are joined by an edge if their Euclidean distance $|x - y| < d$.

Lemma 11.16 *(McDiarmid and Reed [567])*

$$\chi(\Gamma(T,d)) \leq (d+1)^2.$$

Proof Let $\delta = \lceil d \rceil$. Let R denote a $\delta \times \delta$ rhombus made up of triangles of T with one vertex at the origin. This rhombus has δ^2 vertices, if we exclude those at the top and right hand end. We give each of these vertices a distinct color and then tile the plane with copies of R. This is a proper coloring, by construction. □

Armed with this lemma we can easily obtain an upper bound on $\chi(G_{\mathscr{X},r})$. Let $\delta = 1/\omega_r^{1/3}$ and let $s = \delta r$. Let sT be the contraction of the lattice T by a factor s, i.e. $sT = \{sx : x \in T\}$. Then if $v \in sT$ let sC_v be the hexagon with center v, sides parallel to the sides of C_v but reduced by a factor s. $|\mathscr{X} \cap sC_v|$ is distributed as $\text{Bin}(n, s^2\sqrt{3}/2)$. So the Chernoff bounds imply that with probability $1 - o(n^{-1})$,

$$sC_v \text{ contains } \leq \theta = \left\lceil (1 + \omega_r^{-1/8}) n s^2 \sqrt{3}/2 \right\rceil \text{ members of } \mathscr{X}. \tag{11.18}$$

Let $\rho = r + 2s/\sqrt{3}$. We note that if $x \in C_v$ and $y \in C_w$ and $|x - y| \le r$ then $|v - w| \le \rho$. Thus, given a proper coloring φ of $\Gamma(sT, \rho)$ with colors $[q]$ we can w.h.p. extend it to a coloring ψ of $G_{\mathscr{X},r}$ with colors $[q] \times [\theta]$. If $x \in sC_v$ and $\varphi(x) = a$ then we let $\psi(x) = (a, b)$ where b ranges over $[\theta]$ as x ranges over $sC_v \cap \mathscr{X}$. So, w.h.p.

$$\chi(G_{\mathscr{X},r}) \le \theta\chi(\Gamma(sT, \rho)) = \theta\chi(\Gamma(T, \rho/s)) \le \theta\left(\frac{\rho}{s} + 1\right)^2$$
$$\approx \frac{ns^2\sqrt{3}}{2} \times \frac{r^2}{s^2} = \frac{\omega_r\sqrt{3}\log n}{2\pi}. \tag{11.19}$$

For the lower bound we use a classic result on packing disks in the plane.

Lemma 11.17 *Let $A_n = [0, n]^2$ and $\mathscr{C}$ be a collection of disjoint disks of unit area that touch A_n. Then $|\mathscr{C}| \le (1 + o(1))\pi n^2/\sqrt{12}$.*

Proof Thue's theorem states that the densest packing of disjoint same size disks in the plane is the hexagonal packing that has density $\lambda = \pi/\sqrt{12}$. Let $\mathscr{C}'$ denote the disks that are contained entirely in A_n, then we have

$$|\mathscr{C}'| \ge |\mathscr{C}| - O(n) \text{ and } |\mathscr{C}'| \le \frac{\pi n^2}{\sqrt{12}}.$$

The first inequality comes from the fact that if $C \in \mathscr{C} \setminus \mathscr{C}'$ then it is contained in a perimeter of width $O(1)$ surrounding A_n. □

Now consider the subgraph H of $G_{\mathscr{X},r}$ induced by the points of $\mathscr{X}$ that belong to the square with center $(1/2, 1/2)$ and sides $1 - 2r$. It follows from Lemma 11.17 that if $\alpha(H)$ is the size of the largest independent set in H then $\alpha(H) \le (1 + o(1))2/r^2\sqrt{3}$. This is because if S is an independent set of H then the disks $B(x, r/2)$ for $x \in S$ are necessarily disjoint. Now using the Chernoff bounds, we see that w.h.p. H contains at least $(1 - o(1))n$ vertices. Thus

$$\chi(G_{\mathscr{X},r}) \ge \chi(H) \ge \frac{|V(H)|}{\alpha(H)} \ge (1 - o(1))\frac{r^2\sqrt{3}n}{2} = (1 - o(1))\frac{\omega_r\sqrt{3}\log n}{2\pi}.$$

This completes the proof of Theorem 11.15. □

11.3 Exercises

11.3.1 Show that if $p = \omega(n)/\left(n\sqrt{m}\right)$, and $\omega(n) \to \infty$, then $G(n, m, p)$ has w.h.p. at least one edge.

11.3.2 Show that if $p = (2\log n + \omega(n))/m)^{1/2}$ and $\omega(n) \to -\infty$ then w.h.p. $G(n, m, p)$ is not complete.

11.3.3 Prove that the bound (11.2) holds.

11.3.4 Prove that the bound (11.3) holds.

11.3.5 Prove that the bound (11.4) holds.

11.3.6 Prove the claims in Lemma 11.2.

11.3.7 Let X denotes the number of isolated vertices in the Binomial random intersection graph $G(n,m,p)$, where $m = n^\alpha$, $\alpha > 0$. Show that if

$$p = \begin{cases} (\log n + \varphi(n))/m & \text{when } \alpha \le 1 \\ \sqrt{(\log n + \varphi(n))/(nm)} & \text{when } \alpha > 1, \end{cases}$$

then $\mathbb{E} X \to e^{-c}$ if $\lim_{n\to\infty} \varphi(n) \to c$, for any real c.

11.3.8 Find the variance of the random variable X counting isolated vertices in $G(n,m,p)$.

11.3.9 Let Y be a random variable which counts vertices of degree greater than one in $G(n,m,p)$, with $m = n^\alpha$ and $\alpha > 1$. Show that for $p^2m^2n \gg \log n$

$$\lim_{n\to\infty} \mathbb{P}\left(Y > 2p^2m^2n\right) = 0.$$

11.3.10 Suppose that $r \ge (1+\varepsilon)r_0$, as in Theorem 11.8. Show that if $1 \le k = O(1)$ then $G_{\mathscr{X},r}$ is k-connected w.h.p.

11.3.11 Show that if $2 \le k = O(1)$ and $r \gg n^{-\frac{k}{2(k-1)}}$ then w.h.p. $G_{\mathscr{X},r}$ contains a k-clique. On the other hand, show that if $r = o(n^{-\frac{k}{2(k-1)}})$ then $G_{\mathscr{X},r}$ contains no k-clique.

11.3.12 Suppose that $r \gg \sqrt{\frac{\log n}{n}}$. Show that w.h.p. the diameter of $G_{\mathscr{X},r} = \Theta\left(\frac{1}{r}\right)$.

11.3.13 Given $\mathscr{X}$ and an integer k we define the k-nearest neighbor graph $G_{k-NN,\mathscr{X}}$ as follows: We add an edge between x and y of $\mathscr{X}$ if and only if y is one of x's k nearest neighbors, in Euclidean distance or vice versa. Show that if $k \ge C\log n$ for a sufficiently large C then $G_{k-NN,\mathscr{X}}$ is connected w.h.p.

11.3.14 Suppose that we independently deposit n random black points $\mathscr{X}_b$ and n random white points $\mathscr{X}_w$ into D. Let $B_{\mathscr{X}_b,\mathscr{X}_w,r}$ be the bipartite graph where we connect $x \in \mathscr{X}_b$ with $\mathscr{X}_w$ if and only if $|x-y| \le r$. Show that if $r \gg \sqrt{\frac{\log n}{n}}$ then w.h.p. $B_{\mathscr{X}_b,\mathscr{X}_w,r}$ contains a perfect matching.

11.4 Notes

Binomial Random Intersection Graphs

For $G(n,m,p)$ with $m = n^{\alpha}$, α constant, Rybarczyk and Stark [668] provided a condition, called *strictly α-balanced* for the Poisson convergence for the number of induced copies of a fixed subgraph, thus complementing the results of Theorem 11.5 and generalizing Theorem 11.7. (thresholds for small subgraphs in a related model of random intersection digraph are studied by Kurauskas [518].)

Rybarczyk [663] introduced a coupling method to find thresholds for many properties of the Binomial random intersection graph. The method is used to establish *sharp threshold functions* for k-connectivity, the existence of a perfect matching and the existence of a Hamilton cycle.

Stark [690] determined the distribution of the degree of a typical vertex of $G(n,m,p)$, $m = n^{\alpha}$ and showed that it changes sharply between $\alpha < 1, \alpha = 1$ and $\alpha > 1$.

Behrisch [66] studied the evolution of the order of the largest component in $G(n,m,p)$, $m = n^{\alpha}$ when $\alpha \neq 1$. He showed that when $\alpha > 1$ the random graph $G(n,m,p)$ behaves like $\mathbb{G}_{n,p}$ in that a giant component of size order n appears w.h.p. when the expected vertex degree exceeds one. This is not the case when $\alpha < 1$. There is a jump in the order of size of the largest component, but not to one of linear size. Further study of the component structure of $G(n,m,p)$ for $\alpha = 1$ is due to Lageras and Lindholm in [520].

Behrisch, Taraz and Ueckerdt [67] studied the evolution of the chromatic number of a random intersection graph and showed that, in a certain range of parameters, these random graphs can be colored optimally with high probability using various greedy algorithms. Constructions of independent sets in random intersection graphs are given by Rybarczyk in [665].

Uniform Random Intersection Graphs

Uniform random intersection graphs differ from the Binomial random intersection graph in the way a subset of the set M is defined for each vertex of V. Now for every $k = 1, 2, \ldots, n$, each S_k has fixed size r and is randomly chosen from the set M. We use the notation $G(n,m,r)$ for an r-uniform random intersection graph. This version of a random intersection graph was introduced by Eschenauer and Gligor [284] and, independently, by Godehardt and Jaworski [373].

Bloznelis, Jaworski and Rybarczyk [101] determined the emergence of the giant component in $G(n,m,r)$ when $n(\log n)^2 = o(m)$. A precise study of the phase transition of $G(n,m,r)$ is due to Rybarczyk [664]. She proved that if $c > 0$ is a constant, $r = r(n) \geq 2$ and $r(r-1)n/m \approx c$, then if $c < 1$ then w.h.p. the largest component of $G(n,m,r)$ is of size $O(\log n)$, while if $c > 1$ w.h.p. there is a single giant component containing a constant fraction of all vertices, while the second largest component is of size $O(\log n)$.

The connectivity of $G(n,m,r)$ was studied by various authors, among them by Eschenauer and Gligor [284] followed by DiPietro, Mancini, Mei, Panconesi and Radhakrishnan [247], Blackbourn and Gerke [90] and Yagan and Makowski [730]. Finally, Rybarczyk [664] determined the sharp threshold for this property. She proved that if $c > 0$ is a constant, $\omega(n) \to \infty$ as $n \to \infty$ and $r^2 n/m = \log n + \omega(n)$, then similarly as in $\mathbb{G}_{n,p}$, the uniform random intersection graph $G(n,m,r)$ is disconnected w.h.p. if $\omega(n) \to \infty$, is connected w.h.p. if $\omega(n) \to \infty$, while the probability that $G(n,m,r)$ is connected tends to $e^{-e^{-c}}$ if $\omega(n) \to c$. The Hamiltonicity of $G(n,m,r)$ was studied in [104] and by Nicoletseas, Raptopoulos and Spirakis [607].

If in the uniform model we require $|S_i \cap S_j| \geq s$ to connect vertices i and j by an edge, then we denote this random intersection graph by $G_s(n,m,r)$. Bloznelis, Jaworski and Rybarczyk [101] studied phase transition in $G_s(n,m,r)$. Bloznelis and Łuczak [103] proved that w.h.p. for even n the threshold for the property that $G_s(n,m,r)$ contains a perfect matching is the same as that for $G_s(n,m,r)$ being connected. Bloznelis and Rybarczyk [105] showed that w.h.p. the edge density threshold for the property that each vertex of $G_s(n,m,r)$ has degree at least k is the same as that for $G_s(n,m,r)$ being k-connected (for related results see [735]).

Generalized Random Intersection Graphs

Godehardt and Jaworski [373] introduced a model that generalizes both the Binomial and uniform models of random intersection graphs. Let P be a probability measure on the set $\{0,1,2,\ldots,m\}$. Let $V = \{1,2,\ldots,n\}$ be the vertex set. Let $M = \{1,2,\ldots,m\}$ be the set of *attributes*. Let $S_1, S_2, \ldots, S_n$ be independent random subsets of M such that for any $v \in V$ and $S \subseteq M$ we have $\mathbb{P}(S_v = S) = P(|S|)/\binom{m}{|S|}$. If we put an edge between any pair of vertices i and j when $S_i \cap S_j \neq \emptyset$, then we denote such a random intersection graph as $G(n,m,P)$, while if the edge is inserted if $|S_i \cap S_i| \geq s$, $s \geq 1$, the respective graph is denoted as $G_s(n,m,P)$. Bloznelis [94] extended these definitions to random intersection digraphs.

The study of the degree distribution of a typical vertex of $G(n,m,P)$ is given in [446], [230] and [92], see also [447]. Bloznelis (see [93] and [95]) showed that the order of the largest component L_1 of $G(n,m,P)$ is asymptotically equal to $n\rho$, where ρ denotes the non-extinction probability of a related multi-type Poisson branching process. Kurauskas and Bloznelis [519] studied the asymptotic order of the clique number of the sparse random intersection graph $G_s(n,m,P)$.

Finally, a dynamic approach to random intersection graphs is studied by Barbour and Reinert [61], Bloznelis and Karoński [102], Bloznelis and Goetze [99] and Britton, Deijfen, Lageras and Lindholm [159].

One should also notice that some of the results on the connectivity of random intersection graphs can be derived from the corresponding results for random hyperghraphs, see for example [499], [675] and [374].

Inhomogeneous Random Intersection Graphs

Nicoletseas, Raptopoulos and Spirakis [606] introduced a generalization of the Binomial random intersection graph $G(n,m,p)$ in the following way. As before let n,m be positive integers and let $0 \leq p_i \leq 1, i = 1,2,\ldots,m$. Let $V = \{1,2,\ldots,n\}$ be the set of vertices of our graph and for every $1 \leq k \leq n$, let S_k be a random subset of the set $M = \{1,2,\ldots,m\}$ formed by selecting the ith element of M independently with probability p_i. Let $\mathbf{p} = (p_i)_{i=1}^m$. We define the *inhomogeneous random intersection graph* $G(n,m,\mathbf{p})$ as the intersection graph of sets S_k, $k = 1,2,\ldots n$. Here two vertices i and j are adjacent in $G(n,m,\mathbf{p})$ if and only if $S_i \cap S_j \neq \emptyset$. Several asymptotic properties of the random graph $G(n,m,\mathbf{p})$ were studied, such as: large independent sets (in [607]), vertex degree distribution (by Bloznelis and Damarackas in [96]), sharp threshold functions for connectivity, matchings and Hamiltonian cycles (by Rybarczyk in [666]) and the size of the largest component (by Bradonjić, Elsässer, Friedrich, Sauerwald and Stauffer in [156]).

To learn more about different models of random intersection graphs and about other results we refer the reader to recent review papers [97] and [98].

Random Geometric Graphs

McDiarmid and Müller [566] gave the leading constant for the chromatic number when the average degree is $\Theta(\log n)$. The paper also shows a "surprising" phase change for the relation between χ and ω. Also the paper extends the setting to arbitrary dimensions. Müller [592] proved a two-point concentration for the clique number and chromatic number when $nr^2 = o(\log n)$.

Blackwell, Edmonson-Jones and Jordan [91] studied the spectral properties of the adjacency matrix of a random geometric graph (RGG). Rai [641] studied the spectral measure of the transition matrix of a simple random walk. Preciado and Jadbabaie [635] studied the spectrum of RGGs in the context of the spreading of viruses.

Sharp thresholds for monotone properties of RGGs were shown by McColm [560] in the case $d = 1$ viz. a graph defined by the intersection of random sub-intervals. And for all $d \geq 1$ by Goel, Rai and Krishnamachari [375].

First-order expressible properties of random points $\mathscr{X} = \{X_1, X_2, \ldots, X_n\}$ on a unit circle were studied by McColm [559]. The graph has vertex set $\mathscr{X}$ and vertices are joined by an edge if and only if their angular distance is less than some parameter d. He showed among other things that for each fixed d, the set of a.s. first order sentences in this model is a complete non-categorical theory. McColm's results were anticipated in a more precise paper [372] by Godehardt and Jaworski, where the case $d = 1$, i.e. the evolution a random interval graph, was studied.

Diaz, Penrose, Petit and Serna [244] studied the approximability of several layout problems on a family of RGGs. The layout problems that they consider are bandwidth, minimum linear arrangement, minimum cut width, minimum sum cut, vertex separation and edge bisection. Diaz, Grandoni and Marchetti-Spaccemela [241] derived a constant expected approximation algorithm for the β-balanced cut problem on random geometric graphs: find an edge cut of minimum size whose two sides contain at least βn vertices each. Diaz, Perez-Gimenez, Serna and Wormald [243] consider the connectivity of a graph with vertices moving randomly.

Bradonjić, Elsässer, Friedrich, Sauerwald and Stauffer [155] studied the broadcast time of RGGs. They studied a regime where there is likely to be a single giant component and showed that w.h.p. their broadcast algorithm only requires $O(n^{1/2}/r + \log n)$ rounds to pass information from a single vertex, to every vertex of the giant. They showed on the way that the diameter of the giant is $\Theta(n^{1/2}/r)$ w.h.p. Friedrich, Sauerwald and Stauffer [319] extended this to higher dimensions.

A recent interesting development can be described as *Random Hyperbolic Graphs*. These are related to the graphs of Section 11.2 and are posed as models of real world networks. Here points are randomly embedded into hyperbolic, as opposed to Euclidean space. See for example Bode, Fountoulakis and Müller [106], [107]; Candellero and Fountoulakis [170]; Chen, Fang, Hu and Mahoney [179]; Fountoulakis [309]; Friedrich and Krohmer [318]; Gugelmann, Panagiotou and Peter [387]; Kiwi and Mitsche [487]; Krioukov, Papadopolous, Kitsak, Vahdat and Boguñá [502]; and Papadopolous, Krioukov, Boguñá and Vahdat [615]. One version of this model is described in [309]. The models are a little complicated to describe and we refer the reader to the above references.

12

Digraphs

In graph theory, we sometimes orient edges to create a directed graph or digraph. It is natural to consider randomly generated digraphs and this chapter discusses the component size and connectivity of the simplest model $\mathbb{D}_{n,p}$. Hamiltonicity is discussed in the final section.

12.1 Strong Connectivity

In this chapter we study the random digraph $\mathbb{D}_{n,p}$. This has vertex set $[n]$ and each of the $n(n-1)$ possible edges occurs independently with probability p. We first study the size of the strong components of $\mathbb{D}_{n,p}$. Recall the definition of strong components: Given a digraph $D = (V, A)$ we define the relation ρ on V by $x\rho y$ if there is a path from x to y in D and there is a path from y to x in D. It is easy to show that ρ is an equivalence relation and the equivalence classes are called the strong components of D.

Strong component sizes: sub-critical region

Theorem 12.1 *Let $p = c/n$, where c is a constant, $c < 1$, then w.h.p.*

(i) all strong components of $\mathbb{D}_{n,p}$ are either cycles or single vertices
(ii) the number of vertices on cycles is at most ω, for any $\omega = \omega(n) \to \infty$.

Proof The expected number of cycles is

$$\sum_{k=2}^{n} \binom{n}{k} (k-1)! \left(\frac{c}{n}\right)^k \leq \sum_{k=2}^{n} \frac{c^k}{k} = O(1).$$

Part *(ii)* now follows from the Markov inequality.

To tackle (*i*) we observe that if there is a component that is not a cycle or a single vertex then there is a cycle C and vertices $a,b \in C$ and a path P from a to b that is internally disjoint from C.

However, the expected number of such subgraphs is bounded by

$$\sum_{k=2}^{n}\sum_{l=0}^{n-k}\binom{n}{k}(k-1)!\left(\frac{c}{n}\right)^k k^2\binom{n}{l}l!\left(\frac{c}{n}\right)^{l+1}$$
$$\leq \sum_{k=2}^{\infty}\sum_{l=0}^{\infty}\frac{k^2c^{k+l+1}}{kn} = O(1/n).$$

Here l is the number of vertices on the path P, excluding a and b. □

Strong component sizes: super-critical region

We will prove the following beautiful theorem that is a directed analogue of the existence of a giant component in $\mathbb{G}_{n,p}$. It is due to Karp [474].

Theorem 12.2 *Let $p = c/n$, where c is a constant, $c > 1$, and let x be defined by $x < 1$ and $xe^{-x} = ce^{-c}$. Then w.h.p. $\mathbb{D}_{n,p}$ contains a unique strong component of size $\approx \left(1-\frac{x}{c}\right)^2 n$. All other strong components are of logarithmic size.*

We will prove the above theorem through a sequence of lemmas.

For a vertex v let

$$D^+(v) = \{w : \exists \text{ path } v \text{ to } w \text{ in } \mathbb{D}_{n,p}\}$$
$$D^-(v) = \{w : \exists \text{ path } w \text{ to } v \text{ in } \mathbb{D}_{n,p}\}.$$

We will first prove

Lemma 12.3 *There exist constants α, β, dependent only on c, such that w.h.p. $\nexists\ v$ such that $|D^{\pm}(v)| \in [\alpha \log n, \beta n]$.*

Proof If there is a v such that $|D^+(v)| = s$ then $\mathbb{D}_{n,p}$ contains a tree T of size s, rooted at v such that

(i) all arcs are oriented away from v
(ii) there are no arcs oriented from $V(T)$ to $[n] \setminus V(T)$.

The expected number of such trees is bounded above by

$$s\binom{n}{s}s^{s-2}\left(\frac{c}{n}\right)^{s-1}\left(1-\frac{c}{n}\right)^{s(n-s)} \leq \frac{n}{cs}\left(ce^{1-c+s/n}\right)^s.$$

Now $ce^{1-c} < 1$ for $c \neq 1$ and so there exists β such that when $s \leq \beta n$ we can bound $ce^{1-c+s/n}$ by some constant $\gamma < 1$ (γ depends only on c). In which case

$$\frac{n}{cs}\gamma^s \leq n^{-3} \text{ for } \frac{4}{\log 1/\gamma}\log n \leq s \leq \beta n.$$

□

Fix a vertex $v \in [n]$ and consider a directed BFS from v. Let $S_0^+ = S_0^+(v) = \{v\}$ and given $S_0^+, S_1^+ = S_1^+(v), \ldots, S_k^+ = s_k^+(v) \subseteq [n]$ let $T_k^+ = T_k^+(v) = \bigcup_{i=1}^k S_i^+$ and let

$$S_{k+1}^+ = \left\{w \notin T_k^+ : \exists x \in T_k^+ \text{ such that } (x,w) \in E(\mathbb{D}_{n,p})\right\}.$$

We similarly define $S_0^- = S_0^-(v), S_1^- = S_1^-(v), \ldots, S_k^-(v) = S_k^-, T_k^-(v) \subseteq [n]$ with respect to a directed breadth first search into v.

Not surprisingly, we can show that the subgraph Γ_k induced by T_k^+ is close in distribution to the tree defined by the first $k+1$ levels of a Galton–Watson branching process with Po(c) as the distribution of the number of offspring from a single parent. See Chapter 23 for some salient facts about such a process. Here Po(c) is the Poisson random variable with mean c, i.e.

$$\mathbb{P}(\text{Po}(c) = k) = \frac{c^k e^{-c}}{k!} \qquad \text{for } k = 0, 1, 2, \ldots, .$$

Lemma 12.4 *If $\hat{S}_0, \hat{S}_1, \ldots, \hat{S}_k$ and $\hat{T}_k$ are defined with respect to the Galton–Watson branching process and if $k \leq k_0 = (\log n)^3$ and $s_0, s_1, \ldots, s_k \leq (\log n)^4$ then*

$$\mathbb{P}\left(|S_i^+| = s_i, 0 \leq i \leq k\right) = \left(1 + O\left(\frac{1}{n^{1-o(1)}}\right)\right)\mathbb{P}\left(|\hat{S}_i| = s_i, 0 \leq i \leq k\right).$$

Proof We use the fact that if Po(a), Po(b) are independent then Po(a)+Po(b) has the same distribution as Po($a+b$). It follows that

$$\mathbb{P}\left(|\hat{S}_i| = s_i, 0 \leq i \leq k\right) = \prod_{i=1}^k \frac{(cs_{i-1})^{s_i} e^{-cs_{i-1}}}{s_i!}.$$

Furthermore, putting $t_{i-1} = s_0 + s_1 + \ldots + s_{i-1}$ we have for $v \notin T_{i-1}^+$,

$$\mathbb{P}(v \in S_i^+) = 1 - (1-p)^{s_{i-1}} = s_{i-1}p\left(1 + O\left(\frac{(\log n)^7}{n}\right)\right). \tag{12.1}$$

$$\mathbb{P}\left(|S_i^+| = s_i, 0 \le i \le k\right) =$$

$$= \prod_{i=1}^{k} \binom{n-t_{i-1}}{s_i} \left(\frac{s_i c}{n}\left(1+O\left(\frac{(\log n)^7}{n}\right)\right)\right)^{s_i}$$

$$\times \left(1 - \frac{s_{i-1}c}{n}\left(1+O\left(\frac{(\log n)^7}{n}\right)\right)\right)^{n-t_{i-1}-s_i} \tag{12.2}$$

Here we use the fact that given s_{i-1}, t_{i-1}, the distribution of $|S_i^+|$ is the binomial with $n - t_{i-1}$ trials and probability of success given in (12.1). The lemma follows by simple estimations. □

Lemma 12.5 *For* $1 \le i \le (\log n)^3$

(i) $\mathbb{P}\left(|S_i^+| \ge s\log n \,||S_{i-1}^+| = s\right) \le n^{-10}$

(ii) $\mathbb{P}\left(|\hat{S}_i| \ge s\log n \,||\hat{S}_{i-1}| = s\right) \le n^{-10}$.

Proof

$$\begin{aligned}(i)\ \ \mathbb{P}\left(|S_i^+| \ge s\log n \,||S_{i-1}^+| = s\right) &\le \mathbb{P}\left(\mathrm{Bin}(sn, c/n) \ge s\log n\right)\\ &\le \binom{sn}{s\log n}\left(\frac{c}{n}\right)^{s\log n}\\ &\le \left(\frac{snec}{sn\log n}\right)^{s\log n}\\ &\le \left(\frac{ec}{\log n}\right)^{\log n}\\ &\le n^{-10}.\end{aligned}$$

The proof of (ii) is similar. □

Keeping v fixed we next let

$$\begin{aligned}\mathscr{F} &= \left\{\exists\, i : |T_i^+| > (\log n)^2\right\}\\ &= \left\{\exists\, i \le (\log n)^2 : |T_0^+|, |T_1^+|, \ldots, |T_{i-1}^+| < (\log n)^2 < |T_i^+|\right\}.\end{aligned}$$

Lemma 12.6

$$\mathbb{P}(\mathscr{F}) = 1 - \frac{x}{c} + o(1).$$

Proof Applying Lemma 12.4 we see that

$$\mathbb{P}(\mathscr{F}) = \mathbb{P}(\hat{\mathscr{F}}) + o(1), \tag{12.3}$$

where $\hat{\mathscr{F}}$ is defined with respect to the branching process.

Now let $\hat{\mathscr{E}}$ be the event that the branching process eventually becomes extinct. We write

$$\mathbb{P}(\hat{\mathscr{F}}) = \mathbb{P}(\hat{\mathscr{F}} \mid \neg\hat{\mathscr{E}})\,\mathbb{P}(\neg\hat{\mathscr{E}}) + \mathbb{P}(\hat{\mathscr{F}} \cap \hat{\mathscr{E}}). \tag{12.4}$$

To estimate (12.4) we use Theorem 23.1. Let

$$G(z) = \sum_{k=0}^{\infty} \frac{c^k e^{-c}}{k!} z^k = e^{cz-c}$$

be the probability generating function of $Po(c)$. Then Theorem 23.1 implies that $\rho = \mathbb{P}(\hat{\mathscr{E}})$ is the smallest non-negative solution to $G(\rho) = \rho$. Thus

$$\rho = e^{c\rho - c}.$$

Substituting $\rho = \frac{\xi}{c}$ we see that

$$\mathbb{P}(\hat{\mathscr{E}}) = \frac{\xi}{c} \quad \text{where} \quad \frac{\xi}{c} = e^{\xi - c}, \tag{12.5}$$

and so $\xi = x$.

The lemma follows from (12.4) and (12.5) and $\mathbb{P}(\hat{\mathscr{F}} \mid \neg\hat{\mathscr{E}}) = 1 + o(1)$ (which follows from Lemma 12.5) and

$$\mathbb{P}(\hat{\mathscr{F}} \cap \hat{\mathscr{E}}) = o(1).$$

This in turn follows from

$$\mathbb{P}(\hat{\mathscr{E}} \mid \hat{\mathscr{F}}) = o(1), \tag{12.6}$$

which will be established using the following lemma.

Lemma 12.7 *Each member of the branching process has probability at least* $\varepsilon > 0$ *of producing* $(\log n)^2$ *descendants at depth* $\log n$. *Here* $\varepsilon > 0$ *depends only on* c.

Proof If the current population size of the process is s then the probability that it reaches size at least $\frac{c+1}{2}s$ in the next round is

$$\sum_{k \geq \frac{c+1}{2}s} \frac{(cs)^k e^{-cs}}{k!} \geq 1 - e^{-\alpha s}$$

for some constant $\alpha > 0$ provided $s \geq 100$, say.

Now there is a positive probability ε_1, say, that a single member spawns at least 100 descendants and so there is a probability of at least

$$\varepsilon_1 \left(1 - \sum_{s=100}^{\infty} e^{-\alpha s} \right)$$

that a single object spawns

$$\left(\frac{c+1}{2}\right)^{\log n} \gg (\log n)^2$$

descendants at depth $\log n$.

□

Given a population size between $(\log n)^2$ and $(\log n)^3$ at level i_0, let s_i denote the population size at level $i_0 + i \log n$. Then Lemma 12.7 and the Chernoff bounds imply that

$$\mathbb{P}\left(s_{i+1} \leq \frac{1}{2}\varepsilon s_i (\log n)^2\right) \leq \exp\left\{-\frac{1}{8}\varepsilon^2 s_i (\log n)^2\right\}.$$

It follows that

$$\begin{aligned}\mathbb{P}(\hat{\mathscr{E}} \mid \hat{\mathscr{F}}) &\leq \mathbb{P}\left(\exists i:\ s_i \leq \left(\frac{1}{2}\varepsilon(\log n)^2\right)^i s_0 \middle| s_0 \geq (\log n)^2\right)\\ &\leq \sum_{i=1}^{\infty} \exp\left\{-\frac{1}{8}\varepsilon^2\left(\frac{1}{2}\varepsilon(\log n)^2\right)^i (\log n)^2\right\} = o(1).\end{aligned}$$

This completes the proof (12.6) and of Lemma 12.6. □

We must now consider the probability that both $D^+(v)$ and $D^-(v)$ are large.

Lemma 12.8

$$\mathbb{P}\left(|D^-(v)| \geq (\log n)^2 \mid |D^+(v)| \geq (\log n)^2\right) = 1 - \frac{x}{c} + o(1).$$

Proof Expose $S_0^+, S_1^+, \ldots, S_k^+$ until either $S_k^+ = \emptyset$ or we see that $|T_k^+| \in [(\log n)^2, (\log n)^3]$. Now let S denote the set of edges/vertices defined by $S_0^+, S_1^+, \ldots, S_k^+$.

Let $\mathscr{C}$ be the event that there are no edges from T_l^- to S_k^+ where T_l^- is the set of vertices we reach through our BFS into v, up to the point where we first realize that $D^-(v) < (\log n)^2$ (because $S_i^- = \emptyset$ and $|T_i^-| \leq (\log n)^2$) or we realize that $D^-(v) \geq (\log n)^2$. Then

$$\mathbb{P}(\neg\mathscr{C}) = O\left(\frac{(\log n)^4}{n}\right) = \frac{1}{n^{1-o(1)}}$$

and, as in (12.2),

$$\mathbb{P}\left(|S_i^-| = s_i,\ 0 \le i \le k \mid \mathscr{C}\right) =$$

$$= \prod_{i=1}^{k} \binom{n' - t_{i-1}}{s_i} \left(\frac{s_i c}{n}\left(1 + O\left(\frac{(\log n)^7}{n}\right)\right)\right)^{s_i}$$

$$\times \left(1 - \frac{s_{i-1} c}{n}\left(1 + O\left(\frac{(\log n)^7}{n}\right)\right)\right)^{n' - t_{i-1} - s_i}$$

where $n' = n - |T_k^+|$.

Given this we can prove a conditional version of Lemma 12.4 and continue as before. $\square$

We have now shown that if α is as in Lemma 12.3 and if

$$S = \left\{v : |D^+(v)|, |D^-(v)| > \alpha \log n\right\}$$

then the expectation

$$\mathbb{E}(|S|) = (1 + o(1))\left(1 - \frac{x}{c}\right)^2 n.$$

We also claim that for any two vertices v, w

$$\mathbb{P}(v, w \in S) = (1 + o(1))\,\mathbb{P}(v \in S)\,\mathbb{P}(w \in S) \tag{12.7}$$

and therefore the Chebyshev inequality implies that w.h.p.

$$|S| = (1 + o(1))\left(1 - \frac{x}{c}\right)^2 n.$$

But (12.7) follows in a similar manner to the proof of Lemma 12.8.

All that remains of the proof of Theorem 12.2 is to show that

$$S \text{ is a strong component w.h.p.} \tag{12.8}$$

Recall that any $v \notin S$ is in a strong component of size $\le \alpha \log n$ and so the second part of the theorem will also be complete.

We prove (12.8) by arguing that

$$\mathbb{P}\left(\exists\, v, w \in S : w \notin D^+(v)\right) = o(1). \tag{12.9}$$

In which case, we know that w.h.p. there is a path from each $v \in S$ to every other vertex $w \neq v$ in S.

To prove (12.9) we expose $S_0^+, S_1^+, \ldots, S_k^+$ until we find that $|T_k^+(v)| \ge n^{1/2} \log n$. At the same time we expose $S_0^-, S_1^-, \ldots, S_l^-$ until we find that $|T_l^-(w)| \ge n^{1/2} \log n$. If $w \notin D^+(v)$ then this experiment will have tried at least

$\left(n^{1/2}\log n\right)^2$ times to find an edge from $D^+(v)$ to $D^-(w)$ and failed every time. The probability of this is at most

$$\left(1-\frac{c}{n}\right)^{n(\log n)^2}=o(n^{-2}).$$

This completes the proof of Theorem 12.2. □

Threshold for strong connectivity

Here we prove

Theorem 12.9 *Let $\omega=\omega(n)$, $c>0$ be a constant, and let $p=\frac{\log n+\omega}{n}$, then*

$$\lim_{n\to\infty}\mathbb{P}(\mathbb{D}_{n,p}\text{ is strongly connected})=\begin{cases}0 & \text{if } \omega\to-\infty\\ e^{-2e^{-c}} & \text{if } \omega\to c\\ 1 & \text{if } \omega\to\infty.\end{cases}$$
$$=\lim_{n\to\infty}\mathbb{P}(\nexists\, v \text{ s.t. } d^+(v)=0 \text{ or } d^-(v)=0)$$

Proof

We leave as an exercise to prove that

$$\lim_{n\to\infty}\mathbb{P}(\exists\, v\text{ s.t. } d^+(v)=0\text{ or } d^-(v)=0)=\begin{cases}1 & \text{if } \omega\to-\infty\\ 1-e^{-2e^{-c}} & \text{if } \omega\to c\\ 0 & \text{if } \omega\to\infty.\end{cases}$$

Given this, one only has to show that if $\omega\not\to-\infty$ then w.h.p. there does not exist a set S such that (i) $2\le|S|\le n/2$ and (ii) $E(S:\bar S)=\emptyset$ or $E(\bar S:S)=\emptyset$ and (iii) S induces a connected component in the graph obtained by ignoring orientation. But, here with $s=|S|$,

$$\begin{aligned}\mathbb{P}(\exists\, S)&\le 2\sum_{s=2}^{n/2}\binom{n}{s}s^{s-2}(2p)^{s-1}(1-p)^{s(n-s)}\\ &\le\frac{2n}{\log n}\sum_{s=2}^{n/2}\left(\frac{ne}{s}\right)^s s^{s-2}\left(\frac{2\log n}{n}\right)^s n^{-s(1-s/n)}e^{\omega s/n}\\ &\le\frac{2n}{\log n}\sum_{s=2}^{n/2}(2n^{-(1-s/n)}e^{\omega/n}\log n)^s\\ &=o(1).\end{aligned}$$

□

12.2 Hamilton Cycles

Existence of a Hamilton Cycle

Here we prove the following remarkable inequality. It is due to McDiarmid [562]

Theorem 12.10

$$\mathbb{P}(\mathbb{D}_{n,p} \text{ is Hamiltonian}) \geq \mathbb{P}(\mathbb{G}_{n,p} \text{ is Hamiltonian})$$

Proof We consider an ordered sequence of random digraphs $\Gamma_0, \Gamma_1, \Gamma_2, \dots, \Gamma_N$, $N = \binom{n}{2}$ defined as follows: Let $e_1, e_2, \dots, e_N$ be an enumeration of the edges of the complete graph K_n. Each $e_i = \{v_i, w_i\}$ gives rise to two directed edges $\overrightarrow{e_i} = (v_i, w_i)$ and $\overleftarrow{e_i} = (w_i, v_i)$. In Γ_i we include $\overrightarrow{e_j}$ and $\overleftarrow{e_j}$ independently of each other, with probability p, for $j \leq i$. While for $j > i$ we include both or neither with probability p. Thus, Γ_0 is just $\mathbb{G}_{n,p}$ with each edge $\{v, w\}$ replaced by a pair of directed edges $(v, w), (w, v)$ and $\Gamma_N = \mathbb{D}_{n,p}$. Theorem 12.10 follows from

$$\mathbb{P}(\Gamma_i \text{ is Hamiltonian}) \geq \mathbb{P}(\Gamma_{i-1} \text{ is Hamiltonian}).$$

To prove this we condition on the existence or otherwise of directed edges associated with $e_1, \dots, e_{i-1}, e_{i+1}, \dots, e_N$. Let $\mathscr{C}$ denote this conditioning. Either

(i) $\mathscr{C}$ gives us a Hamilton cycle without arcs associated with e_i, or
(ii) $\exists$ a Hamilton cycle if at least one of $\overrightarrow{e_i}, \overleftarrow{e_i}$ is present.

In Γ_{i-1} this happens with probability p, while in Γ_i this happens with probability $1 - (1-p)^2 > p$.

Note that we will never require that **both** $\overrightarrow{e_i}, \overleftarrow{e_i}$ occur. □

Theorem 12.10 was subsequently improved by Frieze [323], who proved the equivalent of Theorem 6.5.

Theorem 12.11 *Let $\omega = \omega(n), C > 0$ be a constant, and let $p = \frac{\log n + \omega}{n}$, then*

$$\lim_{n\to\infty} \mathbb{P}(\mathbb{D}_{n,p} \text{ has a Hamilton cycle}) = \begin{cases} 0 & \text{if } \omega \to -\infty \\ e^{-2e^{-c}} & \text{if } \omega \to c \\ 1 & \text{if } \omega \to \infty. \end{cases}$$

Number of Distinct Hamilton Cycles

Here we give an elegant result of Ferber, Kronenberg and Long [294].

Theorem 12.12 *Let $p = \omega\left(\frac{\log^2 n}{n}\right)$. Then w.h.p. $\mathbb{D}_{n,p}$ contains $e^{o(n)} n!p^n$ directed Hamilton cycles.*

Proof The upper bound follows from the first moment method. Let X_H denote the number of Hamilton cycles in $D = \mathbb{D}_{n,p}$. Now $\mathbb{E}X_H = (n-1)!p^n$, and therefore the Markov inequality implies that w.h.p. we have $X_H \leq e^{o(n)} n!p^n$.

For the lower bound let $\alpha := \alpha(n)$ be a function tending slowly to infinity with n. Let $S \subseteq V(G)$ be a fixed set of size s, where $s \approx \frac{n}{\alpha \log n}$ and let $V' = V \setminus S$. Moreover, assume that s is chosen so that $|V'|$ is divisible by integer $\ell = 2\alpha \log n$. From now on the set S will be fixed and we will use it for closing Hamilton cycles. Our strategy is as follows: we first expose all the edges within V', and show that one can find the "correct" number of distinct families $\mathscr{P}$ consisting of $m := |V'|/\ell$ vertex-disjoint paths which span V'. Then, we expose all the edges with at least one endpoint in S, and show that w.h.p. one can turn "most" of these families into Hamilton cycles and that all of these cycles are distinct.

We take a random partitioning $V' = V_1 \cup \cdots \cup V_\ell$ such that all the V_is are of size m. Let us denote by D_j the bipartite graph with parts V_j and V_{j+1}. Observe that D_j is distributed as $\mathbb{G}_{m,m,p}$, and therefore, since $p = \omega\left(\frac{\log n}{m}\right)$, by Exercise 12.3.2, with probability $1 - n^{-\omega(1)}$ we conclude that D_j contains $(1-o(1))m$ edge-disjoint perfect matchings (in particular, a $(1-o(1))m$ regular subgraph). The Van der Waerden conjecture proved by Egorychev [273] and by Falikman [287] implies the following: Let $G = (A \cup B, E)$ be an r-regular bipartite graph with part sizes $|A| = |B| = n$. Then, the number of perfect matchings in G is at least $\left(\frac{r}{n}\right)^n n!$.

Applying this and the union bound, it follows that w.h.p. each D_j contains at least $(1-o(1))^m m!p^m$ perfect matchings for each j. Taking the union of one perfect matching from each of the D_j's we obtain a family $\mathscr{P}$ of m vertex disjoint paths which spans V'. Therefore, there are

$$((1-o(1))^m m!p^m)^\ell = (1-o(1))^{n-s} (m!)^\ell p^{n-s}$$

distinct families $\mathscr{P}$ obtained from this partitioning in this manner. Since this occurs w.h.p. we conclude (applying the Markov inequality to the number of partitions for which the bound fails) that this bound holds for $(1-o(1))$-fraction of such partitions. Since there are $\frac{(n-s)!}{(m!)^\ell}$ such partitions, one can find at least

$$(1-o(1))\frac{(n-s)!}{(m!)^\ell}(1-o(1))^{n-s} (m!)^\ell p^{n-s}$$
$$= (1-o(1))^{n-s}(n-s)!p^{n-s} = (1-o(1))^n n!p^n$$

distinct families, each of which consists of exactly m vertex-disjoint paths of size ℓ (for the last equality, we used the fact that $s = o(n/\log n)$).

We show next how to close a given family of paths into a Hamilton cycle. For each such family $\mathscr{P}$, let $A := A(\mathscr{P})$ denote the collection of all pairs (s_P, t_P) where s_P is a starting point and t_P is the endpoint of a path $P \in \mathscr{P}$, and define an auxiliary directed graph $D(A)$ as follows. The vertex set of $D(A)$ is $V(A) = S \cup \{z_P = (s_P, t_P) : z_P \in A\}$. Edges of $D(A)$ are determined as follows: if $u, v \in S$ and $(u, v) \in E(D)$ then (u, v) is an edge of $D(A)$. The in-neighbors (out-neighbors) of vertices z_P in S are the in-neighbors of s_P in D (out-neighbors of t_P). Lastly, (z_P, z_Q) is an edge of $D(A)$ if (t_P, s_Q) is an edge D.

Clearly $D(A)$ is distributed as $\mathbb{D}_{s+m,p}$, and a Hamilton cycle in $D(A)$ corresponds to a Hamilton cycle in D after adding the corresponding paths between each s_P and t_P. Now distinct families $\mathscr{P} \neq \mathscr{P}'$ yield distinct Hamilton cycles (to see this, just delete the vertices of S from the Hamilton cycle, to recover the paths). Using Theorem 12.11 we see that for $p = \omega(\log n/(s+m)) = \omega(\log(s+m)/(s+m))$, the probability that $D(A)$ does not have a Hamilton cycle is $o(1)$. Therefore, using the Markov inequality we see that for almost all of the families $\mathscr{P}$, the corresponding auxiliary graph $D(A)$ is indeed Hamiltonian and we have at least $(1-o(1))^n n! p^n$ distinct Hamilton cycles, as desired. □

12.3 Exercises

12.3.1 Let $p = \frac{\log n + (k-1)\log\log n + \omega}{n}$ for a constant $k = 1, 2, \ldots$. Show that w.h.p. $\mathbb{D}_{np}$ is k-strongly connected.

12.3.2 The Gale–Ryser theorem states: Let $G = (A \cup B, E)$ be a bipartite graph with parts of sizes $|A| = |B| = n$. Then, G contains an r-factor if and only if for every two sets $X \subseteq A$ and $Y \subseteq B$, we have

$$e_G(X, Y) \geq r(|X| + |Y| - n).$$

Show that if $p = \omega(\log n/n)$ then with probability $1 - o(1/n)$, $\mathbb{G}_{n,n,p}$ contains $(1 - n^{-\omega(1)}$ edge disjoint perfect matchings.

12.3.3 Show that if $p = \omega((\log n)^2/n)$ then w.h.p. $\mathbb{G}_{n,p}$ contains $e^{o(n)} n! p^n$ distinct Hamilton cycles.

12.3.4 A *tournament* T is an orientation of the complete graph K_n. In a random tournament, edge $\{u, v\}$ is oriented from u to v with probability 1/2 and from v to u with probability 1/2. Show that w.h.p. a random tournament is strongly connected.

12.3.5 Let T be a random tournament. Show that w.h.p. the size of the largest acyclic sub-tournament is asymptotic to $2\log_2 n$. (A tournament is acyclic if it contains no directed cycles.)

12.3.6 Suppose that $0 < p < 1$ is constant. Show that w.h.p. the size of the largest acyclic tournament contained in $\mathbb{D}_{np}$ is asymptotic to $2\log_b n$ where $b = 1/p$.

12.3.7 Let $mas(D)$ denote the number of vertices in the largest acyclic subgraph of a digraph D. Suppose that $0 < p < 1$ is constant. Show that w.h.p. $mas(\mathbb{D}_{n,p}) \leq \frac{4\log n}{\log q}$ where $q = \frac{1}{1-p}$.

12.3.8 Consider the random digraph D_n obtained from $G_{n,1/2}$ by orienting edge (i,j) from i to j when $i < j$. This can be viewed as a partial order on $[n]$ and is called a *Random Graph Order*. Show that w.h.p. D_n contains a path of length at least $0.51n$. (In terms of partial orders, this bounds the height of the order.)

12.4 Notes

Packing

The paper of Frieze [323] was in terms of the hitting time for a digraph process $\mathbb{D}_t$. It proves that the first time that the $\delta^+(G_t), \delta^-(G_t) \geq k$ is w.h.p. the time when G_t has k edge disjoint Hamilton cycles. The paper of Ferber, Kronenberg and Long [294] shows that if $p = \omega((\log n)^4/n)$ then w.h.p. $\mathbb{D}_{n,p}$ contains $(1 - o(1))np$ edge disjoint Hamilton cycles.

Long Cycles

The papers by Hefetz, Steger and Sudakov [405] and by Ferber, Nenadov, Noever, Peter and Škorić [296] study the local resilience of having a Hamilton cycle. In particular, [296] proves that if $p \gg \frac{(\log n)^8}{n}$ then w.h.p. one can delete any subgraph H of $\mathbb{D}_{n,p}$ with maximum degree at most $(\frac{1}{2} - \varepsilon)np$ and still leave a Hamiltonian subgraph.

Krivelevich, Lubetzky and Sudakov [507] proved that w.h.p. the random digraph $\mathbb{D}_{n,p}, p = c/n$ contains a directed cycle of length $(1 - (1 + \varepsilon_c)e^{-c})n$, where $\varepsilon_c \to 0$ as $c \to \infty$.

Cooper, Frieze and Molloy [220] showed that a random regular digraph with indegree = outdegree = r is Hamiltonian w.h.p. if and only if $r \geq 3$.

Connectivity

Cooper and Frieze [209] studied the size of the largest strong component in a random digraph with a given degree sequence. The strong connectivity of an inhomogeneous random digraph was studied by Bloznelis, Götze and Jaworski in [100].

13
Hypergraphs

In this chapter we discuss random k-uniform hypergraphs. We are concerned with the models $\mathbb{H}_{n,p;k}$ and $\mathbb{H}_{n,m;k}$. For $\mathbb{H}_{n,p;k}$ we consider the hypergraph with vertex set $[n]$ in which each possible k-set in $\binom{[n]}{k}$ is included as an edge with probability p. In $\mathbb{H}_{n,m;k}$ the edge set is a random m-subset of $\binom{[n]}{k}$. The parameter k is fixed and independent of n throughout this chapter.

Many of the properties of $\mathbb{G}_{n,p}$ and $\mathbb{G}_{n,m}$ have been generalized without too much difficulty to these models of hypergraphs. Hamilton cycles have only recently been tackled with any success. Surprisingly enough, in some cases it is enough to use the Chebyshev inequality and we will describe these cases. Perfect matchings have also been tackled recently. Here the proof is a remarkable tour de force and we include it as some of the ideas will likely be of use in other places.

13.1 Hamilton Cycles

Suppose that $1 \leq \ell < k$. An *ℓ-overlapping Hamilton cycle C* in a k-uniform hypergraph $H = (V, \mathscr{E})$ on n vertices is a collection of $m_\ell = n/(k-\ell)$ edges of H such that for some cyclic order of $[n]$ every edge consists of k consecutive vertices and for every pair of consecutive edges E_{i-1}, E_i in C (in the natural ordering of the edges) we have $|E_{i-1} \cap E_i| = \ell$. Thus, in every ℓ-overlapping Hamilton cycle the sets $C_i = E_i \setminus E_{i-1}$, $i = 1, 2, \ldots, m_\ell$, are a partition of V into sets of size $k - \ell$. Hence, $m_\ell = n/(k-\ell)$. We thus always assume, when discussing ℓ-overlapping Hamilton cycles, that this necessary condition, $k - \ell$ divides n, is fulfilled. In the literature, when $\ell = k-1$ we have a *tight* Hamilton cycle and when $\ell = 1$ we have a *loose* Hamilton cycle.

In this section we restrict our attention to the case $\ell = k - 1$, i.e. tight Hamilton cycles. So when we say that a hypergraph is Hamiltonian, we mean that it contains a tight Hamilton cycle. The proof extends easily to $\ell \geq 2$. The case $\ell = 1$ poses some more problems and is discussed in Dudek and Frieze [252], Dudek, Frieze, Loh and Speiss [254] and Frieze [326], and in Ferber [292]. The following theorem is from Dudek and Frieze [253]. Furthermore, we assume that $k \geq 3$.

Theorem 13.1

(i) *If $p \leq (1-\varepsilon)e/n$, then w.h.p. $\mathbb{H}_{n,p;k}$ is not Hamiltonian.*

(ii) *If $k = 3$ and $np \to \infty$ then $\mathbb{H}_{n,p;k}$ is Hamiltonian w.h.p.*

(iii) *For all fixed $\varepsilon > 0$, if $k \geq 4$ and $p \geq (1+\varepsilon)e/n$, then w.h.p. $\mathbb{H}_{n,p;k}$ is Hamiltonian.*

Proof We will prove parts (*i*) and (*ii*) and leave the proof of (*iii*) as an exercise, with a hint.

Let $([n], \mathscr{E})$ be a k-uniform hypergraph. A permutation π of $[n]$ is *Hamilton cycle inducing* if

$$E_\pi(i) = \{\pi(i-1+j) : j \in [k]\} \in \mathscr{E} \textit{ for all } i \in [n].$$

(We use the convention $\pi(n+r) = \pi(r)$ for $r > 0$.) Let the term *hamperm* refer to such a permutation.

Let X be the random variable that counts the number of hamperms π for $\mathbb{H}_{n,p;k}$. Every Hamilton cycle induces at least one hamperm and so we can concentrate on estimating $\mathbb{P}(X > 0)$.

Now

$$\mathbb{E}(X) = n!p^n.$$

This is because π induces a Hamilton cycle if and only if a certain n edges are all in $\mathbb{H}_{n,p;k}$.

For part (*i*) we use Stirling's formula to argue that

$$\mathbb{E}(X) \leq 3\sqrt{n}\left(\frac{np}{e}\right)^n \leq 3\sqrt{n}(1-\varepsilon)^n = o(1).$$

This verifies part (*i*).

We see that

$$\mathbb{E}(X) \geq \left(\frac{np}{e}\right)^n \to \infty \tag{13.1}$$

in parts (*ii*) and (*iii*).

Fix a hamperm π. Let $H(\pi) = (E_\pi(1), E_\pi(2), \ldots, E_\pi(n))$ be the Hamilton cycle induced by π. Then let $N(b,a)$ be the number of permutations π' such that $|E(H(\pi)) \cap E(H(\pi'))| = b$ and $E(H(\pi)) \cap E(H(\pi'))$ consists of a edge disjoint paths. Here a path is a maximal sub-sequence $F_1, F_2, \ldots, F_q$ of the edges of $H(\pi)$ such that $F_i \cap F_{i+1} \neq \emptyset$ for $1 \leq i < q$. The set $\bigcup_{j=1}^q F_j$ may contain other edges of $H(\pi)$. Observe that $N(b,a)$ does not depend on π.

Note that

$$\frac{\mathbb{E}(X^2)}{\mathbb{E}(X)^2} = \frac{n!N(0,0)p^{2n}}{\mathbb{E}(X)^2} + \sum_{b=1}^{n}\sum_{a=1}^{b} \frac{n!N(b,a)p^{2n-b}}{\mathbb{E}(X)^2}.$$

Since trivially, $N(0,0) \leq n!$, we obtain,

$$\frac{\mathbb{E}(X^2)}{\mathbb{E}(X)^2} \leq 1 + \sum_{b=1}^{n}\sum_{a=1}^{b} \frac{n!N(b,a)p^{2n-b}}{\mathbb{E}(X)^2}. \tag{13.2}$$

We show that

$$\sum_{b=1}^{n}\sum_{a=1}^{b} \frac{n!N(b,a)p^{2n-b}}{\mathbb{E}(X)^2} = \sum_{b=1}^{n}\sum_{a=1}^{b} \frac{N(b,a)p^{n-b}}{\mathbb{E}(X)} = o(1). \tag{13.3}$$

The Chebyshev inequality implies that

$$\mathbb{P}(X=0) \leq \frac{\mathbb{E}(X^2)}{\mathbb{E}(X)^2} - 1 = o(1),$$

as required.

It remains to show (13.3). First we find an upper bound on $N(b,a)$. Choose a vertices v_i, $1 \leq i \leq a$, on π. We have at most

$$n^a \tag{13.4}$$

choices. Let

$$b_1 + b_2 + \cdots + b_a = b,$$

where $b_i \geq 1$ is an integer for every $1 \leq i \leq a$. Note that this equation has exactly

$$\binom{b-1}{a-1} < 2^b \tag{13.5}$$

solutions. For every i, we choose a path of length b_i in $H(\pi)$ which starts at v_i. Suppose a path consists of edges $F_1, F_2, \ldots, F_q$, $q = b_i$. Assuming that $F_1, \ldots, F_j$ are chosen, we have at most k possibilities for F_{j+1}. Hence, every such a path can be selected in most k^{b_i} ways. Consequently, we have at most

$$\prod_{i=1}^{a} k^{b_i} = k^b$$

choices for all a paths.

Thus, by the above considerations we can find a edge disjoint paths in $H(\pi)$ with the total of b edges in at most

$$n^a(2k)^b \tag{13.6}$$

many ways.

Let $P_1, P_2, \ldots, P_a$ be any collection of the above a paths. Now we count the number of permutations π' containing these paths.

First we choose for every P_i a sequence of vertices inducing this path in π'. We see the vertices in each edge of P_i in at most $k!$ orderings. Crudely, every such sequence can be chosen in at most $(k!)^{b_i}$ ways. Thus, we have

$$\prod_{i=1}^{a}(k!)^{b_i} = (k!)^b \tag{13.7}$$

choices for all a sequences.

Now we bound the number of permutations containing these sequences. First note that

$$|V(P_i)| \geq b_i + k - 1.$$

Thus, we have at most

$$n - \sum_{i=1}^{a}(b_i + k - 1) = n - b - a(k-1) \tag{13.8}$$

vertices not in $V(P_1) \cup \cdots \cup V(P_a)$. We choose a permutation σ of $V \setminus (V(P_1) \cup \cdots \cup V(P_a))$. Here we have at most

$$(n - b - a(k-1))!$$

choices. Now we extend σ to a permutation of $[n]$. We mark a positions on σ and then insert the sequences. We can do it in

$$\binom{n}{a} a! < n^a$$

ways. Therefore, the number of permutations containing $P_1, P_2, \ldots, P_a$ is smaller than

$$(k!)^b(n - b - a(k-1))!n^a. \tag{13.9}$$

Thus, by (13.6) and (13.9) and the Stirling formula we obtain

$$\begin{aligned} N(b,a) &< n^{2a}(2k!k)^b(n - b - a(k-1))! \\ &< n^{2a}(2k!k)^b\sqrt{2\pi n}\left(\frac{n}{e}\right)^{n-b-a(k-1)}. \end{aligned}$$

Since

$$\mathbb{E}(X) = n!p^n = \sqrt{(2+o(1))\pi n}\left(\frac{n}{e}\right)^n p^n,$$

we get

$$\frac{N(b,a)p^{n-b}}{\mathbb{E}(X)} < (1+o(1))n^{2a}(2k!k)^b\left(\frac{e}{n}\right)^{b+a(k-1)} p^{-b}.$$

Finally, since $a \le b$ we estimate $e^{b+a(k-1)} \le e^{kb}$, and consequently,

$$\frac{N(b,a)p^{n-b}}{\mathbb{E}(X)} < \left(\frac{2k!ke^k}{np}\right)^b \frac{1+o(1)}{n^{a(k-3)}}. \tag{13.10}$$

Proof of *(ii)*:
Here $k = 3$ and $np \ge \omega$. Hence, we obtain in (13.10)

$$\frac{N(b,a)p^{n-b}}{\mathbb{E}(X)} \le (1+o(1))\left(\frac{2k!ke^k}{\omega}\right)^b.$$

Thus,

$$\sum_{b=1}^{n}\sum_{a=1}^{b}\frac{N(b,a)p^{n-b}}{\mathbb{E}(X)} < (1+o(1))\sum_{b=1}^{n} b\left(\frac{2k!ke^k}{\omega}\right)^b = o(1). \tag{13.11}$$

This completes the proof of part *(ii)*.

We prove part *(iii)* by estimating $N(b,a)$ more carefully, see Exercise 13.3.2 at the end of the chapter. □

Before leaving Hamilton cycles, we note that Allen, Böttcher, Kohayakawa, and Person [19] described a polynomial time algorithm for finding a tight Hamilton cycle in $\mathbb{H}_{n,p;k}$ w.h.p. when $p = n^{-1+\varepsilon}$ for a constant $\varepsilon > 0$.

There is a weaker notion of Hamilton cycle due to Berge [79] viz. an alternating sequence $v_1, e_1, v_2, \ldots, v_n, e_n$ of vertices and edges such that (i) $v_1, v_2, \ldots, v_n$ are distinct and (ii) $v_i \in e_{i-1} \cap e_i$ for $i = 1, 2, \ldots, n$. The cycle is weak if we do not insist that the edges are distinct. Poole [630] proves that the threshold for the existence of a weak Hamilton cycle in $\mathbb{H}_{n,m;k}$ is equal to the threshold for minimum degree one.

13.2 Perfect Matchings

A 1-factor or perfect matching of $\mathbb{H}_{n,m;k}$ is a set of disjoint edges $e_i, i = 1, 2, \ldots, n/k$ that partition $[n]$. The existence of a perfect matching or 1-factor requires that n is a multiple of k. When $k = 2$ this reduces to the ordinary notion of a perfect matching in a graph. In the literature, the problem of the existence of a perfect matching was first discussed in Schmidt and Shamir [674] and became known as the "Shamir problem."

The following theorem, which is a special case of the result of Johansson, Kahn and Vu [451] constituted a considerable breakthrough. Restricting ourselves to matchings enables us to present a simplified version of the proof in [451]. (In truth it is not so simple, but it avoids a lot of the technical problems faced in [451].)

Theorem 13.2 *Fix $k \geq 2$. Then there exists a constant $K > 0$ such that if $m \geq Kn\log n$ then*

$$\lim_{n\to\infty} \mathbb{P}(\mathbb{H}_{n,m;k} \text{ has a 1-factor}) = 1.$$

Proof In the following, K is taken to be sufficiently large so that all inequalities involving it are valid. Assume from now on that k divides n and let $e_1, e_2, \ldots, e_N, N = \binom{n}{k}$ be a random ordering of the edges of $\mathbb{H}_{n,m;k}$, the complete k-uniform hypergraph $\mathbb{H}_{n,N;k}$ on vertex set $V = [n]$. Let $H_i = \mathbb{H}_{n,N;k} - \{e_1, \ldots, e_i\}$ and $E_i = E(H_i)$ and $m_i = N - i = |E_i|$.

H_i is distributed as $\mathbb{H}_{n,m_i;k}$ and the idea is to show that w.h.p. H_i has many 1-factors as long as $m_i \geq Kn\log n$. Thus, we start with $\mathbb{H}_{n,N;k}$ and remove $N - m$ edges one by one in random order, checking that the number of 1-factors remains large w.h.p.

For a k-uniform hypergraph $H = (V, E)$, where $k \mid |V|$ we let $\mathscr{F}(H)$ denote the set of 1-factors of H and

$$\Phi(H) = |\mathscr{F}(H)|.$$

Let $\Phi_t = \Phi(H_t)$. Then if

$$1 - \xi_i = \frac{\Phi_i}{\Phi_{i-1}} = \frac{\Phi(H_{i-1} - e_i)}{\Phi_{i-1}}$$

we have

$$\Phi_t = \Phi_0 \frac{\Phi_1}{\Phi_0} \cdots \frac{\Phi_t}{\Phi_{t-1}} = \Phi_0(1 - \xi_1) \cdots (1 - \xi_t)$$

or

$$\log \Phi_t = \log \Phi_0 + \sum_{i=1}^{t} \log(1 - \xi_i),$$

where

$$\log \Phi_0 = \log \frac{n!}{(n/k)!(k!)^{n/k}} = \frac{k-1}{k} n \log n - c_1 n, \tag{13.12}$$

where

$$|c_1| \leq 1 + \frac{1}{k} \log k!.$$

We also have

$$\mathbb{E}\xi_i = \gamma_i = \frac{n/k}{N-i+1} \le \frac{1}{kK\log n}. \tag{13.13}$$

for $i \le T = N - Kn\log n$.

Equation (13.13) with

$$p_t = \frac{N-t}{N}, \qquad \text{(so that } |E_t| = Np_t\text{)},$$

gives rise to

$$\sum_{i=1}^{t} \mathbb{E}\xi_i = \sum_{i=1}^{t} \gamma_i = \frac{n}{k}\left(\log\frac{N}{N-t} + O\left(\frac{1}{N-t}\right)\right)$$
$$= \frac{n}{k}\left(\log\frac{1}{p_t} + o\left(\frac{1}{n}\right)\right) \tag{13.14}$$

using the fact that $\sum_{i=1}^{m} \frac{1}{i} = \log m + (\textit{Euler's constant}) + O(1/m)$. Recall that Eulers constant is about $0.577\cdots$.

For $t = T$ this gives

$$p_T = \frac{Kn\log n}{N}$$

and so

$$\sum_{i=1}^{T} \gamma_i = \frac{k-1}{k} n\log n - \frac{n}{k}\log\log n + c_2 n,$$

where

$$|c_2| \le 1 + \log k! + \frac{1}{k}\log K.$$

Our basic goal is to prove that if

$$L = K^{1/4}$$

and

$$\mathscr{A}_t = \left\{\log|\mathscr{F}_t| > \log|\mathscr{F}_0| - \sum_{i=1}^{t} \gamma_i - \frac{n}{L}\right\}$$

then

$$\mathbb{P}(\bar{\mathscr{A}_t}) \le n^{-L/10} \text{ for } t \le T. \tag{13.15}$$

Note that if $\mathscr{A}_T$ occurs then

$$\Phi(H_T) = e^{\Omega(n\log\log n)}. \tag{13.16}$$

Partition of $\mathscr{A}_t$ into sub-events

We need the following notation: suppose $\mathbf{w} : A \to [0, \infty)$ where A is a finite set, then

$$\mathbf{w}(A) = \sum_{a \in A} \mathbf{w}(a),$$

$$\bar{\mathbf{w}}(A) = \frac{\mathbf{w}(A)}{|A|},$$

$$\max \mathbf{w}(A) = \max_{a \in A} \mathbf{w}(a),$$

$$\text{maxr } \mathbf{w}(A) = \frac{\max \mathbf{w}(A)}{\bar{\mathbf{w}}(A)}$$

and

$$\text{med } \mathbf{w}(A) \text{ is the median value of } \mathbf{w}(a), a \in A,$$

i.e the smallest value $\mathbf{w}(a)$ such that at least half of the $a' \in A$ satisfy $\mathbf{w}(a') \leq \mathbf{w}(a)$.

We let $V_r = \binom{V}{r}$. For $Z \in V_k$ we let $H_i - Z$ be the sub-hypergraph of H_i induced by $V \setminus Z$ and let

$$\mathbf{w}_i(Z) = |\Phi(H_i - Z)|.$$

So, if $Z \in E$ then $\mathbf{w}_i(Z)$ is the number of perfect matchings that contain Z as an edge.

Now define property

$$\mathscr{B}_i = \{\text{maxr } \mathbf{w}_i(E_i) \leq L\}.$$

We also define

$$\mathscr{R}_i = \left\{ \left| D(x, H_i) - \binom{n-1}{k-1} p_i \right| \leq \frac{1}{L} \binom{n-1}{k-1} p_i, \text{ for all } x \in V \right\}$$

where

$$D(x, H_i) = |\{e \in E_i : x \in e\}|$$

is the number of edges of H_i that contain x.

We consider the first time $t \leq T$, if any, where $\mathscr{A}_t$ fails. Then,

$$\bar{\mathscr{A}_t} \cap \bigcap_{i<t} \mathscr{A}_i \subseteq \left[\bigcup_{i<t} \bar{\mathscr{R}}_i \right] \cup \left[\bigcup_{i<t} \mathscr{A}_i \mathscr{R}_i \bar{\mathscr{B}}_i \right] \cup \left[\bar{\mathscr{A}_t} \cap \bigcap_{i<t} (\mathscr{B}_i \mathscr{R}_i) \right]$$

Indeed, if the first two events in square brackets fail and $\bigcap_{i<t} \mathscr{A}_i \mathscr{R}_i \mathscr{B}_i$ occurs, then for the LHS to occur, we must have the occurrence of the third event in square brackets.

We can therefore write

$$\mathbb{P}\left(\bar{\mathscr{A}_t} \cap \bigcap_{i<t} \mathscr{A}_i\right) < \sum_{i<t} \mathbb{P}(\bar{\mathscr{R}}_i) + \sum_{i<t} \mathbb{P}(\mathscr{A}_i \mathscr{R}_i \bar{\mathscr{B}}_i) + \mathbb{P}\left(\bar{\mathscr{A}_t} \cap \bigcap_{i<t} (\mathscr{B}_i \mathscr{R}_i)\right). \tag{13.17}$$

Dealing with $\mathscr{R}_i$

The hypergraph H_i is distributed as $\mathbb{H}_{n,m_i;k}$, the random k-uniform hypergraph on vertex set $[n]$ with $m_i = N - i$ edges. It is a little easier to work with $\mathbb{H}_{n,p_i;k}$ where each possible edge occurs independently with probability p_i. Now the probability that $\mathbb{H}_{n,p_i;k}$ has exactly m_i edges is $\Omega(m_i^{-1/2})$ and so we can use $\mathbb{H}_{n,p_i;k}$ as our model if we multiply the probability of unlikely events by $O(m_i^{1/2})$ – using the simple inequality, $\mathbb{P}(A \mid B) \leq \mathbb{P}(A)/\mathbb{P}(B)$.

It then follows that the Chernoff bounds imply that

$$\mathbb{P}(\exists i \leq T : \neg \mathscr{R}_i) = O(n^{-L^2/4}). \tag{13.18}$$

This deals with the first sum in (13.17).

Dealing with the third term in (13.17)

We show next that

$$\mathscr{B}_{i-1} \Rightarrow \xi_i \leq \frac{1}{K^{3/4} \log n}.$$

This enables us to use a standard concentration argument to show that $\mathscr{A}_t$ holds w.h.p., given that $\bigcap_{i<t}(\mathscr{B}_i \mathscr{R}_i)$ holds.

We first compute

$$\begin{aligned} \mathbf{w}_{i-1}(E_{i-1}) &= \sum_{e \in E_{i-1}} \sum_{F \in \mathscr{F}_{i-1}} 1_{e \in F} \\ &= \sum_{F \in \mathscr{F}_{i-1}} \frac{n}{k}. \end{aligned}$$

Hence, for any $e \in E_{i-1}$,

$$\begin{aligned} \Phi_{i-1} &= \frac{k}{n} \mathbf{w}_{i-1}(E_{i-1}) \\ &\geq \frac{k}{Ln} |E_{i-1}| \max \mathbf{w}_{i-1}(E_{i-1}) \\ &\geq \frac{kNp_{i-1}}{Ln} \mathbf{w}_{i-1}(e). \end{aligned}$$

Hence, if the event $\mathscr{E}_i = \{\mathscr{B}_j, \mathscr{R}_j, j < i\}$ holds then

$$\xi_i = \frac{\mathbf{w}_{i-1}(e_i)}{\Phi_{i-1}} \leq \max_{e \in E_{i-1}} \frac{\mathbf{w}_{i-1}(e)}{\Phi_{i-1}} \leq \frac{Ln}{kNp_{i-1}} \leq \frac{L}{kK \log n} \leq \frac{1}{K^{3/4} \log n}. \tag{13.19}$$

Now define

$$Z_i = \begin{cases} \xi_i - \gamma_i & \text{if } \mathscr{B}_j, \mathscr{R}_j \text{ hold for } j < i, \\ 0 & \text{otherwise} \end{cases}$$

and

$$X_t = \sum_{i=1}^{t} Z_i.$$

We show momentarily that

$$\mathbb{P}(X_t \geq n) \leq e^{-\Omega(n)}. \tag{13.20}$$

So if we do have $\mathscr{B}_i, \mathscr{R}_i$ for $i < t \leq T$ (so that $X_t = \sum_{i=1}^{t} (\xi_i - \gamma_i)$) and $X_t \leq n$ then

$$\sum_{i=1}^{t} \xi_i < \sum_{i=1}^{t} \gamma_i + n \leq \frac{k-1}{k} n \log n$$

and hence

$$\sum_{i=1}^{t} \xi_i^2 \leq \frac{1}{K^{3/4} \log n} \sum_{i=1}^{t} \xi_i \leq \frac{n}{K^{3/4}}.$$

So,

$$\log |\mathscr{F}_t| > \log |\mathscr{F}_0| - \sum_{i=1}^{t} (\xi_i + \xi_i^2) > \log |\mathscr{F}_0| - \sum_{i=1}^{t} \gamma_i - \frac{n}{K^{3/4}}.$$

This deals with the third term in (13.17). (If $\bigcap_{i<t} (\mathscr{B}_t \mathscr{R}_t)$ holds then $\mathscr{A}_t$ holds with sufficient probability).

Let us now verify (13.20). Note that $|Z_i| \leq \frac{1}{K^{3/4} \log n}$ and that for any $h > 0$

$$\mathbb{P}(X_t \geq n) = \mathbb{P}(e^{h(Z_1 + \cdots + Z_t)} \geq e^{hn}) \leq E(e^{h(Z_1 + \cdots + Z_t)}) e^{-hn}. \tag{13.21}$$

Now $Z_i = \xi_i - \gamma_i$ (whenever $\mathscr{E}_i$ holds) and $\mathbb{E}(\xi_i \mid \mathscr{E}_i) = \gamma_i$. The conditioning does not affect the expectation since we have the same expectation given any previous history. Also $0 \leq \xi_i \leq \varepsilon = \frac{1}{\log n}$ (whenever $\mathscr{E}_i$ holds). So, with $h \leq 1$, by convexity

$$\mathbb{E}(e^{hZ_i}) = \mathbb{E}(e^{hZ_i} \mid \mathscr{E}_i)\,\mathbb{P}(\mathscr{E}_i) + \mathbb{E}(e^{hZ_i} \mid \neg\mathscr{E}_i)\,\mathbb{P}(\neg\mathscr{E}_i)$$
$$\leq e^{-h\gamma_i}\,\mathbb{E}\left(1 - \frac{\xi_i}{\varepsilon} + \frac{\xi_i}{\varepsilon}e^{h\varepsilon}\middle|\mathscr{E}_i\right)\mathbb{P}(\mathscr{E}_i) + \mathbb{P}(\neg\mathscr{E}_i)$$
$$= e^{-h\gamma_i}\left(1 - \frac{\gamma_i}{\varepsilon} + \frac{\gamma_i}{\varepsilon}e^{h\varepsilon}\right)\mathbb{P}(\mathscr{E}_i) + 1 - \mathbb{P}(\mathscr{E}_i) \leq e^{h^2\varepsilon\gamma_i}.$$

Here we used the fact that in putting $\mathbb{P}(\mathscr{E}_i) = x$, we maximize the LHS of the penultimate expression by putting $x = 1$.

So,

$$\mathbb{E}(e^{h(Z_1+\cdots+Z_t)}) \leq e^{h^2\varepsilon\sum_{i=1}^t \gamma_i}$$

and going back to (13.21) we get

$$\mathbb{P}(X_t \geq n) \leq e^{h^2\varepsilon\sum_{i=1}^t \gamma_i - hn}.$$

Now $\sum_{i=1}^t \gamma_i = O(n\log n)$ and $\varepsilon = 1/\log n$ and so putting h equal to a small enough positive constant makes the RHS of the above less than $e^{-hn/2}$ and (13.20) follows.

Dealing with the second term in (13.17)

It only remains to deal with the second term in (13.17) and show that

$$\mathbb{P}(\mathscr{A}_i\mathscr{R}_i\bar{\mathscr{B}}_i) < n^{-K/4}. \tag{13.22}$$

For $|Y| \leq k$ we let

$$V_{k,Y} = \{Z \in V_k : Z \supset Y\} \tag{13.23}$$

and

$$\mathscr{C}_i = \left\{\max \mathbf{w}_i(V_{k,Y}) \leq \max\left\{n^{-k}\Phi(H_i), 2\mathrm{med}\,\mathbf{w}_i(V_{k,Y})\right\}, \forall Y \in V_{k-1}\right\}$$

This event "replaces" the average of $\mathbf{w}_i$ by the median of $\mathbf{w}_i$. A subtle, but important idea.

We will prove

$$\mathbb{P}(\mathscr{R}_i\mathscr{C}_i\bar{\mathscr{B}}_i) < \varepsilon = n^{-\delta_1 K} \text{ where } \delta_1 = 2^{-(k+3)}. \tag{13.24}$$
$$\mathbb{P}(\mathscr{A}_i\mathscr{R}_i\bar{\mathscr{C}}_i) < n^{-L}. \tag{13.25}$$

And then use

$$\mathscr{A}_i\mathscr{R}_i\bar{\mathscr{B}}_i \subseteq \mathscr{A}_i\mathscr{R}_i\bar{\mathscr{C}}_i \cup \mathscr{R}_i\mathscr{C}_i\bar{\mathscr{B}}_i.$$

Proof of (13.24): We make the following assumption:

$$\mathbb{P}(\mathscr{R}_i\mathscr{C}_i) \geq \varepsilon.$$

For if this doesn't hold, (13.24) will be trivially satisfied.

Suppose that $|V| = n$ and $\mathbf{w} : V_k \to \mathbf{R}^+$. For $X \subseteq V$ with $|X| \le k$ we let $\psi(X) = \max \mathbf{w}(V_{k,X})$. (See (13.23) for the definition of $V_{k,X}$.)

We need the following simple lemma:

Lemma 13.3 *Suppose that for each $Y \in V_{k-1}$ and $\psi(Y) \ge B$ we have*

$$\left|\left\{Z \in V_{k,Y} : \mathbf{w}(Z) \ge \frac{1}{2}\psi(Y)\right\}\right| \ge \frac{n-k}{2}.$$

Then for any $X \subseteq V$ with $|X| = k-j$ and $\psi(X) \ge 2^{j-1}B$ we have

$$\left|\left\{Z \in V_{k,X} : \mathbf{w}(Z) \ge \frac{1}{2^j}\psi(X)\right\}\right| \ge \left(\frac{n-k}{2}\right)^j \frac{1}{(j-1)!}. \tag{13.26}$$

Proof Write N_j for the RHS of (13.26). We proceed by induction on j, with the case $j = 1$ given. Assume that X is as in the statement of the lemma and choose $Z \in V_{k,X}$ with $\mathbf{w}(Z)$ maximum (i.e. $\mathbf{w}(Z) = \psi(X)$). Let $y \in Z \setminus X$ and $Y = X \cup \{y\}$. Then $|Y| = k - (j-1)$ and $\psi(Y) = \psi(X) \ge 2^{j-1}B \ge 2^{j-2}B$. So by our induction hypothesis there are at least N_{j-1} sets $Z' \in V_{k,Y}$ with $\mathbf{w}(Z') \ge 2^{-(j-1)}\psi(Y) = 2^{-(j-1)}\psi(X)$. For each such $Z', Z' \setminus \{y\}$ is a $(k-1)$-subset of V with $\psi(Z' \setminus \{y\}) \ge \mathbf{w}(Z') \ge 2^{-(j-1)}B$. So for each such Z' there are at least $(n-k)/2$ sets $Z'' \in V_{k,Z'\setminus\{y\}}$ with

$$\mathbf{w}(Z'') \ge \frac{1}{2}\psi(Z' \setminus \{y\}) \ge 2^{-j}\psi(X).$$

The number of these pairs (Z', Z'') is thus at least $(n-k)N_{j-1}/2$. On the other hand, each Z' associated with a given Z'' is $(Z'' \setminus \{u\}) \cup \{y\}$ for some $u \in Z'' \setminus (X \cup \{y\})$; so the number of such Z' is at most $j-1$ and the lemma follows. □

Now $\max \mathbf{w}_i(V_{k,Y}) \ge \Phi(H_i)/\binom{n}{k}$. So, applying Lemma 13.3 with $B = n^{-k}\Phi(H_i)$ we see that if $\mathscr{C}_i$ and a fortiori if $\mathscr{R}_i\mathscr{C}_i$ holds then $\max \mathbf{w}_i(V_{k,Y}) > B$ implies that $2\,\mathrm{med}\,\mathbf{w}_i(V_{k,Y}) \ge \max \mathbf{w}_i(V_{k,Y})$ and so

$$\left|\left\{Z \in V_{k,Y} : \mathbf{w}_i(Z) \ge \frac{1}{2}\psi(Y)\right\}\right| \ge \frac{n-k}{2}.$$

Putting $j = k$ so that $X = \emptyset$ and $\psi(\emptyset) = \max \mathbf{w}_i(V_k)$, we see that if $\mathscr{R}_i\mathscr{C}_i$ holds then

$$\left|\left\{K \in V_k = V_{k,\emptyset} : \mathbf{w}_i(K) \ge \frac{\max \mathbf{w}_i(V_k)}{2^k}\right\}\right| \ge \frac{\delta(n-k)^k}{(k-1)!} \tag{13.27}$$

where $\delta = 2^{-k}$.

Let

$$E_i^* = \{e \in E_i : \mathbf{w}_i(e) \ge \delta \max \mathbf{w}_i(E_i)/2\}.$$

We show that (13.27) implies

$$\mathbb{P}\left(|E_i^*| \leq \frac{\delta^2 n^k p_i}{2k!} \;\middle|\; \mathscr{R}_i\mathscr{C}_i\right) \leq n^{-\delta K/10}. \tag{13.28}$$

Now let X_1 denote the set of vertices for which there are at least $\delta n^{k-1}/k!$ choices for $x_2,\dots,x_k$ such that

$$\mathbf{w}_i(x_1,\dots,x_k) > \delta \max \mathbf{w}_i(E_i). \tag{13.29}$$

Now (13.27) implies that

$$|X_1| \times \frac{n^{k-1}}{(k-1)!} + (n - |X_1|) \times \frac{\delta n^{k-1}}{k!} \geq \frac{\delta(n-k)^k}{(k-1)!}$$

which implies that

$$|X_1| \geq \frac{\delta n}{2}.$$

Now fix $0 \leq \ell \leq 2n \log n$ and let $L = 2^\ell$. Fix a vertex $x_1 \in X_1$ and let $A_L = \left\{e \in \binom{[n]}{k} : x_1 \in e \text{ and } \mathbf{w}_i(e) \geq L\right\}$. Here L is an approximation to the random variable $\delta \max \mathbf{w}_i(E_i)$. Using L in place of $\max \mathbf{w}_i(E_i)$ reduces the conditioning. There are not too many choices for L and so we can use the union bound over L.

Note that without the conditioning $\mathscr{R}_i\mathscr{C}_i$, the two events $\{S \subseteq A_L, T \cap A_L = \emptyset\}$ and $\{S \subseteq E_i, T \cap E_i = \emptyset\}$ are independent for all $S, T \in \binom{[n]}{k}$. This is because $\mathbf{w}_i(e)$ depends only on the existence of edges f where $e \cap f = \emptyset$. Hence, if we work with the model $\mathbb{H}_{n,p_i;k}$, without the conditioning, $|A_L \cap E_i|$ will be distributed as $\text{Bin}(|A_L|, p_i)$. Hence, if $|A_L| \geq \Delta = \delta n^{k-1}/k!$ then

$$\mathbb{P}(|A_L \cap E_i| \leq \Delta/2 \mid \mathscr{R}_i\mathscr{C}_i) \leq \frac{\mathbb{P}(|A_L \cap E_i| \leq \Delta/2)}{\mathbb{P}(\mathscr{R}_i\mathscr{C}_i)} \leq \varepsilon^{-1} e^{-\Delta p_i/8} \leq n^{-\delta K/9}.$$

There are at most n choices for x_1. The number of choices for ℓ is $2n \log n$ and for one of these we have $2^\ell \leq \delta \max \mathbf{w}_i(E_i) \leq 2^{\ell+1}$ and so with probability $1 - n^{2+o(1)-\delta K/9}$ we have that for each choice of $x_1 \in X_1$ there are $\frac{\delta}{2k!} n^{k-1} p_i$ choices for $x_2,\dots,x_k$ such that $\{x_1,\dots,x_k\}$ is an edge and $\mathbf{w}_i(x_1,\dots,x_k) > \delta \max \mathbf{w}_i(E_i)$. This verifies (13.28) and we have

$$\frac{\sum_{e \in E_i} \mathbf{w}_i(e)}{\max \mathbf{w}_i(E_i)} \geq \frac{\sum_{e \in E_i^*} \mathbf{w}_i(e)}{\max \mathbf{w}_i(E_i)} \geq \delta |E_i^*| \geq \frac{\delta^3}{2k!} n^k p_i \geq \frac{\delta^3}{2} |E_i|$$

which implies property $\mathscr{B}_i$ if $K \geq 2^{12k+4}$.

Proof of (13.25): Given $y \in V$ we let $X(y,H)$ denote the edge containing y in a uniformly random 1-factor of H. We let

$$h(y,H) = -\sum_{e \ni y} \mathbb{P}(X(y,H) = e) \log \mathbb{P}(X(y,H) = e)$$

denote the entropy of $X(y,H)$. See Chapter 24 for a brief discussion of the properties of entropy that we need.

Lemma 13.4

$$\log \Phi(H) \leq \frac{1}{k} \sum_{y \in V} h(y,H).$$

Proof This follows from Shearer's lemma, Lemma 24.4. To obtain Lemma 13.4 from this, let X be the indicator of a random 1-factor so that we can take $B = \binom{[n]}{k}$ and $\mathscr{A} = (A_v : v \in [n])$, where A_v is the set of edges of $\mathbb{H}_{n,m;k}$ containing v. Finally note that the entropy $h(X)$ of the random variable X satisfies $h(X) = \log \Phi(H)$ since X is (essentially) a random 1-factor. □

For the next lemma let S be a finite set and $\mathbf{w} : S \to \mathbf{R}^+$ and let X be the random variable with

$$\mathbb{P}(X = x) = \frac{\mathbf{w}(x)}{\mathbf{w}(S)}.$$

Lemma 13.5 *If* $h(X) \geq \log |S| - M$ *then there exist* $a, b \in range(\mathbf{w})$ *with*

$$a \leq b \leq 2^{4(M+\log 3)} a$$

such that for $J = \mathbf{w}^{-1}[a,b]$ *we have*

$$|J| \geq e^{-2M-2}|S| \text{ and } \mathbf{w}(J) > 0.7\mathbf{w}(S).$$

Proof Let

$$h(X) = \log |S| - M \tag{13.30}$$

and define C by $\log C = 4(M + \log 3)$. With $\bar{\mathbf{w}} = \mathbf{w}(S)/|S|$, let $a = \bar{\mathbf{w}}/C, b = C\bar{\mathbf{w}}$, $L = \mathbf{w}^{-1}([0,a)), U = \mathbf{w}^{-1}((b,\infty])$, and $J = S \setminus (L \cup U)$. We have

$$h(X) = \sum_{A \in \{L,J,U\}} \frac{\mathbf{w}(A)}{\mathbf{w}(S)} \sum_{x \in A} \frac{\mathbf{w}(x)}{\mathbf{w}(A)} \left(\log \frac{\mathbf{w}(x)}{\mathbf{w}(A)} + \log \frac{\mathbf{w}(A)}{\mathbf{w}(S)} \right).$$

It follows from (24.4) that

$$\sum_{x \in A} \frac{\mathbf{w}(x)}{\mathbf{w}(A)} \log \frac{\mathbf{w}(x)}{\mathbf{w}(A)} \leq \log |A|$$

and that

$$\sum_{A\in\{L,J,U\}} \frac{\mathbf{w}(A)}{\mathbf{w}(S)} \log \frac{\mathbf{w}(A)}{\mathbf{w}(S)} \leq \log 3.$$

So,

$$h(X) \leq \log 3 + \frac{\mathbf{w}(L)}{\mathbf{w}(S)} \log |L| + \frac{\mathbf{w}(J)}{\mathbf{w}(S)} \log |J| + \frac{\mathbf{w}(U)}{\mathbf{w}(S)} \log |U|. \tag{13.31}$$

Then we have a few observations. First, the Markov inequality implies that $|U| < |S|/C$ and this then implies that the RHS of (13.31) is less than

$$\log 3 + \log |S| - \frac{\mathbf{w}(U)}{\mathbf{w}(S)} \log C,$$

which with (13.30) implies

$$\mathbf{w}(U) < \frac{M + \log 3}{\log C} \mathbf{w}(S) = \mathbf{w}(S)/4. \tag{13.32}$$

Of course this also implies $|U| < |S|/4$. Then, combining (13.32) with $\mathbf{w}(L) \leq a|S| = \mathbf{w}(S)/C$, we have $\mathbf{w}(J) \geq 0.7\mathbf{w}(S)$. But then since the RHS of (13.31) is at most

$$\log 3 + \log |S| + \frac{\mathbf{w}(J)}{\mathbf{w}(S)} \log \frac{|J|}{|S|} \leq \log 3 + \log |S| + 0.7 \log \frac{|J|}{|S|}, \tag{13.33}$$

we have from (13.30) and (13.32) that

$$|J| \geq e^{-(0.7)^{-1}(M+\log 3)}|S|.$$

To verify (13.33) observe that it is equivalent to

$$\frac{\mathbf{w}(L)}{\mathbf{w}(S)} \log |L| + \frac{\mathbf{w}(J)}{\mathbf{w}(S)} \log |J| + \frac{\mathbf{w}(U)}{\mathbf{w}(S)} \log |S| \leq \log |S|$$

or

$$\frac{\mathbf{w}(L)}{\mathbf{w}(S)} \log |L| + \frac{\mathbf{w}(J)}{\mathbf{w}(S)} \log |J| \leq \frac{\mathbf{w}(L) + \mathbf{w}(J)}{\mathbf{w}(S)} \log |S|.$$

□

Assume that we have $\mathscr{A}_i$ and $\mathscr{R}_i$ and that $\mathscr{C}_i$ fails at Y. Let $e = Y \cup \{x\} \in V_k$ satisfy $\mathbf{w}_i(e) = \max \mathbf{w}_i(V_{k,Y})$. Note that if $H_i - e$ is the sub-hypergraph of H_i induced by $V \setminus e$ then, using (13.16),

$$\mathbf{w}_i(e) = \Phi(H_i - e) > n^{-k}\Phi(H_i) = e^{\Omega(n \log\log n)}. \tag{13.34}$$

Define for all $u \in \binom{[n]}{k}$

$$\mathscr{R}_{i,u} = \left\{ \left| D(z, H_i - u) - \binom{n-1}{k-1} p_i \right| \leq \frac{1}{L} \binom{n-1}{k-1} p_i, \text{ for all } z \in V \setminus u \right\}.$$

Arguing as in (13.18) we see that

$$\mathbb{P}\left(\exists i \leq T, u \in \binom{[n]}{k} : \neg\mathscr{R}_{i,u}\right) = O(n^{k-L^2/4}). \tag{13.35}$$

Choose $y \in V \setminus Y$ with $\mathbf{w}_i(Y \cup \{y\}) \leq \text{med}\, \mathbf{w}_i(V_{k,Y})$ and with $h(y, H_i - e)$ maximum subject to this restriction and set $f = Y \cup \{y\}$. Note that $y \neq x$ by its definition.

We have

$$\mathbf{w}_i(e) > 2\text{med}\, \mathbf{w}_i(V_{k,Y}) \geq 2\mathbf{w}_i(f).$$

Since we have $\mathscr{A}_i$, we have

$$\log|\Phi(H_i)| > \log|\Phi(H_0)| - \sum_{t=1}^{i} \gamma_t - \frac{n}{L} = \frac{k-1}{k} n \log n + \frac{n}{k} \log p_i - c_3 n,$$

where $c_3 = c_1 + 1/L + o(1)$.

This and (13.34) implies that

$$\log \Phi(H_i - e) = \log \mathbf{w}_i(e) \geq \frac{k-1}{k} n \log n + \frac{n}{k} \log p_i - c_4 n, \tag{13.36}$$

where $c_4 = c_3 + o(1)$.

But Lemma 13.4 with $H = H_i - e$ implies that

$$\log \Phi(H_i - e) \leq \frac{1}{k} \sum_{z \in V \setminus e} h(z, H_i - e)$$

and by our choice of y we have $h(z, H_i - e) \leq h(y, H_i - e)$ for at least half the z's in $V \setminus e$ and that for all $z \in V \setminus e$ we have,

$$h(z, H_i - e) \leq \log D(z, H_i - e) \leq \log\left(\left(1 + \frac{1}{L}\right)\binom{n-1}{k-1} p_i\right).$$

We have used (13.35) for the last inequality.

This implies that

$$\log \Phi(H_i - e) \leq \frac{n}{2k}\left(\log\left(\left(1 + \frac{1}{L}\right)\binom{n-1}{k-1} p_i\right) + h(y, H_i - e)\right). \tag{13.37}$$

Combining (13.36) and (13.37) we obtain

$$\begin{aligned} h(y, H_i - e) &\geq (k-1)\log n + \log p_i - 2kc_4 - \log(1 + 1/L) \\ &= \log D(y, H_i - e) - c_5, \end{aligned} \tag{13.38}$$

where

$$|c_5| \leq 2kc_4 + \log(k-1)! + 2\log(1 + 1/L) + o(1).$$

Recall the definition of Y,f following (13.34). Let $W = V \setminus (Y + \{x,y\})$ and for $Z \in W_{k-1} = \{Z \subseteq W : |Z| = k-1\}$ let

$$\mathbf{w}_i'(Z) = \Phi(H_i - (Y \cup Z \cup \{x,y\})).$$

Then define

$$\mathbf{w}_y \text{ on } W_y = \{K \subseteq V \setminus e : \ |K| = k, y \in K\}$$

and

$$\mathbf{w}_x \text{ on } W_x = \{K \subseteq V \setminus f : \ |K| = k, x \in K\}$$

by

$$\mathbf{w}_y(K) = \mathbf{w}_i'(K \setminus \{y\}) \text{ and } \mathbf{w}_x(K) = \mathbf{w}_i'(K \setminus \{x\}).$$

Recall next that $X(y, H_i - e)$ denotes the edge containing y in a random 1-factor of $H_i - e$. In which case, $X(y, H_i - e)$ is chosen according to the weights $\mathbf{w}_y$ and $X(x, H_i - f)$ is chosen according to the weights $\mathbf{w}_x$. Note also that $\mathbf{w}_y(W_y) = \mathbf{w}_i(e)$ and $\mathbf{w}_x(W_x) = \mathbf{w}_i(f)$.

Let $a, b \in range(\mathbf{w}_y) \subseteq range(\mathbf{w}_i')$ be as defined in Lemma 13.5, using (13.38). Let $J = \mathbf{w}_y^{-1}[a,b]$ and let $Z_1, Z_2, \ldots, Z_M$ be an enumeration of the sets in J. Then

$$M = |J| \geq e^{-2(c_5+1)} \binom{n-k}{k-1}. \tag{13.39}$$

Now

$$\begin{aligned}
\mathbf{w}_y(J) &= \sum_{j=1}^{M} \Phi(H_i - (Y + Z_j + x + y)) 1_{Z_j + y \in E} \\
&= \sum_{j=1}^{M} a_j \zeta_1^{(A)} > 0.7\mathbf{w}_y(W_y) = 0.7\mathbf{w}_i(e). \qquad (13.40) \\
\mathbf{w}_x(J) &= \sum_{j=1}^{M} \Phi(H_i - (Y + Z_j + x + y)) 1_{Z_j + x \in E} \\
&= \sum_{j=1}^{M} a_j \zeta_1^{(B)} \leq \mathbf{w}_x(W_x) = \mathbf{w}_i(f) \leq 0.5\mathbf{w}_i(e). \qquad (13.41)
\end{aligned}$$

For (13.40) we let $a_j = \Phi(H_i - (Y + Z_j + x + y))$ and $\zeta_j^{(A)} = 1_{Z_j + y \in E}$. Similarly for (13.41).

But

$$\mathbb{P}\left(\sum_{j=1}^{M} a_j \zeta_j^{(A)} \leq \frac{5}{7} \sum_{j=1}^{M} a_j \zeta_j^{(B)}\right) \leq \mathbb{P}\left(\sum_{j=1}^{M} a_j \zeta_j^{(A)} \leq \frac{9}{10}\mu\right) + \mathbb{P}\left(\sum_{j=1}^{M} a_j \zeta_j^{(B)} \geq \frac{11}{10}\mu\right)$$

where $\mu = \sum_{j=1}^{M} a_j p_i$.

Let

$$X = \sum_{j=1}^{M} \alpha_j \zeta_j^{(A)}$$

where

$$e^{-4(c_5+\log 3)} \leq \alpha_j = \frac{a_j}{b} \leq 1.$$

Then

$$\mathbb{E}X \geq e^{-4(c_5+\log 3)} M p_i \geq e^{-4(c_5+\log 3)} e^{-2(c_5+1)} \binom{n-k}{k-1} \frac{K \log n}{N} \geq L^2 \log n$$

for K sufficiently large.

It follows from Corollary 21.7 that

$$\mathbb{P}\left(\sum_{j=1}^{M} a_j \zeta_j^{(A)} \leq \frac{9}{10}\mu\right) = \mathbb{P}(X \leq 0.9\,\mathbb{E}X) \leq e^{-\mathbb{E}X/200} \leq n^{-L^2/200}. \qquad (13.42)$$

A similar argument gives

$$\mathbb{P}\left(\sum_{j=1}^{M} a_j \zeta_j^{(B)} \geq \frac{11}{10}\mu\right) \leq n^{-L^2/300}. \qquad (13.43)$$

By conditioning on the values $a'_{i,j}$ we see that we see that the probability that a vertex from the set a_{m+1} participates in σ_{r+1} is $\lambda/(m-r+\lambda)$. (We use the fact that if E_1, E_2 are exponentials with rates λ_1, λ_2 respectively, then $P(E_1 \leq E_2) = \frac{\lambda_1}{\lambda_1+\lambda_2}$, even if $\min\{E_1, E_2\}$ is given). □

13.3 Exercises

13.3.1 Generalize the notion of configuration model to k-uniform hypergraphs. Use it to show that if $r = O(1)$ then the number of r-regular, k-uniform hypergraphs with vertex set $[n]$ is asymptotically equal to

$$\frac{(rkn)!}{(k!)^{rn}(rn)!} e^{-(k-1)(r-1)/2}.$$

13.3.2 Prove part (iii) of Theorem 13.1 by showing that

$$N(b,a) \leq n^{2a} \binom{b-1}{a-1} \sum_{t \geq 0} 2^{t+a} (n-b-a(k-1)-t)!(k!)^{a+t}$$
$$\leq c_k (2k!)^a (n-b-a(k-1))!,$$

where c_k depends only on k. Then use (13.2).

13.3.3 In a *directed* k-uniform hypergraph, the vertices of each edge are totally ordered. Thus, each k-set has $k!$ possible orientations. Given a permutation $i_1, i_2, \ldots, i_n$ of $[n]$ we construct a directed ℓ-overlapping Hamilton cycle $\vec{E}_1 = (i_1, \ldots, i_k), \vec{E}_2 = (i_{k-\ell+1} \ldots, i_{2k-\ell}), \ldots, \vec{E}_{m_\ell} = (i_{n-(k-\ell)+1}, \ldots, i_\ell)$. Let $\vec{\mathbb{H}}_{n,p:k}$ be the directed hypergraph in which each possible directed edge is included with probability p. Use the idea of McDiarmid in Section 12.2 to show (see Ferber [292]) that

$$\mathbb{P}(\vec{\mathbb{H}}_{n,p:k} \text{ contains a directed } \ell\text{-overlapping Hamilton cycle})$$
$$\geq \mathbb{P}(\mathbb{H}_{n,p;k} \text{ contains an } \ell\text{-overlapping Hamilton cycle}).$$

13.3.4 A hypergraph $H = (V, E)$ is 2-colorable if there exists a partition of V into two non-empty sets A, B such that $e \cap A \neq \emptyset, e \cap B \neq \emptyset$ for all $e \in E$. Let $m = \binom{n}{k}p$. Show that if c is sufficiently large and $m = c2^k n$ then w.h.p. $H_{n,p;k}$ is not 2-colorable.

13.3.5 Let $c_k = \frac{1}{k(k-1)}$. Show that if $c < c_k$ then w.h.p. the maximum component in $H_{n,cn:k}$ is of size $O(\log n)$. (Here two vertices u, v are in the same component if there exists a sequence of edges $e_1, e_2, \ldots, e_\ell$ such that $u \in e_1, v \in e_\ell$ and $e_i \cap e_{i+1} \neq \emptyset$ for $1 \leq i < \ell$.)

13.3.6 Show that if $c > c_k$ then w.h.p. $H_{n,cn:k}$ contains a linear size component.

13.3.7 Given a hypergraph H, let a vertex coloring be *strongly proper* if no edge contains two vertices of the same color. The strong chromatic number $\chi_1(H)$ is the minimum number of colors in a strongly proper coloring. Suppose that $k \geq 3$ and $0 < p < 1$ is constant. Show that w.h.p.

$$\chi_1(H_{n,p;k}) \approx \frac{d}{2\log d} \quad \text{where} \quad d = \frac{n^{k-1}p}{(k-2)!}.$$

13.3.8 Let $U_1, U_2, \ldots, U_k$ denote k disjoint sets of size n. Let $\mathscr{H}\mathscr{P}_{n,m,k}$ denote the set of k-partite, k-uniform hypergraphs with vertex set $V = U_1 \cup U_2 \cup \cdots \cup U_k$ and m edges. Here each edge contains exactly one vertex from each $U_i, 1 \leq i \leq k$. The random hypergraph $HP_{n,m,k}$ is sampled uniformly from $\mathscr{H}\mathscr{P}_{n,m,k}$. Prove the k-partite analogue of Theorem 13.2 viz. there exists a constant $K > 0$ such that if $m \geq Kn \log n$ then

$$\lim_{n \to \infty} \mathbb{P}(HP_{n,m,k} \text{ has a 1-factor}) = 1.$$

13.4 Notes

Components and cores

If $H = (V,E)$ is a k-uniform hypergraph and $1 \leq j \leq k-1$ then two sets $J_1, J_2 \in \binom{V}{j}$ are said to be j-connected if there is a sequence of sets $E_1, E_2, \ldots, E_\ell$ such that $J_1 \subseteq E_1, J_2 \subseteq E_\ell$ and $|E_i \cap E_{i+1}| \geq j$ for $1 \leq i < \ell$. This defines an equivalence relation on $\binom{V}{j}$ and the equivalance classes are called *j-components*. Karoński and Łuczak [467] studied the sizes of the 1-components of the random hypergraph $H_{n,m;k}$ and proved the existence of a phase transition at $m \approx \frac{n}{k(k-1)}$. Cooley, Kang and Koch [194] generalized this to j-components and proved the existence of a phase transition at $m \approx \frac{\binom{n}{k}}{\left(\binom{k}{j}-1\right)\binom{n}{k-j}}$. As usual, a phase transition corresponds to the emergence of a unique giant, i.e. one of order $\binom{n}{j}$.

The notion of a core extends simply to hypergraphs and the sizes of cores in random hypergraphs has been considered by Molloy [582]. The r-core is the largest sub-hypergraph with minimum degree r. Molloy proved the existence of a constant $c_{k,r}$ such that if $c < c_{r,k}$ then w.h.p. $H_{n,cn;k}$ has no r-core and that if $c > c_{r,k}$ then w.h.p. $H_{n,cn;k}$ has a r-core. The efficiency of the peeling algorithm for finding a core has been considered by Jiang, Mitzenmacher and Thaler [449]. They showed that w.h.p. the number of rounds in the peeling algorithm is asymptotically $\frac{\log\log n}{\log(k-1)(r-1)}$ if $c < c_{r,k}$ and $\Omega(\log n)$ if $c > c_{r,k}$. Gao and Molloy [357] showed that for $|c - c_{r,k}| \leq n^{-\delta}$, $0 < \delta < 1/2$, the number of rounds grows like $\tilde{\Theta}(n^{\delta/2})$. In this discussion, $(r,k) \neq (2,2)$.

Chromatic number

Krivelevich and Sudakov [510] studied the chromatic number of the random k-uniform hypergraph $H_{n,p;k}$. For $1 \leq \gamma \leq k-1$ we say that a set of vertices S is γ-independent in a hypergraph H if $|S \cap e| \leq \gamma$. The γ-chromatic number of a hypergraph $H = (V,E)$ is the minimum number of sets in a partition of V into γ-independent sets. They showed that if $d^{(\gamma)} = \gamma\binom{k-1}{\gamma}\binom{n-1}{k-1}p$ is sufficiently large then w.h.p. $\frac{d^{(\gamma)}}{(\gamma+1)\log d^{(\gamma)}}$ is a good estimate of the γ-chromatic number of $H_{n,p;k}$.

Dyer, Frieze and Greenhill [267] extended the results of [5] to hypergraphs. Let $u_{k,\ell} = \ell^{k-1}\log\ell$. They showed that if $u_{k,\ell-1} < c < u_{k,\ell}$ then w.h.p. the (weak $(\gamma = k-1)$) chromatic number of $H_{n,cn;k}$ is either k or $k+1$.

Achlioptas, Kim, Krivelevich and Tetali [2] studied the 2-colorability of $H = H_{n,p;k}$. Let $m = \binom{n}{k}p$ be the expected number of edges in H. They showed that if $m = c2^k n$ and $c > \frac{\log 2}{2}$ then w.h.p. H is not 2-colorable. They also showed that if c is a small enough constant then w.h.p. H is 2-colorable.

Orientability

Gao and Wormald [359], Fountoulakis, Khosla and Panagiotou [311] and Lelarge [522] discussed the orientability of random hypergraphs. Suppose that $0 < \ell < k$. To ℓ-orient an edge e of a k-uniform hypergraph $H = (V,E)$, we assign positive signs to ℓ of its vertices and $k-\ell$ negative signs to the rest. An (ℓ,r)-orientation of H consists of an ℓ-orientation of each of its edges so that each vertex receives at most r positive signs due to incident edges. This notion has uses in load balancing. The papers established a threshold for the existence of an (ℓ,r)-orientation. Describing it is somewhat complex and we refer the reader to the papers themselves.

VC-dimension

Ycart and Ratsaby [731] discussed the VC-dimension of $H = H_{n,p;k}$. Let $p = cn{-}\alpha$ for constants c,α. They give the likely VC-dimension of H for various values of α. For example, if $h \in [k]$ and $\alpha = k - \frac{h(h-1)}{h+1}$ then the VC-dimension is h or $h-1$ w.h.p.

Erdős–Ko–Rado

The famous Erdős–Ko–Rado theorem states that if $n > 2k$ then the maximum size of a family of mutually intersecting k-subsets of $[n]$ is $\binom{n-1}{k-1}$ and this is achieved by all the subsets that contain the element 1. Such collections are called *stars*. Balogh, Bohman and Mubayi [52] considered this problem in relation to the random hypergraph $H_{n,p;k}$. They considered for what values of k,p is it true that maximum size intersecting family of edges is w.h.p. a star. More recently Hamm and Kahn [392], [393] have answered some of these questions. For many ranges of k,p the answer is as yet unknown.

Bohman, Cooper, Frieze, Martin and Ruszinko [111] and Bohman, Frieze, Martin, Ruszinko and Smyth [116] studied the k-uniform hypergraph H obtained by adding random k-sets one by one, only adding a set if it intersects all previous sets. They proved that w.h.p. H is a star for $k = o(n^{1/3})$ and were able to analyze the structure of H for $k = o(n^{5/12})$.

Perfect matchings and Hamilton cycles in regular hypergraphs

The perfect matching problem turns out to be a much easier problem than that discussed in Section 13.2. Cooper, Frieze, Molloy and Reed [223] used small subgraph conditioning to prove that $H_{n,r;k}$ has a perfect matching w.h.p. if and only if $k > k_r$ where $k_r = \frac{\log r}{(r-1)\log(\binom{r}{r-1})} + 1$. Dudek, Frieze, Ruciński and Šilekis [255] made some progress on loose Hamilton cycles in random regular hypergraphs. Their approach was to find an embeddding of $\mathbb{H}_{n,m;k}$ in a random regular k-uniform hypergraph.

PART III

Other models

14
Trees

The properties of various kinds of trees are one of the main objects of study in graph theory mainly due to their wide range of application in various areas of science. Here we concentrate our attention on the "average" properties of two important classes of trees: labeled and recursive. The first class plays an important role in both the sub-critical and super-critical phase of the evolution of random graphs. While random recursive trees serve as an example of the very popular random *preferential attachment* models. In particular we will point out, an often overlooked fact, that the first demonstration of a *power law* for the degree distribution in the preferential attachment model was shown in a special class of inhomogeneous random recursive trees.

The families of random trees, whose properties are analyzed in this chapter, fall into two major categories according to the order of their heights: they are either of square root (labeled trees) or logarithmic (recursive trees) height. While most of square-root-trees appear in probability context, most log-trees are encountered in algorithmic applications.

14.1 Labeled Trees

Consider the family $\mathcal{T}_n$ of all n^{n-2} labeled trees on vertex set $[n] = \{1,2,\ldots,n\}$. Let us choose a tree T_n uniformly at random from the family $\mathcal{T}_n$. The tree T_n is called a *random tree* (*random Cayley tree*).

The Prüfer code [638] establishes a bijection between labeled trees on vertex set $[n]$ and the set of sequences $[n]^{n-2}$ of length $n-2$ with items in $[n]$. Such a coding also implies that there is a one-to-one correspondence between the number of labeled trees on n vertices with a given degree sequence $d_1, d_2, \ldots, d_n$ and the number of ways in which one can distribute $n-2$ particles into n cells, such that the ith cell contains exactly $d_i - 1$ particles.

If the positive integers d_i, $i = 1,2,\dots,n$ satisfy

$$d_1 + d_2 + \cdots + d_n = 2(n-1),$$

then there exist

$$\binom{n-2}{d_1-1, d_2-1, \dots, d_n-1} \tag{14.1}$$

trees with n labeled vertices, the ith vertex having degree d_i.

The following observation is a simple consequence of the Prüfer bijection. Namely, there are

$$\binom{n-2}{i-1}(n-1)^{n-i-1} \tag{14.2}$$

trees with n labeled vertices in which the degree of a fixed vertex v is equal to i.

Let X_v be the degree of the vertex v in a random tree T_n, and let $X_v^* = X_v - 1$. Dividing the above formula by n^{n-2}, it follows that, for every i, X_i^* has the $\mathrm{Bin}(n-2, 1/n)$ distribution, which means that the asymptotic distribution of X_i^* tends to the Poisson distribution with mean one.

This observation allows us to obtain an immediate answer to the question of the limiting behavior of the maximum degree of a random tree. Indeed, the proof of Theorem 3.4 yields:

Theorem 14.1 *Denote by $\Delta = \Delta(T_n)$ the maximum degree of a random tree. Then w.h.p.*

$$\Delta(T_n) \approx \frac{\log n}{\log\log n}.$$

□

The classical approach to the study of the properties of labeled trees chosen at random from the family of all labeled trees was purely combinatorial, i.e. via counting trees with certain properties. In this way, Rényi and Szekeres [643], using complex analysis, found the height of a random labeled tree on n vertices (see also Stepanov [693], while for a general probabilistic context of their result, see a survey paper by Biane, Pitman and Yor [89]).

Assume that a tree with vertex set $V = [n]$ is rooted at vertex 1, then there is a unique path connecting the root with any other vertex of the tree. The height of a tree is the length of the longest path from the root to any pendant vertex of the tree. Pendant vertices are the vertices of degree one.

Theorem 14.2 *Let $h(T_n)$ be the height of a random tree T_n, then*

$$\lim_{n\to\infty} \mathbb{P}\left(\frac{h(T_n)}{\sqrt{2n}} < x\right) = \eta(x),$$

where

$$\eta(x) = \frac{4\pi^{5/2}}{x^3} \sum_{k=1}^{\infty} k^2 e^{-(k\pi/x)^2}.$$

Moreover,

$$\mathbb{E}\, h(T_n) \approx \sqrt{2\pi n} \quad \text{and} \quad \operatorname{Var} h(T_n) \approx \frac{\pi(\pi-3)}{3} n.$$

We now introduce a useful relationship between certain characteristics of random trees and branching processes. Consider a Galton–Watson branching process $\mu(t)$, $t = 0,1,\ldots$, starting with M particles, i.e. with $\mu(0) = M$, in which the number of offspring of a single particle is equal to r with probability $p_r, \sum_{r=0}^{\infty} p_r = 1$. Denote by Z_M the total number of offspring in the process $\mu(t)$. Dwass [264] (see also Viskov [715]) proved the following relationship.

Lemma 14.3 *Let $Y_1, Y_2, \ldots, Y_N$ be a sequence of independent identically distributed random variables, such that*

$$\mathbb{P}(Y_1 = r) = p_r \text{ for } r = 1,2,\ldots,N,$$

then

$$\mathbb{P}(Z_M = N) = \frac{M}{N} \mathbb{P}(Y_1 + Y_2 + \ldots + Y_N = N - M).$$

Now, instead of a random tree T_n chosen from the family of all labeled trees $\mathcal{T}_n$ on n vertices, consider a tree chosen at random from the family of all $(n+1)^{n-1}$ trees on $n+1$ vertices, with the root labeled 0 and all other vertices labeled from 1 to n. In such a random tree, with a natural orientation of the edges from the root to pendant vertices, denote by V_t the set of vertices at distance t from the root 0. Let the number of outgoing edges from a given vertex be called its *out-degree* and $X^+_{r,t}$ be the number of vertices of out-degree r in V_t. For our branching process, choose the probabilities p_r, for $r = 0,1,\ldots,$ as equal to

$$p_r = \frac{\lambda^r}{r!} e^{-\lambda},$$

i.e. assume that the number of offspring has the Poisson distribution with mean $\lambda > 0$. Note that λ is arbitrary here.

Let $Z_{r,t}$ be the number of particles in the tth generation of the process, having exactly r offspring. Next let $\underline{X} = [m_{r,t}]$, $r,t = 0,1,\ldots,n$ be a matrix of non-negative integers. Let $s_t = \sum_{r=0}^{n} m_{r,t}$ and suppose that the matrix $\underline{X}$ satisfies the following conditions:

(i) $s_0 = 1$,

$$s_t = m_{1,t-1} + 2m_{2,t-1} + \ldots n m_{n,t-1} \text{ for } t = 1,2,\ldots n.$$

(ii) $s_t = 0$ implies that $s_{t+1} = \ldots = s_n = 0$.
(iii) $s_0 + s_1 + \ldots + s_n = n+1$.

Then, as proved by Kolchin [495], the following relationship holds between the out-degrees of vertices in a random rooted tree and the number of offspring in the Poisson process starting with a single particle.

Theorem 14.4

$$\mathbb{P}([X^+_{r,t}] = \underline{X}) = \mathbb{P}([Z_{r,t}] = \underline{X} | Z = n+1).$$

Proof In Lemma 14.3 let $M = 1$ and $N = n+1$, then,

$$\begin{aligned}\mathbb{P}(Z_1 = n+1) &= \frac{1}{n+1}\mathbb{P}(Y_1 + Y_2 + \ldots + Y_{n+1} = n) \\ &= \frac{1}{n+1}\sum_{r_1+\ldots+r_{n+1}=n}\prod_{i=1}^{n+1}\frac{\lambda^{r_i}}{r_i!}e^{-\lambda} \\ &= \frac{(n+1)^n\lambda^n e^{-\lambda(n+1)}}{(n+1)!}.\end{aligned}$$

Therefore,

$$\begin{aligned}&\mathbb{P}([Z_{r,t}] = \underline{X} | Z = n+1) \\ &\quad= \frac{\prod_{t=0}^{n}\binom{s_t}{m_{0,t},\ldots,m_{n,t}} p_0^{m_{0,t}} \cdots p_n^{m_{n,t}}}{\mathbb{P}(Z = n+1)} \\ &\quad= \frac{(n+1)!\prod_{t=0}^{n}\frac{s_t!}{m_{0,t}!\, m_{1,t}! \,\ldots\, m_{n,t}!}\prod_{r=0}^{n}\left(\frac{\lambda^r}{r!}e^{-\lambda}\right)^{m_{r,t}}}{(n+1)^n\lambda^n e^{-\lambda(n+1)}} \\ &\quad= \frac{(n+1)!\, s_1!\, s_2! \,\ldots\, s_n!}{(n+1)^n}\prod_{t=0}^{n}\prod_{r=0}^{n}\frac{1}{m_{r,t}!\,(r!)^{m_{r,t}}}. \qquad (14.3)\end{aligned}$$

On the other hand, we can construct all rooted trees such that $[X^+_{r,t}] = \underline{X}$ in the following manner. We first layout an unlabeled tree in the plane. We choose a single point $(0,0)$ for the root and then points $S_t = \{(i,t) : i = 1,2,\ldots,s_t\}$ for $t = 1,2,\ldots,n$. Then for each t,r we choose $m_{r,t}$ points of S_t that will be joined to r points in S_{t+1}. Then, for $t = 0,1,\ldots,n-1$ we add edges. Note that S_n, if non-empty, has a single point corresponding to a leaf. We go through S_t in increasing order of the first component. Suppose that we have reached (i,t) and this has been assigned out-degree r. Then we join (i,t) to the first r vertices of S_{t+1} that have not yet been joined by an edge to a point in S_t. Having put in these edges, we assign labels $1,2,\ldots,n$ to $\bigcup_{t=1}^{n} S_t$. The number of ways of doing this is

$$\prod_{t=1}^{n} \frac{s_t!}{\prod_{r=1}^{n} m_{r,t}!} \times n!.$$

The factor $n!$ is an over count. As a set of edges, each tree with $[X_{r,t}^{+}] = \underline{X}$ appears exactly $\prod_{t=0}^{n} \prod_{r=0}^{n} (r!)^{m_{r,t}}$ times, due to permutations of the trees below each vertex. Summarizing, the total number of tree with out-degrees given by the matrix $\underline{X}$ is

$$n!\, s_1!\, s_2! \, \ldots \, s_n! \prod_{t=0}^{n} \prod_{r=0}^{n} \frac{1}{m_{r,t}!\, (r!)^{m_{r,t}}},$$

which, after division by the total number of labeled trees on $n+1$ vertices, i.e. by $(n+1)^{n-1}$, results in an identical formula to that given for the random matrix $[X_{+}^{r,t}]$ in the case of $[Z_{r,t}]$, see (14.3). To complete the proof one has to notice that for those matrices $\underline{X}$ which do not satisfy conditions (i) to (iii) both probabilities in question are equal to zero. □

Hence, roughly speaking, a random rooted labeled tree on n vertices has asymptotically the same shape as a branching process with Poisson, parameter one in terms of family sizes. Grimmett [384] uses this probabilistic representation to deduce the asymptotic distribution of the distance from the root to the nearest pendant vertex in a random labeled tree T_n, $n \geq 2$. Denote this random variable by $d(T_n)$.

Theorem 14.5 *As $n \to \infty$,*

$$\mathbb{P}(d(T_n) \geq k) \to \exp\left\{\sum_{i=1}^{k-1} \alpha_i\right\},$$

where the α_i are given recursively by

$$\alpha_0 = 0, \alpha_{i+1} = e^{\alpha_i} - e^{-1} - 1.$$

Proof Let k be a positive integer and consider the sub-tree of T_n induced by the vertices at distance at most k from the root. Within any level (strata) of T_n, order the vertices in increasing lexicographic order, and then delete all labels, excluding that of the root. Denote the resulting tree by T_n^k.

Now consider the following branching process constructed recursively according to the following rules:

(i) Start with one particle (the unique member of generation zero).
(ii) For $k \geq 0$, the $(k+1)$th generation A_{k+1} is the union of the families of descendants of the kth generation together with one additional member which is allocated at random to one of these families, each of the $|A_k|$

families having equal probability of being chosen for this allocation. As in Theorem 14.4, all family sizes are independent of each other and the past, and are Poisson distributed with mean one.

Lemma 14.6 *As $n \to \infty$ the numerical characteristics of T_n^k have the same distribution as the corresponding characteristics of the tree defined by the first k generations of the branching process described above.*

Proof For a proof of Lemma 14.6, see the proof Theorem 3 of [384].

Let Y_k be the size of the kth generation of our branching process and let N_k be the number of members of the kth generation with no offspring. Let $\mathbf{i} = (i_1, i_2, \ldots, i_k)$ be a sequence of positive integers, and let

$$A_j = \{N_j = 0\} \quad \text{and} \quad B_j = \{Y_j = i_j\} \quad \text{for} \quad j = 1, 2, \ldots, k.$$

Then, by Lemma 14.6, as $n \to \infty$,

$$\mathbb{P}(d(T_n) \geq k) \to \mathbb{P}(A_1 \cap A_2 \cap \ldots \cap A_k) .$$

Now,

$$\mathbb{P}(A_1 \cap A_2 \cap \ldots \cap A_k) = \sum_{\mathbf{i}} \prod_{j=1}^{k} \mathbb{P}(A_j | A_1 \cap \ldots \cap A_{j-1} \cap B_1 \cap \ldots B_j)$$
$$\times \mathbb{P}(B_j | A_1 \cap \ldots \cap A_{j-1} \cap B_1 \cap \ldots B_{j-1}).$$

Using the Markov property,

$$\mathbb{P}(A_1 \cap A_2 \cap \ldots \cap A_k) = \sum_{\mathbf{i}} \prod_{j=1}^{k} \mathbb{P}(A_j | B_j) \mathbb{P}(B_j | A_{j-1} \cap B_{j-1})$$
$$= \sum_{\mathbf{i}} \prod_{j=1}^{k} \left(1 - e^{-1}\right)^{i_j - 1} C_j(i_j), \tag{14.4}$$

where $C_j(i_j) = \mathbb{P}(B_j | A_{j-1} \cap B_{j-1})$ is the coefficient of x^{i_j} in the probability generating function $D_j(x)$ of Y_j conditional upon $Y_{j-1} = i_{j-1}$ and $N_j = 0$. Thus,

$$Y_j = 1 + Z + R_1 + \ldots + R_{i_{j-1}-1},$$

where Z has the Poisson distribution and the R_i are independent random variables with Poisson distribution conditioned on being nonzero. Hence

$$D_j(x) = x e^{x-1} \left(\frac{e^x - 1}{e - 1} \right)^{i_{j-1} - 1} .$$

Now,

$$\sum_{i_k=1}^{\infty}(1-e^{-1})^{i_k-1}C_k(i_k)=\frac{D_k(1-e^{-1})}{1-e^{-1}}.$$

We can use this to eliminate i_k in (14.4) and give

$$\mathbb{P}(A_1\cap A_2\cap\ldots\cap A_k)=\sum_{(i_1,\ldots,i_{k-1})}\prod_{j=1}^{k-1}\beta_1^{i_j-1}C_j(i_j)e^{\beta_1-1}\left(\frac{e^{\beta_1}-1}{e-1}\right)^{i_{k-1}-1}, \tag{14.5}$$

where $\beta_1=1-e^{-1}$. Eliminating i_{k-1} from (14.5) we get

$$\mathbb{P}(A_1\cap A_2\cap\ldots\cap A_k)=$$

$$\sum_{(i_1,\ldots,i_{k-2})}\prod_{j=1}^{k-2}\beta_1^{i_j-1}C_j(i_j)e^{\beta_1+\beta_2-2}\left(\frac{e^{\beta_2}-1}{e-1}\right)^{i_{k-2}-1},$$

where $\beta_2=(e^{\beta_1}-1)$. Continuing we see that, for $k\geq 1$,

$$\mathbb{P}(A_1\cap A_2\cap\ldots\cap A_k)=\exp\left\{\sum_{i=1}^{k}(\beta_i-1)\right\}=\exp\left\{\sum_{i=1}^{k}\alpha_i\right\},$$

where $\beta_0,\beta_1,\ldots$ are given by the recurrence

$$\beta_0=1,\beta_{i+1}=\left(e^{\beta_i}-1\right)e^{-1},$$

and $\alpha_i=\beta_i-1$. One can easily check that β_i remains positive and decreases monotonically as $i\to\infty$, and so $\alpha_i\to -1$. □

Another consequence of Lemma 14.3 is that, for a given N, one can associate with the sequence $Y_1,Y_2,\ldots,Y_N$, a generalized occupancy scheme of distributing n particles into N cells (see [495]). In such a scheme, the joint distribution of the number of particles in each cell $(\nu_1,\nu_2,\ldots,\nu_N)$ is given, for $r=1,2,\ldots,N$ by

$$\mathbb{P}(\nu_r=k_r)=\mathbb{P}\left(Y_r=k_r\left|\sum_{r=1}^{N}Y_r=n\right.\right). \tag{14.6}$$

Now, denote by $X_r^+=\sum_{t=0}^{n}X_{r,t}^+$ the number of vertices of out-degree r in a random tree on $n+1$ vertices, rooted at a vertex labeled 0. Denote by $Z^{(r)}=\sum_{t=0}^{n}Z_{r,t}$, the number of particles with exactly r offspring in the Poisson process $\mu(t)$. Then by Theorem 14.4,

$$\mathbb{P}(X_r^+=k_r,\ r=0,1,\ldots,n)=\mathbb{P}(Z^{(r)}=k_r,\ r=0,1,\ldots,n|Z_1=n+1).$$

Hence, by equation (14.1), the fact that we can choose $\lambda = 1$ in the process $\mu(t)$ and (14.6), the joint distribution of out-degrees of a random tree coincides with the joint distribution of the number of cells containing the given number of particles in the classical model of distributing n particles into $n+1$ cells, where each choice of a cell by a particle is equally likely.

The above relationship, allows us to determine the asymptotic behavior of the expectation of the number X_r of vertices of degree r in a random labeled tree T_n.

Corollary 14.7

$$\mathbb{E}X_r \approx \frac{n}{(r-1)!\, e}.$$

14.2 Recursive Trees

We call a tree on n vertices labeled $1,2,\ldots,n$ a *recursive tree* (or *increasing tree*) if the tree is rooted at vertex 1 and, for $2 \le i \le n$, the labels on the unique path from the root to vertex i form an increasing sequence. It is not difficult to see that any such tree can be constructed "recursively": starting with the vertex labeled 1 and assuming that vertices "arrive" in order of their labels, and connect themselves by an edge to one of the vertices that "arrived" earlier. So the number of recursive (increasing) trees on n vertices is equal to $(n-1)!$.

A *random recursive tree* is a tree chosen uniformly at random from the family of all $(n-1)!$ recursive trees. Or equivalently, it can be generated by a recursive procedure in which each new vertex chooses a neighbor at random from previously arrived vertices. We assume that our tree is rooted at vertex 1 and all edges are directed from the root to the leaves.

Let T_n be a random recursive tree and let $D^+_{n,i}$ be the out-degree of the vertex with label i, i.e the number of "children" of vertex i. We start with the exact probability distribution of these random variables.

Theorem 14.8 *For $i = 1,2,\ldots,n$ and $r = 1,2,\ldots,n-1$,*

$$\mathbb{P}(D^+_{n,i} = r) = \frac{(i-1)!}{(n-1)!} \sum_{k=r}^{n-i} \binom{k}{r} (i-1)^{k-r} |s(n-i,k)|, \tag{14.7}$$

where $s(n-i,k)$ is the Stirling number of the first kind.

Proof Conditioning on tree T_{n-1} we see that, for $r \ge 1$,

$$\mathbb{P}(D^+_{n,i} = r) = \frac{n-2}{n-1} \mathbb{P}(D^+_{n-1,i} = r) + \frac{1}{n-1} \mathbb{P}(D^+_{n-1,i} = r-1). \tag{14.8}$$

Fix i and let

$$\Phi_{n,i}(z) = \sum_{r=0}^{n-i} \mathbb{P}(D^+_{n,i} = r)z^r$$

be the probability generating function of $D^+_{n,i}$.

Multiplying (14.8) by z^r and then summing over $r \geq 1$ we see that

$$\Phi_{n,i}(z) - \mathbb{P}(D^+_{n,i} = 0) = \frac{n-2}{n-1}\left(\Phi_{n-1,i}(z) - \mathbb{P}(D^+_{n-1,i} = 0)\right) + \frac{z}{n-1}\Phi_{n-1,i}(z).$$

Notice, that the probability that vertex i is a leaf equals

$$\mathbb{P}(D^+_{n,i} = 0) = \prod_{j=i}^{n-1}\left(1 - \frac{1}{j}\right) = \frac{i-1}{n-1}. \tag{14.9}$$

Therefore

$$\Phi_{n,i}(z) = \frac{n-2}{n-1}\Phi_{n-1,i}(z) + \frac{z}{n-1}\Phi_{n-1,i}(z).$$

With the boundary condition,

$$\Phi_{i,i}(z) = \mathbb{P}(D^+_{i,i} = 0) = 1.$$

We can verify inductively that

$$\begin{aligned}\Phi_{n,i}(z) &= \prod_{k=1}^{n-i}\left(\frac{z+i+k-2}{i+k-1}\right)\\ &= \frac{(i-1)!}{(n-1)!}(z+i-1)(z+i)\dots(z+n-2).\end{aligned} \tag{14.10}$$

Recall the definition of Stirling numbers of the first kind $s(n,k)$. For non-negative integers n and k

$$z(z-1)\dots(z-n+1) = \sum_{k=1}^{n} s(n,k)z^k.$$

Hence

$$\begin{aligned}\Phi_{n,i}(z) &= \frac{(i-1)!}{(n-1)!}\sum_{k=1}^{n-i}|s(n-i,k)|(z+i-1)^k\\ &= \frac{(i-1)!}{(n-1)!}\sum_{k=1}^{n-i}\sum_{r=0}^{k}\binom{k}{r}z^r(i-1)^{k-r}|s(n-i,k)|\\ &= \sum_{r=0}^{n-i}\left(\frac{(i-1)!}{(n-1)!}\sum_{k=r}^{n-i}\binom{k}{r}(i-1)^{k-r}|s(n-i,k)|\right)z^r.\end{aligned}$$

□

It follows from (14.10), by putting $z = 0$, that the expected number of vertices of out-degree zero is

$$\sum_{i=1}^{n} \frac{i-1}{n-1} = \frac{n}{2}.$$

Then (14.8) with $i = r = 1$ implies that $\mathbb{P}(D^+_{n,1} = 1) = 1/(n-1)$. Hence, if L_n is the number of leaves in T_n, then

$$\mathbb{E} L_n = \frac{n}{2} + \frac{1}{n-1}. \tag{14.11}$$

For a positive integer n, let $\zeta_n(s) = \sum_{k=1}^{n} k^{-s}$ be the incomplete Riemann zeta function, and let $H_n = \zeta(1) = \sum_{k=1}^{n} k^{-1}$ be the nth harmonic number, and let $\delta_{n,k}$ denote the Kronecker function $1_{n=k}$.

Theorem 14.9 *For $1 \le i \le n$, let $D_{n,i}$ be the degree of vertex i in a random recursive tree T_n. Then*

$$\mathbb{E} D_{n,i} = H_{n-1} - H_{i-1} + 1 - \delta_{1,i},$$

while

$$\operatorname{Var} D_{n,i} = H_{n-1} - H_{i-1} - \zeta_{n-1}(2) + \zeta_{i-1}(2).$$

Proof Let N_j be the label of that vertex among vertices $1, 2, \ldots j-1$, which is the parent of vertex j. Then for $j \ge 1$ and $1 \le i < j$

$$D_{n,i} = \sum_{j=i+1}^{n} \delta_{N_j,i}. \tag{14.12}$$

By definition $N_2, N_3, \ldots, N_n$ are independent random variables and for all i,j,

$$\mathbb{P}(N_j = i) = \frac{1}{j-1}. \tag{14.13}$$

The expected value of $D_{n,i}$ follows immediately from (14.12) and (14.13). To compute the variance observe that

$$\operatorname{Var} D_{n,i} = \sum_{j=i+1}^{n} \frac{1}{j-1}\left(1 - \frac{1}{j-1}\right).$$

□

From the above theorem it follows that $\operatorname{Var} D_{n,i} \le \mathbb{E} D_{n,i}$. Moreover, for fixed i and n large, $\mathbb{E} D_{n,i} \approx \log n$, while for i growing with n the expectation $\mathbb{E} D_{n,i} \approx \log n - \log i$. The following theorem, see Kuba and Panholzer [516], shows a standard limit behavior of the distribution of $D_{n,i}$.

Theorem 14.10 *Let $i \geq 1$ be fixed and $n \to \infty$, then*

$$(D_{n,i} - \log n)/\sqrt{\log n} \xrightarrow{D} N(0,1).$$

Now, let $i = i(n) \to \infty$ as $n \to \infty$. If

(i) $i = o(n)$, then

$$(D_{n,i} - (\log n - \log i))/\sqrt{\log n - \log i} \xrightarrow{D} N(0,1),$$

(ii) $i = cn$, $0 < c < 1$, then

$$D_{n,i} \xrightarrow{D} \mathrm{Po}(-\log c),$$

(iii) $n - i = o(n)$, then

$$\mathbb{P}(D^+_{n,i} = 0) \to 1.$$

Now, consider another parameter of a random recursive tree.

Theorem 14.11 *Let $r \geq 1$ be fixed and let $X_{n,r}$ be the number of vertices of degree r in a random recursive tree T_n. Then, w.h.p.*

$$X_{n,r} \approx n/2^r,$$

and

$$\frac{X_{n,r} - n/2^r}{\sqrt{n}} \xrightarrow{d} Y_r,$$

as $n \to \infty$, where Y_r has the $N(0, \sigma_r^2)$ distribution.

□

In place of proving the above theorem we will give a simple proof of its immediate implication, i.e. the asymptotic behavior of the expectation of the random variable $X_{n,r}$. The proof of asymptotic normality of suitably normalized $X_{n,r}$ is due to Janson and can be found in [424]. (In fact, in [424] a stronger statement is proved, namely, that, asymptotically, for all $r \geq 1$, random variables $X_{n,r}$ are jointly normally distributed.)

Corollary 14.12 *Let $r \geq 1$ be fixed, then*

$$\mathbb{E} X_{n,r} \approx n/2^r.$$

Proof Let us introduce a random variable $Y_{n,r}$ counting the number of vertices of degree at least r in T_n. Obviously,

$$X_{n,r} = Y_{n,r} - Y_{n,r+1}. \tag{14.14}$$

Moreover, using a similar argument to that given for formula (14.7), we see that for $2 \le r \le n$,

$$\mathbb{E}[Y_{n,r}|T_{n-1}] = \frac{n-2}{n-1} Y_{n-1,r} + \frac{1}{n-1} Y_{n-1,r-1} \tag{14.15}$$

Notice, that the boundary condition for the recursive formula (14.15) is, trivially given by

$$\mathbb{E}\, Y_{n,1} = n.$$

We will show that $\mathbb{E}\, Y_{n,r}/n \to 2^{-r+1}$, which, by (14.14), implies the theorem. Set

$$a_{n,r} := n2^{-r+1} - \mathbb{E}\, Y_{n,r}. \tag{14.16}$$

$\mathbb{E}\, Y_{n,1} = n$ implies that $a_{n,1} = 0$. We see from (14.11) that the expected number of leaves in a random recursive tree on n vertices is given by

$$\mathbb{E} X_{n,1} = \frac{n}{2} + \frac{1}{n-1}.$$

Hence, $a_{n,2} = 1/(n-1)$ as $\mathbb{E}\, Y_{n,2} = n - \mathbb{E}\, X_{n,1}$.

Now we show that,

$$0 < a_{n,1} < a_{n,2} < \cdots < a_{n,n-1}. \tag{14.17}$$

From the relationships (14.15) and (14.16) we get

$$a_{n,r} = \frac{n-2}{n-1} a_{n-1,r} + \frac{1}{n-1} a_{n-1,r-1}. \tag{14.18}$$

Inductively assume that (14.17) holds for some $n \ge 3$. Now, by (14.18), we get

$$a_{n,r} > \frac{n-2}{n-1} a_{n-1,r-1} + \frac{1}{n-1} a_{n-1,r-1} = a_{n-1,r-1}.$$

Finally, notice that

$$a_{n,n-1} = n2^{2-n} - \frac{2}{(n-1)!},$$

since there are only two recursive trees with n vertices and a vertex of degree $n-1$. So, we conclude that $a(n,r) \to 0$ as $n \to \infty$, for every r, and our theorem follows. □

Finally, consider the maximum degree $\Delta_n = \Delta_n(T_n)$ of a random recursive tree T_n. It is easy to see that for large n, its expected value should exceed $\log n$, since it is as large as the expected degree of the vertex 1, which, by Theorem 14.9, equals $H_{n-1} \approx \log n$. Szymański [699] proved that the upper bound is $O(\log_2 n)$ (see Goh and Schmutz [376] for a strengthening of his result). Finally, Devroye and Lu (see [240]) have shown that in fact $\Delta_n \approx$

$\log_2 n$. This is somewhat surprising. While each vertex in $[1, n^{1-o(1)}]$ only has a small chance of having such a degree, there are enough of these vertices to guarantee one w.h.p..

Theorem 14.13 *In a random recursive tree T_n, w.h.p.*

$$\Delta_n \approx \log_2 n.$$

The next theorem was originally proved by Devroye [235] and Pittel [625]. Both proofs were based on an analysis of certain branching processes. The proof below is related to [235].

Theorem 14.14 *Let $h(T_n)$ be the height of a random recursive tree T_n. Then w.h.p.*

$$h(T_n) \approx e \log n.$$

Proof
Upper Bound: For the upper bound we simply estimate the number ν_1 of vertices at height $h_1 = (1+\varepsilon)e\log n$, where $\varepsilon = o(1)$ but is sufficiently large so that claimed inequalities are valid. Each vertex at this height can be associated with a path $i_0 = 1, i_1, \ldots, i_h$ of length h in T_n. So, if $S = \{i_1, \ldots, i_h\}$ refers to such a path, then

$$\mathbb{E}\nu_1 = \sum_{|S|=h_1} \prod_{i \in S} \frac{1}{i-1} \leq \frac{1}{h_1!}\left(\sum_{i=1}^{n} \frac{1}{i}\right)^{h_1} \leq \left(\frac{(1+\log n)e}{h_1}\right)^{h_1} = o(1), \quad (14.19)$$

assuming that $h_1\varepsilon \to \infty$.

Explanation: If $S = \{i_1 = 1, i_2, \ldots, i_{h_1}\}$ then the term $\prod_{j=1}^{h_1} 1/i_j$ is the probability that i_j chooses i_{j-1} in the construction of T_n.

Lower Bound: The proof of the lower bound is more involved. We consider a different model of tree construction and relate it to T_n. We consider a *Yule* process. We run the process for a specific time t and construct a tree $Y(t)$. We begin by creating a single particle x_1 at time 0 this is the root of a tree $Y(t)$. New particles are generated at various times $\tau_1 = 0, \tau_2, \ldots,$. Then at time τ_k there will be k particles $X_k = \{x_1, x_2, \ldots, x_k\}$ and we will have $Y(t) = Y(\tau_k)$ for $\tau_k \leq t < \tau_{k+1}$. After x_k has been added to $Y(\tau_k)$, each $x \in X_k$ is associated with an exponential random variable E_x with mean one. An exponential random variable Z with mean λ is characterized by $\mathbb{P}(Z \geq x) = e^{-\lambda x}$. If z_k is the particle in X_k that minimizes $E_x, x \in X_k$ then a new particle x_{k+1} is generated at time $\tau_{k+1} = \tau_k + E_{z_k}$ and an edge $\{z_k, x_{k+1}\}$ is added to $Y(\tau_k)$ to create $Y(\tau_{k+1})$. After this we independently generate new random variables $E_x, x \in X_{k+1}$.

Remark 14.15 The *memory-less property* of the exponential random variable, i.e. $\mathbb{P}(Z \geq a+b \mid Z \geq a) = \mathbb{P}(Z \geq b)$, implies that we could equally well think that at time $t \geq \tau_k$ the E_x are independent exponentials conditional on being at least τ_k. In which case the choice of z_k is uniformly random from X_k, even conditional on the process's prior history.

Suppose then that we focus attention on $Y(y;s,t)$, the sub-tree rooted at y containing all descendants of y that are generated after time s and before time t.

We observe three things:

(T1) The tree $Y(\tau_n)$ has the same distribution as T_n. This is because each particle in X_k is equally likely to be z_k.
(T2) If $s < t$ and $y \in Y(s)$ then $Y(y;s,t)$ is distributed as $Y(t-s)$. This follows from Remark 14.15, because when $z_k \notin Y(y;s,t)$ it does not affect any of the the variables $E_x, x \subset Y(y;s,t)$.
(T3) If $x, y \in Y(s)$ then $Y(x;s,t)$ and $Y(y;s,t)$ are independent. This also follows from Remark 14.15 for the same reasons as in (T2).

It is not difficult to prove (see Exercise 14.4.7 or Feller [290]) that if $P_n(t)$ is the probability there are exactly n particles at time t then

$$P_n(t) = e^{-t}(1-e^{-t})^{n-1}. \tag{14.20}$$

Next let

$$t_1 = (1-\varepsilon)\log n.$$

Then it follows from (14.20) that if $\nu(t)$ is the number of particles in our Yule process at time t then

$$\mathbb{P}(\nu(t_1) \geq n) \leq \sum_{k \geq n} e^{-t_1}(1-e^{-t_1})^{k-1} = \left(1 - \frac{1}{n^{1-\varepsilon}}\right)^{n-1} = o(1). \tag{14.21}$$

We will show that w.h.p. the tree $T_{\nu(t_1)}$ has height at least

$$h_0 = (1-\varepsilon)et_1$$

and this completes the proof of the theorem.

We choose $s \to \infty$, $s = O(\log t_1)$. It follows from (14.20) that if $\nu_0 = \varepsilon e^s$ then

$$\mathbb{P}(\nu(s) \leq \nu_0) = \sum_{k=0}^{\nu_0} e^{-s}(1-e^{-s})^{k-1} \leq \varepsilon = o(1). \tag{14.22}$$

Suppose now that $\nu(s) \geq \nu_0$ and that the vertices of $T_{1;0,s}$ are $\{x_1, x_2, \ldots, x_{\nu(s)}\}$. Let $\sigma = \nu_0^{1/2}$ and consider the sub-trees $A_j, j = 1,2,\ldots,\tau$ of $T_{1;0,t_1}$ rooted at

$x_j, j = 1,2,\ldots,\nu(s)$. We will show that

$$\mathbb{P}(T_{x_1;s,t_1} \text{ has height at least } (1-\varepsilon)^3 e\log n) \geq \frac{1}{2\sigma\log\sigma}. \tag{14.23}$$

Assuming that (14.23) holds, since the trees $A_1, A_2, \ldots, A_\tau$ are independent, by T3, we have

$$\mathbb{P}(h(T_n) \leq (1-\varepsilon)^3 e\log n) \leq o(1) + \mathbb{P}(h(T_{1;0,t_1})$$
$$\leq (1-\varepsilon)^3 e\log n) \leq o(1) + \left(1 - \frac{1}{2\sigma\log\sigma}\right)^{\nu_0} = o(1).$$

To prove all this we associate a Galton–Watson branching process with each of $x_1, x_2, \ldots, x_\tau$. Consider for example $x = x_1$ and let $\tau_0 = \log\sigma$. The vertex x is the root of a branching process Π, which we now define. We consider the construction of $Y(x;s,t)$ at times $\tau_i = s + i\tau_0$ for $i = 1,2,\ldots,i_0 = (t_1 - s)/\tau_0$. The children of x in Π are the vertices at depth at least $(1-\varepsilon)e\tau_0$ in $Y(x;s,\tau_1)$. In general, the particles in generation i correspond to particles at depth at least $(1-\varepsilon)e\tau_0$ in the tree $Y(\xi;\tau_{i-1},\tau_i)$, where ξ is a particle of $Y(x;s,t)$ included in generation $i-1$ of Π.

If the process Π does not ultimately become extinct then generation i_0 corresponds to vertices in $Y(t)$ that are at depth

$$i_0 \times (1-\varepsilon)e\tau_0 = (1-\varepsilon)e(t_1 - s) \geq (1-\varepsilon)^3 e\log n.$$

We will prove that

$$\mathbb{P}(\Pi \text{ does not become extinct}) \geq \frac{1}{2\sigma\log\sigma}, \tag{14.24}$$

and this implies (14.23) and the theorem.

To prove (14.24) we first show that μ, the expected number of progeny of a particle in Π satisfies $\mu > 1$ and after that we prove (14.24).

Let $D(h,m)$ denote the expected number of vertices at depth h in the tree T_m. Then for any $\xi \in \Pi$,

$$\mu \geq D((1-\varepsilon)e\tau_0, \sigma) \times \mathbb{P}(\nu(\tau_0) \geq \sigma). \tag{14.25}$$

It follows from (14.20) and $\sigma = e^{\tau_0}$ that

$$\mathbb{P}(|Y(\xi,0,\tau_0)| \geq \sigma) = \sum_{k=\sigma}^{\infty} e^{-\tau_0}(1-e^{-\tau_0})^k = (1-e^{-\tau_0})^\sigma \geq \frac{1}{2e}. \tag{14.26}$$

We show next that for $m \gg h$ we have

$$D(h,m) \geq \frac{(\log m - \log h - 1)^h}{h!}. \tag{14.27}$$

To prove this, we go back to (14.19) and write

$$\begin{aligned}
D(h,m) &= \frac{1}{h}\sum_{i=2}^{m}\frac{1}{i-1}\sum_{S\in\binom{[2,m]\setminus\{i\}}{h-1}}\prod_{j\in S\setminus\{i\}}\frac{1}{j-1} \\
&= \frac{1}{h}\sum_{S\in\binom{[2,m]}{h-1}}\prod_{j\in S}\frac{1}{j-1}\sum_{1\neq k\notin S}^{m}\frac{1}{k-1} \geq \frac{1}{h}\sum_{S\in\binom{[2,m]}{h-1}}\prod_{j\in S}\frac{1}{j-1}\sum_{k=h+1}^{m}\frac{1}{k} \\
&\geq \frac{\log m - \log h - 1}{h}D(h-1,m). \qquad (14.28)
\end{aligned}$$

Equation (14.27) follows by induction since $D(1,m) \geq \log m$.

Explanation of (14.28): We choose a path of length h by first choosing a vertex i and then choosing $S \subseteq [2,m] \setminus \{i\}$. We divide by h because each h-set arises h times in this way. Each choice contributes $\prod_{j\in S\cup\{i\}}\frac{1}{j-1}$. We change the order of summation i, S and then lower bound $\sum_{1\neq k\notin S}^{m}\frac{1}{k-1}$ by $\sum_{k=h+1}^{m}\frac{1}{k}$.

We now see from (14.25), (14.26) and (14.27) that

$$\begin{aligned}
\mu &\geq \frac{(\tau_0 - \log((1-\varepsilon)e\tau_0) - 1)^{(1-\varepsilon)e\tau_0}}{((1-\varepsilon)e\tau_0)!} \times \frac{1}{2e} \\
&\geq \frac{1}{2e\sqrt{2\pi}} \times \frac{1}{(1-\varepsilon/2)^{(1-\varepsilon)e\tau_0}} \gg 1,
\end{aligned}$$

if we take $\varepsilon\tau_0/\log\tau_0 \to \infty$.

We are left to prove (14.24). Let $G(z)$ be the probability generating function for the random variable Z equal to the number of descendants of a single particle. We first observe that for any $\theta \geq 1$,

$$\mathbb{P}(Z \geq \theta\sigma) \leq \mathbb{P}(|Y(\xi,0,\tau_0)| \geq \theta\sigma) = \sum_{k=\theta\sigma}^{\infty} e^{-\tau_0}(1-e^{-\tau_0})^k \leq e^{-\theta}.$$

Note that for $0 \leq x \leq 1$, any $k \geq 0$ and $a \geq k$ it holds that

$$\left(1-\frac{k}{a}\right) + \frac{k}{a}x^a \geq x^k.$$

We then write for $0 \le x \le 1$,

$$\begin{aligned}
G(x) &\le \sum_{k=0}^{\theta\sigma} p_k x^k + \mathbb{P}(Z \ge \theta\sigma) \le \sum_{k=0}^{\theta\sigma} p_k x^k + e^{-\theta} \\
&\le \sum_{k=0}^{\theta\sigma} \left(\left(1 - \frac{k}{\theta\sigma}\right) p_k + \frac{k}{\theta\sigma} p_k x^{\theta\sigma} \right) + e^{-\theta} \\
&\le \sum_{k=0}^{\infty} \left(\left(1 - \frac{k}{\theta\sigma}\right) p_k + \frac{k}{\theta\sigma} p_k x^{\theta\sigma} \right) + e^{-\theta} \\
&= H(x) = 1 - \frac{\mu}{\theta\sigma} + \frac{\mu}{\theta\sigma} x^{\theta\sigma} + e^{-\theta}.
\end{aligned}$$

The function H is monotone increasing in x and so $\rho = \mathbb{P}(\Pi$ becomes extinct) being the smallest non-negative solution to $x = G(x)$ (see Theorem 23.1) implies that ρ is at most the smallest non-negative solution q to $x = H(x)$. The convexity of H and the fact that $H(0) > 0$ implies that q is at most the value ζ satisfying $H'(\zeta) = 1$ or

$$q \le \zeta = \frac{1}{\mu^{1/(\theta\sigma - 1)}} < 1.$$

But $\rho = G(\rho) \le G(q) \le H(q)$ and so

$$1 - \rho \ge \frac{\mu}{\theta\sigma}\left(1 - \frac{1}{\mu^{\theta\sigma/(\theta\sigma-1)}}\right) - e^{-\theta} \ge \frac{\mu - 1}{\theta\sigma} - e^{-\theta} \ge \frac{1}{2\sigma\log\sigma},$$

after putting $\theta = 2\log\sigma$ and using $\mu \gg 1$. □

Devroye, Fawzi and Fraiman [236] give another proof of the above theorem that works for a wider class of random trees called *scaled attachment random recursive trees*, where each vertex i attaches to the random vertex $\lfloor iX_i \rfloor$ and $X_0, \dots, X_n$ is a sequence of independent identically distributed random variables taking values in $[0,1)$.

14.3 Inhomogeneous Recursive Trees

Plane-oriented recursive trees

This section is devoted to the study of the properties of a class of *inhomogeneous recursive trees* that are closely related to the Barabási–Albert "preferential attachment model", see [56]. Bollobás, Riordan, Spencer and Tusnády gave a proper definition of this model and showed how to reduce it to random *plane-oriented recursive trees*, see [149]. In this section we present some results that preceded [56] and created a solid mathematical ground for the further development of general preferential attachment models, which are discussed later in the book (see Chapter 17).

Suppose that we build a recursive tree in the following way. We start as before with a single vertex labeled 1 and add $n-1$ vertices labeled $2,3,\ldots,n$, one by one. We assume that the children of each vertex are ordered (say, from left to right). At each step a new vertex added to the tree is placed in a position "in between" old vertices. A tree built in this way is called a *plane-oriented recursive tree*. To study this model it is convenient to introduce an *extension* of a plane-oriented recursive tree: given a plane-oriented recursive tree we connect each vertex with external nodes, representing a possible insertion position for an incoming new vertex. See Figure 14.3 for a diagram of all plane-oriented recursive trees on $n=3$ vertices, together with their extensions.

Assume now, as before that all the edges of a tree are directed toward the leaves, and denote the out-degree of a vertex v by $d^+(v)$. Then the total number of extensions of a plane-oriented recursive tree on n vertices is equal to

$$\sum_{v\in V}(d^+(v)+1) = 2n-1.$$

So a new vertex can choose of those those $2n-1$ places to join the tree and create a tree on $n+1$ vertices. If we assume that this choice in each step is made uniformly at random then a tree constructed this way is called a *random plane-oriented recursive tree*. Notice that the probability that the vertex labeled $n+1$ is attached to vertex v is equal to $\frac{d^+(v)+1}{2n-1}$, i.e. it is proportional to the degree of v. Such random trees, called *plane-oriented* because of the above *geometric* interpretation, were introduced by Szymański [698] under the name of *non-uniform recursive trees*. Earlier, Prodinger and Urbanek [637] described plane-oriented recursive trees combinatorially, as labeled ordered (or plane) trees with the property that labels along any path down from the root are increasing. Such trees are also known in the literature as *heap-ordered trees* (see Chen and Ni [178], Morris, Panholzer and Prodinger [590]), Prodinger [636] or, more recently, as *scale-free trees*. So, random plane-oriented recursive trees are the simplest example of random preferential attachment graphs.

Denote by a_n the number of plane-oriented recursive trees on n vertices. This number, for $n\geq 2$ satisfies an obvious recurrence relation

$$a_{n+1} = (2n-1)a_n.$$

Solving this equation we get that

$$a_n = 1\cdot 3\cdot 5\ \cdots (2n-3) = (2n-3)!!.$$

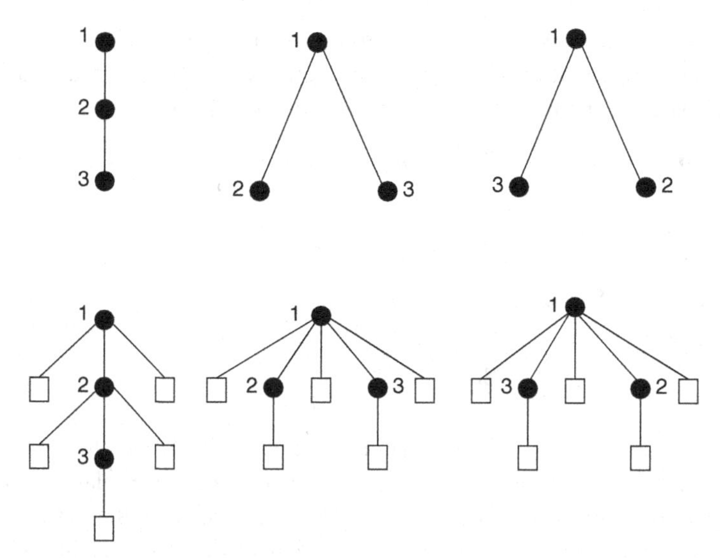

Figure 14.1. Plane-oriented recursive trees and their extensions, $n = 3$

This is also the number of Stirling permutations, introduced by Gessel and Stanley [366], i.e. the number of permutations of the multiset $\{1,1,2,2,3,3,\ldots,n,n\}$, with the additional property that, for each value of $1 \leq i \leq n$, the values lying between the two copies of i are greater than i.

There is a one-to-one correspondence between such permutations and plane-oriented recursive trees, given by Koganov [491] and, independently, by Janson [426]. To see this relationship consider a plane-oriented recursive tree on $n+1$ vertices labeled $0,1,2,\ldots,n$, where the vertex with label 0 is the root of the tree and is connected to the vertex labeled 1 only, and the edges of the tree are oriented in the direction from the root. Now, perform a Depth First Search (DFS) of the tree in which we start from the root. Next we go to the leftmost child of the root, explore that branch recursively, go to the next child in order etc., until we stop at the root. Notice that every edge in such a walk is traversed twice. If every edge of the tree gets a label equal to the label of its end-vertex furthest from the root, then the DFS encodes each tree by a string of length $2n$, where each label $1,2,\ldots,n$ appears twice. So the unique code of each tree is a unique permutation of the multiset $\{1,1,2,2,3,3,\ldots,n,n\}$ with additional property described above. Note also that the insertion of a pair $(n+1,n+1)$ into one of the $2n-1$ gaps between labels of the permutation of this multiset, corresponds to the insertion of the vertex labeled $n+1$ into a plane-oriented recursive tree on n vertices.

Let us start with exact formulas for probability distribution of the out-degree $D^+_{n,i}$ of a vertex with label i, $i = 1,2,\ldots,n$ in a random plane-oriented recursive tree. Kuba and Panholzer [516] proved the following theorem.

Theorem 14.16 *For $i = 1,2,\ldots,n$ and $r = 1,2,\ldots,n-1$,*

$$\mathbb{P}(D^+_{n,i} = r) = \sum_{k=0}^{r} \binom{r}{k} (-1)^k \frac{\Gamma(n-3/2)\Gamma(i-1/2)}{\Gamma(i-1-k/2)\Gamma(n-1/2)},$$

where $\Gamma(z) = \int_0^\infty t^{z-1}e^{-t}dt$ is the Gamma function. Moreover,

$$\mathbb{E}(D^+_{n,i}) = \frac{\binom{2i-2}{i-1}4^{n-i}}{\binom{2n-2}{n-1}} - 1. \tag{14.29}$$

For simplicity, we show below that the formula (14.29) holds for $i = 1$, i.e. the expected value of the out-degree of the root of a random plane-oriented recursive tree, and investigate its behavior as $n \to \infty$. It is then interesting to compare the latter with the asymptotic behavior of the degree of the root of a random recursive tree. Recall that for large n this is roughly $\log n$ (see Theorem 14.10).

The result below was proved by Mahmoud, Smythe and Szymański [554].

Corollary 14.17 *For $n \geq 2$ the expected value of the degree of the root of a random plane-oriented recursive tree is*

$$\mathbb{E}(D^+_{n,1}) = \frac{4^{n-1}}{\binom{2n-2}{n-1}} - 1,$$

and,

$$\mathbb{E}(D^+_{n,1}) \approx \sqrt{\pi n}.$$

Proof Denote by

$$u_n = \frac{4^n}{\binom{2n}{n}} = \prod_{i=1}^{n} \frac{2i}{2i-1} = \frac{(2n)!!}{(2n-1)!!}.$$

Hence, in terms of u_n, we want to prove that $\mathbb{E}(D^+_{n,1}) = u_{n-1} - 1$.

It is easy to see that the claim holds for $n = 1,2$ and that

$$\mathbb{P}(D^+_{n,1} = 1) = \prod_{i=1}^{n-1} \left(1 - \frac{2}{2i-1}\right) = \frac{1}{2n-3},$$

while, for $r > 1$ and $n \geq 1$,

$$\mathbb{P}(D^+_{n+1,1} = r) = \left(1 - \frac{r+1}{2n-1}\right)\mathbb{P}(D^+_{n,1} = r) + \frac{r}{2n-1}\mathbb{P}(D^+_{n,1} = r-1).$$

Hence

$$\begin{aligned}
\mathbb{E}(D^+_{n+1,1}) &= \sum_{r=1}^{n} r\left(\frac{2n-r-2}{2n-1}\,\mathbb{P}(D^+_{n,1}=r) + \frac{r}{2n-1}\,\mathbb{P}(D^+_{n,1}=r-1)\right)\\
&= \frac{1}{2n-1}\left(\sum_{r=1}^{n-1} r(2n-r-2)\,\mathbb{P}(D^+_{n,1}=r) + \sum_{r=1}^{n-1}(r+1)^2\,\mathbb{P}(D^+_{n,1}=r)\right)\\
&= \frac{1}{2n-1}\sum_{r=1}^{n}(2nr+1)\,\mathbb{P}(D^+_{n,1}=r).
\end{aligned}$$

So, we get the following recurrence relation

$$\mathbb{E}(D^+_{n+1,1}) = \frac{2n}{2n-1}\,\mathbb{E}(D^+_{n,1}) + \frac{1}{2n-1}$$

and the first part of the theorem follows by induction.

To see that the second part also holds we have to use the Stirling approximation to check that

$$u_n = \sqrt{\pi n} - 1 + \frac{3}{8}\sqrt{\pi/n} + \cdots.$$

□

The next theorem, due to Kuba and Panholzer [516], summarizes the asymptotic behavior of the suitably normalized random variable $D^+_{n,i}$.

Theorem 14.18 *Let $i \geq 1$ be fixed and let $n \to \infty$. If*

(i) $i = 1$, then

$$n^{-1/2}D^+_{n,1} \xrightarrow{D} D_1, \quad \text{with density} \quad f_{D_1}(x) = (x/2)e^{-x^2/2},$$

i.e. is asymptotically Rayleigh distributed with parameter $\sigma = \sqrt{2}$,

(ii) $i \geq 2$, then $n^{-1/2}D^+_{n,i} \xrightarrow{D} D_i$, with density

$$f_{D_i}(x) = \frac{2i-3}{2^{2i-1}(i-2)!}\int_x^{\infty}(t-x)^{2i-4}e^{-t^2/4}\,dt.$$

Let $i = i(n) \to \infty$ as $n \to \infty$. If

(i) $i = o(n)$, then the normalized random variable $(n/i)^{-1/2}D^+_{n,i}$ is asymptotically Gamma distributed $\gamma(\alpha,\beta)$, with parameters $\alpha = -1/2$ and $\beta = 1$,

(ii) $i = cn$, $0 < c < 1$, then the random variable $D^+_{n,i}$ is asymptotically negative Binomial distributed NegBinom(r,p) with parameters $r = 1$ an $p = \sqrt{c}$,

(iii) $n - i = o(n)$, then $\mathbb{P}(D^+_{n,i} = 0) \to 1$, as $n \to \infty$.

We now turn our attention to the number of vertices of a given out-degree. The next theorem shows a characteristic feature of random graphs built by the preferential attachment rule, where every new vertex prefers to attach to a vertex with high degree (*rich get richer* rule). The proportion of vertices with degree r in such a random graph with n vertices grows like n/r^{α}, for some constant $\alpha > 0$, i.e. its distribution obeys a so-called *power law*. The next result was proved by Szymański [698] (see also [554] and [700]) and it indicates such a behavior for the degrees of the vertices of a random plane-oriented recursive tree, where $\alpha = 3$.

Theorem 14.19 *Let r be fixed and denote by $X^+_{n,r}$ the number of vertices of out-degree r in a random plane-oriented recursive tree T_n. Then,*

$$\mathbb{E}X^+_{n,r} = \frac{4n}{(r+1)(r+2)(r+3)} + O\left(\frac{1}{r}\right).$$

Proof Observe first that conditional on T_n,

$$\mathbb{E}(X^+_{n+1,r}|T_n) = X^+_{n,r} - \frac{r+1}{2n-1}X^+_{n,r} + \frac{r}{2n-1}X^+_{n,r-1} + 1_{r=0}, \tag{14.30}$$

which gives

$$\mathbb{E}X^+_{n+1,r} = \frac{2n-r-2}{2n-1}\mathbb{E}X^+_{n,r} + \frac{r}{2n-1}\mathbb{E}X^+_{n,r-1} + 1_{r=0} \tag{14.31}$$

for $r \geq 1$, ($X^+_{n,-1} = 0$).

We will show that the difference

$$a_{n,r} \stackrel{def}{=} \mathbb{E}X^+_{n,r} - \frac{4n}{(r+1)(r+2)(r+3)}$$

is asymptotically negligible with respect to the leading term in the statement of the theorem. Substitute $a_{n,r}$ in the Equation (14.31) to get that for $r \geq 1$,

$$a_{n+1,r} = \frac{2n-r-2}{2n-1}a_{n,r} + \frac{r}{2n-1}a_{n,r-1} - \frac{1}{2n-1}. \tag{14.32}$$

We want to show that $|a_{n,r}| \leq \frac{2}{\max\{r,1\}}$, for all $n \geq 1, r \geq 0$. Note that this is true for all n and $r = 0, 1$, since from (14.31) it follows (inductively) that for $n \geq 2$

$$\mathbb{E}X^+_{n,0} = \frac{2n-1}{3} \text{ and so } a_{n,0} = -\frac{1}{3}.$$

For $n \geq 2$,

$$\mathbb{E}X^+_{n,1} = \frac{n}{6} - \frac{1}{12} + \frac{3}{4(2n-3)} \text{ and so } a_{n,1} = -\frac{1}{12} + \frac{3}{4(2n-3)}.$$

We proceed by induction on r. By definition

$$a_{r,r} = -\frac{4r}{(r+1)(r+2)(r+3)},$$

and so,

$$|a_{r,r}| < \frac{2}{r}.$$

We then see from (14.32) that for and $r \geq 2$ and $n \geq r$ that

$$\begin{aligned}|a_{n+1,r}| &\leq \frac{2n-r-2}{2n-1} \cdot \frac{2}{r} + \frac{r}{2n-1} \cdot \frac{2}{r-1} - \frac{1}{2n-1}.\\ &= \frac{2}{r} - \frac{2}{(2n-1)r}\left(r+1-\frac{r^2}{r-1}-\frac{r}{2}\right)\\ &\leq \frac{2}{r},\end{aligned}$$

which completes the induction and the proof of the theorem. □

In fact much more can be proved.

Theorem 14.20 *Let $\varepsilon > 0$ and r be fixed, then, w.h.p.*

$$(1-\varepsilon)a_r \leq \frac{X^+_{n,r}}{n} \leq (1+\varepsilon)a_r, \tag{14.33}$$

where

$$a_r = \frac{4}{(r+1)(r+2)(r+3)}.$$

Moreover,

$$\frac{\left(X^+_{n,r} - na_r\right)}{\sqrt{n}} \xrightarrow{d} Y_r, \tag{14.34}$$

as $n \to \infty$, jointly for all $r \geq 0$, where the Y_r are jointly normally distributed with expectations $\mathbb{E}\, Y_r = 0$ and covariances $\sigma_{rs} = \mathrm{Cov}(Y_r, Y_s)$ given by

$$\sigma_{rs} = 2\sum_{k=0}^{r}\sum_{l=0}^{s}\frac{(-1)^{k+l}}{k+l+4}\binom{r}{k}\binom{s}{l}\left(\frac{2(k+l+4)!}{(k+3)!(l+3)!} - 1 - \frac{(k+1)(l+1)}{(k+3)(l+3)}\right).$$

Proof For the proof of asymptotic normality of a suitably normalized random variable $X^+_{n,r}$, i.e. for the proof of statement (14.34) see Janson [424]. We will give a short proof of the first statement (14.33), due to Bollobás, Riordan, Spencer and Tusnády [149] (see also Mori [588]).

Consider a random plane-oriented recursive tree T_n as an element of a process $(T_t)_{t=o}^{\infty}$. Fix $n \geq 1$ and $r \geq 0$ and for $0 \leq t \leq n$ define the martingale

$$Y_t = \mathbb{E}(X^+_{n,r}|T_t) \quad \text{where} \quad Y_0 = \mathbb{E}(X^+_{n,r}) \text{ and } Y_n = X^+_{n,r}.$$

One sees that the differences

$$|Y_{t+1} - Y_t| \le 2.$$

For a proof of this, see the proof of Theorem 17.3. Applying the Hoeffding–Azuma inequality (see Theorem 21.15) we get, for any fixed r,

$$\mathbb{P}(|X^+_{n,r} - \mathbb{E}X^+_{n,r}| \ge \sqrt{n\log n}) \le e^{-(1/8)\log n} = o(1).$$

But Theorem 14.19 shows that for any fixed r, $\mathbb{E}X^+_{n,r} \gg \sqrt{n\log n}$ and (14.33) follows. □

Similarly, as for uniform random recursive trees, Pittel [625] established the asymptotic behavior of the height of a random plane-oriented recursive tree.

Theorem 14.21 *Let h^*_n be the height of a random plane-oriented recursive tree, then w.h.p.*

$$h^*_n \approx \frac{\log n}{2\gamma},$$

where γ is the unique solution of the equation

$$\gamma e^{\gamma+1} = 1,$$

i.e. $\gamma = 0.27846..$, *so* $\frac{1}{2\gamma} = 1.79556....$

Inhomogeneous recursive trees: a general model

As before, consider a tree that grows randomly in time. Each time a new vertex appears, it chooses exactly one of the existing vertices and attaches to it. This way we build a tree T_n of order n with $n+1$ vertices labeled $\{0,1,\ldots,n\}$, where the vertex labeled 0 is the root. Now assume that for every $n \ge 0$ there is a probability distribution

$$P^{(n)} = (p_0, p_1, \ldots, p_n), \quad \sum_{j=0}^{n} p_j = 1.$$

Suppose that T_n has been constructed for some $n \ge 1$. Given T_n we add an edge connecting one of its vertices with a new vertex labeled $n+1$, thus forming a tree T_{n+1}. A vertex $v_n \in \{0,1,2,\ldots,n\}$ is chosen to be a neighbor of the incoming vertex with probability

$$\mathbb{P}(v_n = j|T_n) = p_j, \quad \text{for} \quad j = 0,1,\ldots,n.$$

Note that for the uniform random recursive tree we have

$$p_j = \frac{1}{n+1}, \qquad \text{for } 0 \le j \le n.$$

We say that a random recursive tree is *inhomogeneous* if the attachment rule of new vertices is determined by a non-uniform probability distribution. Most often the probability that a new vertex chooses a vertex $j \in \{0, 1, \ldots, n\}$ is proportional to $w(d_n(j))$, the value of a weight function w applied to the degree $d_n(j)$ of vertex j after n-th step. Then the probability distribution $P^{(n)}$ is defined as

$$p_j = \frac{w(d_n(j))}{\sum_{k=0}^{n} w(d_n(k))}.$$

Consider a special case when the weight function is linear and, for $0 \le j \le n$,

$$w(d_n(j)) = d_n(j) + \beta, \ \beta > -1, \tag{14.35}$$

so that the total weight

$$w_n = \sum_{k=0}^{n} (d_n(k) + \beta) = 2n + (n+1)\beta. \tag{14.36}$$

Obviously the model with such a probability distribution is only a small generalization of plane-oriented random recursive trees and we obtain the latter when we put $\beta = 0$ in (14.35). Inhomogeneous random recursive trees of this type are known in the literature as either *scale free random trees* or *Barabási–Albert random trees*. For obvious reasons, we will call such graphs *generalized random plane-oriented recursive trees.*

Let us focus attention on the asymptotic behavior of the maximum degree of such random trees. We start with some useful notation and observations.

Let $X_{n,j}$ denote the weight of vertex j in a generalized plane-oriented random recursive tree, with initial values $X_{1,0} = X_{j,j} = 1 + \beta$ for $j > 0$. Let

$$c_{n,k} = \frac{\Gamma\left(n + \frac{\beta}{\beta+2}\right)}{\Gamma\left(n + \frac{\beta+k}{\beta+2}\right)}, \quad n \ge 1, \ k \ge 0,$$

be a double sequence of normalizing constants. Note that

$$\frac{c_{n+1,k}}{c_{n,k}} = \frac{w_n}{w_n + k}, \tag{14.37}$$

and, for any fixed k,

$$c_{n,k} = n^{-k/(\beta+2)}(1 + O(n^{-1})).$$

Let k be a positive integer and

$$X_{n,j;k} = c_{n,k} \binom{X_{n,j} + k - 1}{k}.$$

Lemma 14.22 *Let $\mathscr{F}_n$ be the σ-field generated by the first n steps. If $n \geq \max\{1, j\}$, then $(X_{n,j;k}, \mathscr{F}_n)$ is a martingale.*

Proof Because $X_{n+1,j} - X_{n,j} \in \{0,1\}$, we see that

$$\binom{X_{n+1,j}+k-1}{k} = \binom{X_{n,j}+k-1}{k} + \binom{X_{n,j}+k-1}{k-1}\binom{X_{n+1,j}-X_{n,j}}{1}$$
$$= \binom{X_{n,j}+k-1}{k}\left(1+\frac{k(X_{n+1,j}-X_{n,j})}{X_{n,j}}\right).$$

Hence, noting that

$$\mathbb{P}(X_{n+1,j}-X_{n,j}=1|\mathscr{F}_n) = \frac{X_{n,j}}{w_n},$$

and applying (14.37)

$$\mathbb{E}(X_{n+1,j;k}|\mathscr{F}_n) = X_{n,j;k}\frac{c_{n+1,k}}{c_{n,k}}\left(1+\frac{k}{w_n}\right) = X_{n,j;k},$$

we arrive at the lemma. □

Thus, the random variable $X_{n,j;k}$, as a non-negative martingale, is bounded in L_1 and it almost surely converges to $X_j^k/k!$, where X_j is the limit of $X_{n,j;1}$. Since $X_{n,j;k} \leq cX_{n,j;2k}$, where the constant c does not depend on n, it is also bounded in L_2, which implies that it converges in L_1. Therefore we can determine all moments of the random variable X_j. Namely, for $j \geq 1$,

$$\frac{X_j^k}{k!} = \lim_{n\to\infty} \mathbb{E}X_{n,j;k} = X_{j,j;k} = c_{j,k}\binom{\beta+k}{k}. \tag{14.38}$$

Let Δ_n be the maximum degree in a generalized random plane-oriented recursive tree T_n and let, for $j \leq n$,

$$\Delta_{n,j} = \max_{0\leq i\leq j} X_{n,i;1} = \max_{0\leq i\leq j} c_{n,1}X_{n,i}.$$

Note that since $X_{n,i}$ is the weight of vertex i, i.e. its degree plus β, we find that $\Delta_{n,n} = c_{n,1}(\Delta_n+\beta)$. Define

$$\xi_j = \max_{0\leq i\leq j} X_i \quad \text{and} \quad \xi = \xi_\infty = \sup_{j\geq 0} X_j. \tag{14.39}$$

Now we are ready to prove the following result, due to Móri [589].

Theorem 14.23

$$\mathbb{P}\left(\lim_{n\to\infty} n^{-1/(\beta+2)}\Delta_n = \xi\right) = 1.$$

The limiting random variable ξ is almost surely finite and positive and it has an absolutely continuous distribution. The convergence also holds in L_p, for all p, $1 \leq p < \infty$.

Proof In the proof we skip the part dealing with the positivity of ξ and the absolute continuity of its distribution.

By Lemma 14.22, $\Delta_{n,n}$ is the maximum of martingales, therefore $(\Delta_{n,n}|\mathscr{F})$ is a non-negative sub-martingale, and so

$$\mathbb{E}\,\Delta_{n,n}^k \leq \sum_{j=0}^{n} \mathbb{E}X_{n,j;1}^k \leq \sum_{j=0}^{\infty} \mathbb{E}X_j^k = k!\binom{\beta+k}{k}\sum_{j=0}^{\infty} c_{j,k} < \infty,$$

if $k > \beta+2$. (Note $c_{0,k}$ is defined here as equal to $c_{1,k}$.) Hence, $(\Delta_{n,n}|\mathscr{F})$ is bounded in L_k, for every positive integer k, which implies both almost sure convergence and convergence in L_p, for any $p \geq 1$.

Assume that $k > \beta+2$ is fixed, then, for $n \geq k$,

$$\mathbb{E}(\Delta_{n,n} - \Delta_{n,j})^k \leq \sum_{i=j+1}^{n} \mathbb{E}X_{n,i;1}.$$

Take the limit as $n \to \infty$ of both sides of the above inequality. Applying (14.39) and (14.38), we get

$$\mathbb{E}\left(\lim_{n\to\infty} n^{-1/(\beta+2)}\Delta_n - \xi_j\right)^k \leq \sum_{i=j+1}^{\infty} \mathbb{E}\xi_i^k = k!\binom{\beta+k}{k}\sum_{i=j+1}^{\infty} c_{j,k}.$$

The right-hand side tends to 0 as $j \to \infty$, which implies that $n^{-1/(\beta+2)}\Delta_n$ tends to ξ, as claimed. □

To conclude this section, setting $\beta = 0$ in Theorem 14.23, we can obtain the asymptotic behavior of the maximum degree of a plane-oriented random recursive tree.

14.4 Exercises

14.4.1 Use the Prüfer code to show that there is one-to-one correspondence between the family of all labeled trees with vertex set $[n]$ and the family of all ordered sequences of length $n-2$ consisting of elements of $[n]$.

14.4.2 Prove Theorem 14.1.

14.4.3 Let Δ be the maximum degree of a random labeled tree on n vertices. Use (14.1) to show that for every $\varepsilon > 0$, $\mathbb{P}(\Delta > (1+\varepsilon)\log n/\log\log n)$ tends to 0 as $n \to \infty$.

14.4.4 Let Δ be defined as in the previous exercise and let $t(n,k)$ be the number of labeled trees on n vertices with maximum degree at most k. Knowing that $t(n,k) < (n-2)!\left(1+1+\frac{1}{2!}+\ldots+\frac{1}{(k-1)!}\right)^n$, show that for every $\varepsilon > 0$, $\mathbb{P}(\Delta < (1-\varepsilon)\log n/\log\log n)$ tends to 0 as $n \to \infty$.

14.4.5 Determine a one-to-one correspondence between the family of permutations on $\{2,3,\dots,n\}$ and the family of recursive trees on the set $[n]$.

14.4.6 Let L_n denote the number of leaves of a random recursive tree with n vertices. Show that $\mathbb{E}L_n = n/2$ and $\mathrm{Var}\,L_n = n/12$.

14.4.7 Prove (14.20).

14.4.8 Show that $\Phi_{n,i}(z)$ given in Theorem 14.8 is the probability generating function of the convolution of $n-i$ independent Bernoulli random variables with success probabilities equal to $1/(i+k-1)$ for $k = 1,2,\dots,n-i$.

14.4.9 Let L_n^* denotes the number of leaves of a random plane-oriented recursive tree with n vertices. Show that

$$\mathbb{E}L_n^* = \frac{2n-1}{3} \quad \text{and} \quad \mathrm{Var}\,L_n^* = \frac{2n(n-2)}{9(2n-3)}.$$

14.4.10 Prove that L_n^*/n (defined above) converges in probability, to $2/3$.

14.5 Notes

Labeled trees

The literature on random labeled trees and their generalizations is very extensive. For a comprehensive list of publications in this broad area we refer the reader to a recent book of Drmota [250], to a chapter of Bollobás's book [130] on random graphs, as well as to the book by Kolchin [497]. For a review of some classical results, including the most important contributions, forming the foundation of the research on random trees, mainly due to Meir and Moon (see, for example : [574], [575]and [577]), one may also consult a survey by Karoński [465].

Recursive trees

Recursive trees have been introduced as probability models for system generation (Na and Rapoport [597]), spread of infection (Meir and Moon [576]), pyramid schemes (Gastwirth [360]) and stemma construction in philology (Najock and Heyde [601]). Most likely, the first place that such trees were introduced in the literature, is the paper by Tapia and Myers [703], presented there under the name "concave node-weighted trees." Systematic studies of random recursive trees were initiated by Meir and Moon ([576] and [587]) who investigated distances between vertices as well as the process of cutting down such random trees. Observe that there is a bijection between families of recursive trees and binary search trees, and this has opened many interesting directions of research, as shown in a survey by Mahmoud and Smythe [553] and the book by Mahmoud [551].

Early papers on random recursive trees (see, for example, [597], [360] and [249]) were focused on the distribution of the degree of a given vertex and of the number of vertices of a given degree. Later, these studies were extended to the distribution of the number of vertices at each level, which is referred to as the *profile*. Recall, that in a rooted tree, a *level* (*strata*) consists of all those vertices that are at the same distance from the root.

The profile of a random recursive tree is analyzed in many papers. For example, Drmota and Hwang [251] derived asymptotic approximations to the correlation coefficients of two level sizes in random recursive trees and binary search trees. These coefficients undergo sharp sign-changes when one level is fixed and the other is varying. They also proposed a new means of deriving an asymptotic estimate for the expected *width*, which is the number of nodes at the most abundant level.

Devroye and Hwang [238] proposed a new, direct, correlation-free approach based on central moments of profiles to the asymptotics of width in a class of random trees of logarithmic height. This class includes random recursive trees.

Fuchs, Hwang, and Neininger [354] proved convergence in distribution for the profile, normalized by its mean, of random recursive trees when the limit ratio α of the level and the logarithm of tree size lies in $[0,e)$. Convergence of all moments is shown to hold only for $\alpha \in (0,1)$ (with only convergence of finite moments when $\alpha \in (1,e)$).

van der Hofstadt, Hooghiemstra and Van Mieghem [409] studied the covariance structure of the number of nodes k and l steps away from the root in random recursive trees and give an analytic expression valid for all k,l and tree sizes n.

For an arbitrary positive integer $i \leq i_n \leq n-1$, a function of n, Su, Liu and Feng [696] demonstrated the distance between nodes i and n in random recursive trees T_n, is asymptotically normal as $n \to \infty$ by using the classical limit theory method.

Holmgren and Janson [411] proved limit theorems for the sums of functions of sub-trees of binary search trees and random recursive trees. In particular, they give new simple proofs of the fact that the number of fringe trees of size $k = k_n$ in a binary search tree and the random recursive tree (of total size n) asymptotically has a Poisson distribution if $k \to \infty$, and that the distribution is asymptotically normal for $k = o(\sqrt{n})$. Recall that a *fringe tree* is a sub-tree consisting of some vertex of a tree and all its descendants (see Aldous [14]). For other results on that topic see Devroye and Janson [239].

Feng, Mahmoud and Panholzer [291] studied the variety of sub-trees lying on the fringe of recursive trees and binary search trees by analyzing the distributional behavior of $X_{n,k}$, which counts the number of sub-trees of size k in a random tree of size n, with $k = k(n)$. Using analytic methods, they characterized for both tree families the phase change behavior of $X_{n,k}$.

One should also notice interesting applications of random recursive trees. For example, Mehrabian [573] presented a new technique for proving logarithmic upper bounds for diameters of evolving random graph models, which is based on defining a coupling between random graphs and variants of random recursive trees. Goldschmidt and Martin [377] described a representation of the Bolthausen–Sznitman coalescent in terms of the cutting of random recursive trees.

Bergeron, Flajolet and Salvy [80] have defined and studied a wide class of *random increasing trees*. A tree with vertices labeled $\{1,2,\ldots,n\}$ is *increasing* if the sequence of labels along any branch starting at the root is increasing. Obviously, recursive trees and binary search trees (as well as the general class of inhomogeneous trees, including plane-oriented trees) are increasing. Such a general model, which has been intensively studied, yields many important results for random trees discussed in this chapter. Here we restrict ourselves to pointing out just a few papers dealing with random increasing trees authored by Dobrow and Smythe [248], Kuba and Panholzer [516] and Panholzer and Prodinger [613], as well as with their generalizations, i.e. *random increasing k-trees*, published by Zhang, Rong and Comellas [734], Panholzer and Seitz [614] and Darrasse, Hwang and Soria [229].

Inhomogeneous recursive trees

Plane-oriented recursive trees

As we already mentioned in Section 14.3, Prodinger and Urbanek [637] and, independently, Szymański [698] introduced the concept of plane-oriented random trees (more precisely, this notion was introduced in an unpublished paper by Dondajewski and Szymański [249]), and studied the vertex degrees of such random trees. Mahmoud, Smythe and Szymański [554], using Pólya urn models, investigated the exact and limiting distributions of the size and the number of leaves in the branches of the tree (see [426] for a follow up). Lu and Feng [532] considered the strong convergence of the number of vertices of given degree as well as of the degree of a fixed vertex (see also [553]). In Janson's [424] paper, the distribution of vertex degrees in random recursive trees and random plane recursive trees are shown to be asymptotically normal. Brightwell and Luczak [157] investigated the number $D_{n,k}$ of vertices of each degree k at each time n, focusing particularly on the case where $k = k(n)$ is a growing function of n. They showed that $D_{n,k}$ is concentrated around its mean, which is approximately $4n/k^3$, for all $k \leq (n \log n)^{-1/3}$, which is best possible up to a logarithmic factor.

Hwang [413] derived several limit results for the profile of random plane-oriented recursive trees. These include the limit distribution of the normalized profile, asymptotic bimodality of the variance, asymptotic approximation to the expected width and the correlation coefficients of two level sizes.

Fuchs [353] outlined how to derive limit theorems for the number of sub-trees of size k on the fringe of random plane-oriented recursive trees.

Finally, Janson, Kuba and Panholzer [428] considered generalized Stirling permutations and related them to certain families of generalized plane recursive trees.

Generalized recursive trees

Móri [588] proved the strong law of large numbers and central limit theorem for the number of vertices of low degree in a generalized random plane-oriented recursive tree. Szymański [700] gives the rate of concentration of the number of vertices with given degree in such trees. Móri [589] studied the maximum degree of a scale-free trees. Katona [478] showed that the degree distribution is the same on every sufficiently high level of the tree and in [477] investigated the width of scale-free trees.

Rudas, Toth and Valko [661], using results from the theory of general branching processes, give the asymptotic degree distribution for a wide range of weight functions. Backhausz and Móri [44] presented sufficient conditions for the almost sure existence of an asymptotic degree distribution constrained to the set of selected vertices and describe that distribution.

Bertoin and Bravo [81] considered Bernoulli bond percolation on a large scale-free tree in the super-critical regime, i.e. when there exists a giant cluster with high probability. They obtained a weak limit theorem for the sizes of the next largest clusters, extending a result in Bertoin [83] for large random recursive trees.

Devroye, Fawzi and Fraiman [236] studied depth properties of a general class of random recursive trees called *attachment random recursive trees*. They proved that the height of such tree is asymptotically given by $\alpha_{max} \log n$, where α_{max} is a constant. This gives a new elementary proof for the height of uniform random recursive trees that does not use a branching random walk. For further generalizations of random recursive trees see Mahmoud [552].

15
Mappings

In the evolution of the random graph $\mathbb{G}_{n,p}$, during its sub-critical phase, tree components and components with exactly one cycle, i.e. graphs with the same number of vertices and edges, are w.h.p. the only elements of its structure. Similarly, they are the only graphs outside the giant component after the phase transition, until the random graph becomes connected w.h.p. In the previous chapter we studied the properties of random trees. Now we focus our attention on random mappings of a finite set into itself. Such mappings can be represented as digraphs with the same number of vertices and edges. So the study of their "average" properties may help us to better understand the typical structure of classical random graphs. We start the chapter with a short look at the basic properties of random permutations (one-to-one mappings) and then continue to the general theory of random mappings.

15.1 Permutations

Let f be chosen uniformly at random from the set of all $n!$ permutations on the set $[n]$, i.e. from the set of all one-to-one functions $[n] \to [n]$. In this section we concentrate our attention on the properties of a functional digraph representing a random permutation.

Let D_f be the functional digraph $([n],(i,f(i)))$. The digraph D_f consists of vertex disjoint cycles of any length $1,2,\ldots,n$. Loops represent fixed points, see Figure 15.1.

Let $X_{n,t}$ be the number of cycles of length t, $t = 1,2,\ldots,n$ in the digraph D_f. Thus, $X_{n,1}$ counts the number of fixed points of a random permutation. One can easily check that

$$\mathbb{P}(X_{n,t} = k) = \frac{1}{k!t^k} \sum_{i=0}^{\lfloor n/t \rfloor} \frac{(-1)^i}{t^i i!} \to \frac{e^{-1/t}}{t^k k!} \quad \text{as} \quad n \to \infty, \tag{15.1}$$

x	1	2	3	4	5	6	7	8	9	10
f(x)	3	10	1	4	2	6	8	9	7	5

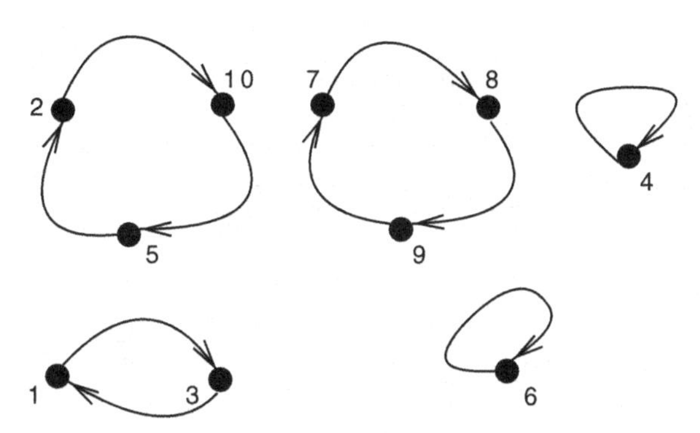

Figure 15.1. A permutation digraph example

for $k = 0,1,2,\ldots,n$. Indeed, convergence in (15.1) follows directly from Lemma 20.10 and the fact that

$$B_i = \mathbb{E}\binom{X_{n,t}}{i} = \frac{1}{n!} \cdot \frac{n!}{(t!)^i(n-ti)!} \frac{((t-1)!)^i(n-ti)!}{i!} = \frac{1}{t^i i!}.$$

This means that $X_{n,t}$ converges in distribution to a random variable with Poisson distribution with mean $1/t$.

Moreover, direct computation gives

$$\begin{aligned}
&\mathbb{P}(X_{n,1} = j_1, X_{n,2} = j_2, \ldots, X_{n,n} = j_n) \\
&= \frac{1}{n!} \frac{n!}{\prod_{t=1}^n j_t!(t!)^{j_t}} \prod_{t=1}^n ((t-1)!)^{j_t} \\
&= \prod_{t=1}^n \left(\frac{1}{t}\right)^{j_t} \frac{1}{j_t!},
\end{aligned}$$

for non-negative integers $j_1, j_2, \ldots, j_n$ satisfying $\sum_{t=1}^n t j_t = n$.

Hence, asymptotically, the random variables $X_{n,t}$ have independent Poisson distributions with expectations $1/t$, respectively (see Goncharov [380] and Kolchin [494]).

Next, consider the random variable $X_n = \sum_{j=1}^{n} X_{n,j}$ counting the total number of cycles in a functional digraph D_f of a random permutation. It is not difficult to show that X_n has the following probability distribution.

Theorem 15.1 *For $k = 1,2,\dots,n$,*

$$\mathbb{P}(X_n = k) = \frac{|s(n,k)|}{n!},$$

where the $s(n,k)$ are Stirling numbers of the first kind, i.e. numbers satisfying the following relation:

$$x(x-1)\cdots(x-n+1) = \sum_{k=0}^{n} s(n,k)x^k .$$

Moreover,

$$\mathbb{E}X_n = H_n = \sum_{j=1}^{n} \frac{1}{j}, \qquad \operatorname{Var} X_n = H_n - \sum_{j=1}^{n} \frac{1}{j^2} .$$

Proof Denote by $c(n,k)$ the number of digraphs D_f (permutations) on n vertices and with exactly k cycles. Consider a vertex n in D_f. It either has a loop (belongs to a unit cycle) or it doesn't. If it does, then D_f is composed of a loop in n and a cyclic digraph (permutation) on $n-1$ vertices with exactly $k-1$ cycles and there are $c(n-1,k-1)$ such digraphs (permutations). Otherwise, the vertex n can be thought as dividing (lying on) one of the $n-1$ arcs that belongs to the cyclic digraph on $n-1$ vertices with k cycles and there are $(n-1)c(n-1,k)$ such permutations (digraphs) of the set $[n]$. Hence,

$$c(n,k) = c(n-1,k-1) + (n-1)c(n-1,k).$$

Now, multiplying both sides by x^k, dividing by $n!$ and summing up over all k, we obtain

$$G_n(x) = (x+n-1)G_{n-1}(x),$$

where $G_n(x)$ is the probability generating function of X_n. But $G_1(x) = x$, so

$$G_n(x) = \frac{x(x+1)\cdots(x+n-1)}{n!},$$

and the first part of the theorem follows. Note that

$$G_n(x) = \binom{x+n-1}{n} = \frac{\Gamma(x+n)}{\Gamma(x)\Gamma(n+1)},$$

where Γ is the Gamma function.

The results for the expectation and variance of X_n can be obtained by calculating the first two derivatives of $G_n(x)$ and evaluating them at $x = 1$ in a standard way but we can also show them using only the fact that the cycles of functional digraphs must be disjoint. Notice, for example, that

$$\mathbb{E} X_n = \sum_{\emptyset \neq S \subset [n]} \mathbb{P}(S \text{ induces a cycle})$$

$$= \sum_{k=1}^{n} \binom{n}{k} \frac{(k-1)!(n-k)!}{n!} = H_n.$$

Similarly, we can derive the second factorial moment of X_n counting ordered pairs of cycles (see Exercises 15.3.2 and 15.3.3), which implies the formula for the variance. □

Goncharov [380] proved a Central Limit Theorem for the number X_n of cycles.

Theorem 15.2

$$\lim_{n\to\infty} \mathbb{P}\left(\frac{X_n - \log n}{\sqrt{\log n}} \leq x \right) = \int_{-\infty}^{x} e^{-t^2/2} dt,$$

i.e. the standardized random variable X_n converges in distribution to the standard normal random variable.

Another numerical characteristic of a digraph D_f is the length L_n of its longest cycle. Shepp and Lloyd [682] established the asymptotic behavior of the expected value of L_n.

Theorem 15.3

$$\lim_{n\to\infty} \frac{\mathbb{E} L_n}{n} = \int_0^\infty \exp\left\{ -x - \int_x^\infty \frac{1}{y} e^{-y} dy \right\} dx = 0.62432965....$$

15.2 Mappings

Let f be chosen uniformly at random from the set of all n^n mappings from $[n] \to [n]$. Let D_f be the functional digraph $([n], (i, f(i)))$ and let G_f be the graph obtained from D_f by ignoring orientation. In general, D_f has unicyclic components only, where each component consists of a directed cycle C with trees rooted at vertices of C, see the Figure 15.2.

Therefore, the study of functional digraphs is based on results for permutations of the set of cyclical vertices (those lying on cycles) and results for forests consisting of trees rooted at these cyclical vertices (we also allow trivial one

x	1	2	3	4	5	6	7	8	9	10
f(x)	3	10	5	4	2	5	8	9	7	5

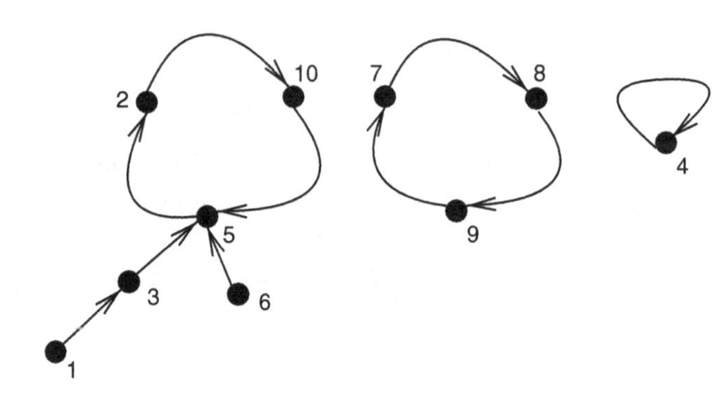

Figure 15.2. A mapping digraph example

vertex trees). For example, to show our first result on the connectivity of G_f we need the following enumerative result for the forests.

Lemma 15.4 *Let $T(n,k)$ denote the number of forests with vertex set $[n]$, consisting of k trees rooted at the vertices $1,2,\ldots,k$, then*

$$T(n,k) = kn^{n-k-1}.$$

Proof Observe first that by (14.2) there are $\binom{n-1}{k-1}n^{n-k}$ trees with $n+1$ labeled vertices in which the degree of a vertex $n+1$ is equal to k. Hence, there are

$$\binom{n-1}{k-1}n^{n-k} \Big/ \binom{n}{k} = kn^{n-k-1}$$

trees with $n+1$ labeled vertices in which the set of neighbors of the vertex $n+1$ is exactly $[k]$. An obvious bijection (obtained by removing the vertex $n+1$ from the tree) between such trees and the considered forests leads directly to the lemma. □

Theorem 15.5

$$\mathbb{P}(G_f \text{ is connected}) = \frac{1}{n}\sum_{k=1}^{n}\frac{(n)_k}{n^k} \approx \sqrt{\frac{\pi}{2n}}.$$

Proof If G_f is connected then there is a cycle with k vertices, say, such that after removing the cycle we have a forest consisting of k trees rooted at the vertices of the cycle. Hence,

$$\begin{aligned}
\mathbb{P}(G_f \text{ is connected }) &= n^{-n}\sum_{k=1}^{n}\binom{n}{k}(k-1)!\,T(n,k) \\
&= \frac{1}{n}\sum_{k=1}^{n}\frac{(n)_k}{n^k} = \frac{1}{n}\sum_{k=1}^{n}\prod_{j=0}^{k-1}\left(1-\frac{j}{n}\right) \\
&= \frac{1}{n}\sum_{k=1}^{n}u_k.
\end{aligned}$$

If $k \geq n^{3/5}$, then

$$u_k \leq \exp\left\{-\frac{k(k-1)}{2n}\right\} \leq \exp\left\{-\frac{1}{3}n^{1/5}\right\},$$

while, if $k < n^{3/5}$,

$$u_k = \exp\left\{-\frac{k^2}{2n} + O\left(\frac{k^3}{n^2}\right)\right\}.$$

So

$$\begin{aligned}
\mathbb{P}(G_f \text{ is connected }) &= \frac{1+o(1)}{n}\sum_{k=1}^{n^{3/5}} e^{-k^2/2n} + O\left(ne^{-n^{1/5}/3}\right) \\
&= \frac{1+o(1)}{n}\int_0^\infty e^{-x^2/2n}dx + O\left(ne^{-n^{1/5}/3}\right) \\
&= \frac{1+o(1)}{\sqrt{n}}\int_0^\infty e^{-y^2/2}dy + O\left(ne^{-n^{1/5}/3}\right) \\
&\approx \sqrt{\frac{\pi}{2n}}.
\end{aligned}$$

□

Let Z_k denote the number of cycles of length k in a random mapping, then

$$\mathbb{E}Z_k = \binom{n}{k}(k-1)!\,n^{-k} = \frac{1}{k}\prod_{j=0}^{k-1}\left(1-\frac{j}{n}\right) = \frac{u_k}{k}.$$

If $Z = Z_1 + Z_2 + \cdots + Z_n$, then

$$\mathbb{E}Z = \sum_{k=1}^{n}\frac{u_k}{k} \approx \int_1^\infty \frac{1}{x}e^{-x^2/2n}dx \approx \frac{1}{2}\log n.$$

Moreover, the expected number of vertices of cycles in a random mapping is equal to

$$\mathbb{E}\left(\sum_{k=1}^{n} kZ_k\right) = \sum_{k=1}^{n} u_k \approx \sqrt{\frac{\pi n}{2}}.$$

$\square$

Note that the functional digraph of a random mapping can be interpreted as a representation of a process in which vertex $i \in [n]$ chooses its image independently with probability $1/n$. So, it is natural to consider a general model of a random mapping $\hat{f} : [n] \to [n]$ where, independently for all $i \in [n]$,

$$\mathbb{P}\left(\hat{f}(i) = j\right) = p_j,\ j = 1, 2, \ldots, n, \tag{15.2}$$

and

$$p_1 + p_2 + \ldots + p_n = 1.$$

This model was introduced (in a slightly more general form) independently by Burtin [168] and Ross [655]. We will first prove a generalization of Theorem 15.5.

Theorem 15.6

$$\mathbb{P}(G_{\hat{f}} \text{ is connected})$$
$$= \sum_i p_i^2 \left(1 + \sum_{j \neq i} p_j + \sum_{j \neq i} \sum_{k \neq i,j} p_j p_k + \sum_{j \neq i} \sum_{k \neq i,j} \sum_{l \neq i,j,k} p_j p_k p_l + \cdots \right).$$

To prove this theorem we use the powerful Burtin–Ross Lemma. The short and elegant proof of this lemma given here is due to Jaworski [439]. His general approach can be applied to study other characteristics of a random mappings, not only their connectedness.

Lemma 15.7 Burtin–Ross Lemma *Let $\hat{f}$ be a generalized random mapping defined above and let $G_{\hat{f}}[U]$ be the subgraph of $G_{\hat{f}}$ induced by $U \subset [n]$, then*

$$\mathbb{P}(G_{\hat{f}}[U] \text{ does not contain a cycle}) = \sum_{k \in [n] \setminus U} p_k.$$

Proof The proof is by induction on $r = |U|$. For $r = 0$ and $r = 1$ it is obvious. Assume that the result holds for all values less than r, $r \geq 2$. Let $\emptyset \neq S \subset U$ and denote by $\mathscr{A}$ the event that $G_{\hat{f}}[S]$ is the union of disjoint cycles and by $\mathscr{B}$ the event that $G_{\hat{f}}[U \setminus S]$ does not contain a cycle . Notice that events $\mathscr{A}$ and $\mathscr{B}$ are independent, since the first one depends on choices of vertices from S, only, while the second depends on choices of vertices from $U \setminus S$. Hence,

$$\mathbb{P}(G_{\hat{f}}[U] \text{ contains a cycle }) = \sum_{\emptyset \neq S \subset U} \mathbb{P}(\mathscr{A})\mathbb{P}(\mathscr{B}).$$

But if $\mathscr{A}$ holds then $\hat{f}$ restricted to S defines a permutation on S. So,

$$\mathbb{P}(\mathscr{A}) = |S|! \prod_{j \in S} p_j.$$

Since $|U \setminus S| < r$, by the induction assumption we obtain

$$\begin{aligned}
\mathbb{P}(G_{\hat{f}}[U] \text{ contains a cycle }) &= \sum_{\emptyset \neq S \subset U} |S|! \prod_{j \in S} p_j \sum_{k \in [n] \setminus (U \setminus S)} p_k \\
&= \sum_{\emptyset \neq S \subset U} |S|! \prod_{j \in S} p_j \left(1 - \sum_{k \in (U \setminus S)} p_k\right) \\
&= \sum_{S \subset U,\ |S| \geq 1} |S|! \prod_{k \in S} p_k - \sum_{S \subset U,\ |S| \geq 2} |S|! \prod_{k \in S} p_k \\
&= \sum_{k \in U} p_k,
\end{aligned}$$

completing the induction. □

Before we prove Theorem 15.6 we will point out that Lemma 15.4 can be immediately derived from the above result. To see this, in Lemma 15.7 choose $p_j = 1/n$, for $j = 1, 2, \cdots n$, and U such that $|U| = r = n - k$. Then, on one hand,

$$\mathbb{P}(G_f[U] \text{ does not contain a cycle}) = \sum_{i \in [n] \setminus U} \frac{1}{n} = \frac{k}{n}.$$

On the other hand,

$$\mathbb{P}(G_f[U] \text{ does not contain a cycle}) = \frac{T(n,k)}{n^{n-k}},$$

where $T(n,k)$ is the number of forests on $[n]$ with k trees rooted in vertices from the set $[n] \setminus U$. Comparing both sides we immediately get the result of Lemma 15.4, i.e. that

$$T(n,k) = kn^{n-k-1}.$$

Proof (of Theorem 15.6) Notice that $G_{\hat{f}}$ is connected if and only if there is a subset $U \subseteq [n]$ such that U spans a single cycle while there is no cycle on

$[n] \setminus U$. Moreover, the events "$U \subseteq [n]$ spans a cycle" and "there is no cycle on $[n] \setminus U$" are independent. Hence, by Lemma 15.7,

$$
\begin{aligned}
&Pr(G_{\hat{f}} \text{ is connected}) \\
&= \sum_{\emptyset \neq U \subseteq [n]} \mathbb{P}(U \subset [n] \text{ spans a cycle}) \mathbb{P}(\text{there is no cycle on } [n] \setminus U) \\
&= \sum_{\emptyset \neq U \subset [n]} (|U|-1)! \prod_{j \in U} p_j \sum_{k \in U} p_k \\
&= \sum_i p_i^2 \left(1 + \sum_{j \neq i} p_j + \sum_{j \neq i} \sum_{k \neq i,j} p_j p_k + \sum_{j \neq i} \sum_{k \neq i,j} \sum_{l \neq i,j,k} p_j p_k p_l + \cdots \right).
\end{aligned}
$$

□

Using the same reasoning as in the above proof, one can show the following result due to Jaworski [439].

Theorem 15.8 *Let X be the number of components in $G_{\hat{f}}$ and Y be the number of its cyclic vertices (vertices belonging to a cycle). Then for $k = 1, 2, \ldots, n$,*

$$
\mathbb{P}(X = k) = \sum_{\substack{U \subset [n] \\ |U| \geq k}} \prod_{j \in U} p_j |s(|U|, k)| - \sum_{\substack{U \subset [n] \\ |U| \geq k+1}} \prod_{j \in U} p_j |s(|U|-1, k)| |U|,
$$

where $s(\cdot, \cdot)$ is the Stirling number of the first kind. On the other hand,

$$
\mathbb{P}(Y = k) = k! \sum_{\substack{U \subset [n] \\ |U| = k}} \prod_{j \in U} p_j - (k+1)! \sum_{\substack{U \subset [n] \\ |U| = k+1}} \prod_{j \in U} p_j.
$$

The Burtin–Ross Lemma has another formulation which we present below.

Lemma 15.9 Burtin–Ross Lemma - the second version *Let $\hat{g} : [n] \to [n] \cup \{0\}$ be a random mapping from the set $[n]$ to the set $[n] \cup \{0\}$, where, independently for all $i \in [n]$,*

$$
\mathbb{P}\left(\hat{g}(i) = j\right) = q_j, \; j = 0, 1, 2, \ldots, n,
$$

and

$$
q_0 + q_1 + q_2 + \ldots + q_n = 1.
$$

Let $D_{\hat{g}}$ be the random directed graph on the vertex set $[n] \cup \{0\}$, generated by the mapping $\hat{g}$ and let $G_{\hat{g}}$ denote its underlying simple graph. Then

$$
\mathbb{P}(G_{\hat{g}} \textit{ is connected}) = q_0.
$$

Notice that the event that $G_{\hat{g}}$ is connected is equivalent to the event that $D_{\hat{g}}$ is a (directed) tree, rooted at vertex $\{0\}$, i.e. there are no cycles in $G_{\hat{g}}[[n]]$.

We use this result and Lemma 15.9 to prove the next theorem (for more general results, see [440]).

Theorem 15.10 *Let $D_{\hat{f}}$ be the functional digraph of a mapping $\hat{f}$ defined in (15.2) and let Z_R be the number of predecessors of a set $R \subset [n]$, $|R| = r$, $r \geq 1$, of vertices of $D_{\hat{f}}$, i.e.*

$$Z_R = |\{j \in [n] : \textit{for some non-negative integer } k,\ \hat{f}^{(k)}(j) \in R\}|,$$

where $\hat{f}^{(0)}(j) = j$ and for $k \geq 1$, $\hat{f}^{(k)}(j) = \hat{f}(\hat{f}^{(k-1)}(j))$. Then, for $k = 0, 1, 2, \ldots, n-r$,

$$\mathbb{P}(Z_R = k + r) = \Sigma_R \sum_{\substack{U \subset [n] \setminus R \\ |U| = k}} (\Sigma_{U \cup R})^{k-1} (1 - \Sigma_{U \cup R})^{n-k},$$

where for $A \subseteq [n]$, $\Sigma_A = \sum_{j \in A} p_j$.

Proof The distribution of Z_R follows immediately from the next observation and the application of Lemma 15.9. Denote by $\mathscr{A}$ the event that there is a forest spanned on the set $W = U \cup R$, where $U \subset [n] \setminus R$, composed of r (directed) trees rooted at vertices of R. Then

$$\mathbb{P}(Z_R = k + r) = \sum_{\substack{U \subset [n] \setminus R \\ |U| = k}} \mathbb{P}(\mathscr{A} | \mathscr{B} \cap \mathscr{C}) \mathbb{P}(\mathscr{B}) \mathbb{P}(\mathscr{C}), \tag{15.3}$$

where $\mathscr{B}$ is the event that all edges that begin in U end in W, while $\mathscr{C}$ denotes the event that all edges that begin in $[n] \setminus W$ end in $[n] \setminus W$. Now notice that

$$\mathbb{P}(\mathscr{B}) = (\Sigma_W)^k, \text{ while } \mathbb{P}(\mathscr{C}) = (1 - \Sigma_W)^{n-k}.$$

Furthermore,

$$\mathbb{P}(\mathscr{A} | \mathscr{B} \cap C) = \mathbb{P}(G_{\hat{g}} \text{ is connected}),$$

where $\hat{g} : U \to U \cup \{0\}$, where $\{0\}$ stands for the set R collapsed to a single vertex, is such that for all $u \in U$ independently,

$$q_j = \mathbb{P}(\hat{g}(u) = j) = \frac{p_j}{\Sigma_W}, \text{ for } j \in U, \quad \text{while} \quad q_0 = \frac{\Sigma_R}{\Sigma_W}.$$

So, applying Lemma 15.9, we arrive at the thesis. □

We finish this section by stating the Central Limit Theorem for the number of components of G_f, where f is a uniform random mapping $f : [n] \to [n]$ (see Stepanov [694]). It is an analogous result to Theorem 15.2 for random permutations.

Theorem 15.11

$$\lim_{n\to\infty} \mathbb{P}\left(\frac{X_n - \frac{1}{2}\log n}{\sqrt{\frac{1}{2}\log n}} \le x\right) = \int_{-\infty}^{x} e^{-t^2/2} dt,$$

the standardized random variable X_n converges in distribution to the standard normal random variable.

15.3 Exercises

15.3.1 Prove directly that if $X_{n,t}$ is the number of cycles of length t in a random permutation then $\mathbb{E}X_{n,t} = 1/t$.

15.3.2 Find the expectation and the variance of the number X_n of cycles in a random permutation using the fact that the rth derivative of the gamma function equals $\frac{d^r}{(dx)^r}\Gamma(x) = \int_0^\infty (\log t)^r t^{x-1} e^{-t} dt$.

15.3.3 Determine the variance of the number X_n of cycles in a random permutation (start with computation of the second factorial moment of X_n, counting ordered pairs of cycles).

15.3.4 Find the probability distribution for the length of a typical cycle in a random permutation, i.e. the cycle that contains a given vertex (say vertex 1). Determine the expectation and variance of this characteristic.

15.3.5 Find the probability distribution of the number of components in a functional digraph D_f of a uniform random mapping $f : [n] \to [n]$.

15.3.6 Determine the expectation and variance of the number of components in a functional digraph $D_{\hat{f}}$ of a generalized random mapping $\hat{f}$ (see Theorem 15.8).

15.3.7 Find the expectation and variance of the number of cyclic vertices in a functional digraph $D_{\hat{f}}$ of a generalized random mapping $\hat{f}$ (see Theorem 15.8).

15.3.8 Prove Theorem 15.8.

15.3.9 Show that Lemmas 15.7 and 15.9 are equivalent.

15.3.10 Prove the Burtin–Ross Lemma for a bipartite random mapping, i.e. a mapping with bipartition $([n], [m])$, where each vertex $i \in [n]$ chooses its unique image in $[m]$ independently with probability $1/m$, and, similarly, each vertex $j \in [m]$ selects its image in $[n]$ with probability $1/n$.

15.3.11 Consider an evolutionary model of a random mapping (see [441], [442]), i.e. a mapping $\hat{f}_q[n] \to [n]$, such that for $i,j \in [n]$, $\mathbb{P}(\hat{f}_q(i) =$

$j) = q$ if $i = j$ while, $\mathbb{P}(\hat{f}_q(i) = j) = (1-q)/(n-1)$ if $i \neq j$, where $0 \leq q \leq 1$. Find the probability that $\hat{f}_q$ is connected.

15.3.12 Show that there is one-to-one correspondence between the family of n^n mappings $f : [n] \to [n]$ and the family of all doubly-rooted trees on the vertex set $[n]$ (Joyal bijection).

15.4 Notes

Permutations

Systematic studies of the properties of random permutations of n objects were initiated by Goncharov in [379] and [380]. Golomb [378] showed that the expected length of the longest cycle of D_f, divided by n is monotone decreasing and gave a numerical value for the limit, while Shepp and Lloyd in [682] found the closed form for this limit (see Theorem 15.3). They also gave the corresponding result for kth moment of the rth longest cycle, for $k, r = 1, 2, \ldots$ and showed the limiting distribution for the length of the rth longest cycle.

Kingman [485] and, independently, Vershik and Schmidt [713], proved that for a random permutation of n objects, as $n \to \infty$, the process giving the proportion of elements in the longest cycle, the second longest cycle, and so on, converges in distribution to the Poisson–Dirichlet process with parameter 1 (for further results in this direction see Arratia, Barbour and Tavaré [39]). Arratia and Tavaré [40] provided explicit bounds on the total variation distance between the process that counts the sizes of cycles in a random permutations and a process of independent Poisson random variables.

For other results, not necessarily of a "graphical" nature, such as, for example, the order of a random permutation, the number of derangements or the number of monotone sub-sequences, we refer the reader to the respective sections of books by Feller [290], Bollobás [131] and Sachkov [669] or, in the case of monotone sub-sequences, to a recent monograph by Romik [654].

Mappings

Uniform random mappings were introduced in the mid 1950s by Rubin and Sitgraves [656], Katz [479] and Folkert [307]. More recently, much attention has been focused on their usefulness as a model for epidemic processes, see for example the papers of Gertsbakh [365], Ball, Mollison and Scalia-Tomba [51], Berg [78], Mutafchiev [596], Pittel [622] and Jaworski [442]. The component structure of a random functional digraph D_f has been studied by Aldous [12]. He has shown, that the joint distribution of the normalized order statistics for the component sizes of D_f converges to the Poisson–Dirichlet distribution with parameter $1/2$. For more results on uniform random mappings we refer the reader to Kolchin's monograph [496] or a chapter of Bollobás' [131].

The general model of a random mapping $\hat{f}$, introduced by Burtin [168] and Ross [655], has been intensively studied by many authors. The crucial Burtin–Ross Lemma (see Lemmas 15.7 and 15.9) has many alternative proofs (see [37]) but the most useful seems to be the one used in this chapter, due to Jaworski [439]. His approach can also be applied to derive the distribution of many other characteristics of a random digraph D_f, it can also be used to prove generalizations of the Burtin–Ross Lemma for models of random mappings with independent choices of images. (For an extensive review of results in that direction see [440].) Aldous, Miermont and Pitman ([17], [18]) studied the asymptotic structure of $D_{\hat{f}}$ using an ingenious coding of the random mapping $\hat{f}$ as a stochastic process on the interval $[0,1]$ (see also the related work of Pitman [621], exploring the relationship between random mappings and random forests).

Hansen and Jaworski (see [394], [395]) introduced a random mapping f^D : $[n] \to [n]$ with an in-degree sequence, which is a collection of exchangeable random variables $(D_1, D_2, \ldots, D_n)$. In particular, they studied predecessors and successors of a given set of vertices, and applied their results to random mappings with preferential and anti-preferential attachment.

16

k-out

Several interesting graph properties require that the minimum degree of a graph be at least a certain amount. For example, having a Hamilton cycle requires that the minimum degree is at least two. In Chapter 6 we saw that $\mathbb{G}_{n,m}$ being Hamiltonian and having minimum degree at least two happen at the same time w.h.p. We are therefore interested in models of a random graph that guarantee a certain minimum degree. We have already seen d-regular graphs in Chapter 10. In this chapter we consider another simple and quite natural model $\mathbb{G}_{k-out}$ that generalizes random mappings. It seems to have first appeared in print as Problem 38 of *The Scottish Book* [558]. We discuss the connectivity of this model and then matchings and Hamilton cycles. We also consider a related model of Nearest Neighbor Graphs.

16.1 Connectivity

For an integer k, $1 \leq k \leq n-1$, let $\vec{\mathbb{G}}_{k-out}$ be a random digraph on vertex set $V = \{1,2,\ldots,n\}$ with arcs (directed edges) generated independently for each $v \in V$ by a random choice of k distinct arcs (v,w), where $w \in V \setminus \{v\}$, so that each of the $\binom{n-1}{k}$ possible sets of arcs is equally likely to be chosen. Let $\mathbb{G}_{k-out}$ be the random graph(multigraph) obtained from $\vec{\mathbb{G}}_{k-out}$ by ignoring the orientation of its arcs, but retaining all edges.

Note that $\vec{\mathbb{G}}_{1-out}$ is a functional digraph of a random mapping $f : [n] \to [n]$, with a restriction that loops (fixed points) are not allowed. So for $k = 1$ the following result holds.

Theorem 16.1

$$\lim_{n\to\infty} \mathbb{P}(\mathbb{G}_{1-out} \textit{ is connected }) = 0.$$

The situation changes when each vertex is allowed to choose more than one neighbor. Denote by $\kappa(G)$ and $\lambda(G)$ the vertex and edge connectivity of a graph G, respectively, i.e. the minimum number of vertices (respectively edges) the deletion of which disconnects G. Let $\delta(G)$ be the minimum degree of G. The well known Whitney's theorem states that, for any graph G,

$$\kappa(G) \leq \lambda(G) \leq \delta(G).$$

In the next theorem we show that for random $k-out$ graphs these parameters are equal w.h.p. It is taken from Fenner and Frieze [298]. *The Scottish Book* [558] contains a proof that $\mathbb{G}_{k-out}$ is connected for $k \geq 2$.

Theorem 16.2 *Let $\kappa = \kappa(\mathbb{G}_{k-out}), \lambda = \lambda(\mathbb{G}_{k-out})$ and $\delta = \delta(\mathbb{G}_{k-out})$. Then, for $2 \leq k = O(1)$,*

$$\lim_{n\to\infty} \mathbb{P}(\kappa = \lambda = \delta = k) = 1.$$

Proof In the light of Whitney's theorem, to prove our theorem we have to show that the following two statements hold:

$$\lim_{n\to\infty} \mathbb{P}(\kappa(\mathbb{G}_{k-out}) \geq k) = 1, \tag{16.1}$$

and

$$\lim_{n\to\infty} \mathbb{P}(\delta(\mathbb{G}_{k-out}) \leq k) = 1. \tag{16.2}$$

Then, w.h.p.

$$k \leq \kappa \leq \lambda \leq \delta \leq k,$$

and the theorem follows.

To prove statement (16.1) consider the deletion of r vertices from the random graph $\mathbb{G}_{k-out}$, where $1 \leq r \leq k-1$. If $\mathbb{G}_{k-out}$ can be disconnected by deleting r vertices, then there exists a partition (R,S,T) of the vertex set V, with $\mid R \mid = r$, $\mid S \mid = s$ and $\mid T \mid = t = n-r-s$, with $k-r+1 \leq s \leq n-k-1$, such that $\mathbb{G}_{k-out}$ has no edge joining a vertex in S with a vertex in T. The probability of such an event, for an arbitrary partition given above, is equal to

$$\left(\frac{\binom{r+s-1}{k}}{\binom{n-1}{k}}\right)^s \left(\frac{\binom{n-s-1}{k}}{\binom{n-1}{k}}\right)^{n-r-s} \leq \left(\frac{r+s}{n}\right)^{sk} \left(\frac{n-s}{n}\right)^{(n-r-s)k}.$$

Thus

$$\mathbb{P}(\kappa(\mathbb{G}_{k-out}) \leq r) \leq \sum_{s=k-r+1}^{\lfloor (n-r)/2 \rfloor} \frac{n!}{s!r!(n-r-s)!} \left(\frac{r+s}{n}\right)^{sk} \left(\frac{n-s}{n}\right)^{(n-r-s)k}.$$

We have replaced $n-k-1$ by $\lfloor (n-r)/2 \rfloor$ because we can always interchange S and T so that $|S| \leq |T|$.

But, by Stirling's formula,

$$\frac{n!}{s!r!(n-r-s)!} \leq \alpha_s \frac{n^n}{s^s(n-r-s)^{n-r-s}},$$

where

$$\alpha_s = \alpha(s,n,r) \leq c\left(\frac{n}{s(n-r-s)}\right)^{1/2} \leq 2c,$$

for some absolute constant $c > 0$.

Thus,

$$\mathbb{P}(\kappa(\mathbb{G}_{k-out}) \leq r) \leq 2c \sum_{s=k-r+1}^{\lfloor (n-r)/2 \rfloor} \left(\frac{r+s}{s}\right)^s \left(\frac{n-s}{n-r-s}\right)^{(n-r-s)} u_s,$$

where

$$u_s = (r+s)^{(k-1)s}(n-s)^{(k-1)(n-r-s)}n^{n-k(n-r)}.$$

Now,

$$\left(\frac{r+s}{s}\right)^s \left(\frac{n-s}{n-r-s}\right)^{n-r-s} \leq e^{2r},$$

and

$$(r+s)^s(n-s)^{n-r-s}$$

decreases monotonically, with increasing s, for $s \leq (n-r)/2$. Therefore,

$$\mathbb{P}(\kappa(\mathbb{G}_{k-out}) \leq r) \leq 2ce^{2r}nu_{k-r+1} \leq 2ce^{2r}an^{1-k(k-r)},$$

where

$$a = (k+1)^{(k-1)(k-r+1)}.$$

It follows that

$$\lim_{n\to\infty} \mathbb{P}(\kappa(\mathbb{G}_{k-out}) \leq r) = \lim_{n\to\infty} \mathbb{P}(\kappa(\mathbb{G}_{k-out}) \leq k-1) = 0,$$

which implies that

$$\lim_{n\to\infty} \mathbb{P}(\kappa(\mathbb{G}_{k-out}) \geq k) = 1,$$

i.e. that equation (16.1) holds.

To complete the proof we have to show that equation (16.2) holds, i.e. that

$$\mathbb{P}(\delta(\mathbb{G}_{k-out}) = k) \to 1 \text{ as } n \to \infty.$$

Since $\delta \geq k$ in $\mathbb{G}_{k-out}$, we have to show that w.h.p. there is a vertex of degree k in $\mathbb{G}_{k-out}$.

Let $\mathscr{E}_v$ be the event that vertex v has indegree zero in $\vec{\mathbb{G}}_{k-out}$. Thus, the degree of v in $\mathbb{G}_{k-out}$ is k if and only if $\mathscr{E}_v$ occurs. Now

$$\mathbb{P}(\mathscr{E}_v) = \left(\frac{\binom{n-2}{k}}{\binom{n-1}{k}}\right)^{n-1} = \left(1 - \frac{k}{n-1}\right)^{n-1} \to e^{-k}.$$

Let Z denote the number of vertices of degree k in $\mathbb{G}_{k-out}$, then we have shown that $\mathbb{E}(Z) \approx ne^{-k}$. Now the random variable Z is determined by kn independent random choices. Changing one of these choices can change the value of Z by at most one. Applying the Azuma–Hoeffding concentration inequality – see Section 21.7, in particular Lemma 21.16 we see that for any $t > 0$

$$\mathbb{P}(Z \leq \mathbb{E}(Z) - t) \leq \exp\left\{-\frac{2t^2}{kn}\right\}.$$

Putting $t = ne^{-k}/2$ we see that $Z > 0$ w.h.p. and the theorem follows. $\square$

16.2 Perfect Matchings

Non-bipartite graphs

Assuming that the number of vertices n of a random graph $\mathbb{G}_{k-out}$ is even, Frieze [322] proved the following result.

Theorem 16.3

$$\lim_{\substack{n\to\infty \\ n \text{ even}}} \mathbb{P}(\mathbb{G}_{k-out} \text{ has a perfect matching}) = \begin{cases} 0 & \text{if } k = 1 \\ 1 & \text{if } k \geq 2. \end{cases}$$

We only prove a weakening of the above result to where $k \geq 15$. We follow the ideas of Section 6.1. So, we begin by examining the expansion properties of $G = \mathbb{G}_{a-out}, a \geq 3$.

Lemma 16.4 *W.h.p.* $|N_G(S)| \geq |S|$ *for all* $S \subseteq [n], |S| \leq \kappa_a n$ *where* $\kappa_a = \frac{1}{2}\left(\frac{1}{30}\right)^{1/(a-2)}$.

Proof The probability there exists a set S with insufficient expansion is at most

$$\sum_{s=3}^{\kappa_a n} \binom{n}{s}\binom{n}{s-1}\left(\frac{2s}{n}\right)^{as} \leq \sum_{s=3}^{\kappa_a n} \left(\frac{ne}{s}\right)^{2s}\left(\frac{2s}{n}\right)^{as}$$
$$= \sum_{s=3}^{\kappa_a n} \left(\left(\frac{s}{n}\right)^{a-2} e^2 2^a\right)^s = o(1). \tag{16.3}$$

Lemma 16.5 *Let $b = \lceil (1+\kappa_a^{-2})/2 \rceil$. Then as $n \to \infty$, n even, $G_{(a+b)-out}$ has a perfect matching w.h.p.*

Proof First note that $G_{(a+b)-out}$ contains $H = \mathbb{G}_{a-out} \cup \mathbb{G}_{b-out}$ in the following sense. Start the construction of $\mathbb{G}_{(a+b)-out}$ with H. If there is a $v \in [n]$ that chooses edge $\{v,w\}$ in both $\mathbb{G}_{a-out}$ and $\mathbb{G}_{b-out}$ then add another random choice for v.

Let us show that H has a perfect matching w.h.p. Enumerate the edges of $\mathbb{G}_{b-out}$ as $e_1, e_2, \ldots, e_{bn}$. Here $e_{(i-1)n+j}$ is the ith edge chosen by vertex j. Let $\mathbb{G}_0 = \mathbb{G}_{a-out}$ and let $\mathbb{G}_i = \mathbb{G}_0 + \{e_1, e_2, \ldots, e_i\}$. If $\mathbb{G}_i$ does not have a perfect matching, consider the sets $A, A(x), x \in A$ defined prior to (6.5). It follows from Lemma 16.4 that w.h.p. all of these sets are of size at least $\kappa_a n$. Furthermore, if $i+1 \mod n = x$ and $x \in A$ and $e_i = \{x,y\}$ then $\mathbb{P}(y \in A(x)) \geq \frac{\kappa_a n - i}{n}$. (We subtract i to account for the i previously inspected edges associated with the choices of x.)

It follows that

$$
\begin{aligned}
&\mathbb{P}(\mathbb{G}_{(a+b)-out} \text{ does not have a perfect matching}) \\
&\leq \mathbb{P}(H \text{ does not have a perfect matching}) \\
&\leq \mathbb{P}(\mathrm{Bin}(b\kappa_a n, \kappa_a - b/n) \leq n/2) = o(1).
\end{aligned}
$$

□

Putting $a = 8$ gives $b = 7$ and a proof that $\mathbb{G}_{15-out}$, n even, has a perfect matching w.h.p.

Bipartite graphs

We now consider the related problem of the existence of a perfect matching in a random k-out bipartite graph.

Let $U = \{u_1, u_2, \ldots, u_n\}, V = \{v_1, v_2, \ldots, v_n\}$ and let each vertex from U choose independently and without repetition, k neighbors in V, and let each vertex from V choose independently and without repetition k neighbors in U. Denote by $\vec{\mathbb{B}}_{k-out}$ the digraphs generated by the above procedure and let $\mathbb{B}_{k-out}$ be its underlying simple bipartite graph.

Theorem 16.6

$$
\lim_{n\to\infty} \mathbb{P}(\mathbb{B}_{k-out} \textit{ has a perfect matching}) = \begin{cases} 0 & \textit{if } k=1 \\ 1 & \textit{if } k \geq 2. \end{cases}
$$

We will give two different proofs. The first one – existential – of a combinatorial nature is due to Walkup [718]. The second one – constructive – of an algorithmic nature, is due to Karp, Rinnooy-Kan and Vohra [475]. We start with the combinatorial approach.

Existence proof

Let X denote the number of perfect matchings in $\mathbb{B}_{k-out}$, then

$$\mathbb{P}(X > 0) \leq \mathbb{E}(X) \leq n!\ 2^n (k/n)^n.$$

The above bound follows from the following observations. There are $n!$ ways of pairing the vertices of U with the vertices of V. For each such pairing there are 2^n ways to assign directions for the connecting edges, and then each possible oriented probability has probability at most $(k/n)^n$ of appearing in $\mathbb{B}_{k-out}$.

So, by Stirling's formula,

$$\mathbb{P}(X > 0) \leq 3n^{1/2}(2k/e)^n,$$

which, for $k = 1$, tends to 0 as $n \to \infty$, and the first statement of our theorem follows.

To show that $\mathbb{B}_{k-out}$ has a perfect matching w.h.p. notice that since this is an increasing graph property, it is enough to show that it is true for $k = 2$. Note also, that if there is no perfect matching in $\mathbb{B}_{k-out}$, then there must exist a set $R \subset U$ (or $R \subset V$) such that the cardinality of neighborhood of $S = N(R)$ of R in V (respectively, in V) is smaller that the cardinality of the set R itself, i.e. $\mid S \mid < \mid R \mid$. We call such a pair (R,S) a *bad pair*, and, in particular, we restrict our attention to the "minimal bad pairs," i.e. such that there is no $R' \subset R$ for which $(R', N(R'))$ is bad.

If (R,S) is a bad pair with $R \subseteq U$ then $(V \setminus S, U \setminus R)$ is also a bad pair. Given this, we can concentrate on showing that w.h.p. there are no bad pairs (R,S) with $2 \leq |R| \leq (n+1)/2$.

Every minimal bad pair has to have the following two properties:

(i) $\mid S \mid = \mid R \mid -1$,
(ii) every vertex in S has at least two neighbors in R.

The first property is obvious. To see why property (ii) holds, suppose that there is a vertex $v \in S$ with at most one neighbor u in R. Then the pair $(R \setminus \{u\}, S \setminus \{v\})$ is also "bad pair" and so the pair (R,S) is not minimal.

Let $r \in [2, (n+1)/2]$ and let Y_r be the number of minimal bad pairs (R,S), with $\mid R \mid = r$ in $\mathbb{B}_{k-out}$. To complete the proof of the theorem we have to show that $\sum_r \mathbb{E} Y_r \to 0$ as $n \to \infty$. By symmetry, choose (R,S), such that $R = \{u_1, u_2, \dots u_r\} \subset U$ and $S = \{v_1, v_2, \dots v_{r-1}\} \subset V$ is a minimal "bad pair." Then

$$\mathbb{E} X_r = 2\binom{n}{r}\binom{n}{r-1} P_r Q_r, \tag{16.4}$$

where

$$P_r = \mathbb{P}((R,S) \text{ is bad})$$

and

$$Q_r = \mathbb{P}((R,S) \text{ is minimal} \mid (R,S) \text{ is bad}).$$

We observe that, for any fixed k,

$$P_r = \left(\frac{\binom{r-1}{k}}{\binom{n}{k}}\right)^r \left(\frac{\binom{n-r}{k}}{\binom{n}{k}}\right)^{n-r+1}.$$

Hence, for $k = 2$,

$$P_r \le \left(\frac{r}{n}\right)^{2r} \left(\frac{n-r}{n}\right)^{2(n-r)}. \tag{16.5}$$

Then we use Stirling's formula to show,

$$\binom{n}{r}\binom{n}{r-1} = \frac{r}{n-r+1}\binom{n}{r}^2 \le \frac{r}{n-r+1}\frac{n}{r(n-r)}\left(\frac{n}{2r}\right)^{2r}\left(\frac{n}{n-r}\right)^{2(n-r)}. \tag{16.6}$$

To estimate Q_r we have to consider condition (ii), which a minimal bad pair has to satisfy. So a vertex $v \in S = N(R)$ is chosen by at least one vertex from R (denote this event by A_v), or it chooses both its neighbors in R (denote this event by B_v). Then the events $A_v^c, v \in R$ are negatively correlated (see Section 21.2) and the events $B_v, v \in S$ are independent of other events in this collection.

Let $S = \{v_1, v_2, \ldots, v_{r-1}\}$. Then we can write

$$
\begin{aligned}
Q_r &= \mathbb{P}\left(\bigcup_{i=1}^{r-1}(A_{v_i} \cup B_{v_i})\right) \\
&= \prod_{i=1}^{r-1} \mathbb{P}\left(A_{v_i} \cup B_{v_i} \,\middle|\, \bigcup_{j=1}^{i-1}(A_{v_j} \cup B_{v_j})\right) \\
&\leq \prod_{i=1}^{r-1} \mathbb{P}\left(A_{v_i} \cup B_{v_i}\right) \\
&\leq \left(1 - \mathbb{P}(A_v^c)\,\mathbb{P}(B_v^c)\right)^{r-1} \\
&\leq \left(1 - \left(\frac{r-2}{r-1}\right)^{2r}\left(1 - \frac{\binom{r}{2}}{\binom{n}{2}}\right)\right)^{r-1} \\
&\leq \eta^{r-1} \qquad (16.7)
\end{aligned}
$$

for some absolute constant $0 < \eta < 1$ when $r \leq (n+1)/2$.

Going back to (16.4), and using (16.5), (16.6)–(16.7)

$$
\sum_{r=3}^{(n+1)/2} \mathbb{E} X_r \leq 2 \sum_{r=3}^{(n+1)/2} \frac{\eta^{r-1} n}{(n-r)(n-r+1)} = o(1).
$$

Hence, $\sum_r \mathbb{E} X_r \to 0$ as $n \to \infty$, which means that w.h.p. there are no bad pairs, implying that $\mathbb{B}_{k-out}$ has a perfect matching w.h.p. $\square$

Frieze and Melsted [343] considered the related question. Suppose that M, N are disjoint sets of size m, n and that each $v \in M$ chooses $d \geq 3$ neighbors in N. Suppose that we condition on each vertex in N being chosen at least twice. They showed that w.h.p. there is a matching of size equal to $\min\{m, n\}$. Fountoulakis and Panagiotou [310] proved a slightly weaker result, in the same vein.

Algorithmic Proof

We now give a rather elegant algorithmic proof of Theorem 16.6. It is due to Karp, Rinnooy-Kan and Vohra [475]. We do this for two reasons. First, because it is a lovely proof and second this proof is the basis of the proof that 2-in, 2-out is Hamiltonian in [206]. In particular, this latter example shows that constructive proofs can sometimes be used to achieve results not obtainable through existence proofs alone.

Start with the random digraph $\vec{\mathbb{B}}_{2-out}$ and consider two multigraphs, G_U and G_V with labeled vertices and edges, generated by $\vec{\mathbb{B}}_{2-out}$ on the sets of the bipartition (U,V) in the following way. The vertex set of the graph G_U is U and two vertices, u and u', are connected by an edge, labeled v, if a vertex $v \in V$ chooses u and u' as its two neighbors in U. Similarly, the graph G_U has vertex set V and we put an edge labeled u between two vertices v and v', if a vertex $u \in U$ chooses v and v' as its two neighbors in V. Hence, graphs G_U and G_V are random multigraphs with exactly n labeled vertices and n labeled edges.

We describe below, a randomized algorithm that w.h.p. finds a perfect matching in $\mathbb{B}_{2-out}$ in $O(n)$ expected number of steps.

Algorithm PAIR

Step 0: Set $H_U = G_U$ and let H_V be the empty graph on vertex set V. Initially, all vertices in H_U are *unmarked* and all vertices in G_V are *unchecked*. Let *CORE* denote the set of edges of G_U that lie on cycles in G_U, i.e. the edges of the 2-core of G_U.

Step 1: If every isolated tree in H_U contains a marked vertex, go to Step 5. Otherwise, select any isolated tree T in H_U in which all vertices are unmarked. Pick a random vertex u in T and mark it.

Step 2: Add the edge $\{x,y\}, x,y \in V$ that has label u to the graph H_V.

Step 3: Let C_x, C_y be the components of H_V just before the edge labeled u is added. Let $C = C_x \cup C_y$. If all vertices in C are checked, go to Step 6. Otherwise, select an unchecked vertex v in C. If possible, select an unchecked vertex v for which the edge labeled v in H_U belongs to *CORE*.

Step 4: Delete the edge labeled v from H_U, return to Step 1.

Step 5: STOP and declare success.

Step 6: STOP and declare failure.

We next argue that Algorithm PAIR, when it finishes at Step 5, does indeed produce a perfect matching in $\mathbb{B}_{2-out}$. There are two simple invariants of this process that explain this:

(I1) The number of marked vertices plus the number of edges in H_U is equal to n.

(I2) The number of checked vertices is equal to the number of edges in H_V.

For **I1**, we observe that each round marks one vertex and deletes one edge of H_U. Similarly, for **I2**, we observe that each round checks one vertex and adds one edge to H_V.

Lemma 16.7 *Up until (possible) failure in Step 6, the components of H_V are either trees with a unique unchecked vertex or are unicyclic components with all vertices checked. Also, failure in Step 6 means that PAIR tries to add an edge to a unicyclic component.*

Proof This is true initially, as initially H_V has no edges and all vertices are unchecked. Assume this to be the case when we add an edge $\{x,y\}$ to H_V. If $C_x \neq C_y$ are both trees then we will have a choice of two unchecked vertices in $C = C_x \cup C_y$ and C will be a tree. After checking one vertex, our claim will still hold. The other possibilities are that C_x is a tree and C_y is unicyclic. In this case there is one unchecked vertex and this will be checked and C will be unicyclic. The other possibility is that $C = C_x = C_y$ is a tree. Again there is only one unchecked vertex and adding $\{x,y\}$ will make C unicyclic. □

Lemma 16.8 *If H_U consists of trees and unicyclic components then all the trees in H_U contain a marked vertex.*

Proof Suppose that H_U contains k trees with marked vertices and ℓ trees with no marked vertices and that the rest of the components are unicyclic. It follows that H_U contains $n-k-\ell$ edges and then **(I1)** implies that $\ell = 0$. □

Lemma 16.9 *If the algorithm stops in Step 5, then we can extract a perfect matching from H_U, H_V.*

Proof Suppose that we arrive at Step 5 after k rounds. Suppose that there are k trees with a marked vertex. Let the component sizes in H_U be $n_1, n_2, \dots, n_k$ for the trees and $m_1, m_2, \dots, m_\ell$ for the remaining components. Then,

$$n_1 + n_2 + \cdots + n_k + m_1 + m_2 + \cdots + m_\ell = |V(H_U)| = n.$$

$$|E(H_U)| = n - k,$$

from **I1** and so

$$(n_1 - 1) + (n_2 - 1) + \cdots + (n_k - 1)$$
$$+ (\geq m_1) + (\geq m_2) + (\geq m_\ell) = n - k.$$

It follows that the components of H_U that are not trees with a marked vertex have as many edges as vertices and so are unicyclic.

We now show, given that H_U, H_V only contain trees and unicyclic components, that we can extract a perfect matching. The edges of H_U define a matching of $\mathbb{B}_{2-out}$ of size $n-k$. Consider a tree T component with marked vertex ρ. Orient the edges of T away from ρ. Now consider an edge $\{x,y\}$ of T, oriented from x to y. Suppose that this edge has label $z \in V$. We add the edge

$\{y,z\}$ to M_1. These edges are disjoint; z appears as the label of exactly one edge and y is the head of exactly one oriented edge.

For the unicyclic components, we orient the unique cycle $C = (u_1, u_2, \dots, u_s)$ arbitrarily in one of two ways. We then consider the trees attached to each of the u_i and orient them away from the u_i. An oriented edge $\{x,y\}$ with label z yields a matching edge $\{y,z\}$ as before.

The remaining k edges needed for a perfect matching come from H_V. We extract a set of k matching edges out of H_V in the same way we extracted $n-k$ edges from H_U. We only need to check that these k edges are disjoint from those chosen from H_U. Let $\{y,z\}$ be such an edge, obtained from the edge $\{x,y\}$ of H_V, which has label z. z is marked in H_U and so is the root of a tree and does not appear in any matching edge of M_1. y is a checked vertex and so the edge labeled y has been deleted from H_U and this prevents y appearing in a matching edge of M_1. □

Lemma 16.10 *W.h.p. Algorithm PAIR cannot reach Step 6 in fewer than* $0.49n$ *iterations.*

Proof It follows from Lemma 2.10 that w.h.p. after $\leq 0.499n$ rounds, H_V only contains trees and unicyclic components. The lemma now follows from Lemma 16.7. □

To complete our analysis, it only remains to show

Lemma 16.11 *W.h.p., at most* $0.49n$ *rounds are needed to make H_U the union of trees and unicyclic components.*

Proof Recall that each edge of H_U corresponds to an unchecked vertex of H_V, the edges corresponding to checked vertices having been deleted. Moreover, each tree component T of H_V has one unchecked vertex, u_T say. If u_T is the label of an edge of H_U belonging to *CORE* then due to the choice rule for vertex checking in Step 3, every vertex of T must be the label of an edge of *CORE*. Hence the number of edges left in *CORE*, after a given iteration of the algorithm, is equal to the number of tree components of H_V, where every vertex labels an edge of *CORE*. We use this to estimate the number of edges of CORE that remain in H_U after $.49n$ iterations.

Let $xe^{-x} = 2e^{-2}$, where $0 < x < 1$. One can easily check that $0.40 < x < 0.41$. It follows from Lemma 2.16 that w.h.p. $|CORE| \approx \left(1 - \frac{x}{2}\right)^2 n$, which implies, that $0.63\,n \leq |CORE| \leq 0.64\,n$.

Let Z be the number of tree components in H_V made up of vertices that are the labels of edges belong to *CORE*. Then, after at most $0.49n$ rounds,

$$\begin{aligned}
\mathbb{E}Z \leq\; & o(1) + \sum_{k=1}^{(\log n)^2} \binom{n}{k} k^{k-2} \binom{0.49n}{k-1} (k-1)! \frac{(0.64)^k}{\binom{n}{2}^{k-1}} \times \\
& \times \left(1 - \frac{k(n-k)}{\binom{n}{2}}\right)^{.49n-(k-1)} \qquad (16.8) \\
& \leq (1+o(1))n \sum_{k=1}^{(\log n)^2} \frac{k^{k-2}}{k!} (0.64)^k e^{-0.98k} \\
& \leq (1+o(1))n \times \\
& \left[\left(0.64\theta + \frac{(0.64\theta)^2}{2} + \frac{(0.64\theta)^3}{2} + \frac{2(0.64\theta)^4}{3}\right) + \sum_{k=5}^{\infty} \frac{\left((0.64)e^{.02}\right)^k}{2k^{5/2}}\right]
\end{aligned}$$

where $\theta = e^{-0.98}$

$$\begin{aligned}
& \leq (1+o(1))n \left[0.279 + \frac{1}{2 \times 5^{5/2}\left(1-(0.64)e^{.02}\right)}\right] \\
& \leq (1+o(1))n\,[0.279 + 0.026] \\
& \leq (0.305)n.
\end{aligned}$$

Explanation of (16.8): The $o(1)$ term corresponds to components of size greater than $(\log n)^2$ and w.h.p. there are none of these. For the summand, we choose k vertices and a tree on these k vertices in $\binom{n}{k}k^{k-2}$ ways. The term $\binom{0.49n}{k-1}(k-1)!$ gives the number of sequences of edge choices that lead to a given tree. The term $\binom{n}{2}^{-(k-1)}$ is the probability that these edges exist and $(0.64)^k$ bounds the probability that the vertices of the tree correspond to edges in *CORE*. The final term is the probability that the tree is actually a component.

So after $0.49n$ rounds, in expectation, the number of edges left in *CORE*, is at most $\frac{0.305}{0.63} < 0.485$ of its original size, and the Chebyshev inequality (applied to Z) can be used to show that w.h.p. it is at most 0.49 of its original size. However, randomly deleting approximately 0.51 fraction of *CORE*s will w.h.p. leave just trees and unicyclic components in H_U. To see this, observe that if we delete $0.505n$ random edges from G_U then we will have a random graph in the sub-critical stage and so w.h.p. it will consist of trees and unicyclic components. But deleting $0.505n$ random edges will w.h.p. delete less than a 0.51 fraction of CORE. □

This completes the proof that w.h.p. Algorithm PAIR finishes before $0.49n$ rounds with a perfect matching. In summary,

Theorem 16.12 *W.h.p. the algorithm* **PAIR** *finds a perfect matching in the random graph* $\mathbb{B}_{2-out}$ *in at most* $.49n$ *steps.*

One can ask whether one can w.h.p. secure a perfect matching in a bipartite random graph having more edges then $\mathbb{B}_{1-out}$, but less than $\mathbb{B}_{2-out}$. To see that it is possible, consider the following two-round procedure. In the first round assume that each vertex from the set U chooses exactly one neighbor in V and, likewise, every vertex from the set V chooses exactly one neighbor in U. In the next round, only those vertices from U and V that have not been selected in the first round get a second chance to make yet another random selection. It is easy to see that, for large n, such a second chance is, on average, given to approximately n/e vertices on each side. That is, that the average out-degree of vertices in U and V is approximately $1+1/e$. Therefore, the underlying simple graph is denoted as $\mathbb{B}_{(1+1/e)-out}$, and Karoński and Pittel [468] proved that the following result holds.

Theorem 16.13 *With probability* $1-O(n^{-1/2})$ *a random graph* $\mathbb{B}_{(1+1/e)-out}$ *contains a perfect matching.*

16.3 Hamilton Cycles

Bohman and Frieze [113] proved the following:

Theorem 16.14

$$\lim_{n\to\infty} \mathbb{P}(\mathbb{G}_{k-out} \text{ has a Hamiltonian cycle}) = \begin{cases} 0 & \text{if } k \le 2 \\ 1 & \text{if } k \ge 3. \end{cases}$$

To see that this result is best possible note that one can show that w.h.p. the random graph $\mathbb{G}_{2-out}$ contains a vertex adjacent to three vertices of degree two, which prevents the existence of a Hamiltonian cycle. The proof that $\mathbb{G}_{3-out}$ w.h.p. contains a Hamiltonian cycle is long and complicated, we therefore prove the weaker result given below which has a straightforward proof, using the ideas of Section 6.2. It is taken from Frieze and Łuczak [338].

Theorem 16.15

$$\lim_{n\to\infty} \mathbb{P}(\mathbb{G}_{k-out} \text{ has a Hamiltonian cycle}) = 1, \text{ if } k \ge 5.$$

Proof Let $H = G_0 \cup G_1 \cup G_2$ where $G_i = \mathbb{G}_{k_i-out}$, where (i) $k_0 = 1, k_1 = k_2 = 2$ and (ii) G_0, G_1, G_2 are generated independently of each other. Then we can couple the construction of H and $\mathbb{G}_{5-out}$ so that $H \subseteq \mathbb{G}_{5-out}$. This is because in the construction of H, some random choices in the construction of the associated digraphs might be repeated. In which case, having constructed H, we can give $\mathbb{G}_{5-out}$ some more edges.

It follows from Theorem 16.3 that w.h.p. $G_i, i = 1,2$ contain perfect matchings $M_i, i = 1,2$. Here we allow n to be odd and so a perfect matching may leave one vertex isolated. By symmetry M_1, M_2 are uniform random matchings. Let $M = M_1 \cup M_2$. The components of M are cycles. There could be degenerate 2-cycles consisting of two copies of the same edge and there may be a path in the case n is odd.

Lemma 16.16 *Let X be the number of components of M. Then w.h.p.*

$$X \leq 3 \log n.$$

Proof Let C be the cycle containing vertex 1. We show that

$$\mathbb{P}\left(|C| \geq \frac{n}{2}\right) \geq \frac{1}{2}. \tag{16.9}$$

To see this note that

$$\mathbb{P}(|C| = 2k) = \prod_{i=1}^{k-1} \left(\frac{n-2i}{n-2i+1}\right) \frac{1}{n-2k+1} < \frac{1}{n-2k+1}.$$

Indeed, consider the M_1-edge $\{1 = i_1, i_2\} \in C$ containing vertex 1. Let $\{i_2, i_3\} \in C$ be the M_2-edge containing i_2, Now, $\mathbb{P}(i_3 \neq 1) = (n-2)/(n-1)$. Assume that $i_3 \neq 1$ and let $\{i_3, i_4\} \in C$ be the M_1-edge containing i_3. Let $\{i_4, i_5\} \in C$ be the M_2-edge containing i_4. Then $\mathbb{P}(i_5 \neq 1) = (n-4)/(n-3)$, and so on until we close the cycle with probability $1/(n-2k+1)$. Hence,

$$\mathbb{P}\left(|C| < \frac{n}{2}\right) < \sum_{k=1}^{\lfloor n/4 \rfloor} \frac{1}{n-2k+1} < \frac{1}{2},$$

and the bound given in (16.9) follows.

Consider next the following experiment. Choose the size s of the cycle containing vertex 1. Next choose the size of the cycle containing a particular vertex from the remaining $n-s$ vertices. Continue until the cycle chosen contains all remaining vertices. Observe now, that deleting any cycle from M leaves a random pair of matchings of the remaining vertices. So, by this observation and the fact that the bound (16.9) holds, whatever the currently chosen cycle sizes, with probability at least 1/2, the size of the remaining vertex set halves, at least. Thus,

$$\mathbb{P}(X \geq 3 \log n) \leq \mathbb{P}(\text{Bin}(3 \log n, 1/2) \leq \log_2 n) = o(1).$$

□

We use rotations as in Section 6.2. Lemma 16.16 enables us to argue that we only need to add random edges trying to find x,y, where $y \in END(x)$, at most $O(\log n)$ times. We show next that $H_1 = G_1 \cup G_2$ has sufficient expansion.

Lemma 16.17 *W.h.p.* $S \subseteq [n], |S| \leq n/1000$ *implies that* $|N_{H_1}(S)| \geq 2|S|$.

Proof Let X be the number of vertex sets that violate the claim. Then,

$$\begin{aligned}\mathbb{E}X &\leq \sum_{k=1}^{n/1000} \binom{n}{k}\binom{n}{2k}\left(\left(\frac{\binom{3k}{2}}{\binom{n-1}{2}}\right)^2\right)^k \\ &\leq \sum_{k=1}^{n/1000} \left(\frac{e^3n^3}{4k^3}\frac{81k^4}{n^4}\right)^k \\ &= \sum_{k=1}^{n/1000} \left(\frac{81e^3k}{4n}\right)^k \\ &= o(1).\end{aligned}$$

□

If n is even then we begin our search for a Hamilton cycle by choosing a cycle of H_1 and removing an edge. This gives us our current path P. If n is odd we use the path P joining the two vertices of degree one in $M_1 \cup M_2$. We can ignore the case where the isolated vertex is the same in M_1 and M_2 because this only happens with probability $1/n$. We run Algorithm Pósa of Section 6.2 and observe the following: at each point of the algorithm we have a path P plus a collection of vertex disjoint cycles spanning the vertices not in P. This is because in Step (iv) the edge $\{u,v\}$ joins two cycles, one is the newly closed cycle and the other is a cycle of M. It follows that w.h.p. we only need to execute Step (iv) at most $3\log n$ times.

We now estimate the probability that we reach the start of Step (iv) and fail to close a cycle. Let the edges of G_0 be $\{e_1,e_2,\dots,e_n\}$, where e_i is the edge chosen by vertex i. Suppose that at the beginning of Step (iv) we have identified END. We can go through the vertices of END until we find $x \in END$ such that $e_x = \{x,y\}$, where $y \in END(x)$. Because G_0 and H_1 are independent, we see by Lemma 16.17 that we can assume $\mathbb{P}(y \in END(x)) \geq 1/1000$. Here we use the fact that adding edges to H_1 will not decrease the size of neighborhoods. It follows that with probability $1 - o(1/n)$ we will examine fewer than $(\log n)^2$ edges of G_0 before we succeed in closing a cycle.

Now we try closing cycles $O(\log n)$ times and w.h.p. each time we look at $O((\log n)^2)$ edges of G_0. So, if we only examine an edge of G_0 once, we will w.h.p. still always have $n/1000 - O((\log n)^3)$ edges to try. The probability that we fail to find a Hamilton cycle this way, given that H_1 has sufficient expansion, can therefore be bounded by $\mathbb{P}(\text{Bin}(n/1000 - O((\log n)^3), 1/1000) \leq 3\log n) = o(1)$. □

16.4 Nearest Neighbor Graphs

Consider the complete graph K_n, on vertex set $V = \{1, 2, , \ldots, n\}$, in which each edge is assigned a cost $C_{i,j}, i \neq j$, and the costs are independent identically distributed continuous random variables. Color an edge *green* if it is one of the k shortest edges incident to either end vertex, and color it *blue* otherwise. The graph made up of the *green* edges only is called the *k-th nearest neighbor graph* and is denoted by $G_{k-nearest}$. Note that in the random graph $G_{k-nearest}$ the edges are no longer independent, as in the case of G_{k-out} or in the classical model $\mathbb{G}_{n,p}$. Assume without loss of generality that the $C_{i,j}$ are exponential random variables of mean one. Cooper and Frieze [205] proved

Theorem 16.18

$$\lim_{n\to\infty} \mathbb{P}(G_{k-nearest} \text{ is connected}) = \begin{cases} 0 & \text{if } k = 1, \\ \gamma & \text{if } k = 2, \\ 1 & \text{if } k \geq 3, \end{cases}$$

where $0.99081 \leq \gamma \leq 0.99586$.

A similar result holds for a random *bipartite k-th nearest neighbor graph*, generated in a similar way as $G_{k-nearest}$ but starting with the complete bipartite graph $K_{n,n}$ with vertex sets $V_1, V_2 = \{1, 2, , \ldots, n\}$, and denoted by $B_{k-nearest}$. The following result is from Pittel and Weishar [629].

Theorem 16.19

$$\lim_{n\to\infty} \mathbb{P}(B_{k-nearest} \text{ is connected}) = \begin{cases} 0 & \text{if } k = 1, \\ \gamma & \text{if } k = 2, \\ 1 & \text{if } k \geq 3, \end{cases}$$

where $0.996636 \leq \gamma$.

The paper [629] contains an explicit formula for γ.

Consider the related problem of the existence of a perfect matching in the bipartite k-th nearest neighbor graph $B_{k-nearest}$. For convenience, to simplify computations, we assume here that the $C_{i,j}$ are iid exponential random variables with rate $1/n$. Coppersmith and Sorkin [226] showed that the expected size of the largest matching in $B_{1-nearest}$ (which itself is a forest) is w.h.p. asymptotic to

$$\left(2 - e^{-e^{-1}} - e^{-e^{-e^{-1}}}\right) n \approx 0.807n.$$

The same expression was obtained independently in [629]. Also, w.h.p., $B_{2-nearest}$ does not have a perfect matching. Moreover, w.h.p., in a maximal matching there are at least $\frac{2\log n}{13\log\log n}$ unmatched vertices, see [629]. The situation changes when each vertex chooses three, instead of one or two, of its "green" edges. Then the following theorem was proved in [629]:

Theorem 16.20 *$B_{3-nearest}$ has a perfect matching, w.h.p.*

Proof The proof is analogous to the proof of Theorem 16.6 and uses Hall's theorem. We use the same terminology. We can, as in Theorem 16.6, consider only bad pairs of "size" $k \le n/2$. Consider first the case when $k < \varepsilon n$, where $\varepsilon < 1/(2e^2)$, i.e. "small" bad pairs. Notice, that in a bad pair, each of the k vertices from V_1 must choose its neighbors from the set of $k-1$ vertices from V_2. Let A_k be the number of such sets, then,

$$\mathbb{E}A_k \le 2\binom{n}{k}\binom{n}{k-1}\left(\frac{k}{n}\right)^{3k} \le 2\frac{n^{2k}}{(k!)^2}\left(\frac{k}{n}\right)^{3k} \le 2\left(\frac{ke^2}{n}\right)^k.$$

(The factor 2 arises from allowing R to be chosen from V_1 or V_2.)

Let P_k be the probability that there is a bad pair of size k in $B_{3-nearest}$. Then the probability that $B_{3-nearest}$ contains a bad pair of size less than $t = \lfloor \varepsilon n \rfloor$ is, letting $l = \lfloor (\log n)^2 \rfloor$, at most

$$\begin{aligned}
\sum_{k=4}^{t} P_k &\le 2\sum_{k=4}^{t}\left(\frac{ke^2}{n}\right)^k \\
&= 2\sum_{k=4}^{l}\left(\frac{ke^2}{n}\right)^k + 2\sum_{k=l+1}^{t}\left(\frac{ke^2}{n}\right)^k \\
&\le 2\sum_{k=4}^{l}\left(\frac{le^2}{n}\right)^k + 2\sum_{k=l+1}^{t}\left(\varepsilon e^2\right)^k \\
&\le \frac{2l^2e^8}{n^4} + \left(\varepsilon e^2\right)^l.
\end{aligned}$$

So, if $\varepsilon < 1/(2e^2)$, then

$$\sum_{k=4}^{\lfloor \varepsilon n \rfloor} P_k \to 0.$$

It suffices to show that

$$\sum_{k=\lfloor \varepsilon n \rfloor+1}^{n/2} P_k \to 0.$$

To prove that there are no "large" bad pairs, note that for a pair to be bad it must be the case that there is a set of $n-k+1$ vertices of V_2 that do not choose any of the k vertices from V_1. Let $R \subset V_1, |R| = k$ and $S \subset V_2, |S| = k-1$. Without loss of generality, assume that $R = \{1,2,\ldots k\}, S = \{1,2,\ldots k-1\}$. Then let $Y_i, i = 1,2,\ldots k$ be the smallest weight in $K_{n,n}$ among the weights of edges connecting vertex $i \in R$ with vertices from $V_2 \setminus S$, and let $Z_j, j = k, k+1, \ldots n$ be the smallest weight among the weights of edges connecting vertex $j \in V_2 \setminus S$ with vertices from R. Then, each Y_i has an exponential distribution with rate $(n-k+1)/n$ and each Z_j has the exponential distribution with rate k/n.

Notice that for there to be no edges in $B_{3-nearest}$ between respective sets R and $V_2 \setminus S$ the following property should be satisfied: each vertex $i \in R$ has at least three neighbors in $K_{n,n}$ with weights smaller than Y_i and each vertex $j \in V_2 \setminus S$ has at least three neighbors in $K_{n,n}$ with weights smaller than the corresponding Z_j. If we condition on the value $Y_i = y$, then the probability that vertex i has at least three neighbors with respective edge weight smaller than Y_i, is given by

$$P_{n,k}(y) = 1 - \left(e^{-y/n}\right)^{k-1} - (k-1)\left(1 - e^{-y/n}\right)\left(e^{-y/n}\right)^{k-2} - \binom{k-1}{2}\left(1 - e^{-y/n}\right)^2\left(e^{-y/n}\right)^{k-3}.$$

Putting $a = k/n$

$$P_{n,k}(y) \approx f(a,y) = 1 - e^{-ay} - aye^{-ay} - \frac{1}{2}a^2y^2e^{-ay}.$$

Similarly, the probability that there are three neighbors of vertex $j \in V_2 \setminus S$ with edge weights smaller than Z_j is $\approx f(1-a, Z_j)$.

So, the probability that there is a bad pair in $B_{3-nearest}$ can be bounded by

$$P_k \leq 2\binom{n}{k}\binom{n}{k-1}E_k,$$

where, by the Cauchy–Schwarz inequality and independence separately of $Y_1, \ldots, Y_n$ and $Z_1, \ldots, Z_n$,

$$
\begin{aligned}
E_k &= \mathbb{E}\left(\prod_{i=1}^{k} f(a, Y_i) \prod_{j=k}^{n} f(1-a, Z_j)\right) \\
&\leq \left(\mathbb{E}\left(\prod_{i=1}^{k} f^2(a, Y_i)\right)\right)^{1/2} \left(\mathbb{E}\left(\prod_{j=k}^{n} f^2(1-a, Z_j)\right)\right)^{1/2} \\
&= \prod_{i=1}^{k} \mathbb{E}(f^2(a, Y_i))^{1/2} \prod_{j=k}^{n} \mathbb{E}(f^2(1-a, Z_j))^{1/2} \\
&= \mathbb{E}(f^2(a, Y_1))^{k/2} \mathbb{E}(f^2(1-a, Z_n))^{(n-k+1)/2}.
\end{aligned}
$$

Asymptotically, Y_1 has an exponential $(1-a)$ distribution, so

$$
\begin{aligned}
&\mathbb{E}(f^2(a, Y_1)) \\
&\approx \int_0^\infty \left(1 - e^{-ay} - aye^{-ay} - \frac{1}{2}a^2y^2e^{-ay}\right)^2 (1-a)e^{-(1-a)y} dy \\
&= (1-a)\int_0^\infty (e^{-(1-a)y} - 2e^{y} - 2aye^{-y} - a^2y^2e^{-y} + e^{-(1+a)y} \\
&\qquad + 2aye^{-(1+a)y} + 2a^2y^2e^{-(1+a)y} + a^3y^3e^{-(1+a)y} + \frac{1}{4}a^4y^4e^{-(1+a)y})dy.
\end{aligned}
$$

Now using

$$\int_0^\infty y^i e^{-cy} dy = \frac{i!}{c^{i+1}},$$

we obtain

$$
\begin{aligned}
\mathbb{E}(f^2(a, Y_1)) &= (1-a)\left(\frac{1}{1-a} - 2 - 2a - 2a^2 + \frac{1}{1+a}\right. \\
&\qquad \left. + \frac{2a}{(1+a)^2} + \frac{4a^2}{(1+a)^3} + \frac{6a^3}{(1+a)^4} + \frac{6a^4}{(1+a)^5}\right) \\
&= \frac{2a^6(10 + 5a + a^2)}{(1+a)^5}.
\end{aligned}
$$

Letting

$$g(a) = \mathbb{E}(f^2(a, Y_1))^{a/2},$$

we have

$$P_k \leq 2\binom{n}{k}\binom{n}{k-1}(g(a)g(1-a))^n \approx 2\left(\frac{g(a)g(1-a)}{a^{2a}(1-a)^{2(1-a)}}\right)^n = 2h(a)^n.$$

Numerical examination of the function $h(a)$ shows that it is bounded below 1 for a in the interval $[\delta, 0.5]$, which implies that the expected number of bad pairs is exponentially small for any $k > \delta n$, with $k \leq n/2$. Taking $\delta < \varepsilon < 1/(2e^2)$, we conclude that, w.h.p., there are no bad pairs in $B_{3-nearest}$, and so we arrive at the theorem. $\square$

16.5 Exercises

16.5.1 Let $p = \frac{\log n + (m-1)\log\log n + \omega}{n}$ where $\omega \to \infty$. Show that w.h.p. it is possible to orient the edges of $\mathbb{G}_{n,p}$ to obtain a digraph D such that the minimum out-degree $\delta^+(D) \geq m$.

16.5.2 The random digraph $D_{k-in,\ell-out}$ is defined as follows: each vertex $v \subset [n]$ independently randomly chooses k-in-neighbors and ℓ-out-neighbors. Show that w.h.p. $D_{m-in,m-out}$ is m-strongly connected for $m \geq 2$, i.e. to destroy strong connectivity, one must delete at least m vertices.

16.5.3 Show that w.h.p. the diameter of G_{k-out} is asymptotically equal to $\log_{2k} n$ for $k \geq 2$.

16.5.4 For a graph $G = (V, E)$ let $f : V \to V$ be a G-mapping if $(v, f(v))$ is an edge of G for all $v \in V$. Let G be a connected graph with maximum degree d. Let $H = \bigcup_{i=0}^{k} H_i$, where (i) $k \geq 1$, (ii) H_0 is an arbitrary spanning tree of G and (iii) $H_1, H_2, \ldots, H_k$ are independent uniform random G-mappings. Let $\theta_k = 1 - \left(1 - \frac{1}{d}\right)^{2k}$ and let $\alpha = 16/\theta_k$. Show that w.h.p. for every $A \subset V$, we have

$$|e_H(A)| \geq \frac{\theta_k}{16 \log n} \cdot |e_G(A)|,$$

where $e_G(A)$ (resp. $e_H(A)$) is the number of edges of G (resp. H) with exactly one endpoint in A.

16.5.5 Let G be a graph with n vertices and minimum degree $(\frac{1}{2} + \varepsilon)n$ for some fixed $\varepsilon > 0$. Let $H = \bigcup_{i=1}^{k} H_i$, where (i) $k \geq 2$ and (ii) $H_1, H_2, \ldots, H_k$ are independent uniform random G-mappings. Show that w.h.p. H is connected.

16.5.6 Show that w.h.p. G_{k-out} contains k edge disjoint spanning trees. (Hint: use the Nash–Williams condition [602] – see Frieze and Łuczak [339]).

16.6 Notes

k-out process

Jaworski and Łuczak [443] studied the following process that generates G_{k-out} along the way. Starting with the empty graph, a vertex v is chosen uniformly at random from the set of vertices of minimum out-degree. We then add the arc (v,w), where w is chosen uniformly at random from the set of vertices that are not out-neighbors of v. After kn steps the digraph in question is precisely $\vec{\mathbb{G}}_{k-out}$. Ignoring orientation, we denote the graph obtained after m steps by $U(n,m)$. The paper [443] studied the structure of $U(n,m)$ for $n \leq m \leq 2m$. These graphs sit between random mappings and G_{2-out}.

Nearest neighbor graphs

There has been some considerable research on the nearest neighbor graph generated by n points $\mathscr{X} = \{X_1, X_2, \ldots, X_n\}$ chosen randomly in the unit square. Given a positive integer k we define the k-nearest neighbor graph $\mathbb{G}_{\mathscr{X},k}$ by joining vertex $X \in \mathscr{X}$ to its k nearest neighbors in Euclidean distance. We first consider the existence of a giant component. Teng and Yao [704] showed that if $k \geq 213$ then there is a giant component w.h.p. Balister and Bollobás [48] reduced this number to 11. Now consider connectivity. Balister, Bollobás, Sarkar and Walters [50] proved that there exists a critical constant c^* such that if $k \leq c \log n$ and $c < c^*$ then w.h.p. $\mathbb{G}_{\mathscr{X},k}$ is not connected and if $k \geq c \log n$ and if $c > c^*$ then w.h.p. $\mathbb{G}_{\mathscr{X},k}$ is connected. The best estimates for c^* are given in Balister, Bollobás, Sarkar and Walters [49], i.e. $0.3043 < c^* < 0.5139$.

When distances are independently generated then the situation is much clearer. Cooper and Frieze [205] proved that if $k = 1$ then the k-nearest neighbor graph $\mathscr{O}_1$ is not connected; the graph $\mathscr{O}_2$ is connected with probability $\gamma \in [.99081, .99586]$; for $k \geq 2$, the graph $\mathscr{O}_k$ is k-connected w.h.p.

Directed k-in, ℓ-out

There is a natural directed version of G_{k-out} called $D_{k-in,\ell-out}$ where each vertex randomly chooses k in-neighbors and ℓ out-neighbors.

Cooper and Frieze [203] studied the connectivity of such graphs. They proved for example that if $1 \leq k, \ell \leq 2$ then

$$\lim_{n\to\infty} \mathbb{P}(D_{k-in,\ell-out} \text{ is strongly connected}) = (1-(2-k)e^{-\ell})(1-(2-\ell)e^{-k}).$$

In this result, one can in a natural way allow $k,\ell \in [1,2]$. Hamiltonicity was discussed in [206], where it was shown that w.h.p. $D_{2-in,2-out}$ is Hamiltonian. The random digraph $\mathbb{D}_{n,p}$ as well as $\vec{\mathbb{G}}_{k-out}$ are special cases of a random digraph where each vertex, independently of others, first chooses its out-degree d according to some probability distribution and then the set of its images – uniformly from all d-element subsets of the vertex set. If d is chosen according to the binomial distribution then it is $\mathbb{D}_{n,p}$ while if d equals k with probability 1, then it is $\vec{\mathbb{G}}_{k-out}$. Basic properties of the model (monotone properties, k-connectivity), were studied in Jaworski and Smit [445] and in Jaworski and Palka [444] .

k-out subgraphs of large graphs

Just as in Section 6.5, we can considered replacing the host graph K_n by graphs of large degree. Let an n vertex graph G be *strongly Dirac* if its minimum degree is at least cn for some constant $c > 0$. Frieze and Johansson [332] considered the subgraph G_k obtained from G by letting each vertex independently choose k neighbors in G. They showed that w.h.p. G_k is k-connected for $k \geq 2$ and that G_k is Hamiltonian for k sufficiently large. The paper by Frieze, Goyal, Rademacher and Vempala [328] shows the use of G_k as a cut-sparsifier.

k-out with preferential attachment

Peterson and Pittel [620] considered the following model: vertices $1,2,\dots,n$ in this order, each choose k random out-neighbors one at a time, subject to a "preferential attachment" rule: the current vertex selects vertex i with probability proportional to a given parameter $\alpha = \alpha(n)$ plus the number of times i has already been selected. Intuitively, the larger α gets, the closer the resulting k-out mapping is to the uniformly random k-out mapping. They proved that $\alpha = \Theta(n^{1/2})$ is the threshold for α growing "fast enough" to make the random digraph approach the uniformly random digraph in terms of the total variation distance. They also determined an exact limit of this distance for $\alpha = \beta n^{1/2}$.

17
Real World Networks

There has recently been an increased interest in the networks that we see around us in our every day lives. Most prominent are the Internet or the World Wide Web or social networks like Facebook and Linked In. The networks are constructed by some random process. At least we do not properly understand their construction. It is natural to model such networks by random graphs. When first studying so-called "real world networks," it was observed that often the degree sequence exhibits a tail that decays polynomially, as opposed to classical random graphs, whose tails decay exponentially. See, for example, Faloutsos, Faloutsos and Faloutsos [288]. This has led to the development of other models of random graphs such as the ones described below.

17.1 Preferential Attachment Graph

Fix an integer $m > 0$, constant and define a sequence of graphs $G_1, G_2, \ldots, G_t$. The graph G_t has vertex set $[t]$ and G_1 consists of m loops on vertex 1. Suppose we have constructed G_t. To obtain G_{t+1} we apply the following rule. We add vertex $t+1$ and connect it to m randomly chosen vertices $y_1, y_2, \ldots, y_m \in [t]$ in such a way that for $i = 1, 2, \ldots, m$,

$$\mathbb{P}(y_i = w) = \frac{\deg(w, G_t)}{2mt}.$$

In this way, G_{t+1} is obtained from G_t by adding vertex $t+1$ and m randomly chosen edges, in such a way that the neighbors of $t+1$ are biased towards higher degree vertices. When $m = 1$, G_t is a tree and this is basically a plane-oriented recursive tree as considered in Section 14.5.

This model was considered by Barabási and Albert [56]. This was followed by a rigorous analysis of a marginally different model in Bollobás, Riordan, Spencer and Tusnády [149].

Expected Degree Sequence: Power Law

Fix t and let $V_k(t)$ denote the set of vertices of degree k in G_t, where $m \leq k = \tilde{O}(t^{1/2})$. Let $D_k(t) = |V_k(t)|$. Then (compare with (14.30) when $m = 1$)

$$\mathbb{E}(D_k(t+1)|G_t) = D_k(t) + m\left(\frac{(k-1)D_{k-1}(t)}{2mt} - \frac{kD_k(t)}{2mt}\right) + 1_{k=m} + \varepsilon(k,t). \tag{17.1}$$

Explanation of (17.1): The total degree of G_t is $2mt$ and so $\frac{(k-1)D_{k-1}(t)}{2mt}$ is the probability that y_i is a vertex of degree $k-1$, creating a new vertex of degree k. Similarly, $\frac{kD_k(t)}{2mt}$ is the probability that y_i is a vertex of degree k, destroying a vertex of degree k. At this point $t+1$ has degree m and this accounts for the term $1_{k=m}$. The term $\varepsilon(k,t)$ is an error term that accounts for the possibility that $y_i = y_j$ for some $i \neq j$.

Thus,

$$\varepsilon(k,t) = O\left(\binom{m}{2}\frac{k}{mt}\right) = \tilde{O}(t^{-1/2}). \tag{17.2}$$

Taking expectations over G_t, we obtain

$$\bar{D}_k(t+1) = \bar{D}_k(t) + 1_{k=m} + m\left(\frac{(k-1)\bar{D}_{k-1}(t)}{2mt} - \frac{k\bar{D}_k(t)}{2mt}\right) + \varepsilon(k,t). \tag{17.3}$$

Under the assumption $\bar{D}_k(t) \approx d_k t$ (justified below) we are led to consider the recurrence

$$d_k = \begin{cases} 1_{k=m} + \frac{(k-1)d_{k-1} - kd_k}{2} & \text{if } k \geq m, \\ 0 & \text{if } k < m, \end{cases} \tag{17.4}$$

or

$$d_k = \begin{cases} \frac{k-1}{k+2}d_{k-1} + \frac{2\cdot 1_{k=m}}{k+2} & \text{if } k \geq m, \\ 0 & \text{if } k < m. \end{cases}$$

Therefore,

$$d_m = \frac{2}{m+2}$$
$$d_k = d_m \prod_{l=m+1}^{k} \frac{l+2}{l-2} = \frac{2m(m+1)}{k(k+1)(k+2)}. \tag{17.5}$$

So for large k, under our assumption $\bar{D}_k(t) \approx d_k t$, we see that

$$\bar{D}_k(t) \approx \frac{2m(m+1)}{k^3}.$$

We now show that the assumption $\bar{D}_k(t) \approx d_k t$ can be justified. Note that the following theorem is vacuous for $k \gg t^{1/6}$.

Theorem 17.1

$$|\bar{D}_k(t) - d_k t| = \tilde{O}(t^{1/2}) \text{ for } k = \tilde{O}(t^{1/2}).$$

Proof Let

$$\Delta_k(t) = \bar{D}_k(t) - d_k t.$$

Then, replacing $\bar{D}_k(t)$ by $\Delta_k(t) + d_k t$ in (17.3) and using (17.2) and (17.4) we obtain

$$\Delta_k(t+1) = \frac{k-1}{2t}\Delta_{k-1}(t) + \left(1 - \frac{k}{2t}\right)\Delta_k(t) + \tilde{O}(t^{-1/2}). \tag{17.6}$$

Now assume inductively on t that for every $k \geq 0$

$$|\Delta_k(t)| \leq Ct^{1/2}(\log t)^{\beta},$$

where $(\log t)^{\beta}$ is the hidden power of logarithm in $\tilde{O}(t^{-1/2})$ of (17.6) and C is an unspecified constant.

This is trivially true for $k < m$ also for small t if we make A large enough. So, replacing $\tilde{O}(t^{-1/2})$ in (17.6) by the more explicit $\alpha t^{-1/2}(\log t)^{\beta}$ we obtain

$$\begin{aligned}
\Delta_k(t+1) &\leq \left|\frac{k-1}{2t}\Delta_{k-1}(t)\right| + \left|\left(1 - \frac{k}{2t}\right)\Delta_k(t)\right| + \alpha t^{-1/2}(\log t)^{\beta} \\
&\leq \frac{k-1}{2t} A t^{1/2}(\log t)^{\beta} + \left(1 - \frac{k}{2t}\right) A t^{1/2}(\log t)^{\beta} + \alpha t^{-1/2}(\log t)^{\beta} \\
&\leq (\log t)^{\beta}(At^{1/2} + \alpha t^{-1/2}).
\end{aligned}$$

Note that if t is sufficiently large then

$$(t+1)^{1/2} = t^{1/2}\left(1 + \frac{1}{t}\right)^{1/2} \geq t^{1/2} + \frac{1}{3t^{1/2}},$$

and so

$$\Delta_k(t+1) \leq (\log(t+1))^\beta \left(A\left((t+1)^{1/2} - \frac{1}{3t^{1/2}}\right) + \frac{\alpha}{t^{1/2}}\right)$$
$$\leq A\,(\log(t+1))^\beta\,(t+1)^{1/2},$$

assuming that $A \geq 3\alpha$. □

In the next section, we justify our bound of $\tilde{O}(t^{1/2})$ for vertex degrees. After that we prove concentration of the number of vertices of degree k, for small k.

Maximum Degree

Fix $s \leq t$ and let X_l be the degree of vertex s in G_l for $s \leq l \leq t$.

Lemma 17.2

$$\mathbb{P}(X_t \geq Ae^m(t/s)^{1/2}(\log(t+1))^2) = O(t^{-A}).$$

Proof Note first that $X_s = m$. If $0 < \lambda < \varepsilon_t = \frac{1}{\log(t+1)}$ then,

$$\mathbb{E}\left(e^{\lambda X_{l+1}}|X_l\right)$$
$$= e^{\lambda X_l}\sum_{k=0}^{m}\binom{m}{k}\left(\frac{X_l}{2ml}\right)^k\left(1-\frac{X_l}{2ml}\right)^{m-k} e^{\lambda k}$$
$$\leq e^{\lambda X_l}\sum_{k=0}^{m}\binom{m}{k}\left(\frac{X_l}{2ml}\right)^k\left(1-\frac{X_l}{2ml}\right)^{m-k} (1+k\lambda(1+k\lambda))$$
$$= e^{\lambda X_l}\left(1+\frac{m\lambda X_l}{2m}+\frac{(m-1)\lambda^2X_l^2}{4ml^2}\right)$$
$$\leq e^{\lambda X_l}\left(1+\frac{X_l}{2l}\lambda(1+m\lambda)\right), \qquad \text{since } X_l < ml,$$
$$\leq e^{\lambda\left(1+\frac{(1+m\lambda)}{2l}\right)X_l}.$$

We define a sequence $\lambda = (\lambda_l, \lambda_{l+1}, \ldots, \lambda_t)$, where

$$\lambda_{j+1} = \left(1+\frac{1+m\lambda_j}{2j}\right)\lambda_j < \varepsilon_t.$$

Here our only choice is λ_l. We show below that we can find a suitable value for this, but first observe that if we manage this then

$$\mathbb{E}\left(e^{\lambda X_t}\right) \leq \mathbb{E}\left(e^{\lambda_{l+1}X_{t-1}}\right)\cdots \leq \mathbb{E}\left(e^{\lambda_t X_l}\right) \leq 1+o(1).$$

Now

$$\lambda_{j+1} \leq \left(1 + \frac{1+m\varepsilon_t}{2j}\right)\lambda_j,$$

implies that

$$\lambda_t = \lambda_l \prod_{j=l}^{t}\left(1 + \frac{1+m\varepsilon_t}{2j}\right) \leq \lambda_l \exp\left\{\sum_{j=l}^{t}\frac{1+m\varepsilon_t}{2j}\right\} \leq e^m\left(\frac{t}{l}\right)^{1/2}\lambda_l.$$

So a suitable choice for $\lambda = \lambda_l$ is

$$\lambda_l = e^{-m}\varepsilon_t\left(\frac{l}{t}\right)^{1/2}.$$

This gives

$$\mathbb{E}\left(\exp\left\{e^{-m}\varepsilon_t(l/t)^{1/2}X_t\right\}\right) \leq 1 + o(1).$$

So,

$$\mathbb{P}\left(X_t \geq Ae^m(t/l)^{1/2}(\log(t+1))^2)\right) \leq e^{-\lambda Ae^m(t/l)^{1/2}(\log(t+1)^2)}\,\mathbb{E}\left(e^{\lambda X_t}\right) = O(t^{-A}).$$

□

Thus, with probability $1-o(1)$ as $t \to \infty$ we have that the maximum degree in G_t is $O(t^{1/2}(\log t)^2)$. This is not best possible. One can prove that w.h.p. the maximum degree is $O(t^{1/2}\omega(t))$ and $\Omega(t^{1/2}/\omega(t))$ for any $\omega(t) \to \infty$, see for example Flaxman, Frieze and Fenner [303].

Concentration of Degree Sequence

Fix a value k for a vertex degree. We show that $D_k(t)$ is concentrated around its mean $\bar{D}_k(t)$.

Theorem 17.3

$$\mathbb{P}(|D_k(t) - \bar{D}_k(t)| \geq u) \leq 2\exp\left\{-\frac{u^2}{8mt}\right\}. \tag{17.7}$$

Proof Let $Y_1, Y_2, \ldots, Y_{mt}$ be the sequence of edge choices made in the construction of G_t, and for $Y_1, Y_2, \ldots, Y_i$ let

$$Z_i = Z_i(Y_1, Y_2, \ldots, Y_i) = \mathbb{E}(D_k(t) \mid Y_1, Y_2, \ldots, Y_i).$$

We prove next that $|Z_i - Z_{i-1}| \leq 4$ and then (17.7) follows directly from the Azuma–Hoeffding inequality, see Section 21.7, in particular Lemma 21.16. Fix $Y_1, Y_2, \ldots, Y_i$ and $\hat{Y}_i \neq Y_i$. We define a map (measure preserving projection) φ of

$$Y_1, Y_2, \ldots, Y_{i-1}, Y_i, Y_{i+1}, \ldots, Y_{mt}$$

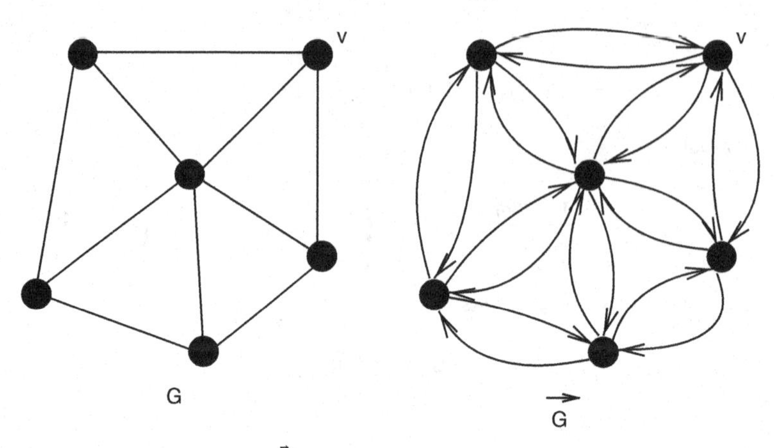

Figure 17.1. Constructing $\vec{G}$ from G

to

$$Y_1, Y_2, \ldots, Y_{i-1}, \hat{Y}_i, \hat{Y}_{i+1}, \ldots, \hat{Y}_{mt}$$

such that

$$|Z_i(Y_1, Y_2, \ldots, Y_i) - Z_i(Y_1, Y_2, \ldots, \hat{Y}_i)| \leq 4.$$

In the preferential attachment model we can view vertex choices in the graph G as random choices of arcs in a digraph $\vec{G}$, which is obtained by replacing every edge of G by a directed 2-cycle (see Figure 17.1).

Indeed, if we choose a random arc and choose its head then v will be chosen with probability proportional to the number of arcs with v as head, i.e. its degree. Hence, $Y_1, Y_2, \ldots$ can be viewed as a sequence of arc choices. Let

$$Y_i = (x, y) \text{ where } x > y$$
$$\hat{Y}_i = (\hat{x}, \hat{y}) \text{ where } \hat{x} > \hat{y}.$$

Note that $x = \hat{x}$ if $i \bmod m \neq 1$.

Now suppose $j > i$ and $Y_j = (u, v)$ arises from choosing (w, v). Then we define

$$\varphi(Y_j) = \begin{cases} Y_j & (w,v) \neq Y_i \\ (w, \hat{y}) & (w,v) = Y_i. \end{cases}$$

This map is measure preserving since each sequence $\varphi(Y_1, Y_2, \ldots, Y_t)$ occurs with probability $\prod_{j=i+1}^{tm} j^{-1}$. Only $x, \hat{x}, y, \hat{y}$ change degree under the map φ so $D_k(t)$ changes by at most four. □

17.2 A General Model of Web Graphs

In the model presented in the previous section a new vertex is added at time t and this vertex chooses m random neighbors, with probability proportional to their current degree. Cooper and Frieze [208] generalize this in the following ways: they allow (a) new edges to be inserted between existing vertices, (b) a variable number of edges to be added at each step and (c) endpoint vertices to be chosen by a mixture of uniform selection and copying. This results in a large number of parameters, which are described below. We first give a precise description of the process.

Initially, at step $t = 0$, there is a single vertex v_0. At any step $t = 1, 2, \ldots, T, \ldots$, there is a birth process in which either new vertices or new edges are added. Specifically, either a procedure NEW is followed with probability $1 - \alpha$ or a procedure OLD is followed with probability α. In procedure NEW, a new vertex v is added to G_{t-1} with one or more edges added between v and G_{t-1}. In procedure OLD, an existing vertex v is selected and extra edges are added at v.

The method for adding edges at step t typically permits the choice of initial vertex v (in the case of OLD) and of terminal vertices (in both cases) to be made from G_{t-1} either u.a.r. (uniformly at random) or according to vertex degree, or a mixture of these two based on further sampling. The number of edges added to vertex v at step t by the procedures (NEW, OLD) is given by distributions specific to the procedure.

Notice that the edges have an intrinsic direction, arising from the way they are inserted, which one can ignore or not. Here the undirected model is considered with a sampling procedure based on vertex degree. The process allows multiple edges, and self-loops can arise from the OLD procedure. The NEW procedure, as described, does not generate self-loops, although this could easily be modified.

Sampling parameters, notation and main properties

Our undirected model G_t has sampling parameters $\alpha, \beta, \gamma, \delta, \boldsymbol{p}, \boldsymbol{q}$ whose meanings are given below:

Choice of procedure at step t.

α: probability that an OLD vertex generates edges.

$1 - \alpha$: probability that a NEW vertex is created.

Procedure NEW

$\boldsymbol{p} = (p_i : i \geq 1)$: probability that the new node generates i new edges.

β: probability that choices of terminal vertices are made uniformly.

$1-\beta$: probability that choices of terminal vertices are made according to degree.

Procedure OLD

$\boldsymbol{q} = (q_i : i \geq 1)$: probability that the old node generates i new edges.

δ: probability that the initial node is selected uniformly.

$1-\delta$: probability that the initial node is selected according to degree.

γ : probability that choices of terminal vertices are made uniformly.

$1-\gamma$: probability that choices of terminal vertices are made according to degree.

The models require $\alpha < 1$ and $p_0 = q_0 = 0$. It is convenient to assume a *finiteness* condition for the distributions $\{p_j\}$, $\{q_j\}$. This means that there exist j_0, j_1 such that $p_j = 0, j > j_0$ and $q_j = 0, j > j_1$. Imposing the finiteness condition helps simplify the difference equations used in the analysis.

The model creates edges in the following way: an initial vertex v is selected. If the terminal vertex w is chosen u.a.r., we say v is *assigned uniformly* to w. If the terminal vertex w is chosen according to its vertex degree, we say v is *copied* to w. In either case the edge has an intrinsic direction (v,w), which we may choose to ignore. Note that sampling according to vertex degree is equivalent to selecting an edge u.a.r and then selecting an endpoint u.a.r.

Let

$$\mu_p = \sum_{j=1}^{j_0} jp_j, \ \mu_q = \sum_{j=1}^{j_1} jq_j$$

be the expected number of edges added by NEW or OLD and let

$$\theta = 2((1-\alpha)\mu_p + \alpha\mu_q).$$

To simplify subsequent notation, we introduce new parameters as follows:

$$\begin{aligned}
a &= 1 + \beta\mu_p + \frac{\alpha\gamma\mu_q}{1-\alpha} + \frac{\alpha\delta}{1-\alpha}, \\
b &= \frac{(1-\alpha)(1-\beta)\mu_p}{\theta} + \frac{\alpha(1-\gamma)\mu_q}{\theta} + \frac{\alpha(1-\delta)}{\theta}, \\
c &= \beta\mu_p + \frac{\alpha\gamma\mu_q}{1-\alpha}, \\
d &= \frac{(1-\alpha)(1-\beta)\mu_p}{\theta} + \frac{\alpha(1-\gamma)\mu_q}{\theta}, \\
e &= \frac{\alpha\delta}{1-\alpha}, \\
f &= \frac{\alpha(1-\delta)}{\theta}.
\end{aligned}$$

We note that

$$c+e=a-1 \text{ and } b=d+f. \tag{17.8}$$

Now define the sequence $(d_0, d_1, \ldots, d_k, \ldots)$ by $d_0 = 0$, and for $k \geq 1$

$$d_k(a+bk) = (1-\alpha)p_k + (c+d(k-1))d_{k-1} + \sum_{j=1}^{k-1}(e+f(k-j))q_j d_{k-j}. \tag{17.9}$$

For convenience we define $d_k = 0$ for $k < 0$. Since $a \geq 1$, this system of equations has a unique solution.

The main quantity we study is the random variable $D_k(t)$, the number of vertices of degree k at step t. Cooper and Frieze [208] prove that, as $t \to \infty$, for small k, $\overline{D}_k(t) \approx td_k$.

Theorem 17.4 *There exists a constant $M > 0$ such that almost surely for all t, $k \geq 1$*

$$|\overline{D}_k(t) - td_k| \leq Mt^{1/2}\log t.$$

This is proved in Section 17.2.

It is shown in (17.10), that the number of vertices $\nu(t)$ at step t is w.h.p. asymptotic to $(1-\alpha)t$. It follows that the proportion of vertices of degree k is w.h.p. asymptotic to

$$\bar{d}_k = \frac{d_k}{1-\alpha}.$$

The next theorem summarizes what is known about the sequence (d_k) defined by (17.9).

Theorem 17.5 *There exist constants $C_1, C_2, C_3, C_4 > 0$ such that*

(i) $C_1 k^{-\zeta} \leq d_k \leq C_2 \min\{k^{-1}, k^{-\zeta/j_1}\}$ *where* $\zeta = (1+d+f\mu_q)/(d+f)$.
(ii) If $j_1 = 1$ *then* $d_k \approx C_3 k^{-(1+1/(d+f))}$.
(iii) If $f = 0$ *then* $d_k \approx C_4 k^{-(1+1/d)}$.

Evolution of the degree sequence of G_t

Let $\nu(t) = |V(t)|$ be the number of vertices and let $\eta(t) = |2E(t)|$ be the total degree of the graph at the end of step t. $\mathbb{E}\,\nu(t) = (1-\alpha)t$ and $\mathbb{E}\,\eta(t) = \theta t$. The random variables $\nu(t)$, $\eta(t)$ are sharply concentrated provided $t \to \infty$. Indeed $\nu(t)$ has distribution $\mathrm{Bin}(t, 1-\alpha)$ and so by Theorem 21.6 and its corollaries,

$$\mathbb{P}(|\nu(t) - (1-\alpha)t| \geq t^{1/2}\log t) = O(t^{-K}) \tag{17.10}$$

for any constant $K > 0$.

Similarly, $\eta(t)$ has expectation θt and is the sum of t independent random variables, each bounded by $\max\{j_0, j_1\}$. Hence, by Theorem 21.6 and its corollaries,

$$\mathbb{P}(|\eta(t) - \theta t| \geq t^{1/2}\log t) = O(t^{-K}) \tag{17.11}$$

for any constant $K > 0$.

These results are almost sure in the sense that they hold for all $t \geq t_0$ with probability $1 - O(t_0^{-K+1})$. Thus we can focus on processes such that this is true.

We remind the reader that $D_k(t)$ is the number of vertices of degree k at step t and that $\overline{D}_k(t)$ is its expectation. Here $\overline{D}_j(t) = 0$ for all $j \leq 0, t \geq 0$, $\overline{D}_1(0) = 1$, $\overline{D}_k(0) = 0$, $k \geq 2$.

Using (17.10) and (17.11) we see that

$$\overline{D}_k(t+1) = \overline{D}_k(t) + (1-\alpha)p_k + O(t^{-1/2}\log t) \tag{17.12}$$

$$+ (1-\alpha)\sum_{j=1}^{j_0} p_j \left(\beta\left(\frac{j\overline{D}_{k-1}(t)}{(1-\alpha)t} - \frac{j\overline{D}_k(t)}{(1-\alpha)t}\right)\right.$$

$$\left.+ (1-\beta)\left(\frac{j(k-1)\overline{D}_{k-1}(t)}{\theta t} - \frac{jk\overline{D}_k(t)}{\theta t}\right)\right) \tag{17.13}$$

$$-\alpha\left(\frac{\delta\overline{D}_k(t)}{(1-\alpha)t} + \frac{(1-\delta)k\overline{D}_k(t)}{\theta t}\right)$$

$$+\alpha\sum_{j=1}^{j_1} q_j\left(\frac{\delta\overline{D}_{k-j}(t)}{(1-\alpha)t} + \frac{(1-\delta)(k-j)\overline{D}_{k-j}(t)}{\theta t}\right) \tag{17.14}$$

$$+\alpha\sum_{j=1}^{j_1} jq_j\left(\gamma\left(\frac{\overline{D}_{k-1}(t)}{(1-\alpha)t} - \frac{\overline{D}_k(t)}{(1-\alpha)t}\right)\right.$$

$$\left.+ (1-\gamma)\left(\frac{(k-1)\overline{D}_{k-1}(t)}{\theta t} - \frac{k\overline{D}_k(t)}{\theta t}\right)\right). \tag{17.15}$$

Here (17.13), (17.14) and (17.15) are the main terms of the change in the expected number of vertices of degree k due to the effect on: terminal vertices in NEW, the initial vertex in OLD and the terminal vertices in OLD, respectively.

Rearranging the RHS, we find:

$$
\begin{aligned}
\overline{D}_k(t+1) =& \overline{D}_k(t) + (1-\alpha)p_k + O(t^{-1/2}\log t) \\
&- \frac{\overline{D}_k(t)}{t}\left(\beta\mu_p + \frac{\alpha\gamma\,\mu_q}{1-\alpha} + \frac{\alpha\delta}{1-\alpha} + \frac{(1-\alpha)(1-\beta)\mu_p k}{\theta}\right. \\
&\left. + \frac{\alpha(1-\gamma)\mu_q k}{\theta} + \frac{\alpha(1-\delta)k}{\theta}\right) \\
&+ \frac{\overline{D}_{k-1(t)}}{t}\left(\beta\mu_p + \frac{\alpha\gamma\,\mu_q}{1-\alpha} + \frac{(1-\alpha)(1-\beta)\mu_p(k-1)}{\theta}\right. \\
&\left. + \frac{\alpha(1-\gamma)\mu_q(k-1)}{\theta}\right) \\
&+ \sum_{j=1}^{j_1} q_j \frac{\overline{D}_{k-j}(t)}{t}\left(\frac{\alpha\delta}{1-\alpha} + \frac{\alpha(1-\delta)(k-j)}{\theta}\right).
\end{aligned}
$$

Thus, for all $k \geq 1$ and almost surely for all $t \geq 1$,

$$
\begin{aligned}
\overline{D}_k(t+1) =& \overline{D}_k(t) + (1-\alpha)p_k + O(t^{-1/2}\log t) \\
&+ \frac{1}{t}((1-(a+bk))\overline{D}_k(t) + (c+d(k-1))\overline{D}_{k-1}(t) \\
&+ \sum_{j=1}^{j_1} q_j(e+f(k-j))\overline{D}_{k-j}(t)).
\end{aligned}
\tag{17.16}
$$

The following Lemma establishes an upper bound on d_k given in Theorem 17.5(i).

Lemma 17.6 *The solution of (17.9) satisfies* $d_k \leq \frac{C_2}{k}$.

Proof Assume that $k > k_0$, where k_0 is sufficiently large, and thus $p_k = 0$. Smaller values of k can be dealt with by adjusting C_2. We proceed by induction on k. From (17.9),

$$
\begin{aligned}
(a+bk)d_k &\leq (c+d(k-1))\frac{C_2}{k-1} + \sum_{j=1}^{j_1}(e+f(k-j))q_j\frac{C_2}{k-j} \\
&\leq C_2(d+f) + \frac{C_2(c+e)}{k-j_1} \\
&= C_2 b + \frac{C_2(a-1)}{k-j_1},
\end{aligned}
$$

from (17.8). So

$$
\begin{aligned}
d_k - \frac{C_2}{k} &\le \frac{C_2 b}{a+bk} + \frac{C_2(a-1)}{(k-j_1)(a+bk)} - \frac{C_2}{k} \\
&= \frac{C_2(a-1)}{(k-j_1)(a+bk)} - \frac{C_2 a}{k(a+bk)} \\
&\le 0,
\end{aligned}
$$

for $k \ge j_1 a$. □

We can now prove Theorem 17.4, which is restated here for convenience.

Theorem 17.7 *There exists a constant $M > 0$ such that almost surely for $t, k \ge 1$,*

$$|\overline{D}_k(t) - td_k| \le Mt^{1/2}\log t. \tag{17.17}$$

Proof Let $\Delta_k(t) = \overline{D}_k(t) - td_k$. It follows from (17.9) and (17.16) that

$$
\begin{aligned}
\Delta_k(t+1) = \Delta_k(t)\left(1 - \frac{a+bk-1}{t}\right) + O(t^{-1/2}\log t) \\
+ \frac{1}{t}\left((c+d(k-1))\Delta_{k-1}(t) + \sum_{j=1}^{j_1}(e+f(k-j))q_j\Delta_{k-j}(t)\right).
\end{aligned} \tag{17.18}
$$

Let L denote the hidden constant in $O(t^{-1/2}\log t)$. We can adjust M to deal with small values of t, so we assume that t is sufficiently large. Let $k_0(t) = \left\lfloor \frac{t+1-b}{a} \right\rfloor$. If $k > k_0(t)$ then we observe that (i) $D_k(t) \le \frac{t\max\{j_0,j_1\}}{k_0(t)} = O(1)$ and (ii) $td_k \le t\frac{C_2}{k_0(t)} = O(1)$ follows from Lemma 17.6, and so (17.17) holds trivially.

Assume inductively that $\Delta_\kappa(\tau) \le M\tau^{1/2}\log\tau$ for $\kappa + \tau \le k + t$ and that $k \le k_0(t)$. Then (17.18) and $k \le k_0$ implies that for M large,

$$
\begin{aligned}
|\Delta_k(t+1)| &\le L\frac{\log t}{t^{1/2}} + Mt^{1/2}\log t \times \\
&\quad \left(1 + \frac{1}{t}\left(c + dk + \sum_{j=1}^{j_1}(e+fk)q_j - (a+bk-1)\right)\right) \\
&= L\frac{\log t}{t^{1/2}} + Mt^{1/2}\log t \\
&\le M(t+1)^{1/2}\log(t+1)
\end{aligned}
$$

provided $M \gg 2L$. We have used (17.8) to obtain the second line.

This completes the proof by induction. □

A general power law bound for d_k

The following lemma completes the proof of Theorem 17.5(i).

Lemma 17.8 *For $k > j_0$ we have,*

(i) $d_k > 0$.
(ii) $C_1 k^{-(1+d+f\mu_q)/b} \leq d_k \leq C_2 k^{-(1+d+f\mu_q)/bj_1}$.

Proof *(i)* Let κ be the first index such that $p_\kappa > 0$, so that, from (17.9), $d_\kappa > 0$. It is not possible for both c and d to be zero. Therefore the coefficient of d_{k-1} in (17.9) is non-zero and thus $d_k > 0$ for $k \geq \kappa$.

(ii) Re-writing (17.9) we see that for $k > j_0$, $p_k = 0$ and then d_k satisfies

$$d_k = d_{k-1}\frac{c+d(k-1)}{a+bk} + \sum_{j=1}^{j_1} d_{k-j}q_j \frac{e+f(k-j)}{a+bk}, \tag{17.19}$$

which is a linear difference equation with rational coefficients (see [581]).

We let $d_i = 0$ for $i < 0$ to handle the cases where $k-j < 0$ in the above sum.

Let $y = 1+d+f\mu_q$, then

$$\frac{c+d(k-1)}{a+bk} + \sum_{j=1}^{j_1} q_j \frac{e+f(k-j)}{a+bk} = 1 - \frac{y}{a+bk} \geq 0$$

and thus

$$\left(1-\frac{y}{a+bk}\right)\min\{d_{k-1},\ldots,d_{k-j_1}\} \leq d_k \leq \left(1-\frac{y}{a+bk}\right)\max\{d_{k-1},\ldots,d_{k-j_1}\}. \tag{17.20}$$

It follows that

$$d_{j_0}\prod_{j=j_0+1}^{k}\left(1-\frac{y}{a+bj}\right) \leq d_k \leq \max\{d_1,d_2,\ldots,d_{j_0}\}\prod_{s=0}^{\lfloor (k-j_0)/j_1 \rfloor}\left(1-\frac{y}{a+b(k-sj_1)}\right). \tag{17.21}$$

The lower bound in (17.21) is proved by induction on k. It is trivial for $k = j_0$ and for the inductive step we have

$$\begin{aligned} d_k &\geq d_{j_0}\left(1-\frac{y}{a+bk}\right)\min_{i=j_0,\ldots,k-1}\left\{\prod_{j=j_0+1}^{i}\left(1-\frac{y}{a+bj}\right)\right\} \\ &= d_{j_0}\prod_{j=j_0+1}^{k}\left(1-\frac{y}{a+bj}\right). \end{aligned}$$

The upper bound in (17.21) is proved as follows: let $d_{i_1} = \max\{d_{k-1}, \ldots, d_{k-j_1}\}$, and in general, let $d_{i_{t+1}} = \max\{d_{i_t-1}, \ldots, d_{i_t-j_1}\}$. Using (17.20) we see there is a sequence $k-1 \geq i_1 > i_2 > \cdots > i_p > j_0 \geq i_{p+1}$ such that $|i_t - i_{t-1}| \leq j_1$ for all t, and $p \geq \lfloor (k-j_0)/j_1 \rfloor$. Thus,

$$d_k \leq d_{i_{p+1}} \prod_{t=0}^{p} \left(1 - \frac{y}{a+bi_t}\right),$$

and the RHS of (17.21) now follows.

Now consider the product in the LHS of (17.21).

$$\begin{aligned}
\prod_{j=j_0+1}^{k} \left(1 - \frac{y}{a+bj}\right) &= \exp\left\{ \sum_{j=j_0+1}^{k} \left(-\frac{y}{a+bj} - \frac{1}{2}\left(\frac{y}{a+bj}\right)^2 - \cdots \right) \right\} \\
&= \exp\left\{ O(1) - \sum_{j=j_0+1}^{k} \frac{y}{a+bj} \right\} \\
&= C_1 k^{-y/b}.
\end{aligned}$$

This establishes the lower bound of the lemma. The upper bound follows similarly, from the upper bound in (17.21). □

The case $j_1 = 1$

We prove Theorem 17.5(ii). When $q_1 = 1$, $p_j = 0, j > j_0 = \Theta(1)$, the general value of d_k, $k > j_0$ can be found directly, by iterating the recurrence (17.9). Thus,

$$\begin{aligned}
d_k &= \frac{1}{a+bk} \left(d_{k-1}\left((a-1) + b(k-1)\right)\right) \\
&= d_{k-1} \left(1 - \frac{1+b}{a+bk}\right) \\
&= d_{j_0} \prod_{j=j_0+1}^{k} \left(1 - \frac{1+b}{a+jb}\right).
\end{aligned}$$

Thus, for some constant $C_6 > 0$,

$$d_k \approx C_6 (a+bk)^{-x}$$

where

$$x = 1 + \frac{1}{b} = 1 + \frac{2}{\alpha(1-\delta) + (1-\alpha)(1-\beta) + \alpha(1-\gamma)}.$$

The case $f = 0$

We prove Theorem 17.5(iii). The case ($f = 0$) arises in two ways. Firstly, if $\alpha = 0$ so that a new vertex is added at each step. Secondly, if $\alpha \neq 0$ but $\delta = 1$ so that the initial vertex of an OLD choice is sampled u.a.r. Observe that $b = d$ now, see (17.8).

We first prove that for a sufficiently large absolute constant $A_2 > 0$ and for all sufficiently large k, that

$$\frac{d_k}{d_{k-1}} = 1 - \frac{1+d}{a+dk} + \frac{\xi(k)}{k^2} \tag{17.22}$$

where $|\xi(k)| \leq A_2$.

We first re-write (17.9) as

$$\frac{d_k}{d_{k-1}} = \frac{c+d(k-1)}{a+dk} + \sum_{j=1}^{j_1} \frac{eq_j}{a+dk} \prod_{t=k-j+1}^{k-1} \frac{d_{t-1}}{d_t}. \tag{17.23}$$

(We assume here that $k > j_0$, so that $p_k = 0$.)

Now use induction to write

$$\prod_{t=k-j+1}^{k-1} \frac{d_{t-1}}{d_t} = 1 + (j-1)\frac{d+1}{a+dk} + \frac{\xi^*(j,k)}{k^2} \tag{17.24}$$

where $|\xi^*(j,k)| \leq A_3$ for some constant $A_3 > 0$. (We use the fact that j_1 is constant here.)

Substituting (17.24) into (17.23) gives

$$\frac{d_k}{d_{k-1}} = \frac{c+d(k-1)}{a+dk} + \frac{e}{a+dk} + \frac{e(\mu_q - 1)(d+1)}{(a+dk)^2} + \frac{\xi^{**}(k)}{(a+dk)k^2}$$

where $|\xi^{**}(k)| \leq eA_3$.

Equation (17.22) follows immediately from this and $c + e = a - 1$. On iterating (17.22) we see that for some constant $C_7 > 0$,

$$d_k \approx C_7 k^{-\left(1+\frac{1}{d}\right)}.$$

□

17.3 Small World

In an influential paper Milgram [580] describes the following experiment. He chose a person X to receive mail and then randomly chose a person Y to send it. If Y did not know X then Y was to send the mail to someone he/she thought more likely to know X and so on. Surprisingly, the mail got through in 64 out of 296 attempts and the number of links in the chain was relatively small, between 5 and 6. More recently, Kleinberg [488] described a model that attempts to explain this phenomenon.

Kleinberg's Model

The model can be generalized significantly, but to be specific we consider the following. We start with the $n \times n$ grid G_0 that has vertex set $[n]^2$ and where (i,j) is adjacent to (i',j') if and only if $d((i,j),(i',j')) = 1$, where $d((i,j),(k,\ell)) = |i-k| + |j-\ell|$. In addition, each vertex $u = (i,j)$ will choose another random neighbor $\varphi(u)$, where

$$\mathbb{P}(\varphi(u) = v = (k,\ell)) = \frac{d(u,v)^{-2}}{D_u}$$

where

$$D_x = \sum_{y \neq x} d(x,y)^{-2}.$$

The random neighbors model "long range contacts." Let the grid G_0 plus the extra random edges be denoted by G.

It is not difficult to show that w.h.p. these random contacts reduce the diameter of G to order $\log n$. This, however, would not explain Milgram's success. Instead, Kleinberg proposed the following decentralized algorithm $\mathscr{A}$ for finding a path from an initial vertex $u_0 = (i_0, j_0)$ to a target vertex $u_\tau = (i_\tau, j_\tau)$: when at u move to the neighbor closest in distance to u_τ.

Theorem 17.9 *Algorithm $\mathscr{A}$ finds a path from initial to target vertex of order $O((\log n)^2)$, in expectation.*

Proof Note that each step of $\mathscr{A}$ finds a node closer to the target than the current node and so the algorithm must terminate with a path.

Observe next that for any vertex x of G we have

$$D_x \leq \sum_{j=1}^{2n-2} 4j \times j^{-2} = 4 \sum_{j=1}^{2n-2} j^{-1} \leq 4 \log(3n).$$

As a consequence, v is the long-range contact of vertex u, with probability at least $(4\log(3n)d(u,v)^2)^{-1}$.

For $0 < j \leq \log_2 n$, we say that the execution of $\mathscr{A}$ is in phase j if the distance of the current vertex u to the target is greater than 2^j, but at most 2^{j+1}. We say that $\mathscr{A}$ is in phase 0 if the distance from u to the target is at most 2.

Let B_j denote the set of nodes at distance 2^j or less from the target. Then

$$|B_j| \geq 1 + \sum_{i=1}^{2^j} i > 2^{2j-1}.$$

Note that by the triangle inequality, each member of B_j is within distance $2^{j+1} + 2^j < 2^{2j+2}$ of u.

Let $X_j \leq 2^{j+1}$ be the time spent in Phase j. Assume first that $\log_2 \log_2 n \leq j \leq \log_2 n$. Phase j will end if the long-range contact of the current vertex lies in B_j. The probability of this is at least

$$\frac{2^{2j-1}}{4\log(3n)2^{2j+4}} = \frac{1}{128\log(3n)}.$$

We can reveal the long-range contacts as the algorithm progresses. In this way, the long-range contact of the current vertex will be independent of the previous contacts of the path. Thus,

$$\mathbb{E}X_j = \sum_{i=1}^{\infty} \mathbb{P}(X_j \geq i) \leq \sum_{i=1}^{\infty} \left(1 - \frac{1}{128\log(3n)}\right)^i < 128\log(3n).$$

Now if $0 \leq j \leq \log_2 \log_2 n$ then $X_j \leq 2^{j+1} \leq 2\log_2 n$. Thus the expected length of the path found by $\mathscr{A}$ is at most $128\log(3n) \times \log_2 n$. □

In the same paper, Kleinberg showed that replacing $d(u,v)^{-2}$ by $d(u,v)^{-r}$ for $r \neq 2$ led to non-polylogarithmic path length.

17.4 Exercises

17.4.1 Show that w.h.p. the preferential attachment graph of Section 17.1 has diameter $O(\log n)$. (Hint: using the idea that vertex t chooses a random edge of the current graph, observe that half of these edges appeared at time $t/2$ or less.)

17.4.2 For the next few questions we modify the preferential attachment graph of Section 17.1 in the following way: first let $m = 1$ and preferentially generate a sequence of graphs $\Gamma_1, \Gamma_2, \ldots, \Gamma_{mn}$. Then if the edges of Γ_{mn} are $(u_i, v_i), i = 1,2,\ldots,mn$ let the edges of $\mathbb{G}_n$ be $(u_{\lceil i/m \rceil}, v_{\lceil i/m \rceil}), i = 1,2,\ldots,mn$. Show that (17.1) continues to hold.

17.4.3 Show that $\mathbb{G}_n$ of the previous question can also be generated in the following way:

(a) Let π be a random permutation of $[2mn]$. Let $X = \{(a_i, b_i), i = 1, 2, \dots, mn\}$ where $a_i = \min\{\pi(2i-1), \pi(2i)\}$ and $b_i = \max\{\pi(2i-1), \pi(2i)\}$.

(b) Let the edges of $\mathbb{G}_n$ be $(a_{\lceil i/m \rceil}, b_{\lceil i/m \rceil}), i = 1, 2, \dots, mn$.

This model was introduced in [149].

17.4.4 Show that the edges of the graph in the previous question can be generated as follows:

(a) Let $\zeta_1, \zeta_2, \dots, \zeta_{2mn}$ be independent uniform $[0,1]$ random variables. Let $\{x_i < y_i\} = \{\zeta_{2i-1}, \zeta_{2i}\}$ for $i = 1, 2, \dots, mn$. Sort the y_i in increasing order $R_1 < R_2 < \cdots < R_{mn}$ and let $R_0 = 0$. Then let

$$W_j = R_{mj} \text{ and } I_j = (W_{j-1}, W_j] \text{ for } j = 1, 2, \dots, n.$$

This model was introduced in [148].

(b) The edges of $\mathbb{G}_n$ are $(u_i, v_i), i = 1, 2, \dots, mn$ where $x_i \in I_{u_i}, y_i \in I_{v_i}$.

17.4.5 Prove that $(R_1, R_2, \dots, R_{mn})$ can be generated as

$$R_i = \left(\frac{\Upsilon_i}{\Upsilon_{mn+1}}\right)^{1/2}$$

where $\Upsilon_N = \xi_1 + \xi_2 + \cdots + \xi_N$ for $N \geq 1$ and $\xi_1, \xi_2, \dots, \xi_{mn+1}$ are independent exponential copies of $EXP(1)$.

17.4.6 Let L be a large constant and let $\omega = \omega(n) \to \infty$ arbitrarily slowly. Then let $\mathcal{E}$ be the event that

$$\Upsilon_k \approx k \text{ for } \frac{k}{m} \in [\omega, n] \text{ or } k = mn + 1.$$

Show that

(a) $\mathbb{P}(\neg\mathcal{E}) = o(1)$.

(b) Let $\eta_i = \xi_{(i-1)m+1} + \xi_{(i-1)m+2} + \cdots + \xi_{im}$. If $\mathcal{E}$ occurs then

(1) $W_i \approx \left(\frac{i}{n}\right)^{1/2}$ for $\omega \leq i \leq n$, and

(2) $w_i = W_i - W_{i-1} \approx \dfrac{\eta_i}{2m(in)^{1/2}}$ for $\omega \leq i \leq n$.

(c) $\eta_i \leq \log n$ for $i \in [n]$ w.h.p.

(d) $\eta_i \leq \log\log n$ for $i \in [(\log n)^{10}]$ w.h.p.

(e) If $\omega \leq i < j \leq n$ then $\mathbb{P}(\text{edge } ij \text{ exists}) \approx \frac{\eta_i}{2(ij)^{1/2}}$.

(f) $\eta_i \geq \frac{1}{\log\log n}$ and $i \leq \frac{n}{\omega(\log n)^3}$ implies the degree $d_n(i) \approx \eta_i \left(\frac{n}{i}\right)^{1/2}$.

17.5 Notes

There are by now a vast number of papers on different models of Real World Networks. We point out a few additional results in the area. The books by Durrett [262] and Bollobás, Kozma and Miklós [146] cover the area. See also van der Hofstadt [407].

Preferential Attachment Graph

Perhaps the most striking result is due to Bollobás and Riordan [148]. There they proved that the diameter of $\mathbb{G}_n$ is asymptotic to $\frac{\log n}{\log\log n}$ w.h.p. To prove this they introduced the model in question 17.4.4 above. Cooper [200] and Peköz, Röllin and Ross [617] discussed the degree distribution of $\mathbb{G}_n$ in some detail. Flaxman, Frieze and Fenner [303] showed that the if Δ_k, λ_k are the kth largest degree and eigenvalue, respectively, then $k \approx \Delta_k^{1/2}$ for $k = O(1)$. The proof follows ideas from Mihail and Papadimitriou [579] and Chung, Lu and Vu [184], [185]. Cooper and Frieze [208] discussed the likely proportion of vertices visited by a random walk on a growing preferential attachment graph. They showed that w.h.p. this is just over 40% at all times. Borgs, Brautbar, Chayes, Khanna and Lucier [152] discussed "local algorithms" for finding a specific vertex or the largest degree vertex. Frieze and Pegden [346] described an algorithm for the same problem, but with reduced storage requirements.

Geometric models

Some real world graphs have a geometric constraint. Flaxman, Frieze and Vera [304], [305] considered a geometric version of the preferential attachment model. Here the vertices $X_1, X_2, \dots, X_n$ are randomly chosen points on the unit sphere in $\mathbb{R}^3$. X_{i+1} chooses m neighbors and these vertices are chosen with probability $P(deg, dist)$ dependent on (i) their current degree and (ii) their distance from X_{i+1}. van den Esker [285] added *fitness* to the models in [304] and [305]. Jordan [452] considered more general spaces than $\mathbb{R}^3$. Jordan and Wade [453] considered the case of $m = 1$ and a variety of definitions of P that enable one to interpolate between the preferential attachment graph and the on-line nearest neighbor graph.

Aiello, Bonato, Cooper, Janssen and Pralat [7] introduced the SPA model. Here the vertices are points in the unit hyper-cube D in $\mathbb{R}^m$, equipped with a toroidal metric. At time t each vertex v has a domain of attraction $S(v,t)$ of volume $\frac{A_1 \deg^-(v,t) + A_2}{t}$. Then at time t we generate a uniform random point X_{t+1} as a new vertex. If the new point lies in the domain $S(v,t)$ then we join X_{t+1} to

v by an edge directed to v, with probability p. The paper by [7] deals mainly with the degree distribution. The papers by Jannsen, Pralat and Wilson [437], [438] show that for graphs formed according to the SPA model it is possible to infer the metric distance between vertices from the link structure of the graph. The paper by Cooper, Frieze and Pralat [221] shows that w.h.p. the directed diameter at time t lies between $\frac{c_1 \log t}{\log \log t}$ and $c_2 \log t$.

Zhou, Yan and Wang [736] introduced random Apollonian networks. Here we build a random triangulation by inserting a vertex into a randomly chosen face. Frieze and Tsourakakis [350] studied their degree sequence and eigenvalue structure. Ebrahimzadeh, Farczadi, Gao, Mehrabian, Sato, Wormald and Zung [270] studied their diameter and length of the longest path. Cooper and Frieze [216] gave an improved longest path estimate and this was further improved by Collevecchio, Mehrabian and Wormald [225].

Interpolating between Erdős–Rényi and Preferential Attachment

Pittel [626] considered the following model: $G_0, G_1, \dots, G_m$ is a random (multi) graph growth process G_m on a vertex set $[n]$. G_{m+1} is obtained from G_m by inserting a new edge e at random. Specifically, the conditional probability that e joins two currently disjoint vertices, i and j, is proportional to $(d_i+\alpha)(d_j+\alpha)$, where d_i, d_j are the degrees of i,j in G_m, and $\alpha > 0$ is a fixed parameter. The limiting case $\alpha = \infty$ is the Erdős–Rényi graph process. He showed that w.h.p. G_m contains a unique giant component if and only if $c := 2m/n > c_\alpha = \alpha/(1+\alpha)$, and the size of this giant is asymptotic to $n\big[1 - \big(\frac{\alpha+c^*}{\alpha+c}\big)^\alpha\big]$, where $c^* < c_\alpha$ is the root of $\frac{c}{(\alpha+c)^{2+\alpha}} = \frac{c^*}{(\alpha+c^*)^{2+\alpha}}$. A phase transition window is proved to be contained, essentially, in $[c_\alpha - An^{-1/3}, c_\alpha + Bn^{-1/4}]$, and he conjectured that $1/4$ may be replaced with $1/3$. For the multigraph version, MG_m, he showed that MG_m is connected w.h.p. if and only if $m \gg m_n := n^{1+\alpha^{-1}}$. He conjectured that, for $\alpha > 1$, m_n is the threshold for connectedness of G_m itself.

18
Weighted Graphs

There are many cases in which we put weights $X_e, e \in E$ on the edges of a graph or digraph and ask for the minimum or maximum weight object. The optimization questions that arise from this are the backbone of Combinatorial Optimization. When the X_e are random variables we can ask for properties of the optimum value, which will be a random variable. In this chapter we consider three of the most basic optimization problems viz. minimum weight spanning trees, shortest paths and minimum weight matchings in bipartite graphs.

18.1 Minimum Spanning Tree

Let $X_e, e \in E(K_n)$ be a collection of independent uniform $[0,1]$ random variables. Consider X_e to be the length of edge e and let L_n be the length of the minimum spanning tree (MST) of K_n with these edge lengths.

Frieze [320] proved the following theorem. The proof we give uses the rather lovely integral formula (18.1) due to Janson [421], (see also the related equation (7) from [341]).

Theorem 18.1

$$\lim_{n\to\infty} \mathbb{E} L_n = \zeta(3) = \sum_{k=1}^{\infty} \frac{1}{k^3} = 1.202\cdots$$

Proof Suppose that $T = T(\{X_e\})$ is the MST, unique with probability one. We use the identity

$$a = \int_0^1 1_{\{x \le a\}}\, dx.$$

Therefore,

$$\begin{aligned}
L_n &= \sum_{e\in T} X_e \\
&= \sum_{e\in T} \int_{p=0}^{1} 1_{\{p\le X_e\}} dp \\
&= \int_{p=0}^{1} \sum_{e\in T} 1_{\{p\le X_e\}} dp \\
&= \int_{p=0}^{1} |\{e\in T : X_e \ge p\}| dp \\
&= \int_{p=0}^{1} (\kappa(G_p) - 1) dp,
\end{aligned}$$

where $\kappa(\mathbb{G}_p)$ denote the number of components of graph $\mathbb{G}_p$. Here $\mathbb{G}_p$ is the graph induced by the edges e with $X_e \le p$, i.e. $\mathbb{G}_p \equiv \mathbb{G}_{n,p}$. The last line may be considered to be a consequence of the fact that the greedy algorithm solves the minimum spanning tree problem. This algorithm examines edges in increasing order of edge weight. It builds a tree, adding one edge at a time. It adds the edge to the forest F of edges accepted so far, only if the two endpoints lie in distinct components of F. Otherwise it moves onto the next edge. Thus the number of edges to be added given F, is $\kappa(F) - 1$ and if the longest edge in $e \in F$ has $X_e = p$ then $\kappa(F) = \kappa(G_p)$, which follows by an easy induction. Hence,

$$\mathbb{E} L_n = \int_{p=0}^{1} (\mathbb{E}\kappa(\mathbb{G}_p) - 1) dp. \tag{18.1}$$

We therefore estimate $\mathbb{E}\kappa(\mathbb{G}_p)$. We observe first that

$$p \ge \frac{6\log n}{n} \Rightarrow \mathbb{E}\kappa(G_p) = 1 + o(1).$$

Indeed, $1 \le \mathbb{E}\kappa(G_p)$ and

$$\begin{aligned}
\mathbb{E}\kappa(G_p) &\le 1 + n\mathbb{P}(G_p \text{ is not connected}) \\
&\le 1 + n\sum_{k=1}^{n/2} \binom{n}{k} k^{k-2} p^{k-1} (1-p)^{k(n-k)} \\
&\le 1 + \frac{n}{p} \sum_{k=1}^{n/2} \left(\frac{ne}{k} \frac{6k\log n}{n} \frac{1}{n^3} \right)^k \\
&= 1 + o(1).
\end{aligned}$$

Hence, if $p_0 = \frac{6\log n}{n}$ then

$$\mathbb{E} L_n = \int_{p=0}^{p_0} (\mathbb{E}\kappa(G_p) - 1)dp + o(1)$$
$$= \int_{p=0}^{p_0} \mathbb{E}\kappa(G_p)dp + o(1).$$

Write

$$\kappa(G_p) = \sum_{k=1}^{(\log n)^2} A_k + \sum_{k=1}^{(\log n)^2} B_k + C,$$

where A_k stands for the number of components that are k vertex trees, B_k is the number of k vertex components that are not trees and C denotes the number of components on at least $(\log n)^2$ vertices. Then, for $1 \leq k \leq (\log n)^2$ and $p \leq p_0$,

$$\mathbb{E} A_k = \binom{n}{k} k^{k-2} p^{k-1} (1-p)^{k(n-k)+\binom{k}{2}-k+1}$$
$$= (1+o(1)) n^k \frac{k^{k-2}}{k!} p^{k-1} (1-p)^{kn}.$$

$$\mathbb{E} B_k \leq \binom{n}{k} k^{k-2} \binom{k}{2} p^k (1-p)^{k(n-k)}$$
$$\leq (1+o(1))(npe^{1-np})^k$$
$$\leq 1+o(1).$$

$$C \leq \frac{n}{(\log n)^2}.$$

Hence,

$$\int_{p=0}^{\frac{6\log n}{n}} \sum_{k=1}^{(\log n)^2} \mathbb{E} B_k dp \leq \frac{6\log n}{n} (\log n)^2 (1+o(1)) = o(1),$$

and

$$\int_{p=0}^{\frac{6\log n}{n}} C dp \leq \frac{6\log n}{n} \frac{n}{(\log n)^2} = o(1).$$

So

$$\mathbb{E}L_n = o(1) + (1+o(1)) \sum_{k=1}^{(\log n)^2} n^k \frac{k^{k-2}}{k!} \int_{p=0}^{\frac{6\log n}{n}} p^{k-1}(1-p)^{kn} dp.$$

But

$$\begin{aligned}
&\sum_{k=1}^{(\log n)^2} n^k \frac{k^{k-2}}{k!} \int_{p=\frac{6\log n}{n}}^{1} p^{k-1}(1-p)^{kn} dp \\
&\quad \leq \sum_{k=1}^{(\log n)^2} n^k \frac{k^{k-2}}{k!} \int_{p=\frac{6\log n}{n}}^{1} n^{-6k} dp \\
&\quad = o(1).
\end{aligned}$$

Therefore,

$$\begin{aligned}
\mathbb{E}L_n &= o(1) + (1+o(1)) \sum_{k=1}^{(\log n)^2} n^k \frac{k^{k-2}}{k!} \int_{p=0}^{1} p^{k-1}(1-p)^{kn} dp \\
&= o(1) + (1+o(1)) \sum_{k=1}^{(\log n)^2} n^k \frac{k^{k-2}}{k!} \frac{(k-1)!(kn)!}{(k(n+1))!} \\
&= o(1) + (1+o(1)) \sum_{k=1}^{(\log n)^2} n^k k^{k-3} \prod_{i=1}^{k} \frac{1}{kn+i} \\
&= o(1) + (1+o(1)) \sum_{k=1}^{(\log n)^2} \frac{1}{k^3} \\
&= o(1) + (1+o(1)) \sum_{k=1}^{\infty} \frac{1}{k^3}.
\end{aligned}$$

□

One can obtain the same result if the uniform $[0,1]$ random variable is replaced by any random non-negative random variable with distribution F having a derivative equal to one at the origin, e.g. an exponential variable with mean one, see Steele [691].

18.2 Shortest Paths

Let the edges of the complete graph K_n on $[n]$ be given independent lengths X_e, $e \in [n]^2$. Here X_e is exponentially distributed with mean 1. The following theorem was proved by Janson [423]:

Theorem 18.2 *Let X_{ij} be the distance from vertex i to vertex j in the complete graph with edge weights independent $EXP(1)$ random variables. Then, for every $\varepsilon > 0$, as $n \to \infty$,*

(i) For any fixed i,j,

$$\mathbb{P}\left(\left|\frac{X_{ij}}{\log n/n} - 1\right| \geq \varepsilon\right) \to 0.$$

(ii) For any fixed i,

$$\mathbb{P}\left(\left|\frac{\max_j X_{ij}}{\log n/n} - 2\right| \geq \varepsilon\right) \to 0.$$

(iii)

$$\mathbb{P}\left(\left|\frac{\max_{i,j} X_{ij}}{\log n/n} - 3\right| \geq \varepsilon\right) \to 0.$$

Proof We will prove statements (i) and (ii), only. First, recall the following two properties of the exponential X:

(P1) $\mathbb{P}(X > \alpha + \beta | X > \alpha) = \mathbb{P}(X > \beta)$.
(P2) If $X_1, X_2, \ldots, X_m$ are independent $EXP(1)$ exponential random variables then $\min\{X_1, X_2, \ldots, X_m\}$ is an exponential with mean $1/m$.

Suppose that we want to find shortest paths from a vertex s to all other vertices in a digraph with non-negative arc-lengths. Recall Dijkstra's algorithm. After several iterations there is a rooted tree T such that if v is a vertex of T then the tree path from s to v is a shortest path. Let $d(v)$ be its length. For $x \notin T$ let $d(x)$ be the minimum length of a path P that goes from s to v to x, where $v \in T$ and the sub-path of P that goes to v is the tree path from s to v. If $d(y) = \min\{d(x) : x \notin T\}$ then $d(y)$ is the length of a shortest path from s to y and y can be added to the tree.

Suppose that vertices are added to the tree in the order $v_1, v_2, \ldots, v_n$ and that $Y_j = dist(v_1, v_j)$ for $j = 1,2,\ldots,n$. It follows from property P1 that

$$Y_{k+1} = \min_{\substack{i=1,2,\ldots,k \\ v \neq v_1,\ldots,v_k}} [Y_i + X_{v_i,v}] = Y_k + E_k$$

where E_k is exponential with mean $\frac{1}{k(n-k)}$ and is independent of Y_k.

This is because X_{v_i,v_j} is distributed as an independent exponential X conditioned on $X \geq Y_k - Y_i$. Hence,

$$\begin{aligned}
\mathbb{E}\, Y_n &= \sum_{k=1}^{n-1} \frac{1}{k(n-k)} \\
&= \frac{1}{n} \sum_{k=1}^{n-1} \left(\frac{1}{k} + \frac{1}{n-k} \right) \\
&= \frac{2}{n} \sum_{k=1}^{n-1} \frac{1}{k} \\
&= \frac{2 \log n}{n} + O(n^{-1}).
\end{aligned}$$

Also, from the independence of E_k, Y_k,

$$\begin{aligned}
\operatorname{Var} Y_n &= \sum_{k=1}^{n-1} \operatorname{Var} E_k \\
&= \sum_{k=1}^{n-1} \left(\frac{1}{k(n-k)} \right)^2 \\
&\leq 2 \sum_{k=1}^{n/2} \left(\frac{1}{k(n-k)} \right)^2 \\
&\leq \frac{8}{n^2} \sum_{k=1}^{n/2} \frac{1}{k^2} \\
&= O(n^{-2})
\end{aligned}$$

and we can use the Chebyshev inequality Lemma 20.3 to prove (ii).

Now fix $j = 2$. Then if i is defined by $v_i = 2$, we see that i is uniform over $\{2,3,\dots,n\}$. So

$$\begin{aligned}
\mathbb{E} X_{1,2} &= \frac{1}{n-1}\sum_{i=2}^{n}\sum_{k=1}^{i-1}\frac{1}{k(n-k)} \\
&= \frac{1}{n-1}\sum_{k=1}^{n-1}\frac{n-k}{k(n-k)} \\
&= \frac{1}{n-1}\sum_{k=1}^{n-1}\frac{1}{k} \\
&= \frac{\log n}{n} + O(n^{-1}).
\end{aligned}$$

For the variance of $X_{1,2}$ we have

$$X_{1,2} = \delta_2 Y_2 + \delta_3 Y_3 + \cdots + \delta_n Y_n,$$

where

$$\delta_i \in \{0,1\}; \quad \delta_2 + \delta_3 + \cdots + \delta_n = 1; \quad \mathbb{P}(\delta_i = 1) = \frac{1}{n-1}.$$

$$\begin{aligned}
\operatorname{Var} X_{1,2} &= \sum_{i=2}^{n} \operatorname{Var}(\delta_i Y_i) + \sum_{i \neq j} \operatorname{Cov}(\delta_i Y_i, \delta_j Y_j) \\
&\le \sum_{i=2}^{n} \operatorname{Var}(\delta_i Y_i).
\end{aligned}$$

The last inequality holds since

$$\operatorname{Cov}(\delta_i Y_i, \delta_j Y_j) = \mathbb{E}(\delta_i Y_i \delta_j Y_j) - \mathbb{E}(\delta_i Y_i)\,\mathbb{E}(\delta_j Y_j) = -\mathbb{E}(\delta_i Y_i)\,\mathbb{E}(\delta_j Y_j) \le 0.$$

So

$$\begin{aligned}
\operatorname{Var} X_{1,2} &\le \sum_{i=2}^{n} \operatorname{Var}(\delta_i Y_i) \\
&\le \sum_{i=2}^{n} \frac{1}{n-1}\sum_{k=1}^{i-1}\left(\frac{1}{k(n-k)}\right)^2 \\
&= O(n^{-2}).
\end{aligned}$$

We can now use the Chebyshev inequality. □

We can, as for spanning trees, replace the exponential random variables by random variables that behave like the exponential close to the origin. The paper of Janson [423] allows for any random variable X satisfying $\mathbb{P}(X \leq t) = t + o(t)$ as $t \to 0$.

18.3 Minimum Weight Assignment

Consider the complete bipartite graph $K_{n,n}$ and suppose that its edges are assigned independent exponentially distributed weights, with rate 1. (The rate of an exponential variable is one over its mean.) Denote the minimum total weight of a perfect matching in $K_{n,n}$ by C_n. Aldous [13], [16] proved that $\lim_{n\to\infty} \mathbb{E}\, C_n = \zeta(2) = \sum_{k=1}^{\infty} \frac{1}{k^2}$. The following theorem was conjectured by Parisi [616]. It was proved independently by Linusson and Wästlund [531] and Nair, Prabhakar and Sharma [600]. The proof given here is from Wästlund [720].

Theorem 18.3

$$\mathbb{E}\, C_n = \sum_{k=1}^{n} \frac{1}{k^2} = 1 + \frac{1}{4} + \frac{1}{9} + \frac{1}{16} + \cdots + \frac{1}{n^2}. \tag{18.2}$$

From the above theorem we immediately get the following corollary, first proved by Aldous [16].

Corollary 18.4

$$\lim_{n\to\infty} \mathbb{E}\, C_n = \zeta(2) = \sum_{k=1}^{\infty} \frac{1}{k^2} = \frac{\pi^2}{6} = 1.6449\cdots$$

Let us introduce a more general model of minimum weight assignment. Consider an m by n complete bipartite graph $K_{m,n}$, with bipartition (A,B), where $A = \{a_1, a_2, \ldots, a_m\}$ and $B = \{b_1, b_2, \ldots, b_n\}$, and with edge weights that are independent and exponentially distributed, with rate 1. A k-assignment is defined to be a set of k independent edges, i.e. a set of k edges, no two of them having end vertices in common. The weight of an assignment is the sum of the weights of its edges. Let $C_{k,m,n}$ denote the minimum weight of a k-assignment in such an edge weighted $K_{m,n}$. Then, as conjectured by Coppersmith and Sorkin [226], the following holds:

Theorem 18.5

$$\mathbb{E}\, C_{k,m,n} = \sum_{\substack{i,j \geq 0 \\ i+j<k}} \frac{1}{(m-i)(n-j)}. \tag{18.3}$$

We first prove the above theorem and then show that Equation (18.3) reduces to Equation (18.2) for $k = m = n$.

Proof (of Theorem 18.5).

In order to establish (18.3) inductively it suffices to show that

$$\mathbb{E}\,C_{k,m,n} - \mathbb{E}\,C_{k-1,m,n-1} = \frac{1}{mn} + \frac{1}{(m-1)n} + \cdots + \frac{1}{(m-k+1)n}. \tag{18.4}$$

For then we obtain (18.3) by summing $\mathbb{E}\,C_{\ell,m,n-k+\ell} - \mathbb{E}\,C_{\ell-1,m,n-k+\ell-1}$ for $\ell = 1,2,\ldots,k$.

Let σ_r, $0 \le r < k$, be the minimum weight r-assignment in $K_{m,n}$. First notice that since the edge weights of $K_{m,n}$ are exponentially distributed, we have that with probability 1, no two disjoint sets of edges have the same total weight. It implies that every vertex v that *participates* in σ_r (i.e. is incident to an edge from σ_r and denoted $v \in \sigma_r$) also participates in σ_{r+1}. Briefly,

$$v \in \sigma_r \Rightarrow v \in \sigma_{r+1}. \tag{18.5}$$

To see this, let H be a subgraph of $K_{m,n}$ induced by the symmetric difference $\sigma_r \,\triangle\, \sigma_{r+1}$, i.e. by those edges of minimum r and $r+1$ assignments that do not belong to both of them. Observe that each vertex of H has degree at most two and so H consists of vertex disjoint paths and cycles. We claim that with probability one, H is in fact a single path. If this is not the case then there would exist a subgraph H_1 of H, being either a cycle or two paths, containing the same number of edges from σ_r as from σ_{r+1}. With probability one the edge sets in $H_1 \cap \sigma_r$ and $H_1 \cap \sigma_{r+1}$ have different total weight. But then either $H_1 \,\triangle\, \sigma_r$ has smaller weight than σ_r, or $H_1 \,\triangle\, \sigma_{r+1}$ has smaller weight than σ_{r+1}, a contradiction.

To proceed with the proof, let us introduce a slightly modified model. Namely, consider a weighted complete graph $K_{m+1,n}$, which is obtained from $K_{m,n}$ by adding a new vertex a_{m+1} to A and connecting it with all n vertices of B by edges with independent, exponentially distributed weights with rate $\lambda > 0$ (recall that all weights of $K_{m,n}$ are exponential with rate one).

We will show that, in $K_{m+1,n}$, as $\lambda \to 0$,

$$\mathbb{P}(a_{m+1} \in \sigma_k) = \left(\frac{1}{m} + \frac{1}{m-1} + \cdots + \frac{1}{m-k+1}\right)\lambda + O(\lambda^2). \tag{18.6}$$

We will show that (18.6) is an immediate consequence of the fact that

$$\mathbb{P}(a_{m+1} \in \sigma_{r+1} | a_{m+1} \notin \sigma_r) = \frac{\lambda}{m-r+\lambda}. \tag{18.7}$$

Indeed, (18.7) implies that

$$\mathbb{P}(a_{m+1} \in \sigma_k) = 1 - \frac{m}{m+\lambda} \cdot \frac{m-1}{m-1+\lambda} \cdots \frac{m-k+1}{m-k+1+\lambda}$$
$$= 1 - \left(1 + \frac{\lambda}{m}\right)^{-1} \cdots \left(1 + \frac{\lambda}{m-k+1}\right)^{-1}$$
$$= \left(\frac{1}{m} + \frac{1}{m-1} + \cdots + \frac{1}{m-k+1}\right)\lambda + O(\lambda^2),$$

as $\lambda \to 0$, and (18.6) follows.

To see that (18.7) holds, suppose, without loss of generality that the vertices of A participating in σ_r are $a_1, a_2, \ldots, a_r$. Let K' be a multi-graph obtained from $K_{m+1,n}$ by identifying vertices $a_{r+1}, \ldots, a_m, a_{m+1}$ to a single "super-vertex" a'_{r+1}, of degree $(m-r+1)n$, where $a'_{r+1,j} = \min\left\{a_{r+1,j}, \ldots, a_{m+1,j}\right\}$.

Now, by (18.5), σ_{r+1} cannot contain two edges incident with $a_{r+1}, \ldots, a_{m+1}$. Hence, once we know the edge incident with a'_{r+1} that belongs to σ'_{r+1} in K', we know that it corresponds to a unique edge emanating from one of vertices from $a_{r+1}, \ldots, a_{m+1}$ that belongs to σ_{r+1}.

By conditioning on the values $a'_{i,j}$ we see that the probability that a vertex from the set a_{m+1} participates in σ_{r+1} is $\lambda/(m-r+\lambda)$. (We use the fact that if E_1, E_2 are exponentials with rates λ_1, λ_2 respectively, then $\mathbb{P}(E_1 \leq E_2) = \frac{\lambda_1}{\lambda_1+\lambda_2}$, even if $\min\{E_1, E_2\}$ is given.)

Now let W denote the cost of the edge (a_{m+1}, b_n), and let $X = C_{k,m,n}$ of $K_{m,n}$, while $Y = C_{k-1,m,n-1}$ of $K_{m,n-1}$, with bipartition $(A, B \setminus \{b_n\})$. Let I be the indicator variable for the event that the cost of the cheapest k-assignment that contains (a_{m+1}, b_n) is smaller than the cost of the cheapest k-assignment that does not use a_{m+1}. In other words, I is the indicator variable for the event that $Y + W < X$.

Lemma 18.6 *As $\lambda \to 0$, we have*

$$\mathbb{E} I = \left(\frac{1}{mn} + \frac{1}{(m-1)n} + \cdots + \frac{1}{(m-k+1)n}\right)\lambda + O(\lambda^2).$$

Proof Observe first that if $(a_{m+1}, b_n) \in \sigma_k$ then we have $Y + W < X$. Conversely, if $Y + W < X$ and no other edge incident with a_{m+1} has weight smaller than X then $(a_{m+1}, b_n) \in \sigma_k$. But, when $\lambda \to 0$, the probability that there are two distinct edges incident to a_{m+1} of weight smaller than X is of order $O(\lambda^2)$ (the hidden constant depends on n here). Hence,

$$\mathbb{E} I = \mathbb{P}((a_{m+1}, b_n) \in \sigma_k) + O(\lambda^2) = \frac{1}{n}\mathbb{P}(a_{m+1} \in \sigma_k) + O(\lambda^2), \tag{18.8}$$

where $\mathbb{P}(a_{m+1} \in \sigma_k)$ is given by (18.6). □

Since W is exponentially distributed with rate λ, we have

$$\mathbb{E} I = \mathbb{P}(W < X - Y) = \mathbb{E}\left(1 - e^{-\lambda(X-Y)}\right) = 1 - \mathbb{E}\left(e^{-\lambda(X-Y)}\right). \tag{18.9}$$

Hence,

$$\mathbb{E}(X - Y) = \frac{d}{d\lambda}\left[-\mathbb{E}\left(e^{-\lambda(X-Y)}\right)\right]_{\lambda=0} = \frac{d}{d\lambda}(\mathbb{E} I - 1)_{\lambda=0} = \frac{d}{d\lambda}\mathbb{E} I|_{\lambda=0}.$$

It follows from Lemma 18.6 that

$$\begin{aligned}\mathbb{E}(X - Y) &= \frac{d}{d\lambda}\left[\frac{1}{n}\left(\frac{1}{m} + \frac{1}{m-1} + \cdots + \frac{1}{m-k+1}\right)\lambda + O(\lambda^2)\right]_{\lambda=0} \\ &= \frac{1}{mn} + \frac{1}{(m-1)n} + \cdots + \frac{1}{(m-k+1)n},\end{aligned}$$

arriving at the formula (18.4) and so proving Theorem 18.5. □

Now we will show that with $k = m = n$ the statement of Theorem 18.5 reduces to the statement of Theorem 18.3. So, let

$$\begin{aligned}S_n = \mathbb{E} C_{n,n,n} &= \sum_{\substack{i,j\geq 0 \\ i+j<n}} \frac{1}{(n-i)(n-j)} \\ &= \sum_{\substack{i',j'\geq 1 \\ i'+j'<n+2}} \frac{1}{(n+1-i')(n+1-j')}.\end{aligned}$$

It follows that

$$\begin{aligned}S_{n+1} - S_n &= \sum_{\substack{i=0,j\leq n \\ or \\ i\leq n,j=0}} \frac{1}{(n+1-i)(n+1-j)} - \sum_{\substack{i,j\geq 1 \\ i+j=n+1}} \frac{1}{(n+1-i)(n+1-j)} \\ &= \frac{1}{(n+1)^2} + \sum_{i=1}^{n} \frac{2}{(n+1)(n+1-i)} - \sum_{i=1}^{n} \frac{1}{i(n+1-i)} \\ &= \frac{1}{(n+1)^2},\end{aligned} \tag{18.10}$$

since

$$\frac{1}{i(n+1-i)} = \frac{1}{n+1}\left(\frac{1}{i} + \frac{1}{n+1-i}\right).$$

$S_1 = 1$ and so (18.2) follows from (18.10). □

18.4 Exercises

18.4.1 Suppose that the edges of the complete bipartite graph $K_{n,n}$ are given independent uniform $[0,1]$ edge weights. Show that if $L_n^{(b)}$ is the length of the minimum spanning tree, then

$$\lim_{n\to\infty} \mathbb{E} L_n^{(b)} = 2\zeta(3).$$

18.4.2 Tighten Theorem 18.1 and prove that

$$\mathbb{E} L_n = \zeta(3) + O\left(\frac{1}{n}\right).$$

18.4.3 Suppose that the edges of K_n are given independent uniform $[0,1]$ edge weights. Let Z_k denote the minimum total edge cost of the union of k edge-disjoint spanning trees. Show that $\lim_{k\to\infty} Z_k/k^2 = 1$.

18.4.4 Suppose that the edges of $G_{n,p}$ where $0 < p \leq 1$ is a constant, are given exponentially distributed weights with rate 1. Show that if X_{ij} is the shortest distance from i to j then

(i) For any fixed i,j,

$$\mathbb{P}\left(\left|\frac{X_{ij}}{\log n/n} - \frac{1}{p}\right| \geq \varepsilon\right) \to 0.$$

(ii)

$$\mathbb{P}\left(\left|\frac{\max_j X_{ij}}{\log n/n} - \frac{2}{p}\right| \geq \varepsilon\right) \to 0.$$

18.4.5 The quadratic assignment problem is to

Minimize

$$Z = \sum_{i,j,p,q=1}^{n} a_{ijpq} x_{ip} x_{jq}$$

Subject to

$$\sum_{i=1}^{n} x_{ip} = 1 \qquad p = 1,2,\ldots,n$$

$$\sum_{p=1}^{n} x_{ip} = 1 \qquad i = 1,2,\ldots,n$$

$$x_{ip} = 0/1.$$

Suppose now that the a_{ijpq} are independent uniform $[0,1]$ random variables. Show that w.h.p. $Z_{\min} \approx Z_{\max}$ where $Z_{\min}$ (resp. $Z_{\max}$) denotes the minimum (resp. maximum) value of Z, subject to the assignment constraints.

18.4.6 The 0/1 knapsack problem is to

Maximize

$$Z = \sum_{i=1}^{n} a_i x_i$$

Subject to

$$\sum_{i=1}^{n} b_i x_i \leq L$$

$$x_i = 0/1 \quad \text{for } i = 1, 2, \ldots, n.$$

Suppose that the (a_i, b_i) are chosen independently and uniformly from $[0,1]^2$ and that $L = \alpha n$. Show that w.h.p. the maximum value of Z, $Z_{\max}$, satisfies

$$Z_{\max} \approx \begin{cases} \frac{\alpha^{1/2} n}{2} & \alpha \leq \frac{1}{4}. \\ \frac{(8\alpha - 8\alpha^2 - 1)n}{2} & \frac{1}{4} \leq \alpha \leq \frac{1}{2}. \\ \frac{n}{2} & \alpha \geq \frac{1}{2} \end{cases}$$

18.4.7 Suppose that $X_1, X_2, \ldots, X_n$ are points chosen independently and uniformly at random from $[0,1]^2$. Let Z_n denote the total Euclidean length of the shortest tour (Hamilton cycle) through each point. Show that there exist constants c_1, c_2 such that $c_1 n^{1/2} \leq Z_n \leq c_2 n^{1/2}$ w.h.p.

18.4.8 Prove Equation (18.11) below.

18.4.9 Prove Equation (18.12) below.

18.5 Notes

Shortest paths

There have been some strengthenings and generalizations of Theorem 18.2. For example, Bhamidi and van der Hofstad [86] found the (random) second-order term in (i), i.e. convergence in distribution with the correct norming. They also studied the number of edges in the shortest path.

Spanning trees

Beveridge, Frieze and McDiarmid [85] considered the length of the minimum spanning tree in regular graphs other than complete graphs. For graphs G of large degree r they proved that the length $MST(G)$ of an n-vertex randomly edge weighted graph G satisfies $MST(G) = \frac{n}{r}(\zeta(3) + o_r(1))$ w.h.p., provided some mild expansion condition holds. For r regular graphs of large girth g they proved that if

$$c_r = \frac{r}{(r-1)^2} \sum_{k=1}^{\infty} \frac{1}{k(k+\rho)(k+2\rho)},$$

then w.h.p. $|MST(G) - c_r n| \leq \frac{3n}{2g}$.

Frieze, Ruszinko and Thoma [348] replaced the expansion in [85] by connectivity and in addition proved that $MST(G) \leq \frac{n}{r}(\zeta(3) + 1 + o_r(1))$ for any r-regular graph.

Cooper, Frieze, Ince, Janson and Spencer [217] showed that Theorem 18.1 can be improved to yield $\mathbb{E} L_n = \zeta(3) + \frac{c_1}{n} + \frac{c_2+o(1)}{n^{4/3}}$ for explicit constants c_1, c_2.

Bollobás, Gamarnik, Riordan and Sudakov [142] considered the Steiner Tree problem on K_n with independent random edge weights, $X_e, e \in E(K_n)$. Here they assumed that the X_e have the same distribution $X \geq 0$, where $\mathbb{P}(X \leq x) = x + o(x)$ as $x \to 0$. The main result is that if one fixes $k = o(n)$ vertices then w.h.p. the minimum length W of a sub-tree of K_n that includes these k points satisfies $W \approx \frac{k-1}{n} \log \frac{n}{k}$.

Angel, Flaxman and Wilson [36] considered the minimum length of a spanning tree of K_n that has a fixed root and bounded depth k. The edge weights X_e are independent exponential mean one. They proved that if $k \geq \log_2 \log n + \omega(1)$ then w.h.p. the minimum length tends to $\zeta(3)$ as in the unbounded case. On the other hand, if $k \leq \log_2 \log n - \omega(1)$ then w.h.p. the weight is doubly exponential in $\log_2 \log n - k$. They also considered bounded depth Steiner trees.

Using Talagrand's inequality, McDiarmid [563] proved that for any real $t > 0$ we have $\mathbb{P}(|L_n - \zeta(3)| \geq t) \leq e^{-\delta_1 n}$ where $\delta_1 = \delta_2(t)$. Flaxman [302] proved that $\mathbb{P}(|L_n - \zeta(3)| \leq \varepsilon) \geq e^{-\delta_2 n}$ where $\delta_1 = \delta_2(\varepsilon)$.

Assignment problem

Walkup [719] proved that $\mathbb{E} C_n \leq 3$ (see (18.3)) and later Karp [473] proved that $\mathbb{E} C_n \leq 2$. Dyer, Frieze and McDiarmid [268] adapted Karp's proof to something more general: let Z be the optimum value to the linear program:

$$\text{Minimise} \sum_{j=1}^{n} c_j x_j, \text{subject to } x \in P = \{x \in \mathbb{R}^n : Ax = b, x \geq 0\},$$

where A is an $m \times n$ matrix. As a special case of [268], we have that if $c_1, c_2, \ldots, c_n$ are independent uniform $[0,1]$ random variables and x^* is any member of P, then $\mathbb{E}(Z) \leq m(\max_j x_j^*)$. Karp's result can easily be deduced from this.

The assignment problem can be generalized to multi-dimensional versions: we replace the complete bipartite graph $K_{n,n}$ by the complete k-partite hypergraph $K_n^{(k)}$ with vertex partition $V = V_1 \sqcup V_2 \sqcup \cdots \sqcup V_k$ where each V_i is of size n. We give each edge of $K_n^{(k)}$ an independent exponential mean one value. Assume for example that $k = 3$. In one version of the 3-dimensional assignment problem we ask for a minimum weight collection of hyperedges such that each vertex $v \in V$ appears in exactly one edge. The optimal total weight Z of this collection satisfies

$$\Omega\left(\frac{1}{n}\right) \leq Z \leq O\left(\frac{\log n}{n}\right) \; w.h.p. \tag{18.11}$$

(The upper bound uses the result of [451], see Section 13.2.)

Frieze and Sorkin [349] give an $O(n^3)$ algorithm that w.h.p. finds a solution of value $\frac{1}{n^{1-o(1)}}$.

In another version of the 3-dimensional assignment problem we ask for a minimum weight collection of hyperedges such that each pair of vertices $v, w \in V$ from different sets in the partition appear in exactly one edge. The optimal total weight Z of this collection satisfies

$$\Omega(n) \leq Z \leq O(n \log n) \; w.h.p. \tag{18.12}$$

The upper bound is also from [349] and uses the result of [268] to greedily solve a sequence of restricted assignment problems.

19

Brief Notes on Uncovered Topics

There are several topics that we have not been able to cover and that might be of interest to the reader. For these topics, we provide some short synopses and some references that the reader may find useful.

19.1 Contiguity

Suppose that we have two sequences of probability models on graphs $\mathscr{G}_{1,n}, \mathscr{G}_{2,n}$ on the set of graphs with vertex set $[n]$. We say that the two sequences are *contiguous* if for any sequence of events $\mathscr{A}_n$ we have

$$\lim_{n\to\infty} \mathbb{P}(\mathscr{G}_{1,n} \in \mathscr{A}_n) = 0 \Leftrightarrow \lim_{n\to\infty} \mathbb{P}(\mathscr{G}_{2,n} \in \mathscr{A}_n) = 0.$$

This is useful for us if, for example, we want to see what happens w.h.p. in the model $\mathscr{G}_{1,n}$, but find it easier to work with $\mathscr{G}_{2,n}$. In this context, $\mathbb{G}_{n,p}$ and $\mathbb{G}_{n,m=n^2p/2}$ are almost contiguous.

Interest in this notion in random graphs was stimulated by the results of Robinson and Wormald [650], [651] that random r-regular graphs, $r \geq 3, r = O(1)$ are Hamiltonian. As a result, we find that other non-uniform models of random regular graphs are contiguous to $G_{n,r}$, e.g. the union rM_n of r random perfect matchings when n is even. (There is an implicit conditioning on rM_n being simple here.) The most general result in this line is given by Wormald [728], improving on earlier results of Janson [422] and Molloy, Robalewska, Robinson and Wormald [584] and Kim and Wormald [484]. Suppose that $r = 2j + \sum_{i=1}^{r-1} ik_i$, with all terms non-negative. Then $\mathbb{G}_{n,r}$ is contiguous to the sum $jH_n + \sum_{i=1}^{r-1} k_i\mathbb{G}_{n,i}$, where n is restricted to even integers if $k_i \neq 0$ for any odd i. Here jH_n is the union of j edge disjoint Hamilton cycles etc.

Chapter 8 of [432] is devoted to this subject.

19.2 Edge Colored Random Graphs

Suppose that we color the edges of a graph G. A set of edges S is said to be *rainbow* colored if each edge of S has a different color. Consider first the existence of a rainbow spanning tree. We consider the graph process where the edges are randomly colored using $q \geq n-1$ colors. Let τ_a be the hitting time for $n-1$ colors to appear in the process and let τ_c be the hitting time for connectivity and let $\tau^* = \max\{\tau_a, \tau_c\}$. Frieze and McKay [342] showed that w.h.p. G_{τ^*} contains a rainbow spanning tree. This is clearly best possible. Bal, Bennett, Frieze and Pralat [46] considered the case where each edge has a choice of k random colors. This reduces τ_a, but the result still holds.

The existence of rainbow Hamilton cycles is different. The existence of a rainbow spanning tree can be checked in polynomial time and this leads to a simple criterion for non-existence. This is clearly not likely for Hamilton cycles. Cooper and Frieze [207] proved that if $m \geq Kn\log n$ and $q \geq Kn$ then w.h.p. $\mathbb{G}_{n,m}$ contains a rainbow Hamilton cycle. This was improved to $m \geq \frac{1+o(1)}{2} n\log n$ and $q \geq (1+o(1))n$ by Frieze and Loh [337]. Bal and Frieze [47] showed that if $m \geq Kn\log n$ and $q = n$ and n is even there is a rainbow Hamilton cycle w.h.p. Ferber [292] removed the requirement that n be even. Bal and Frieze also considered rainbow perfect matchings in k-uniform hypergraphs. Janson and Wormald [435] considered random colorings of r-regular graphs. They proved that if $r \geq 4, r = O(1)$ and the edges of $G_{n,2r}$ are randomly colored so that each color is used r times, then w.h.p. there is a rainbow Hamilton cycle. Ferber, Kronenberg, Mousset and Shikhelman [295] give results on packing rainbow structures such as Hamilton cycles. Ferber, Nenadov and Peter [296] proved that if $p \gg n^{-1/d}(\log n)^{1/d}$ and H is a fixed graph of density at most d then w.h.p. $\mathbb{G}_{n,p}$ contains a rainbow copy of H if it is randomly colored with $(1+\varepsilon)|E(H)|$ colors, for any fixed $\varepsilon > 0$.

Cooper and Frieze [204] found the threshold for the following property: if $k = O(1)$ and $\mathbb{G}_{n,m}$ is arbitrarily edge colored so that no color is used more than k times, then $\mathbb{G}_{n,m}$ contains a rainbow Hamilton cycle.

19.3 Games

Positional games can be considered to be a generalization of the game of Noughts and Crosses or Tic-Tac-Toe. There are two players A (Maker) and B (Breaker) and in the context for this section, the board is a graph G. Each player in turn chooses an edge and at the end of the game, the winner is determined by the partition of the edges claimed by the players. As a typical example, in the connectivity game, player A is trying to ensure that the edges she collects contain a spanning tree of G and player B is trying to prevent this. See Chvátal and Erdős [188] for one of the earliest papers on the subject and books by Beck [64] and Hefetz, Krivelevich, Stojaković and Szabó [403]. Most of the analyses have considered $G = K_n$ and to make the problem interesting [188] introduced the notion of *bias*. Thus in the connectivity game, player B is allowed to collect b edges for each edge of A. Now the question becomes what is the largest value of b for which A has a winning strategy. There is a striking though somewhat mysterious connection between the optimal values of b for various games and thresholds for associated properties in random graphs. For example, in the connectivity game, the threshold bias $b \approx \frac{n}{\log n}$, i.e. player A collects about $\frac{1}{2}n \log n$ edges, see Gebauer and Szabó [361]. Another example is the biased H-game where Maker wins if she can create a copy of some fixed graph H with at least two adjacent edges. The optimal threshold bias b for this game is of order $\Theta\left(n^{1/m_2(H)}\right)$, Bednarska and Łuczak [65]. For a sufficiently small constant $c > 0$, if $b \leq cn^{1/m_2(H)}$, then Maker can create $\Theta(EX_H)$ copies of H in K_n, where X_H is the number of copies of H in $G_{n,1/b}$. Furthermore, if Maker plays randomly, she achieves this goal w.h.p.

Recently Stojaković and Szabó [695] began research on random boards, i.e. where G is a random graph. Ben-Shimon, Ferber, Hefetz and Krivelevich [75] proved a hitting time result for the $b = 1$ Hamilton cycle game on the graph process. Assuming that player A wants to build a Hamilton cycle and player B starts first, player A will have a winning strategy in G_m if and only if $m \geq m_4^*$. This is best possible. Biased Hamiltonicity games on $\mathbb{G}_{n,p}$ were considered in Ferber, Glebov, Krivelevich and Naor [293], where it was shown that for $p \gg \frac{\log n}{n}$, the threshold bias b_{HAM} satisfies $b_{HAM} \approx \frac{np}{\log n}$ w.h.p. The H-game where A wins if she can create a copy of some fixed graph H was first studied by Stojakovic and Szabo [695] for the case where H is a clique on k vertices. This was strengthened by Müller and Stojaković [595]. They showed that if $p \leq cn^{-2/(k+1)}$, then w.h.p. B can win this game. For $p \geq Cn^{-2/(k+1)}$ one can use the results of [653] to argue that A wins w.h.p. This result was generalized to arbitrary graphs H (satisfying certain mild conditions) by Nenadov, Steger and Stojaković [605], where they showed that the threshold is where one would expect it to be – at the 2-density of H. As we have seen there are other models of random graphs and Beveridge, Dudek, Frieze, Müller and Stojaković [84] studied these games on random geometric graphs.

The *game chromatic number* $\chi_g(G)$ of a graph G can be defined as follows. Once again there are two players A and B and they take it in turns to properly color vertices of G with one of q colors. Thus if $\{u, v\}$ is an edge and u is colored with color c and v is uncolored at the start of any turn, then v may not be colored with c by either player. The goal of A is to ensure that the game ends with every vertex colored and the goal of B is to prevent this by using all q colors in the neighborhood of some uncolored vertex. The game chromatic number is the minimum q for which A can win. For a survey on results on this parameter see Bartnicki, Grytczuk, Kierstead and and Zhu [62]. Bohman, Frieze and Sudakov [118] studied χ_g for dense random graphs and proved that for such graphs, χ_g is within a constant factor of the chromatic number. Keusch and Steger [481] proved that this factor is asymptotically equal to two. Frieze, Haber and Lavrov [330] extended the results of [118] to sparse random graphs.

19.4 Graph Searching

Cops and Robbers

A collection of cops are placed on the vertices of a graph by player C and then a robber is placed on a vertex by player R. The players take turns. C can move all cops to a neighboring vertex and R can move the robber. The cop number of a graph is the minimum number of cops needed so that C can win. The basic rule being that if there is a cop occupying the same vertex as the robber, then C wins. Łuczak and Pralat [546] proved a remarkable "zigzag" theorem giving the cop number of a random graph. This number being n^α where $\alpha = \alpha(p)$ follows a saw-toothed curve. Pralat and Wormald [634] proved that the cop number of the random regular graph $\mathbb{G}_{n,r}$ is $O(n^{1/2})$. It is worth noting that Meyniel has conjectured $O(n^{1/2})$ as a bound on the cop number of any connected n-vertex graph. There are many variations on this game and the reader is referred to the monograph by Bonato and Pralat [151].

Graph Cleaning

Initially, every edge and vertex of a graph G is dirty, and a fixed number of brushes start on a set of vertices. At each time-step, a vertex v and all its incident edges that are dirty may be cleaned if there are at least as many brushes on v as there are incident dirty edges. When a vertex is cleaned, every incident dirty edge is traversed (i.e. cleaned) by one and only one brush, and brushes cannot traverse a clean edge. The *brush number* $b(G)$ is the minimum number of brushes needed to clean G. Pralat [632], [633] proved that w.h.p. $b(\mathbb{G}_{n,p}) \approx \frac{1-e^{-2d}}{4} n$ for $p = \frac{d}{n}$ where $d < 1$ and w.h.p. $b(\mathbb{G}_{n,p}) \leq (1+o(1))\left(d+1-\frac{1-e^{-2d}}{2d}\right)\frac{n}{4}$ for $d > 1$. For the random d-regular graph $\mathbb{G}_{n,d}$, Alon, Pralat and Wormald [29] proved that w.h.p. $b(\mathbb{G}_{n,d}) \geq \frac{dn}{4}\left(1-\frac{2^{3/2}}{d^{1/2}}\right)$.

Acquaintance Time

Let $G = (V,E)$ be a finite connected graph. We start the process by placing one agent on each vertex of G. Every pair of agents sharing an edge are declared to be acquainted, and remain so throughout the process. In each round of the process, we choose some matching M in G. The matching M need not be maximal; perhaps it is a single edge. For each edge of M, we swap the agents occupying its endpoints, which may cause more agents to become acquainted.

We may view the process as a graph searching game with one player, where the player's strategy consists of a sequence of matchings that allow all agents to become acquainted. Some strategies may be better than others, which leads to a graph optimization parameter. The acquaintance time of G, denoted by $\mathscr{A}(G)$, is the minimum number of rounds required for all agents to become acquainted with one another. The parameter $\mathscr{A}(G)$ was introduced by Benjamini, Shinkar and Tsur [69], who showed that $\mathscr{A}(G) = O\left(\frac{n^2 \log\log n}{\log n}\right)$ for an n vertex graph. The $\log\log n$ factor was removed by Kinnersley, Mitsche and Pralat [486]. The paper [486] also showed that w.h.p. $\mathscr{A}(\mathbb{G}_{n,p}) = O\left(\frac{\log n}{p}\right)$ for $\frac{(1+\varepsilon)\log n}{n} \leq p \leq 1-\varepsilon$. The lower bound here was relaxed to $np - \log n \to \infty$ in Dudek and Pralat [260]. A lower bound, $\Omega\left(\frac{\log n}{p}\right)$ for $\mathbb{G}_{n,p}$ and $p \geq n^{-1/2+\varepsilon}$ was proved in [486].

19.5 *H*-free process

In an early attempt to estimate the Ramsey number $R(3,t)$, Erdős, Suen and Winkler [283] considered the following process for generating a triangle-free graph. Let $e_1, e_2, \ldots, e_N, N = \binom{n}{2}$ be a random ordering of the complete graph K_n. Let $\mathscr{P}$ be a graph property, e.g. being triangle free. We generate a sequence of random graphs $\Gamma_0, \Gamma_1, \ldots, \Gamma_N$, where $\Gamma_{i+1} = \Gamma_i + e_{i+1}$ if adding e_{i+1} does not destroy $\mathscr{P}$, otherwise $\Gamma_{i+1} = \Gamma_i$. In this way we can generate a random graph that is guaranteed to have property $\mathscr{P}$.

For $\mathscr{P}$ is bipartite they showed in [283] that Γ_N has expected size greater than $(n^2 - n)/4$. When $\mathscr{P}$ is triangle free they showed that w.h.p. that Γ_N has size $\Omega(n^{3/2})$ w.h.p. Bollobás and Riordan [147] studied the general H-free process. More recently, Bohman [110] showed in the case of the triangle-free process, that w.h.p. Γ_N has size $\Theta(n^{3/2}(\log n)^{1/2})$. This provides an alternative proof to that of Kim [482] that $R(3,t) = \Omega\left(\frac{t^2}{\log t}\right)$. He made careful use of the differential equations method, see Chapter 22. Bohman and Keevash [119] and Flz Pontiveros, Griffiths and Morris [301] have improved this result and shown that w.h.p. Γ_N has size asymptotically equal to $\frac{1}{2\sqrt{2}}n^{3/2}(\log n)^{1/2}$. They also showed that the independence number of Γ_N is bounded by $(1+o(1))(2n\log n)^{1/2}$. This shows that $R(3,t) > \left(\frac{1}{4}-o(1)\right)t^2/\log t$.

Bohman, Mubayi and Picolleli [121] considered an r-uniform hypergraph version. In particular they studied the $T^{(r)}$-free process, where $T^{(r)}$ generalizes a triangle in a graph. It consists of $S \cup \{a_i\}, i = 1,2,\ldots,r$, where $|S| = r-1$ and a further edge $\{a_1,a_2,\ldots,a_r\}$. Here hyperedges are randomly added one by one until one is forced to create a copy of T^r. They showed that w.h.p. the final hypergraph produced has independence number $O((n\log n)^{1/r})$. This proves a lower bound of $\Omega\left(\frac{s^r}{\log s}\right)$ for the Ramsey number $R(T^{(r)},K_s^{(r)})$. The analysis is based on a paper on the random greedy hypergraph independent set process by Bennett and Bohman [74].

There has also been work on the related triangle removal process. Here we start with K_n and repeatedly remove a random triangle until the graph is triangle free. The main question is how many edges are there in the final triangle-free graph. A proof of a bound of $O(n^{7/4+o(1)})$ was outlined by Grable [381]. A simple proof of $O(n^{7/4+o(1)})$ was proved in Bohman, Frieze and Lubetzky [114]. Furthermore, Bohman, Frieze and Lubetzky [115] have proved a tight result of $n^{3/2+o(1)}$ for the number of edges left. This is close to the $\Theta(n^{3/2})$ bound conjectured by Bollobás and Erdős in 1990.

An earlier paper by Ruciński and Wormald [660] considered the d-process. Edges were now rejected if they raised the degree of some vertex above d. Answering a question of Erdős, they proved that the resulting graph was w.h.p. d-regular.

19.6 Logic and Random Graphs

The *first order theory of graphs* is a language in which we can describe some, but certainly not all, properties of graphs. It can describe G has a triangle, but not G is connected. Fagin [286] and Glebskii, Kogan, Liagonkii and Talanov [371] proved that for any property $\mathscr{A}$ that can be described by a first-order sentence, $\lim_{n\to\infty}\mathbb{P}(\mathbb{G}_{n,1/2} \in \mathscr{A}) \in \{0,1\}$. We say that $p = 1/2$ obeys a 0-1 law. We do not need to restrict ourselves to $\mathbb{G}_{n,1/2}$. Shelah and Spencer [681] proved that if α is irrational then $p = n^{-\alpha}$ also obeys a 0-1 law. See the book by Spencer [688] for much more on this subject.

19.7 Planarity

We have said very little about random planar graphs. This is partially because there is no simple way of generating a random planar graph. The study begins with the seminal work of Tutte [710], [711] on counting planar maps. The number of rooted maps on surfaces was found by Bender and Canfield [71]. The size of the largest components were studied by Banderier, Flajolet, Schaeffer and Soria [55].

When it comes to random labeled planar graphs, McDiarmid, Steger and Welsh [568] showed that if $pl(n)$ denotes the number of labeled planar graphs with n vertices, then $(pl(n)/n!)^{1/n}$ tends to a limit γ as $n \to \infty$. Osthus, Prömel and Taraz [609] found an upper bound for γ, Bender, Gao and Wormald [73] found a lower bound for γ. Finally, Giménez and Noy [368] proved that $pl(n) \approx cn^{-7/2}\gamma^n n!$ for explicit values of c, γ.

Next let $pl(n,m)$ denote the number of labeled planar graphs with n vertices and m edges. Gerke, Schlatter, Steger and Taraz [364] proved that if $0 \leq a \leq 3$ then $(pl(n,an)/n!)^{1/n}$ tends to a limit γ_a as $n \to \infty$. Giménez and Noy [368] showed that if $1 < a < 3$ then $pl(n,an) \approx c_a n^{-4}\gamma_a^n n!$. Kang and Łuczak [462] proved the existence of two critical ranges for the sizes of complex components.

19.8 Planted Cliques, Cuts and Hamilton cycles

The question here is the following: suppose that we plant an unusual object into a random graph. Can someone else find it? One motivation being that if finding the planted object is hard for someone who does not know where it is planted, then this modified graph can be used as a signature. To make this more precise, consider starting with $\mathbb{G}_{n,1/2}$, choosing an s-subset S of $[n]$ and then making S into a clique. Let the modified graph be denoted by Γ. Here we assume that $s \gg \log n$ so that S should stand out. Can we find S, if we are given Γ, but we are not told S? Kucera [517] proved that if $s \geq C(n \log n)^{1/2}$ for a sufficiently large C then w.h.p. one can find S by looking at vertex degrees. Alon, Krivelevich and Sudakov [26] improved this to $s = \Omega(n^{1/2})$. They show that the second eigenvector of the adjacency matrix of Γ contains enough information so that w.h.p. S can be found. Frieze and Kannan [333] related this to a problem involving the optimization of a tensor product. Recently, Feldman, Grigorescu, Reyzin, Vempala and Xiao [289] showed that a large class of algorithms will fail w.h.p. if $s \leq n^{1/2-\delta}$ for some positive constant δ.

There has also been a considerable amount of research on planted cuts. Beginning with the paper by Bui, Chaudhuri, Leighton and Sipser [164] there have been many papers that deal with the problem of finding a cut in a random graph of unusual size. By this we mean that starting with $\mathbb{G}_{n,p}$, someone selects a partition of the vertex set into $k \geq 2$ sets of large size and then alters the edges between the subsets of the partition so that it is larger or smaller than can be usually found in $\mathbb{G}_{n,p}$. See Coja-Oghlan [189] for a recent paper with many pertinent references.

As a final note on this subject of planted objects. Suppose that we start with a Hamilton cycle C and then add a copy of $\mathbb{G}_{n,p}$, where $p = \frac{c}{n}$ to create Γ. Broder, Frieze and Shamir [161] showed that if c is sufficiently large then w.h.p. one can in polynomial time find a Hamilton cycle H in Γ. While H may not necessarily be C, this rules out a simple use of Hamilton cycles for a signature scheme.

19.9 Random Lifts

For a graph K, an *n-lift* G of K has vertex set $V(K) \times [n]$, where for each vertex $v \in V(K)$, $\{v\} \times [n]$ is called the *fiber* above v and is denoted by Π_v. The edge set of an n-lift G consists of a perfect matching between fibers Π_u and Π_w for each edge $\{u, w\} \in E(K)$. The set of n-lifts is denoted $\Lambda_n(K)$. In a random n-lift, the matchings between fibers are chosen independently and uniformly at random.

Lifts of graphs were introduced by Amit and Linial in [33], where they proved that if K is a connected, simple graph with minimum degree $\delta \geq 3$ and G is a random n-lift of K, then G is $\delta(G)$-connected w.h.p., where the asymptotics are for $n \to \infty$. They continued the study of random lifts in [34], where they proved expansion properties of lifts. Together with Matoušek, they gave bounds on the independence number and chromatic number of random lifts in [35]. Linial and Rozenman [530] gave a tight analysis for when a random n-lift has a perfect matching. Greenhill, Janson and Ruciński [383] considered the number of perfect matchings in a random lift.

Łuczak, Witkowski and Witkowski [549] proved that a random lift of H is Hamiltonian w.h.p. if H has minimum degree at least 5 and contains two disjoint Hamiltonian cycles whose union is not a bipartite graph. Chebolu and Frieze [175] considered a directed version of lifts and showed that a random lift of the complete digraph $\vec{K}_h$ is Hamiltonian w.h.p. provided h is sufficiently large.

19.10 Random Simplicial Complexes

Linial and Meshulam [529] pioneered the extension of the analysis of $\mathbb{G}_{n,p}$ to higher dimensional complexes. We are at the beginning of research in this area and can look forward to exciting connections with algebraic topology. For more details see the survey of Kahle [455].

19.11 Random Subgraphs of the n-cube

While most work on random graphs has been on random subgraphs of K_n, it is true to say that has also been a good deal of work on random subgraphs of the n-cube, $\mathbb{Q}_n$. This has vertex set $V_n = \{0,1\}^n$ and an edge between $\mathbf{x},\mathbf{y} \in V_n$ if and only if they differ in exactly one coordinate. To obtain a subgraph, we can either randomly delete vertices with probability $1 - p_v$ or edges with probability $1 - p_e$ or both. If only edges are deleted then the connectivity threshold is around $p_e = 1/2$, see Burtin [167] or Saposhenko [670] or Erdős and Spencer [282]. If only vertices are deleted then the connectivity threshold is around $p_v = 1/2$, see Saposhenko [671] or Weber [721]. If both edges and vertices are deleted then the connectivity threshold is around $p_e p_v = 1/2$, see Dyer, Frieze and Foulds [266].

Ajtai, Komlós and Szemerédi [8] showed that if $p_e = (1 + \varepsilon)/n$ then w.h.p. there will be a unique giant component of order 2^n. Their results were tightened in Bollobás, Kohayakawa and Łuczak [144], where the case $\varepsilon = o(1)$ was considered. In further analysis, Bollobás, Kohayakawa and Łuczak [145] proved the existence of giant components in the case $p_v = (1+\varepsilon)/n$.

The threshold for the existence of a perfect matching at around $p_e = 1/2$ was established by Bollobás [133]. The threshold for the existence of a Hamilton cycle remains an open question.

19.12 Random Walks on Random Graphs

For a random walk, two of the most interesting parameters, are the mixing time and the cover time.

Mixing Time

Generally speaking, the probability that a random walk is at a particular vertex tends to a steady state probability $\frac{\deg(v)}{2m}$. The *mixing time* is the time taken for the k-step distribution to get to within variation distance $1/4$, say, of the steady state. Above the threshold for connectivity, the mixing time of $\mathbb{G}_{n,p}$ is certainly $O(\log n)$ w.h.p. For sparser graphs, the focus has been on finding the mixing time for a random walk on the giant component. Fountoulakis and Reed [313] and Benjamini, Kozma and Wormald [68] showed that w.h.p. the mixing time of a random walk on the giant component of $\mathbb{G}_{n,p}, p = c/n, c > 1$ is $O((\log n)^2)$. Nachmias and Peres [598] showed that the mixing time of the largest component of $\mathbb{G}_{n,p}, p = 1/n$ is in $[\varepsilon n, (1-\varepsilon)n]$ with probability $1 - p(\varepsilon)$, where $p(\varepsilon) \to 0$ as $\varepsilon \to 0$. Ding, Lubetzky and Peres [246] showed that the mixing time for the emerging giant at $p = (1+\varepsilon)/n$, where $\lambda = \varepsilon^3 n \to \infty$ is of order $(n/\lambda)(\log \lambda)^2$. For random regular graphs, the mixing time is $O(\log n)$ and Lubetzky and Sly [533] proved that the mixing time exhibits a *cut-off* phenomenon, i.e. the variation distance goes from near one to near zero very rapidly.

Cover Time

The cover time C_G of a graph G is the maximum over starting vertex of the expected time for a random walk to visit every vertex of G. For $G = \mathbb{G}_{n,p}$ with $p = \frac{c \log n}{n}$, where $c > 1$, Jonasson [450] showed that w.h.p. $C_G = \Theta(n \log n)$. Cooper and Frieze [211] proved that $C_G \approx A(c) n \log n$ where $A(c) = c \log\left(\frac{c}{c-1}\right)$. Then in [210] they showed that the cover time of a random r-regular graph is w.h.p. asymptotic to $\frac{r-1}{r-2} n \log n$, for $r \geq 3$. Then in a series of papers they established the asymptotic cover time for preferential attachment graphs [211]; the giant component of $\mathbb{G}_{n,p}, p = c/n$, where $c > 1$ is constant [212]; random geometric graphs of dimension $d \geq 3$, [213]; random directed graphs [214]; random graphs with a fixed degree sequence [1], [219] and random hypergraphs [222]. The asymptotic cover time of random geometric graphs for $d = 2$ is still unknown. Avin and Ercal [41] proved that w.h.p. it is $\Theta(n \log n)$. The paper [215] deals with the structure of the subgraph H_t induced by the unvisited vertices in a random walk on a random graph after t steps. It gives tight results on a phase transition, i.e. a point where H breaks up into small components. Cerny and Teixeira [172] refined the result of [215] near the phase transition.

19.13 Stable Matching

In the stable matching problem we have a complete bipartite graph on vertex sets A, B, where $A = \{a_1, a_2, \ldots, a_n\}$ and $B = \{b_1, b_2, \ldots, b_n\}$. If we think of A as a set of women and B as a set of men, then we refer to this as the *stable marriage* problem. Each $a \in A$ has a total ordering p_a of B and each $b \in B$ has a total ordering p_b of B. The problem is to find a perfect matching $(a_i, b_i), i = 1, 2, \ldots, n$ such that there *does not exist* a pair i,j such that $b_j > b_i$ in the order p_{a_i} and $a_i > b_j$ in the order p_{b_j}. The existence of i,j leads to an unstable matching. Gale and Shapley [356] proved that there is always a stable matching and gave an algorithm for finding one. We focus on the case where p_a, p_b are uniformly random for all $a \in A, b \in B$. Wilson [724] showed that the expected number of proposals in a sequential version of the Gale–Shapley algorithm is asymptotically equal to $n \log n$. Knuth, Motwani and Pittel [489] studied the likely number of stable husbands for an element of $A \cup B$, i.e. they showed that w.h.p. there are constants $c < C$ such that for a fixed $a \in A$ there are between $c \log n$ and $C \log n$ choices $b \in B$ such that a and b are matched together in some stable matching. The question of how many distinct stable matchings there are likely to be was raised in Pittel [624] who showed that w.h.p. there are at least $n^{1/2-o(1)}$. More recently, Lennon and Pittel [523] showed that there are at least $n \log n$ with probability at least 0.45. Thus the precise growth rate of the number of stable matchings is not clear at the moment. Pittel, Shepp and Veklerov [627] considered the number $Z_{n,m}$ of $a \in A$ that have exactly m choices of stable husband. They show that $\lim_{n\to\infty} \frac{\mathbb{E}(Z_{n,m})}{(\log n)^{m+1}} = \frac{1}{(m-1)!}$.

19.14 Universal graphs

A graph G is *universal* for a class of graphs $\mathscr{H}$ if G contains a copy of every $H \in \mathscr{H}$. In particular, let $\mathscr{H}(n,d)$ denote the set of graphs with vertex set $[n]$ and maximum degree at most d. One question that has concerned researchers, is to find the threshold for $\mathbb{G}_{n,p}$ being universal for $\mathscr{H}(n,d)$. A counting argument shows that any $\mathscr{H}(n,d)$ universal graph has $\Omega(n^{2-2/d})$ edges. For random graphs this can be improved to $\Omega(n^{2-2/(d+1)}(\log n)^{O(1)})$. This is because to contain the union of $\left\lfloor \frac{n}{d+1} \right\rfloor$ disjoint copies of K_{d+1}, all but at most d vertices must lie in a copy of K_{d+1}. This problem was first considered in Alon, Capalbo, Kohayakawa, Rödl, Ruciński and Szemerédi [23]. Currently the best upper bound on the value of p needed to make $\mathbb{G}_{n,m}$ $\mathscr{H}(n,d)$ universal is $O(n^{2-1/d}(\log n)^{1/d})$ in Dellamonica, Kohayakawa, Rödl, and Ruciński [231]. Ferber, Nenadov and Peter [297] proved that if $p \gg \Delta^8 n^{-1/2} \log n$ then $\mathbb{G}_{n,p}$ is universal for the set of trees with maximum degree Δ.

PART IV

Tools and Methods

20

Moments

20.1 First and Second Moment Method

Lemma 20.1 Markov's inequality *Let X be a non-negative random variable, then, for all $t > 0$,*

$$\mathbb{P}(X \geq t) \leq \frac{\mathbb{E}X}{t}.$$

Proof Let

$$I_A = \begin{cases} 1 & \text{if event } A \text{ occurs,} \\ 0 & \text{otherwise.} \end{cases}$$

Notice that

$$X = XI_{\{X \geq t\}} + XI_{\{X < t\}} \geq XI_{\{X \geq t\}} \geq tI_{\{X \geq t\}}.$$

Hence,

$$\mathbb{E}X \geq t\mathbb{E}I_{\{X \geq t\}} = t\mathbb{P}(X \geq t).$$

□

As an immediate corollary, we obtain

Lemma 20.2 First Moment Method *Let X be a non-negative integer valued random variable, then*

$$\mathbb{P}(X > 0) \leq \mathbb{E}X.$$

Proof Put $t = 1$ in Markov's inequality. □

The following inequality is a simple consequence of Lemma 20.1.

Lemma 20.3 Chebyshev inequality *If X is a random variable with a finite mean and variance, then, for $t > 0$,*

$$\mathbb{P}(|X - \mathbb{E}X| \geq t) \leq \frac{\operatorname{Var} X}{t^2}.$$

Proof

$$\mathbb{P}(|X-\mathbb{E}X|\geq t)=\mathbb{P}((X-\mathbb{E}X)^2\geq t^2)\leq\frac{\mathbb{E}(X-\mathbb{E}X)^2}{t^2}=\frac{\operatorname{Var}X}{t^2}.$$

□

Throughout the book the following consequence of the Chebyshev inequality plays a particularly important role.

Lemma 20.4 Second Moment Method *If X is a non-negative integer valued random variable then*

$$\mathbb{P}(X=0)\leq\frac{\operatorname{Var}X}{(\mathbb{E}X)^2}=\frac{\mathbb{E}X^2}{(\mathbb{E}X)^2}-1.$$

Proof Set $t=\mathbb{E}X$ in the Chebyshev inequality, then

$$\mathbb{P}(X=0)\leq\mathbb{P}(|X-\mathbb{E}X|\geq\mathbb{E}X)\leq\frac{\operatorname{Var}X}{(\mathbb{E}X)^2}.$$

□

Lemma 20.5 (Strong) Second Moment Method *If X is a non-negative integer valued random variable then*

$$\mathbb{P}(X=0)\leq\frac{\operatorname{Var}X}{\mathbb{E}X^2}=1-\frac{(\mathbb{E}X)^2}{\mathbb{E}X^2}.$$

Proof Notice that

$$X=X\cdot I_{\{X\geq 1\}}.$$

Then, by the Cauchy–Schwarz inequality,

$$(\mathbb{E}X)^2=\left(\mathbb{E}(X\cdot I_{\{X\geq 1\}})\right)^2\leq\mathbb{E}I^2_{\{X\geq 1\}}\,\mathbb{E}X^2=\mathbb{P}(X\geq 1)\,\mathbb{E}X^2.$$

□

The bound in Lemma 20.5 is stronger than the bound in Lemma 20.4, since $\mathbb{E}X^2\geq(\mathbb{E}X)^2$. However, for many applications, these bounds are equally useful since the Second Moment Method can be applied if

$$\frac{\operatorname{Var}X}{(\mathbb{E}X)^2}\to 0, \tag{20.1}$$

or, equivalently,

$$\frac{\mathbb{E}X^2}{(\mathbb{E}X)^2}\to 1, \tag{20.2}$$

as $n \to \infty$. In fact if (20.1) holds, then much more than $\mathbb{P}(X > 0) \to 1$ is true. Note that

$$\frac{\operatorname{Var} X}{(\mathbb{E}X)^2} = \operatorname{Var}\left(\frac{X}{\mathbb{E}X}\right) = \mathbb{E}\left(\frac{X}{\mathbb{E}X}\right)^2 - \left(\mathbb{E}\left(\frac{X}{\mathbb{E}X}\right)\right)^2$$
$$= \mathbb{E}\left(\frac{X}{\mathbb{E}X} - 1\right)^2.$$

Hence,

$$\mathbb{E}\left(\frac{X}{\mathbb{E}X} - 1\right)^2 \to 0 \text{ if } \frac{\operatorname{Var} X}{(\mathbb{E}X)^2} \to 0.$$

It simply means that

$$\frac{X}{\mathbb{E}X} \xrightarrow{L^2} 1. \tag{20.3}$$

In particular, it implies (as does the Chebyshev inequality) that

$$\frac{X}{\mathbb{E}X} \xrightarrow{\mathbb{P}} 1, \tag{20.4}$$

i.e. for every $\varepsilon > 0$,

$$\mathbb{P}((1-\varepsilon)\mathbb{E}X < X < (1+\varepsilon)\mathbb{E}X) \to 1. \tag{20.5}$$

So, we can only apply the Second Moment Method, if the random variable X has its distribution asymptotically concentrated at a single value (X can be approximated by the non-random value $\mathbb{E}X$, as stated by (20.3), (20.4) and (20.5)).

We complete this section with another lower bound on the probability $\mathbb{P}(X_n \geq 1)$, when X_n is a sum of (asymptotically) negatively correlated indicators. Notice that in this case we do not need to compute the second moment of X_n.

Lemma 20.6 *Let $X_n = I_1 + I_2 + \cdots + I_n$, where $\{I_i\}_{i=1}^n$ is a collection of 0–1 random variables, such that*

$$\mathbb{P}(I_i = I_j = 1) \leq (1+\varepsilon_n)\mathbb{P}(I_i = 1)\mathbb{P}(I_j = 1)$$

for $i,j = 1,2,\ldots,n$. Here $\varepsilon_n \to 0$ as $n \to \infty$. Then

$$\mathbb{P}(X_n \geq 1) \geq \frac{1}{1+\varepsilon_n + 1/\mathbb{E}X_n}.$$

Proof By the (Strong) Second Moment Method (see Lemma 20.5)

$$\mathbb{P}(X_n \geq 1) \geq \frac{(\mathbb{E}X_n)^2}{\mathbb{E}X_n^2}.$$

Now

$$
\begin{aligned}
\mathbb{E}X_n^2 &= \sum_{i=1}^{n}\sum_{j=1}^{n}\mathbb{E}(I_iI_j) \\
&\leq \mathbb{E}X_n + (1+\varepsilon_n)\sum_{i\neq j}\mathbb{E}I_i\mathbb{E}I_j \\
&= \mathbb{E}X_n + (1+\varepsilon_n)\left(\left(\sum_{i=1}^{n}\mathbb{E}I_i\right)^2 - \sum_{i=1}^{n}(\mathbb{E}I_i)^2\right) \\
&\leq \mathbb{E}X_n + (1+\varepsilon_n)(\mathbb{E}X_n)^2.
\end{aligned}
$$

□

20.2 Convergence of Moments

Let X be a random variable such that $\mathbb{E}|X|^k < \infty$, $k \geq 1$, i.e. all *k-th moments* $\mathbb{E}X^k$ exist and are finite. Let the distribution of X be completely determined by its moments. This means that all random variables with the same moments as X have the same distribution as X. In particular, this is true when X has the Normal or the Poisson distribution.

The Method of Moments provides a tool to prove the convergence in distribution of a sequence of random variables with finite moments (see Durrett [263] for details).

Lemma 20.7 Method of Moments *Let X be a random variable with probability distribution completely determined by its moments. If $X_1, X_2, \ldots X_n, \ldots$ are random variables with finite moments such that $\mathbb{E}X_n^k \to \mathbb{E}X^k$ as $n \to \infty$, for every integer $k \geq 1$, then the sequence of random variables $\{X_n\}$ converges in distribution to random variable X, denoted as $X_n \xrightarrow{D} X$.*

The next result, which can be deduced from Theorem 20.7, provides a tool to prove asymptotic normality.

Corollary 20.8 *Let $X_1, X_2, \ldots, X_n, \ldots$ be a sequence of random variables with finite moments and let $a_1, a_2, \ldots, a_n, \ldots$ be a sequence of positive numbers, such that*

$$
\mathbb{E}(X_n - \mathbb{E}X_n)^k = \begin{cases} \frac{(2m)!}{2^m m!}a_n^k + o(a_n^k), & \textit{when } k = 2m,\ m \geq 1, \\ o(a_n^k), & \textit{when } k = 2m-1,\ m \geq 2, \end{cases}
$$

as $n \to \infty$. *Then*

$$\frac{X_n - \mathbb{E}X_n}{a_n} \xrightarrow{D} Z, \text{ and } \tilde{X}_n = \frac{X_n - \mathbb{E}X_n}{\sqrt{\operatorname{Var} X_n}} \xrightarrow{D} Z,$$

where Z *is a random variable with the standard normal distribution* $N(0,1)$.

A similar result for convergence to the Poisson distribution can also be deduced from Theorem 20.7. Instead, we will show how to derive it directly from the *Inclusion–Exclusion Principle.*

The following lemma sometimes simplifies the proof of some probabilistic inequalities:

Lemma 20.9 (Rényi's lemma) *Suppose that* $A_1, A_2, \ldots, A_r$ *are events in some probability space* Ω, $f_1, f_2, \ldots, f_s$ *are Boolean functions of* $A_1, A_2, \ldots, A_r$, *and* $\alpha_1, \alpha_2, \ldots, \alpha_s$ *are reals. Then, if*

$$\sum_{i=1}^{s} \alpha_i \mathbb{P}(f_i(A_1, A_2, \ldots, A_r)) \geq 0, \tag{20.6}$$

whenever $\mathbb{P}(A_i) = 0$ *or* 1, *then (20.6) holds in general.*

Proof Write

$$f_i = \bigcup_{S \in \mathscr{S}_i} \left(\left(\bigcap_{i \in S} A_i \right) \cap \left(\bigcap_{i \notin S} A_i^c \right) \right),$$

for some collection of $\mathscr{S}_i$ of subset of $[r] = \{1, 2, \ldots, r\}$.

Then,

$$\mathbb{P}(f_i) = \sum_{S \in \mathscr{S}_i} \mathbb{P}\left(\left(\bigcap_{i \in S} A_i \right) \cap \left(\bigcap_{i \notin S} A_i^c \right) \right),$$

and then the LHS of (20.6) becomes

$$\sum_{S \subseteq [r]} \beta_S \mathbb{P}\left(\left(\bigcap_{i \in S} A_i \right) \cap \left(\bigcap_{i \notin S} A_i^c \right) \right),$$

for some real β_S. If (20.6) holds, then $\beta_S \geq 0$ for every S, since we can choose $A_i = \Omega$ if $i \in S$, and $A_i = \emptyset$ for $i \notin S$. □

For $J \subseteq [r]$ let $A_J = \bigcap_{i \in J} A_i$, and let $S = |\{j : A_j \text{ occurs}\}|$ denote the number of events that occur. Then let

$$B_k = \sum_{J:|J|=k} \mathbb{P}(A_J) = \mathbb{E}\binom{S}{k}.$$

Let $\mathscr{E}_j$ be the event that exactly j among the events $A_1, A_2, \ldots, A_r$ occur. Then,

Lemma 20.10

$$\mathbb{P}(\mathcal{E}_j) \begin{cases} \leq \sum_{k=j}^{s} (-1)^{k-j} \binom{k}{j} B_k & s-j \text{ even.} \\ \geq \sum_{k=j}^{s} (-1)^{k-j} \binom{k}{j} B_k & s-j \text{ odd} \\ = \sum_{k=j}^{s} (-1)^{k-j} \binom{k}{j} B_k & s=r. \end{cases}$$

Proof It follows from Lemma 20.9 that we only need to check the truth of the statement for

$$\mathbb{P}(A_i) = 1 \quad 1 \leq i \leq \ell,$$
$$\mathbb{P}(A_i) = 0 \quad \ell < i \leq r,$$

where $0 \leq \ell \leq r$ is arbitrary.

Now,

$$\mathbb{P}(S=j) = \begin{cases} 1 & \text{if } j = l, \\ 0 & \text{if } j \neq l, \end{cases}$$

and

$$B_k = \binom{\ell}{k}.$$

So,

$$\sum_{k=j}^{s} (-1)^{k-j} \binom{k}{j} B_k = \sum_{k=j}^{s} (-1)^{k-j} \binom{k}{j} \binom{\ell}{k}$$
$$= \binom{\ell}{j} \sum_{k=j}^{s} (-1)^{k-j} \binom{\ell-j}{k-j}. \tag{20.7}$$

If $\ell < j$ then $\mathbb{P}(\mathcal{E}_j) = 0$ and the sum in (20.7) reduces to zero. If $\ell = j$ then $\mathbb{P}(\mathcal{E}_j) = 1$ and the sum in (20.7) reduces to one. Thus in this case, the sum is exact for all s. Assume then that $r \geq \ell > j$. Then $\mathbb{P}(\mathcal{E}_j) = 0$ and

$$\sum_{k=j}^{s} (-1)^{k-j} \binom{\ell-j}{k-j} = \sum_{t=0}^{s-j} (-1)^{t} \binom{\ell-j}{t} = (-1)^{s-j} \binom{\ell-j-1}{s-j}.$$

This explains the alternating signs of the theorem. Finally, observe that $\binom{\ell-j-1}{r-j} = 0$, as required. □

Now we are ready to state the main tool for proving convergence to the Poisson distribution.

Theorem 20.11 *Let $S_n = \sum_{i\geq 1} I_i$ be a sequence of random variables, $n \geq 1$ and let $B_k^{(n)} = \mathbb{E}\binom{S_n}{k}$. Suppose that there exists $\lambda \geq 0$, such that for every fixed $k \geq 1$,*

$$\lim_{n\to\infty} B_k^{(n)} = \frac{\lambda^k}{k!}.$$

Then, for every $j \geq 0$,

$$\lim_{n\to\infty} \mathbb{P}(S_n = j) = e^{-\lambda}\frac{\lambda^j}{j!},$$

i.e. S_n converges in distribution to the Poisson distributed random variable with expectation λ ($S_n \xrightarrow{D} \mathrm{Po}(\lambda)$).

Proof By Lemma 20.10, for $l \geq 0$,

$$\sum_{k=j}^{j+2l+1} (-1)^{k-j}\binom{k}{j} B_k^{(n)} \leq \mathbb{P}(S_n = j) \leq \sum_{k=j}^{j+2l} (-1)^{k-j}\binom{k}{j} B_k^{(n)}.$$

So, as n grows to ∞,

$$\sum_{k=j}^{j+2l+1} (-1)^{k-j}\binom{k}{j} B_k^{(n)} \leq \liminf_{n\to\infty} \mathbb{P}(S_n = j)$$

$$\leq \limsup_{n\to\infty} \mathbb{P}(S_n = j) \leq \sum_{k=j}^{j+2l} (-1)^{k-j}\binom{k}{j} B_k^{(n)}.$$

So, for $m \geq 0$,

$$\sum_{k=j}^{j+m} (-1)^{k-j}\binom{k}{j}\frac{\lambda^k}{k!} = \frac{\lambda^k}{k!}\sum_{t=0}^{m}(-1)^t\frac{\lambda^t}{t!} \to \frac{\lambda^k}{k!}e^{-\lambda},$$

as $m \to \infty$. □

Notice that the falling factorial

$$(S_n)_k = S_n(S_n-1)\cdots(S_n-k+1)$$

counts number of ordered k-tuples of events with $I_i = 1$. Hence, the binomial moments of S_n can be replaced in Theorem 20.11 by the factorial moments, defined as

$$\mathbb{E}(S_n)_k = \mathbb{E}[S_n(S_n-1)\cdots(S_n-k+1)],$$

and one has to check whether,for every $k \geq 1$,

$$\lim_{n\to\infty} \mathbb{E}(S_n)_k = \lambda^k.$$

20.3 Stein–Chen Method

Stein in [692] introduced a powerful technique for obtaining estimates of the rate of convergence to the standard normal distribution. His approach was subsequently extended to cover convergence to the Poisson distribution by Chen [177], while Barbour [57] ingeniously adapted both methods to random graphs.

The Stein–Chen approach has some advantages over the Method of Moments. The principal advantage is that a rate of convergence is automatically obtained. Also the computations are often easier and fewer moment assumptions are required. Moreover, it frequently leads to conditions for convergence weaker than those obtainable by the Method of Moments.

Consider a sequence of random variables $(X_n)_{n=1}^{\infty}$ and let $(\lambda_n)_{n=1}^{\infty}$ be a sequence of positive integers, and let $\mathrm{Po}(\lambda)$ denote, as before, the Poisson distribution with expectation λ. We say that X_n is *Poisson convergent* if the *total variation distance* between the distribution $\mathscr{L}(X_n)$ of X_n and $\mathrm{Po}(\lambda_n)$, $\lambda_n = \mathbb{E} X_n$, distribution, tends to zero as n tends to infinity. So, we ask for

$$d_{TV}(\mathscr{L}(X_n), \mathrm{Po}(\lambda_n)) = \sup_{A \subseteq \mathbb{Z}^+} \left| \mathbb{P}(X_n \in A) - \sum_{k \in A} \frac{\lambda_n^k}{k!} e^{-\lambda_n} \right| \to 0, \tag{20.8}$$

as $n \to \infty$, where $\mathbb{Z}^+ = \{0, 1, \ldots\}$.

Notice, that if X_n is Poisson convergent and $\lambda_n \to \lambda$, then X_n converges in distribution to the $\mathrm{Po}(\lambda)$ distributed random variable. Furthermore, if $\lambda_n \to 0$, then X_n converges to a random variable with distribution degenerated at 0. More importantly, if $\lambda_n \to \infty$, then the Central Limit Theorem for Poisson distributed random variables implies that $\tilde{X}_n = (X_n - \lambda_n)/\sqrt{\lambda_n}$ converges in distribution to a random variable with the standard normal random distribution $N(0,1)$.

The basic feature and advantage of the Stein–Chen approach is that it gives computationally tractable bounds for the distance d_{TV}, when the random variables in question are sums of indicators with a fairly general dependence structure.

Let $\{I_a\}_{a \in \Gamma}$, be a family of indicator random variables, where Γ is some index set. To describe the relationship between these random variables we define a *dependency graph* $L = (V(L), E(L))$, where $V(L) = \Gamma$. Graph L has the property that whenever there are no edges between A and B, $A, B \subseteq \Gamma$, then $\{I_a\}_{a \in A}$ and $\{I_b\}_{b \in B}$ are mutually independent families of random variables. The following general bound on the total variation distance was proved by Barbour, Holst and Janson [58] via the Stein–Chen method.

Theorem 20.12 *Let $X = \sum_{a \in \Gamma} I_a$ where the I_a are indicator random variables with a dependency graph L. Then, with $\pi_a = \mathbb{E} I_a$ and $\lambda = \mathbb{E} X = \sum_{a \in \Gamma} \pi_a$,*

$$d_{TV}(\mathscr{L}(X), \mathrm{Po}(\lambda)) \leq$$

$$\min(\lambda^{-1}, 1) \left(\sum_{a \in V(L)} \pi_a^2 + \sum_{ab \in E(L)} \{\mathbb{E}(I_a I_b) + \pi_a \pi_b\} \right),$$

where $\sum_{ab \in E(L)}$ means summing over all ordered pairs (a,b), such that $\{a,b\} \in E(L)$.

Finally, let us briefly mention, that the original Stein method investigates the convergence to the normal distribution in the following metric

$$d_S(\mathscr{L}(X_n), N(0,1)) = \sup_h ||h||^{-1} \left| \int h(x) dF_n(x) - \int h(x) d\Phi(x) \right|, \qquad (20.9)$$

where the supremum is taken over all bounded test functions h with bounded derivative, $||h|| = \sup|h(x)| + \sup|h'(x)|$.

Here F_n is the distribution function of X_n, while Φ denotes the distribution function of the standard normal distribution. So, if $d_S(\mathscr{L}(X_n), N(0,1)) \to 0$ as $n \to \infty$, then $\tilde{X}_n$ converges in distribution to $N(0,1)$ distributed random variable.

Barbour, Karoński and Ruciński [60] obtained an effective upper bound on $d_S(\mathscr{L}(X_n), N(0,1))$ if S belongs to a general class of *decomposable* random variables. This bound involves the first three moments of S only.

For a detailed and comprehensive account of the Stein–Chen method the reader is referred to the book by Barbour, Holst and Janson [58], or to Chapter 6 of the book by Janson, Łuczak and Ruciński [432], where other interesting approaches to studying asymptotic distributions of random graph characteristics are also discussed. For some applications of the Stein–Chen method in random graphs, one can look at a survey by Karoński [466].

21
Inequalities

21.1 Binomial Coefficient Approximation

We state some important inequalities. The proofs of all but (g) are left as exercises:

Lemma 21.1

$$(a)\quad 1+x \le e^x, \qquad \forall x.$$

$$(b)\quad 1-x \ge e^{-x/(1-x)}, \qquad 0 \le x < 1.$$

$$(c)\quad \binom{n}{k} \le \left(\frac{ne}{k}\right)^k, \qquad \forall n,k.$$

$$(d)\quad \binom{n}{k} \le \frac{n^k}{k!}\left(1-\frac{k}{2n}\right)^{k-1}, \qquad \forall n,k.$$

$$(e)\quad \frac{n^k}{k!}\left(1-\frac{k(k-1)}{2n}\right) \le \binom{n}{k} \le \frac{n^k}{k!}e^{-k(k-1)/(2n)}, \qquad \forall n,k.$$

$$(f)\quad \binom{n}{k} \approx \frac{n^k}{k!}, \qquad \textit{if } k^2 = o(n).$$

$$(g)\quad \textit{If } a \ge b \textit{ then } \frac{\binom{n-a}{t-b}}{\binom{n}{t}} \le \left(\frac{t}{n}\right)^b \left(\frac{n-t}{n-b}\right)^{a-b}.$$

Proof (of (g))

$$
\begin{aligned}
\frac{\binom{n-a}{t-b}}{\binom{n}{t}} &= \frac{(n-a)!t!(n-t)!}{n!(t-b)!(n-t-a+b)!} \\
&= \frac{t(t-1)\cdots(t-b+1)}{n(n-1)\cdots(n-b+1)} \times \frac{(n-t)(n-t-1)\cdots(n-t-a+b+1)}{(n-b)(n-b-1)\cdots(n-a+1)} \\
&\leq \left(\frac{t}{n}\right)^b \times \left(\frac{n-t}{n-b}\right)^{a-b}.
\end{aligned}
$$

□

We also need the following estimate for binomial coefficients. It is a little more precise than those given in Lemma 21.1.

Lemma 21.2 *Let* $k = o(n^{3/4})$, *then*

$$\binom{n}{k} \approx \frac{n^k}{k!} \exp\left\{-\frac{k^2}{2n} - \frac{k^3}{6n^2}\right\}.$$

Proof

$$
\begin{aligned}
\binom{n}{k} &= \frac{n^k}{k!} \prod_{i=0}^{k-1} \left(1 - \frac{i}{n}\right) \\
&= \frac{n^k}{k!} \exp\left\{\sum_{i=0}^{k-1} \log\left(1 - \frac{i}{n}\right)\right\} \\
&= \frac{n^k}{k!} \exp\left\{-\sum_{i=0}^{k-1} \left(\frac{i}{n} + \frac{i^2}{2n^2}\right) + O\left(\frac{k^4}{n^3}\right)\right\} \\
&= (1+o(1)) \frac{n^k}{k!} \exp\left\{-\frac{k^2}{2n} - \frac{k^3}{6n^2}\right\}.
\end{aligned}
$$

□

21.2 Balls in Boxes

Suppose that we have M boxes and we independently place N distinguishable balls into them. Let us assume that a ball goes into box i with probability p_i where $p_1 + \cdots + p_M = 1$. Let W_i denote the number of balls that are placed in box i and for $S \subseteq [M]$, let $W_S = \sum_{i \in S} W_i$. The following looks obvious and is extremely useful.

Theorem 21.3 *Let S, T be disjoint subsets of $[M]$ and let s, t be non-negative integers, then*

$$\mathbb{P}(W_S \le s \mid W_T \le t) \le \mathbb{P}(W_S \le s). \tag{21.1}$$

$$\mathbb{P}(W_S \ge s \mid W_T \le t) \ge \mathbb{P}(W_S \ge s). \tag{21.2}$$

$$\mathbb{P}(W_S \ge s \mid W_T \ge t) \le \mathbb{P}(W_S \ge s). \tag{21.3}$$

$$\mathbb{P}(W_S \le s \mid W_T \ge t) \ge \mathbb{P}(W_S \le s). \tag{21.4}$$

Proof Equation (21.2) follows immediately from (21.1). Also, Equation (21.4) follows immediately from (21.3). The proof of (21.3) is very similar to that of (21.1) and so we will only prove (21.1).

Let

$$\pi_i = \mathbb{P}(W_S \le s \mid W_T = i).$$

Given $W_T = i$, we are looking at throwing $N - i$ balls into $M - 1$ boxes. It is clear therefore that π_i is monotone increasing in i. Now, let $q_i = \mathbb{P}(W_T = i)$. Then,

$$\mathbb{P}(W_S \le s) = \sum_{i=0}^{N} \pi_i q_i.$$

$$\mathbb{P}(W_S \le s \mid W_T \le t) = \sum_{i=0}^{t} \pi_i \frac{q_i}{q_0 + \cdots + q_t}.$$

So, (21.1) reduces to

$$(q_0 + \cdots + q_t) \sum_{i=0}^{N} \pi_i q_i \le (q_0 + \cdots + q_N) \sum_{i=t+1}^{N} \pi_i q_i,$$

or

$$(q_{t+1} + \cdots + q_N) \sum_{i=0}^{t} \pi_i q_i \le (q_0 + \cdots + q_t) \sum_{i=t+1}^{N} \pi_i q_i,$$

or

$$\sum_{i=0}^{t} \sum_{j=t+1}^{N} q_i q_j \pi_i \le \sum_{i=0}^{t} \sum_{j=t+1}^{N} q_i q_j \pi_j.$$

The result now follows from the monotonicity of π_i. □

The following is an immediate corollary:

Corollary 21.4 *Let $S_1,S_2,\dots,S_k$ be disjoint subsets of $[M]$ and let $s_1,s_2,\dots,s_k$ be non-negative integers, then*

$$\mathbb{P}\left(\bigcap_{i=1}^{k}\{W_{S_i}\leq s_i\}\right)\leq\prod_{i=1}^{k}\mathbb{P}(\{W_{S_i}\leq s_i\}).$$

$$\mathbb{P}\left(\bigcap_{i=1}^{k}\{W_{S_i}\geq s_i\}\right)\leq\prod_{i=1}^{k}\mathbb{P}(\{W_{S_i}\geq s_i\}).$$

21.3 FKG Inequality

A function $f: C_N=\{0,1\}^{[N]}\to\mathbb{R}$ is said to be *monotone increasing* if whenever $x=(x_1,x_2,\dots,x_N), y=(y_1,y_2,\dots,y_N)\in C_N$ and $x\leq y\in C_N$ (i.e. $x_j\leq y_j, j=1,2,\dots,N$) then $f(x)\leq f(y)$. Similarly, f is said to be *monotone decreasing* if $-f$ is monotone increasing.

An important example for us is the case where f is the indicator function of some subset $\mathscr{A}$ of $2^{[N]}$. Then

$$f(x)=\begin{cases}1 & x\in\mathscr{A}\\ 0 & x\notin\mathscr{A}.\end{cases}$$

A typical example for us would be $N=\binom{n}{2}$ and then each $G\in 2^{[N]}$ corresponds to a graph with vertex set $[n]$. Then $\mathscr{A}$ will be a set of graphs, i.e. a graph property. Suppose that f is the indicator function for $\mathscr{A}$. Then f is monotone increasing, if whenever $G\in\mathscr{A}$ and $e\notin E(G)$ we have $G+e\in\mathscr{A}$, i.e. adding an edge does not destroy the property. We say that the set/property is monotone increasing. For example, if $\mathscr{H}$ is the set of Hamiltonian graphs then $\mathscr{H}$ is monotone increasing. If $\mathscr{P}$ is the set of planar graphs then $\mathscr{P}$ is monotone decreasing. In other words a property is monotone increasing if and only if its indicator function is monotone increasing.

Suppose that next we turn C_N into a probability space by choosing some $p_1,p_2,\dots,p_N\in[0,1]$ and then for $\mathbf{x}=(x_1,x_2,\dots,x_N)\in C_N$ letting

$$\mathbb{P}(\mathbf{x})=\prod_{j:x_j=1}p_j\prod_{j:x_j=0}(1-p_j). \tag{21.5}$$

If $N=\binom{n}{2}$ and $p_j=p, j=1,2,\dots,N$, then this model corresponds to $\mathbb{G}_{n,p}$.

The following is a special case of the FKG inequality, Harris [396] and Fortuin, Kasteleyn and Ginibre [308]:

Theorem 21.5 *If f,g are monotone increasing functions on C_N, then $\mathbb{E}(fg)\geq\mathbb{E}(f)\,\mathbb{E}(g)$.*

Proof We prove this by induction on N. If $N = 0$ then $\mathbb{E}(f) = a, \mathbb{E}(g) = b$ and $\mathbb{E}(fg) = ab$ for some constants a, b.

So assume the truth for $N-1$. Suppose that $\mathbb{E}(f \mid x_N = 0) = a$ and $\mathbb{E}(g \mid x_N = 0) = b$, then

$$\mathbb{E}((f-a)(g-b)) - \mathbb{E}(f-a)\,\mathbb{E}(g-b) = \mathbb{E}(fg) - \mathbb{E}(f)\,\mathbb{E}(g).$$

By replacing f by $f-a$ and g by $g-b$ we may therefore assume that $\mathbb{E}(f \mid x_N = 0) = \mathbb{E}(g \mid x_N = 0) = 0$. By monotonicity, we see that $\mathbb{E}(f \mid x_N = 1), \mathbb{E}(g \mid x_N = 1) \geq 0$.

We observe that by the induction hypothesis

$$\mathbb{E}(fg \mid x_N = 0) \geq \mathbb{E}(f \mid x_N = 0)\,\mathbb{E}(g \mid x_N = 0) = 0$$
$$\mathbb{E}(fg \mid x_N = 1) \geq \mathbb{E}(f \mid x_N = 1)\,\mathbb{E}(g \mid x_N = 1) \geq 0.$$

Now, by the above inequalities,

$$\begin{aligned}\mathbb{E}(fg) &= \mathbb{E}(fg \mid x_N = 0)(1-p_N) + \mathbb{E}(fg \mid x_N = 1)p_N \\ &\geq \mathbb{E}(f \mid x_N = 1)\,\mathbb{E}(g \mid x_N = 1)p_N. \end{aligned} \tag{21.6}$$

Furthermore,

$$\begin{aligned}\mathbb{E}(f)\,\mathbb{E}(g) &= (\mathbb{E}(f \mid x_N = 0)(1-p_N) + \mathbb{E}(f \mid x_N = 1)p_N)\times \\ &\qquad (\mathbb{E}(g \mid x_N = 0)(1-p_N) + \mathbb{E}(g \mid x_N = 1)p_N) \\ &= \mathbb{E}(f \mid x_N = 1)\,\mathbb{E}(g \mid x_N = 1)p_N^2. \end{aligned} \tag{21.7}$$

The result follows by comparing (21.6) and (21.7) and using the fact that $\mathbb{E}(f \mid x_N = 1), \mathbb{E}(g \mid x_N = 1) \geq 0$ and $0 \leq p_N \leq 1$. □

In terms of monotone increasing sets $\mathscr{A}, \mathscr{B}$ and the same probability (21.5) we can express the FKG inequality as

$$\mathbb{P}(\mathscr{A} \mid \mathscr{B}) \geq \mathbb{P}(\mathscr{A}). \tag{21.8}$$

21.4 Sums of Independent Bounded Random Variables

Suppose that S is a random variable and $t > 0$ is a real number. We are concerned here with bounds on the *upper and lower tail* of the distribution of S, i.e. on $\mathbb{P}(S \geq \mu + t)$ and $\mathbb{P}(S \leq \mu - t)$, respectively, where $\mu = \mathbb{E}\,S$.

The basic observation that leads to the construction of such bounds is due to Bernstein [82]. Let $\lambda \geq 0$, then

$$\mathbb{P}(S \geq \mu + t) = \mathbb{P}(e^{\lambda S} \geq e^{\lambda(\mu+t)}) \leq e^{-\lambda(\mu+t)}\,\mathbb{E}(e^{\lambda S}), \tag{21.9}$$

by Markov's inequality (see Lemma 20.1). Similarly for $\lambda \leq 0$,

$$\mathbb{P}(S \leq \mu - t) \leq e^{-\lambda(\mu-t)}\mathbb{E}(e^{\lambda S}). \tag{21.10}$$

Combining (21.9) and (21.10) one can obtain a bound for $\mathbb{P}(|S-\mu| \geq t)$.

Now let $S_n = X_1 + X_2 + \cdots + X_n$, where $X_i, i = 1,\ldots,n$ are independent random variables. Assume that $0 \leq X_i \leq 1$ and $\mathbb{E}X_i = \mu_i$ for $i = 1,2,\ldots,n$. Let $\mu = \mu_1 + \mu_2 + \cdots + \mu_n$. Then for $\lambda \geq 0$

$$\mathbb{P}(S_n \geq \mu + t) \leq e^{-\lambda(\mu+t)}\prod_{i=1}^{n}\mathbb{E}(e^{\lambda X_i}) \tag{21.11}$$

and for $\lambda \leq 0$

$$\mathbb{P}(S_n \leq \mu - t) \leq e^{-\lambda(\mu-t)}\prod_{i=1}^{n}\mathbb{E}(e^{\lambda X_i}). \tag{21.12}$$

Note that $\mathbb{E}(e^{\lambda X_i})$ in (21.11) and (21.12), likewise $\mathbb{E}(e^{\lambda S})$ in (21.9) and (21.10) are the moment generating functions of the X_is and S, respectively. So finding bounds boils down to the estimation of these functions.

Now the convexity of e^x and $0 \leq X_i \leq 1$ implies that

$$e^{\lambda X_i} \leq 1 - X_i + X_i e^{\lambda}.$$

Taking expectations we obtain

$$\mathbb{E}(e^{\lambda X_i}) \leq 1 - \mu_i + \mu_i e^{\lambda}.$$

Equation (21.11) becomes, for $\lambda \geq 0$,

$$\mathbb{P}(S_n \geq \mu + t) \leq e^{-\lambda(\mu+t)}\prod_{i=1}^{n}(1 - \mu_i + \mu_i e^{\lambda}) \leq e^{-\lambda(\mu+t)}\left(\frac{n - \mu + \mu e^{\lambda}}{n}\right)^n. \tag{21.13}$$

The second inequality follows from the fact that the geometric mean is at most the arithmetic mean, i.e. $(x_1 x_2 \cdots x_n)^{1/n} \leq (x_1 + x_2 + \cdots + x_n)/n$ for non-negative $x_1, x_2, \ldots, x_n$. This in turn follows from Jensen's inequality and the concavity of $\log x$.

The RHS of (21.13) attains its minimum, as a function of λ, at

$$e^{\lambda} = \frac{(\mu+t)(n-\mu)}{(n-\mu-t)\mu}. \tag{21.14}$$

Hence, by (21.13) and(21.14), assuming that $\mu + t < n$,

$$\mathbb{P}(S_n \geq \mu + t) \leq \left(\frac{\mu}{\mu+t}\right)^{\mu+t}\left(\frac{n-\mu}{n-\mu-t}\right)^{n-\mu-t}, \tag{21.15}$$

while for $t > n - \mu$ this probability is zero.

Now let

$$\varphi(x) = (1+x)\log(1+x) - x, \quad x \geq -1,$$

and let $\varphi(x) = \infty$ for $x < -1$. Now, for $0 \leq t \leq n - \mu$, we can rewrite the bound (21.15) as

$$\mathbb{P}(S_n \geq \mu + t) \leq \exp\left\{-\mu\varphi\left(\frac{t}{\mu}\right) - (n-\mu)\varphi\left(\frac{-t}{n-\mu}\right)\right\}. \tag{21.16}$$

Since $\varphi(x) \geq 0$ for every x, we obtain

$$\mathbb{P}(S_n \geq \mu + t) \leq e^{-\mu\varphi(t/\mu)}. \tag{21.17}$$

Similarly, putting $n - S_n$ for S_n, or by an analogous argument, using (21.12), we obtain for $0 \leq t \leq \mu$,

$$\mathbb{P}(S_n \leq \mu - t) \leq \exp\left\{-\mu\varphi\left(\frac{-t}{\mu}\right) - (n-\mu)\varphi\left(\frac{t}{n-\mu}\right)\right\}. \tag{21.18}$$

Hence,

$$\mathbb{P}(S_n \leq \mu - t) \leq e^{-\mu\varphi(-t/\mu)}. \tag{21.19}$$

We can simplify the expressions (21.17) and (21.19) by observing that

$$\varphi(x) \geq \frac{x^2}{2(1+x/3)}. \tag{21.20}$$

To see this, observe that for $|x| \leq 1$ we have

$$\varphi(x) - \frac{x^2}{2(1+x/3)} = \sum_{k=2}^{\infty}(-1)^k\left(\frac{1}{k(k-1)} - \frac{1}{2\cdot 3^{k-2}}\right)x^k.$$

Equation (21.20) for $|x| \leq 1$ follows from $\frac{1}{k(k-1)} - \frac{1}{2\cdot 3^{k-2}} \geq 0$ for $k \geq 2$. We leave it as an exercise to check that (21.20) remains true for $x > 1$.

Taking this into account we arrive at the following theorem, see Hoeffding [406].

Theorem 21.6 (Chernoff/Hoeffding inequality) *Suppose that $S_n = X_1 + X_2 + \cdots + X_n$, where (i) $0 \leq X_i \leq 1$ and $\mathbb{E}X_i = \mu_i$ for $i = 1, 2, \ldots, n$, (ii) $X_1, X_2, \ldots, X_n$ are independent. Let $\mu = \mu_1 + \mu_2 + \cdots + \mu_n$. Then for $t \geq 0$,*

$$\mathbb{P}(S_n \geq \mu + t) \leq \exp\left\{-\frac{t^2}{2(\mu + t/3)}\right\} \tag{21.21}$$

and for $t \leq \mu$,

$$\mathbb{P}(S_n \leq \mu - t) \leq \exp\left\{-\frac{t^2}{2(\mu - t/3)}\right\}. \tag{21.22}$$

Putting $t = \varepsilon\mu$, for $0 < \varepsilon < 1$, one can immediately obtain the following bounds.

Corollary 21.7 *Let* $0 < \varepsilon < 1$, *then*

$$\mathbb{P}(S_n \geq (1+\varepsilon)\mu) \leq \left(\frac{e^{\varepsilon}}{(1+\varepsilon)^{1+\varepsilon}}\right)^{\mu} \leq \exp\left\{-\frac{\mu\varepsilon^2}{3}\right\}, \tag{21.23}$$

while

$$\mathbb{P}(S_n \leq (1-\varepsilon)\mu) \leq \exp\left\{-\frac{\mu\varepsilon^2}{2}\right\}. \tag{21.24}$$

Proof The formula (21.24) follows directly from (21.22) and (21.23) follows from (21.16). □

One can "tailor" Chernoff bounds with respect to specific needs. For example, for small ratios t/μ, the exponent in (21.21) is close to $t^2/2\mu$, and the following bound holds.

Corollary 21.8

$$\mathbb{P}(S_n \geq \mu + t) \leq \exp\left\{-\frac{t^2}{2\mu} + \frac{t^3}{6\mu^2}\right\} \tag{21.25}$$

$$\leq \exp\left\{-\frac{t^2}{3\mu}\right\} \qquad \textit{for } t \leq \mu. \tag{21.26}$$

Proof Use (21.21) and note that

$$(\mu + t/3)^{-1} \geq (\mu - t/3)/\mu^2.$$

□

For large deviations we have the following result.

Corollary 21.9 *If* $c > 1$ *then*

$$\mathbb{P}(S_n \geq c\mu) \leq \left\{\frac{e}{ce^{1/c}}\right\}^{c\mu}. \tag{21.27}$$

Proof Put $t = (c-1)\mu$ into (21.17). □

Our next bound incorporates the variance of the X_is.

Theorem 21.10 (Bernstein's theorem) *Suppose that* $S_n = X_1 + X_2 + \cdots + X_n$ *where (i)* $|X_i| \leq 1$ *and* $\mathbb{E}X_i = 0$ *and* $\operatorname{Var} X_i = \sigma_i^2$ *for* $i = 1, 2, \ldots, n$, *(ii)* $X_1, X_2, \ldots, X_n$ *are independent. Let* $\sigma^2 = \sigma_1^2 + \sigma_2^2 + \cdots + \sigma_n^2$, *then for* $t \geq 0$,

$$\mathbb{P}(S_n \geq t) \leq \exp\left\{-\frac{t^2}{2(\sigma^2 + t/3)}\right\} \tag{21.28}$$

and

$$\mathbb{P}(S_n \le -t) \le \exp\left\{-\frac{t^2}{2(\sigma^2 + t/3)}\right\}. \tag{21.29}$$

Proof The strategy is once again to bound the moment generating function. Let

$$F_i = \sum_{r=2}^{\infty} \frac{\lambda^{r-2}\mathbb{E}X_i^r}{r!\sigma_i^2} \le \sum_{r=2}^{\infty} \frac{\lambda^{r-2}\sigma_i^2}{r!\sigma_i^2} = \frac{e^\lambda - 1 - \lambda}{\lambda^2}.$$

Here $\mathbb{E}X_i^r \le \sigma_i^2$, since $|X_i| \le 1$.

We then observe that

$$\begin{aligned}
\mathbb{E}(e^{\lambda X_i}) &= 1 + \sum_{r=2}^{\infty} \frac{\lambda^r \mathbb{E}X_i^r}{r!} \\
&= 1 + \lambda^2 \sigma_i^2 F_i \\
&\le e^{\lambda^2 \sigma_i^2 F_i} \\
&\le \exp\left\{(e^\lambda - \lambda - 1)\sigma_i^2\right\}.
\end{aligned}$$

So,

$$\begin{aligned}
\mathbb{P}(S_n \ge t) &\le e^{-\lambda t} \prod_{i=1}^{n} \exp\left\{(e^\lambda - \lambda - 1)\sigma_i^2\right\} \\
&= e^{\sigma^2(e^\lambda - \lambda - 1) - \lambda t} \\
&= \exp\left\{-\sigma^2 \varphi\left(\frac{t}{\sigma^2}\right)\right\}.
\end{aligned}$$

after assigning

$$\lambda = \log\left(1 + \frac{t}{\sigma^2}\right).$$

To obtain (21.28) we use (21.20). To obtain (21.29) we apply (21.28) to $Y_i = -X_i, i = 1, 2, \ldots, n$. □

21.5 Sampling Without Replacement

Let a multi-set $A = \{a_1, a_2, \ldots, a_N\} \subseteq \mathbf{R}$ be given. We consider two random variables. For the first let $X = a_i$, where i is chosen uniformly at random from $[N]$. Let

$$\mu = \mathbb{E}X = \frac{1}{N}\sum_{i=1}^{N} a_i \text{ and } \sigma^2 = \operatorname{Var} X = \frac{1}{N}\sum_{i=1}^{N}(a_i - \mu)^2.$$

Now let $S_n = X_1 + X_2 + \cdots + X_n$ be the sum of n independent copies of X. Next let $W_n = \sum_{i \in \mathbf{X}} a_i$ where $\mathbf{X}$ is a uniformly random n-subset of $[N]$. We have $\mathbb{E} S_n = \mathbb{E} W_n = n\mu$ but as shown in Hoeffding [406], W_n is more tightly concentrated around its mean than S_n. This follows from the following:

Lemma 21.11 *Let $f : \mathbf{R} \to \mathbf{R}$ be continuous and convex, then*

$$\mathbb{E} f(W_n) \leq \mathbb{E} f(S_n).$$

Proof We write, where $(A)_n$ denotes the set of sequences of n distinct members of A and $(N)_n = N(N-1)\cdots(N-n+1) = |(A)_n|$,

$$\begin{aligned} \mathbb{E} f(S_n) &= \frac{1}{N^n} \sum_{\mathbf{y} \in A^n} f(y_1 + \cdots + y_n) \\ &= \frac{1}{(N)_n} \sum_{\mathbf{x} \in (A)_n} g(x_1, x_2, \ldots, x_n) = \mathbb{E} g(\mathbf{X}), \end{aligned} \tag{21.30}$$

where g is a symmetric function of $\mathbf{x}$ and

$$g(x_1, x_2, \ldots, x_n) = \sum_{k, \mathbf{i}, \mathbf{r}} \psi(k, \mathbf{i}, \mathbf{r}) f(r_{i_1} x_{i_1} + \cdots + r_{i_k} x_{i_k}).$$

Here $\mathbf{i}$ ranges over sequences of k distinct values $i_1, i_2, \ldots, i_k \in [n]$ and $r_{i_1} + \cdots + r_{i_k} = n$. The factors $\psi(k, \mathbf{i}, \mathbf{r})$ are independent of the function f.

Putting $f = 1$ we see that $\frac{1}{(N)_n} \sum_{k, \mathbf{i}, \mathbf{r}} \psi(k, \mathbf{i}, \mathbf{r}) = 1$. Putting $f(x) = x$ we see that g is a linear symmetric function and so

$$\sum_{k, \mathbf{i}, \mathbf{r}} \psi(k, \mathbf{i}, \mathbf{r})(r_{i_1} x_{i_1} + \cdots + r_{i_k} x_{i_k}) = K(x_1 + \cdots + x_n),$$

for some K. Equation (21.30) implies that $K = 1$.

Applying Jensen's inequality we see that

$$g(\mathbf{x}) \geq f(x_1 + \cdots + x_n).$$

It follows that

$$\mathbb{E} g(\mathbf{X}) \geq \mathbb{E} f(W_n)$$

and the Lemma follows from (21.30). □

As a consequence, we have that (i) $\operatorname{Var} W_n \leq \operatorname{Var} S_n$ and (ii) $\mathbb{E} e^{\lambda W_n} \leq \mathbb{E} e^{\lambda S_n}$ for any $\lambda \in \mathbf{R}$.

Thus all the inequalities developed in Section 21.4 can a fortiori be applied to W_n in place of S_n. Of particular importance in this context, is the *hypergeometric distribution*: here we are given a set of $S \subseteq [N]$, $|S| = m$ and we choose a random set X of size k from $[N]$. Let $Z = |X \cap S|$, then

$$\mathbb{P}(Z = t) = \frac{\binom{m}{t}\binom{N-m}{k-t}}{\binom{N}{k}}, \quad \text{for } 0 \le t \le k.$$

21.6 Janson's Inequality

In Section 21.4 we found bounds for the upper and lower tails of the distribution of a random variable S_n composed of n *independent* summands. In the previous section we allowed some dependence between the summands. We consider another case where the random variables in question are not necessarily independent. In this section we prove an inequality of Janson [418]. This generalized an earlier inequality of Janson, Łuczak and Ruciński [431], see Corollary 21.13.

Fix a family of n subsets $D_i, i \in [n]$. Let R be a random subset of $[N]$ such that for $s \in [N]$ we have $0 < \mathbb{P}(s \in R) = q_s < 1$. The elements of R are chosen independently of each other and the sets $D_i, i = 1, 2, \ldots, n$. Let $\mathscr{A}_i$ be the event that D_i is a subset of R. Moreover, let I_i be the indicator of the event $\mathscr{A}_i$. Note that, I_i and I_j are independent if and only if $s \in D_i \cap D_j = \emptyset$. We can easily see that the I_is are increasing.

We let

$$S_n = I_1 + I_2 + \cdots + I_n,$$

and

$$\mu = \mathbb{E}S_n = \sum_{i=1}^{n} \mathbb{E}(I_i).$$

We write $i \sim j$ if $D_i \cap D_j \neq \emptyset$. Then, let

$$\overline{\Delta} = \sum_{\{i,j\}: i \sim j} \mathbb{E}(I_i I_j) = \mu + \Delta, \tag{21.31}$$

where

$$\Delta = \sum_{\substack{\{i,j\}: i \sim j \\ i \neq j}} \mathbb{E}(I_i I_j). \tag{21.32}$$

As before, let $\varphi(x) = (1+x)\log(1+x) - x$. Now, with $S_n, \overline{\Delta}, \varphi$ given above one can establish the following upper bound on the lower tail of the distribution of S_n.

Theorem 21.12 (Janson's inequality) *For any real t, $0 \le t \le \mu$,*

$$\mathbb{P}(S_n \le \mu - t) \le \exp\left\{-\frac{\varphi(-t/\mu)\mu^2}{\overline{\Delta}}\right\} \le \exp\left\{-\frac{t^2}{2\overline{\Delta}}\right\}. \tag{21.33}$$

Proof We begin as we did in Section 21.4. Put $\psi(\lambda) = \mathbb{E}(e^{-\lambda S_n})$, $\lambda \ge 0$. By Markov's inequality we have

$$\mathbb{P}(S_n \le \mu - t) \le e^{\lambda(\mu-t)}\mathbb{E}e^{-\lambda S_n}.$$

Therefore,

$$\log \mathbb{P}(S_n \le \mu - t) \le \log \psi(\lambda) + \lambda(\mu - t). \tag{21.34}$$

Now let us estimate $\log \psi(\lambda)$ and minimize the RHS of (21.34) with respect to λ.

Note that

$$-\psi'(\lambda) = \mathbb{E}(S_n e^{-\lambda S_n}) = \sum_{i=1}^{n} \mathbb{E}(I_i e^{-\lambda S_n}). \tag{21.35}$$

Now for every $i \in [n]$, split S_n into Y_i and Z_i, where

$$Y_i = \sum_{j:j\sim i} I_j, \quad Z_i = \sum_{j:j\not\sim i} I_j, \quad S_n = Y_i + Z_i.$$

Then by the FKG inequality (applied to the random set R and conditioned on $I_i = 1$) we get, setting $p_i = \mathbb{E}(I_i) = \prod_{s\in D_i} q_s$,

$$\mathbb{E}(I_i e^{-\lambda S_n}) = p_i \mathbb{E}(e^{-\lambda Y_i} e^{-\lambda Z_i} \mid I_i = 1) \ge p_i \mathbb{E}(e^{-\lambda Y_i} \mid I_i = 1)\mathbb{E}(e^{-\lambda Z_i} \mid I_i = 1).$$

Since, Z_i and I_i are independent we get

$$\mathbb{E}(I_i e^{-\lambda S_n}) \ge p_i \mathbb{E}(e^{-\lambda Y_i} \mid I_i = 1)\mathbb{E}(e^{-\lambda Z_i}) \ge p_i \mathbb{E}(e^{-\lambda Y_i} \mid I_i = 1)\psi(\lambda). \tag{21.36}$$

From (21.35) and (21.36), applying Jensen's inequality to get (21.37) and remembering that $\mu = \mathbb{E}S_n = \sum_{i=1}^n p_i$, we get

$$
\begin{aligned}
-(\log \psi(\lambda))' &= -\frac{\psi'(\lambda)}{\psi(\lambda)} \\
&\geq \sum_{i=1}^{n} p_i \mathbb{E}(e^{-\lambda Y_i} \mid I_i = 1) \\
&\geq \mu \sum_{i=1}^{n} \frac{p_i}{\mu} \exp\{-\mathbb{E}(\lambda Y_i \mid I_i = 1)\} \\
&\geq \mu \exp\left\{-\frac{1}{\mu}\sum_{i=1}^{n} p_i \mathbb{E}(\lambda Y_i \mid I_i = 1)\right\} \qquad (21.37) \\
&= \mu \exp\left\{-\frac{\lambda}{\mu}\sum_{i=1}^{n} \mathbb{E}(Y_i I_i)\right\} \\
&= \mu e^{-\lambda\overline{\Delta}/\mu}.
\end{aligned}
$$

So

$$
-(\log \psi(\lambda))' \geq \mu e^{-\lambda\overline{\Delta}/\mu}, \qquad (21.38)
$$

which implies that

$$
-\log \psi(\lambda) \geq \int_0^{\lambda} \mu e^{-z\overline{\Delta}/\mu} dz = \frac{\mu^2}{\overline{\Delta}}(1 - e^{-\lambda\overline{\Delta}/\mu}). \qquad (21.39)
$$

Hence by (21.39) and (21.34)

$$
\log \mathbb{P}(S_n \leq \mu - t) \leq -\frac{\mu^2}{\overline{\Delta}}(1 - e^{-\lambda\overline{\Delta}/\mu}) + \lambda(\mu - t), \qquad (21.40)
$$

which is minimized by choosing $\lambda = -\log(1 - t/\mu)\mu/\overline{\Delta}$. It yields the first bound in (21.33), while the final bound in (21.33) follows from the fact that $\varphi(x) \geq x^2/2$ for $x \leq 0$. □

The following corollary is very useful:

Corollary 21.13 (Janson, Łuczak, Ruciński inequality)

$$
\mathbb{P}(S_n = 0) \leq e^{-\mu+\Delta}.
$$

Proof We put $t = \mu$ into (21.33) giving $\mathbb{P}(S_n = 0) \leq \exp\left\{-\frac{\varphi(-1)\mu^2}{\overline{\Delta}}\right\}$. Now note that $\varphi(-1) = 1$ and $\frac{\mu^2}{\overline{\Delta}} \geq \frac{\mu^2}{\mu+\Delta} \geq \mu - \Delta$. □

21.7 Martingales. Azuma–Hoeffding Bounds

Before we present the basic results of this chapter we have to briefly introduce *martingales* and concentration inequalities for martingales. Historically, martingales were applied to random graphs for the first time in the context of the chromatic number of $\mathbb{G}_{n,p}$.

Let $(\Omega, \mathscr{F}, \mathbb{P})$ be a probability space. If the sample space Ω is finite, then $\mathscr{F}$ is the algebra of all subsets of Ω. For simplicity, let us assume that we deal with this case.

Recall that if $\mathscr{D} = \{D_1, D_2, \ldots, D_m\}$ is a *partition* of Ω, i.e. $\bigcup_{i=1}^m D_i = \Omega$ and $D_i \cap D_j = \emptyset$ if $i \neq j$, then it generates an algebra of subsets $\mathscr{A}(\mathscr{D})$ of Ω. The algebra generated by the partition $\mathscr{D}$ and denoted by $\mathscr{A}(\mathscr{D})$ is the family of all unions of the events (sets) from $\mathscr{D}$, with $\emptyset$ obtained by taking an empty union. Let $\mathscr{D} = \{D_1, D_2, \ldots, D_m\}$ be a partition of Ω and A be any event, $A \subset \Omega$ and let $\mathbb{P}(A|\mathscr{D})$ be the random variable defined by

$$\begin{aligned}\mathbb{P}(A|\mathscr{D})(\omega) &= \sum_{i=1}^{m} \mathbb{P}(A|D_i) I_{D_i}(\omega) \\ &= \mathbb{P}(A|D_{i(\omega)}) \text{ where } \omega \in D_{i(\omega)}.\end{aligned}$$

Note that if $\mathscr{D}$ a trivial partition, i.e. $\mathscr{D} = \mathscr{D}_0 = \{\Omega\}$ then $\mathbb{P}(A|\mathscr{D}_0) = \mathbb{P}(A)$, while, in general,

$$\mathbb{P}(A) = \mathbb{E}\,\mathbb{P}(A|\mathscr{D}). \tag{21.41}$$

Suppose that X is a discrete random variable taking values $\{x_1, x_2, \ldots, x_l\}$ and write X as

$$X = \sum_{j=1}^{l} x_j I_{A_j}, \tag{21.42}$$

where $A_j = \{\omega : X(\omega) = x_j\}$. Notice that the random variable X generates a partition $\mathscr{D}_X = \{A_1, A_2, \ldots, A_l\}$.

Now the conditional expectation of X with respect to a partition $\mathscr{D}$ of Ω is given as

$$\mathbb{E}(X|\mathscr{D}) = \sum_{j=1}^{l} x_j \mathbb{P}(A_j|\mathscr{D}). \tag{21.43}$$

Hence, $\mathbb{E}(X|\mathscr{D})(\omega_1)$ is the expected value of X conditional on the event $\left\{\omega \in D_{i(\omega_1)}\right\}$.

Suppose that $\mathscr{D}$ and $\mathscr{D}'$ are two partitions of Ω. We say that $\mathscr{D}'$ is *finer* than $\mathscr{D}$ if $\mathscr{A}(\mathscr{D}) \subseteq \mathscr{A}(\mathscr{D}')$ and denote this as $\mathscr{D} \prec \mathscr{D}'$.

If $\mathscr{D}$ is a partition of Ω and Y is a discrete random variable defined on Ω, then Y is $\mathscr{D}$-measurable if $\mathscr{D}_Y \prec \mathscr{D}$, i.e. if the partition $\mathscr{D}$ is *finer* than the partition induced by Y. It simply means that Y takes constant values y_i on the atoms D_i of $\mathscr{D}$, so Y can be written as $Y = \sum_{i=1}^{m} y_i I_{D_i}$, where some y_i may be equal. Note that a random variable Y is $\mathscr{D}_0$-measurable if Y has a degenerate distribution, i.e. it takes a constant value on all $\omega \in \Omega$. Also, trivially, the random variable Y is $\mathscr{D}_Y$-measurable.

Note that if $\mathscr{D}'$ is finer than $\mathscr{D}$ then

$$\mathbb{E}(\mathbb{E}(X \mid \mathscr{D}') \mid \mathscr{D}) = \mathbb{E}(X \mid \mathscr{D}). \tag{21.44}$$

Indeed, if $\omega \in \Omega$ then

$$\begin{aligned}
\mathbb{E}(\mathbb{E}(X \mid \mathscr{D}') \mid \mathscr{D})(\omega) &= \sum_{\omega' \in D_{i(\omega)}} \left(\sum_{\omega'' \in D'_{i(\omega')}} X(\omega'') \frac{\mathbb{P}(\omega'')}{\mathbb{P}(D'_{i(\omega')})} \right) \frac{\mathbb{P}(\omega')}{\mathbb{P}(D_{i(\omega)})} \\
&= \sum_{\omega'' \in D_{i(w)}} Z(\omega'')\mathbb{P}(\omega'') \sum_{\omega' \in D'_{i(\omega'')}} \frac{\mathbb{P}(\omega')}{\mathbb{P}(D'_{i(\omega')})\mathbb{P}(D_{i(\omega)})} \\
&= \sum_{\omega'' \in D_{i(w)}} Z(\omega'')\mathbb{P}(\omega'') \sum_{\omega' \in D'_{i(\omega'')}} \frac{\mathbb{P}(\omega')}{\mathbb{P}(D'_{i(\omega'')})\mathbb{P}(D_{i(\omega)})} \\
&= \sum_{\omega'' \in D_{i(w)}} Z(\omega'') \frac{\mathbb{P}(\omega'')}{\mathbb{P}(D_{i(\omega)})} \\
&= \mathbb{E}(X \mid \mathscr{D})(\omega).
\end{aligned}$$

Note that despite all the algebra, (21.44) just boils down to saying that the properly weighted average of averages is just the average.

Finally, suppose a partition $\mathscr{D}$ of Ω is induced by a sequence of random variables $\{Y_1, Y_2, \ldots, Y_n\}$. We denote such partition as $\mathscr{D}_{Y_1,Y_2,\ldots,Y_n}$, then the atoms of this partition are defined as

$$D_{y_1,y_2,\ldots,y_n} = \{\omega : Y_1(\omega) = y_1, Y_2(\omega) = y_2, \ldots, Y_n(\omega) = y_n\},$$

where the y_i range over all possible values of the Y_is. $\mathscr{D}_{Y_1,Y_2,\ldots,Y_n}$ is then the *coarsest* partition such that $Y_1, Y_2, \ldots, Y_n$ are all constant over the atoms of the partition. For convenience, we simply write $\mathbb{E}(X|Y_1, Y_2, \ldots, Y_n)$, instead of $\mathbb{E}(X|\mathscr{D}_{Y_1,Y_2,\ldots,Y_n})$.

Now we are ready to introduce an important class of *dependent* random variables called *martingales*.

Let $(\Omega, \mathscr{F}, \mathbb{P})$ be a finite probability space and $\mathscr{D}_0 \prec \mathscr{D}_1 \prec \mathscr{D}_2 \prec \ldots \prec \mathscr{D}_n = \mathscr{D}^*$ be a nested sequence of partitions of Ω (a *filtration* of Ω), where $\mathscr{D}_0$ is a trivial partition, while $\mathscr{D}^*$ stands for the discrete partition (i.e. $\mathscr{A}(\mathscr{D}_0) = \{\emptyset, \Omega\}$, while $\mathscr{A}(\mathscr{D}^*) = 2^{\Omega} = \mathscr{F}$).

A sequence of random variables $X_0, X_1, \ldots, X_n$ is called (a) a *martingale*, (b) a *super-martingale* and (c) a *sub-martingale*, with respect to the partition $\mathscr{D}_0 \prec \mathscr{D}_1 \prec \mathscr{D}_2 \prec \ldots \prec \mathscr{D}_n = \mathscr{D}^*$ if

$$X_k \text{ is } \mathscr{D}_k\text{-measurable}$$

and

(a) $\mathbb{E}(X_{k+1} \mid \mathscr{D}_k) = X_k \quad k = 0, 1, \ldots, n-1.$
(b) $\mathbb{E}(X_{k+1} \mid \mathscr{D}_k) \leq X_k \quad k = 0, 1, \ldots, n-1.$
(c) $\mathbb{E}(X_{k+1} \mid \mathscr{D}_k) \geq X_k \quad k = 0, 1, \ldots, n-1.$

If the partition $\mathscr{D}$ of Ω is generated by a sequence of random variables $Y_1, \ldots, Y_n$, then the sequence $X_1, \ldots, X_n$ is called a martingale with respect to the sequence $Y_1, \ldots, Y_n$. In particular, when $Y_1 = X_1, \ldots, Y_n = X_n$, i.e. when $\mathscr{D}_k = \mathscr{D}_{X_1, \ldots, X_k}$, then we simply say that X is a martingale with respect to itself. Observe also that $\mathbb{E}X_k = \mathbb{E}X_1 = X_0$, for every k. Analogous statements hold for super- and sub-martingales.

Martingales are ubiquitous, we can obtain a martingale from essentially any random variable. Let $Z = Z(Y_1, Y_2, \ldots, Y_n)$ be a random variable defined on the random variables $Y_1, Y_2, \ldots, Y_n$. The sequence of random variables

$$X_k = \mathbb{E}(Z \mid Y_1, Y_2, \ldots, Y_k), \quad k = 0, 1, \ldots, n$$

is called the *Doob Martingale* of Z.

Theorem 21.14 *We have (i) $X_0 = \mathbb{E}Z$, (ii) $X_n = Z$ and (iii) the sequence $X_0, X_1, \ldots, X_n$ is a martingale with respect to (the partition defined by) $Y_1, Y_2, \ldots, Y_n$.*

Proof Only (*iii*) needs to be explicitly checked.

$$\begin{aligned}\mathbb{E}(X_k \mid Y_1, \ldots, Y_{k-1}) &= \mathbb{E}(\mathbb{E}(Z \mid Y_1, \ldots, Y_k) \mid Y_1, \ldots, Y_{k-1}) \\ &= \mathbb{E}(Z \mid Y_1, \ldots, Y_{k-1}) \\ &= X_{k-1}.\end{aligned}$$

Here the second equality comes from (21.44).

□

We next show how one can define the so-called, *vertex and edge exposure martingales*, on the space of random graphs. Consider the binomial random graph $\mathbb{G}_{n,p}$. Let us view $\mathbb{G}_{n,p}$ as a vector of random variables $(I_1, I_2, \ldots, I_{\binom{n}{2}})$, where I_i is the indicator of the event that the ith edge is present, with

$\mathbb{P}(I_i = 1) = p$ and $\mathbb{P}(I_i = 0) = 1 - p$ for $i = 1,2,\ldots,\binom{n}{2}$. These random variables are independent of each other. Hence, in this case, Ω consists of all $(0,1)$-sequences of length $\binom{n}{2}$.

Now given any graph invariant (a random variable) $X : \Omega \to \mathbb{R}$, (e.g., the chromatic number, the number of vertices of given degree or the size of the largest clique), we will define a martingale generated by X and certain sequences of partitions of Ω.

Let the random variables $I_1, I_2, \ldots, I_{\binom{n}{2}}$ be listed in a lexicographic order. Define $\mathscr{D}_0 \prec \mathscr{D}_1 \prec \mathscr{D}_2 \prec \ldots \prec \mathscr{D}_n = \mathscr{D}^*$ in the following way: $\mathscr{D}_k$ is the partition of Ω induced by the sequence of random variables $I_1, \ldots, I_{\binom{k}{2}}$ and $\mathscr{D}_0$ is the trivial partition. Finally, for $k = 1, \ldots, n$,

$$X_k = \mathbb{E}(X \mid \mathscr{D}_k) = \mathbb{E}(X \mid \mathscr{D}_{I_1, I_2, \ldots, I_{\binom{k}{2}}}).$$

Hence, X_k is the conditional expectation of X, given that we "uncovered" the set of edges induced by the first k vertices of our random graph $\mathbb{G}_{n,p}$. A martingale determined through such a sequence of nested partitions is called a *vertex exposure martingale*.

An *edge exposure martingale* is defined in a similar way. The martingale sequence is defined as follows

$$X_k = \mathbb{E}(X \mid \mathscr{D}_k) = \mathbb{E}(X \mid \mathscr{D}_{I_1, I_2, \ldots, I_k}),$$

where $k = 1, 2, \ldots, \binom{n}{2}$, i.e. we uncover the edges of $\mathbb{G}_{n,p}$ one by one.

We next give upper bounds for both the lower and upper tails of the probability distributions of certain classes of martingales.

Theorem 21.15 Azuma–Hoeffding bound *Let $\{X_k\}_0^n$ be a sequence of random variables such that $|X_k - X_{k-1}| \le c_k$, $k = 1, \ldots, n$ and X_0 is constant.*

(a) If $\{X_k\}_0^n$ is a super-martingale then for all $t > 0$ we have

$$\mathbb{P}(X_n \ge \mathbb{E}X_n + t) \le \exp\left\{-\frac{t^2}{2\sum_{i=1}^n c_i^2}\right\}.$$

(b) If $\{X_k\}_0^n$ is a sub-martingale then for all $t > 0$ we have

$$\mathbb{P}(X_n \le \mathbb{E}X_n - t) \le \exp\left\{-\frac{t^2}{2\sum_{i=1}^n c_i^2}\right\}.$$

(c) If $\{X_k\}_0^n$ is a martingale then for all $t > 0$ we have

$$\mathbb{P}(|X_n - \mathbb{E}X_n| \ge t) \le 2\exp\left\{-\frac{t^2}{2\sum_{i=1}^n c_i^2}\right\}.$$

Proof We only need to prove (a), since (b) and (c) will then follow easily, since $\{X_k\}_0^n$ is a sub-martingale if and only if $-\{X_k\}_0^n$ is a super-martingale and $\{X_k\}_0^n$ is a martingale if and only if it is a super-martingale and a sub-martingale.

Define the *martingale difference sequence* by $Y_1 = 0$ and

$$Y_k = X_k - X_{k-1} \quad , \quad k = 1, \ldots, n.$$

Then

$$\sum_{k=1}^{n} Y_k = X_n - X_0 = X_n - \mathbb{E}X_n,$$

and

$$\mathbb{E}(Y_{k+1} \mid Y_0, Y_1, \ldots, Y_k) \leq 0. \tag{21.45}$$

Let $\lambda > 0$, then

$$\begin{aligned}\mathbb{P}(X_n \geq t) &= \mathbb{P}\left(\exp\left\{\lambda \sum_{i=1}^{n} Y_i\right\} \geq e^{\lambda t}\right) \\ &\leq e^{-\lambda t}\,\mathbb{E}\left(\exp\left\{\lambda \sum_{i=1}^{n} Y_i\right\}\right),\end{aligned}$$

by Markov's inequality.

Note that $e^{\lambda x}$ is a convex function of x, and since $-c_i \leq Y_i \leq c_i$, we have

$$\begin{aligned}e^{\lambda Y_i} &\leq \frac{1 - Y_i/c_i}{2} e^{-\lambda c_i} + \frac{1 + Y_i/c_i}{2} e^{\lambda c_i} \\ &= \cosh(\lambda c_i) + \frac{Y_i}{c_i} \sinh(\lambda c_i).\end{aligned}$$

It follows from (21.45) that

$$\mathbb{E}(e^{\lambda Y_n} \mid Y_0, Y_1, \ldots, Y_k) \leq \cosh(\lambda c_i). \tag{21.46}$$

We then see that

$$\begin{aligned}\mathbb{E}\left(\exp\left\{\lambda \sum_{i=1}^{n} Y_i\right\}\right) &= \mathbb{E}\left(\mathbb{E}(e^{\lambda Y_n} \mid Y_0, Y_1, \ldots, Y_n) \times \mathbb{E}\left(\exp\left\{\lambda \sum_{i=1}^{n-1} Y_i\right\}\right)\right) \\ &\leq \cosh(\lambda c_n)\,\mathbb{E}\left(\exp\left\{\lambda \sum_{i=1}^{n-1} Y_i\right\}\right) \leq \prod_{i=1}^{n} \cosh(\lambda c_i).\end{aligned}$$

The expectation in the middle term is over $Y_0, Y_1, \ldots, Y_{n-1}$ and the last inequality follows by induction on n.

By the above equality and the Taylor expansion, we obtain

$$\begin{aligned}
e^{\lambda t}\mathbb{P}(X_n \geq t) &\leq \prod_{i=1}^{n}\cosh(\lambda c_i) = \prod_{i=1}^{n}\sum_{m=0}^{\infty}\frac{(\lambda c_i)^{2m}}{(2m)!} \\
&\leq \prod_{i=1}^{n}\sum_{m=0}^{\infty}\frac{(\lambda c_i)^{2m}}{2^m m!} = \exp\left\{\frac{1}{2}\lambda^2\sum_{i=1}^{n}c_i^2\right\}.
\end{aligned}$$

Putting $\lambda = t/\sum_{i=1}^{n} c_i^2$ we arrive at the theorem. □

We end by describing a simple situation where we can apply these inequalities.

Lemma 21.16 (McDiarmid's inequality) *Let $Z = Z(W_1, W_2, \ldots, W_n)$ be a random variable that depends on n independent random variables $W_1, W_2, \ldots, W_n$. Suppose that*

$$|Z(W_1, \ldots, W_i, \ldots, W_n) - Z(W_1, \ldots, W_i', \ldots, W_n)| \leq c_i$$

for all $i = 1, 2, \ldots, n$ and $W_1, W_2, \ldots, W_n, W_i'$. Then for all $t > 0$ we have

$$\mathbb{P}(Z \geq \mathbb{E}Z + t) \leq \exp\left\{-\frac{t^2}{2\sum_{i=1}^{n} c_i^2}\right\},$$

and

$$\mathbb{P}(Z \leq \mathbb{E}Z - t) \leq \exp\left\{-\frac{t^2}{2\sum_{i=1}^{n} c_i^2}\right\}.$$

Proof We consider the martingale

$$X_k = X_k(W_1, W_2, \ldots, W_k) = \mathbb{E}(Z \mid W_1, W_2, \ldots, W_k),$$

then

$$X_0 = \mathbb{E}Z \text{ and } X_n = Z.$$

We only have to show that the martingale differences $Y_k = X_k - X_{k-1}$ are bounded. But,

$$
\begin{aligned}
&|X_k(W_1,\ldots,W_k) - X_{k-1}(W_1,\ldots,W_{k-1})| \\
&\quad \leq \sum_{W_k',W_{k+1},\ldots,W_n} |Z(W_1,\ldots,W_k,\ldots,W_n) - Z(W_1,\ldots,W_k',\ldots,W_n))| \\
&\quad \times \mathbb{P}(W_k') \prod_{i=k+1}^{n} \mathbb{P}(W_i) \\
&\quad \leq \sum_{W_k',W_{k+1},\ldots,W_n} c_i \mathbb{P}(W_k') \prod_{i=k+1}^{n} \mathbb{P}(W_i) \\
&\quad = c_i.
\end{aligned}
$$

□

21.8 Talagrand's Inequality

In this section we describe a concentration inequality that is due to Talagrand [702] that has proved to be very useful. It can often overcome the following problem with using Theorems 21.15, 21.16: if $\mathbb{E} X_n = O(n^{1/2})$ then the bounds they give are weak. Our treatment is a re-arrangement of the treatment in Alon and Spencer [30].

Let $\Omega = \prod_{i=1}^n \Omega_i$, where each Ω_i is a probability space and Ω has the product measure. Let $A \subseteq \Omega$ and let $\mathbf{x} = (x_1, x_2, \ldots, x_n) \in \Omega$.

For $\boldsymbol{\alpha} = (\alpha_1, \alpha_2, \ldots, \alpha_n)$ we let

$$d_{\boldsymbol{\alpha}}(A, \mathbf{x}) = \inf_{\mathbf{y} \in A} \sum_{i: y_i \neq x_i} \alpha_i.$$

Then we define

$$\rho(A, \mathbf{x}) = \sup_{|\alpha|=1} d_{\boldsymbol{\alpha}}(A, \mathbf{x}),$$

where $|\alpha|$ denotes the Euclidean norm, $(\alpha_1^2 + \cdots + \alpha_n^2)^{1/2}$.

We then define, for $t \geq 0$,

$$A_t = \{\mathbf{x} \in \Omega : \rho(A, \mathbf{x}) \leq t\}.$$

The following theorem is due to Talagrand [702]:

Theorem 21.17

$$\mathbb{P}(A)(1 - \mathbb{P}(A_t)) \leq e^{-t^2/4}.$$

Theorem 21.17 follows from

Lemma 21.18

$$\int_\Omega \exp\left\{\frac{1}{4}\rho^2(A,\mathbf{x})\right\} d\mathbf{x} \le \frac{1}{\mathbb{P}(A)}.$$

Proof Indeed, fix A and consider $X = \rho(A,\mathbf{x})$, then,

$$1 - \mathbb{P}(A_t) = \mathbb{P}(X > t) = \mathbb{P}(e^{X^2/4} > e^{t^2/4}) \le \mathbb{E}(e^{X^2/4})e^{-t^2/4}.$$

The lemma states that $\mathbb{E}(e^{X^2/4}) \le \frac{1}{\mathbb{P}(A)}$. □

The following alternative description of ρ is important. Let

$$U(A,\mathbf{x}) = \{\mathbf{s} \in \{0,1\}^n : \exists \mathbf{y} \in A \text{ s.t. } s_i = 0 \text{ implies } x_i = y_i\}$$

and let $V(A,\mathbf{x})$ be the convex hull of $U(A,\mathbf{x})$, then

Lemma 21.19

$$\rho(A,\mathbf{x}) = \min_{\mathbf{v} \in V(A,\mathbf{x})} |\mathbf{v}|.$$

Here $|v|$ denotes the Euclidean norm of v. We leave the proof of this lemma as a simple exercise in convex analysis.

We now give the proof of Lemma 21.18.

Proof We use induction on the dimension n. For $n = 1, \rho(A,\mathbf{x}) = 1_{\mathbf{x} \notin A}$ so that

$$\int_\Omega \exp\left\{\frac{1}{4}\rho^2(A,\mathbf{x})\right\} = \mathbb{P}(A) + (1 - \mathbb{P}(A))e^{1/4} \le \frac{1}{\mathbb{P}(A)},$$

which follows from $u + (1-u)e^{1/4} \le u^{-1}$ for $0 < u \le 1$.

Assume the result for n. Write $\Psi = \prod_{i=1}^n \Omega_i$ so that $\Omega = \Psi \times \Omega_{n+1}$. Any $\mathbf{z} \in \Omega$ can be written uniquely as $\mathbf{z} = (\mathbf{x},\omega)$ where $\mathbf{x} \in \Psi$ and $\omega \in \Omega_{n+1}$. Set

$$B = \{\mathbf{x} \in \Psi : (\mathbf{x},\omega) \in A \text{ for some } \omega \in \Omega_{n+1}\}$$

and for $\omega \in \Omega_{n+1}$ set

$$A_\omega = \{\mathbf{x} \in \Psi : (\mathbf{x},\omega) \in A\}.$$

Then

$$\mathbf{s} \in U(B,\mathbf{x}) \Longrightarrow (\mathbf{s},1) \in U(A,(\mathbf{x},\omega)).$$

$$\mathbf{t} \in U(A_\omega,\mathbf{x}) \Longrightarrow (\mathbf{t},0) \in U(A,(\mathbf{x},\omega)).$$

If $\mathbf{s} \in V(B,\mathbf{x})$ and $\mathbf{t} \in V(A_\omega,\mathbf{x})$ then $(\mathbf{s},1)$ and $(\mathbf{t},0)$ are both in $V(A,(\mathbf{x},\omega))$ and hence for any $\lambda \in [0,1]$,

$$((1-\lambda)\mathbf{s} + \lambda\mathbf{t}, 1-\lambda) \in V(A,(\mathbf{x},\omega)).$$

Then,

$$\rho^2(A,(\mathbf{x},\omega)) \le (1-\lambda)^2 + |(1-\lambda)\mathbf{s} + \lambda\mathbf{t}|^2 \le (1-\lambda)^2 + (1-\lambda)|\mathbf{s}|^2 + \lambda|\mathbf{t}|^2,$$

where the second inequality uses the convexity of $|\cdot|^2$.

Selecting, $\mathbf{s},\mathbf{t}$ with minimal norms yields the critical inequality,

$$\rho^2(A,(\mathbf{x},\omega)) \le (1-\lambda)^2 + \lambda\rho^2(A_\omega,\mathbf{x}) + (1-\lambda)\rho^2(B,\mathbf{x}).$$

Now fix ω and bound,

$$\int_{\mathbf{x}\in\Psi} \exp\left\{\frac{1}{4}\rho^2(A,(\mathbf{x},\omega))\right\} \le$$

$$e^{(1-\lambda)^2/4}\int_{\mathbf{x}\in\Psi} \exp\left\{\frac{1}{4}\rho^2(A_\omega,\mathbf{x}))\right\}^{\lambda} \exp\left\{\frac{1}{4}\rho^2(B,\mathbf{x}))\right\}^{1-\lambda}.$$

By Hölder's inequality this is at most

$$e^{(1-\lambda)^2/4}\left(\int_{\mathbf{x}\in\Psi} \exp\left\{\frac{1}{4}\rho^2(A_\omega,\mathbf{x}))\right\}\right)^{\lambda} \left(\int_{\mathbf{x}\in\Psi} \exp\left\{\frac{1}{4}\rho^2(B,\mathbf{x}))\right\}\right)^{1-\lambda},$$

which by induction is at most

$$e^{(1-\lambda)^2/4}\frac{1}{\mathbb{P}(A_\omega)^\lambda}\cdot\frac{1}{\mathbb{P}(B)^{1-\lambda}} = \frac{1}{\mathbb{P}(B)}e^{(1-\lambda)^2/4}r^{-\lambda}$$

where $r = \mathbb{P}(A_\omega)/\mathbb{P}(B) \le 1$.

Using calculus, we minimize $e^{(1-\lambda)^2/4}r^{-\lambda}$ by choosing $\lambda = 1+2\log r$ for $e^{-1/2} \le r \le 1$, $\lambda = 0$ otherwise. Further calculation shows that $e^{(1-\lambda)^2/4}r^{-\lambda} \le 2-r$ for this value of λ. Thus,

$$\int_{\mathbf{x}\in\Psi} \exp\left\{\frac{1}{4}\rho^2(A,(\mathbf{x},\omega))\right\} \le \frac{1}{\mathbb{P}(B)}\left(2-\frac{\mathbb{P}(A_\omega)}{\mathbb{P}(B)}\right).$$

We integrate over ω to give

$$\int_{\omega\in\Omega_{n+1}}\int_{\mathbf{x}\in\Psi} \exp\left\{\frac{1}{4}\rho^2(A,(\mathbf{x},\omega))\right\} \le \frac{1}{\mathbb{P}(B)}\left(2-\frac{\mathbb{P}(A)}{\mathbb{P}(B)}\right) = \frac{1}{\mathbb{P}(A)}x(2-x),$$

where $x = \mathbb{P}(A)/\mathbb{P}(B) \le 1$. But $x(2-x) \le 1$, completing the induction and hence the theorem. □

We call $h:\Omega\to\mathbb{R}$ *Lipschitz* if $|h(\mathbf{x})-h(\mathbf{y})| \le 1$ whenever $\mathbf{x},\mathbf{y}$ differ in at most one coordinate.

Definition 21.20 Let $f:\mathbb{N}\to\mathbb{N}$. h is *f-certifiable* if whenever $h(\mathbf{x}) \ge s$ then there exists $I \subseteq [n]$ with $|I| \le f(s)$ so that if $\mathbf{y}\in\Omega$ agrees with $\mathbf{x}$ on coordinates I then $h(\mathbf{y}) \ge s$.

Theorem 21.21 *Suppose that h is Lipschitz and f-certifiable. Then if $X = h(\mathbf{x})$ for $\mathbf{x}\in\Omega$, then for all b,t,*

$$\mathbb{P}(X \le b - t\sqrt{f(b)})\,\mathbb{P}(X \ge b) \le e^{-t^2/4}.$$

Proof Set $A = \{\mathbf{x} : h(\mathbf{x}) < b - t\sqrt{f(b)}\}$. Now suppose that $h(\mathbf{y}) \geq b$. We claim that $\mathbf{y} \notin A_t$. Let I be a set of indices of size at most $f(b)$ that certifies $h(y) \geq b$ as given above. Define $\alpha_i = 0$ when $i \notin I$ and $\alpha_i = |I|^{-1/2}$ when $i \in I$. Using Lemma 21.19 we see that if $\mathbf{y} \in A_t$ then there exists a $\mathbf{z} \in A$ that differs from $\mathbf{y}$ in at most $t|I|^{1/2} \leq t\sqrt{f(b)}$ coordinates of I, though at arbitrary coordinates outside I. Let $\mathbf{y}'$ agree with $\mathbf{y}$ on I and agree with $\mathbf{z}$ outside I. By the certification $h(\mathbf{y}') \geq b$. Now $\mathbf{y}', \mathbf{z}$ differ in at most $t\sqrt{f(b)}$ coordinates and so, by Lipschitz,

$$h(\mathbf{z}) \geq h(\mathbf{y}') - t\sqrt{f(b)} \geq b - t\sqrt{f(b)},$$

but then $\mathbf{z} \notin A$, a contradiction. So, $\mathbb{P}(X \geq b) \leq 1 - \mathbb{P}(A_t)$ and from Theorem 21.17,

$$\mathbb{P}(X < b - t\sqrt{f(b)})\,\mathbb{P}(X \geq b) \leq e^{-t^2/4}.$$

As the RHS is continuous in t, we may replace "$<$" by "$\leq$" giving Theorem 21.21. □

21.9 Dominance

We say that a random variable X *stochastically dominates* a random variable Y if

$$\mathbb{P}(X \geq t) \geq \mathbb{P}(Y \geq t) \quad \text{for all real } t.$$

There are many cases when we want to use our inequalities to bound the upper tail of some random variable Y and (i) Y does not satisfy the necessary conditions to apply the relevant inequality, but (ii) Y is dominated by some random variable X that does. Clearly, we can use X as a surrogate for Y.

The following case arises quite often. Suppose that $Y = Y_1 + Y_2 + \cdots + Y_n$, where $0 \leq Y_i \leq 1$ for $i = 1, 2, \ldots, n$. Suppose that $Y_1, Y_2, \ldots, Y_n$ are not independent, but instead we have

$$\mathbb{P}(Y_i \geq t \mid Y_1, Y_2, \ldots, Y_{i-1}) \leq \mathbb{P}(X_i \geq t),$$

where $t \geq 0$ and X_i is a random variable taking values in $[0,1]$. Let $X = X_1 + \cdots + X_n$, where $X_1, X_2, \ldots, X_n$ are independent of each other and $Y_1, Y_2, \ldots, Y_n$. Then we have

Lemma 21.22 *X stochastically dominates Y.*

Proof Let $X^{(i)} = X_1 + \cdots + X_i$ and $Y^{(i)} = Y_1 + \cdots + Y_i$ for $i = 1, 2, \ldots, n$. We will show by induction that $X^{(i)}$ dominates $Y^{(i)}$ for $i = 1, 2, \ldots, n$. This is trivially true for $i = 1$ and for $i > 1$ we have

$$\begin{aligned}\mathbb{P}(Y^{(i)} \geq t \mid Y_1 \ldots, Y_{i-1}) &= \mathbb{P}(Y_i \geq t - (Y_1 + \cdots + Y_{i-1}) \mid Y_1 \ldots, Y_{i-1}) \\ &\leq \mathbb{P}(X_i \geq t - (Y_1 + \cdots + Y_{i-1}) \mid Y_1 \ldots, Y_{i-1}).\end{aligned}$$

Removing the conditioning we have

$$\mathbb{P}(Y^{(i)} \geq t) \leq \mathbb{P}(Y^{(i-1)} \geq t - X_i) \leq \mathbb{P}(X^{(i-1)} \geq t - X_i) = \mathbb{P}(X^{(i)} \geq t),$$

where the second inequality follows by induction. □

22

Differential Equations Method

Let $D \subseteq \mathbb{R}^2$ be open and connected. Consider a general random process

$$X(0), X(1), \dots, X(t), \dots, X(n) \in \mathbb{Z}.$$

where $X(0)$ is fixed and $\left(0, \frac{X(0)}{n}\right) \in D$.

Let H_t denote the history $X(0), X(1), \dots, X(t)$ of the process to time t. We further assume

(P1) $|X(t)| \le C_0 n, \ \forall t < T_D$, where C_0 is a constant.
(P2) $|X(t+1) - X(t)| \le \beta = \beta(n) \ge 1, \ \forall t < T_D$.
(P3) $|\mathbb{E}(X(t+1) - X(t)|H_t, \mathscr{E}) - f(t/n, X(t)/n)| \le \lambda_0, \forall t < T_D$. Here $\mathscr{E}$ is some likely event that holds with probability at least $1 - \gamma$.
(P4) $f(t,x)$ is continuous and satisfies a Lipschitz condition
$|f(t,x) - f(t',x')| \le L\|(t,x) - (t',x)'\|_\infty$
for $(t,x), (t',x') \in D \cap \{(t,x) : t \ge 0\}$.

Theorem 22.1 *Suppose that*

$$\lambda_0 \le \lambda = o(1) \quad \textit{and} \quad \alpha = \frac{n\lambda^3}{\beta^3} \gg 1.$$

$$\sigma = \inf\{\tau : (\tau, z(\tau)) \notin D_0 = \{(t,z) \in D : l^\infty \textit{ distance of } (t,z) \textit{ from the boundary of } D \ge 2\lambda\}\}.$$

Let $z(\tau)$, $0 \le \tau \le \sigma$ be the unique solution to the differential equation

$$z'(\tau) = f(\tau, z(\tau)) \tag{22.1}$$

$$z(0) = \frac{X(0)}{n}. \tag{22.2}$$

Then,

$$X(t) = nz(t/n) + O(\lambda n),$$

uniformly in $0 \leq t \leq \sigma n$, *with probability* $1 - O(\gamma + \beta e^{-\alpha}/\lambda)$.

Proof The γ in the probability of success will be handled by conditioning on $\mathscr{E}$. Now let

$$\omega = \left\lceil \frac{n\lambda}{\beta} \right\rceil.$$

We study the difference $X(t+\omega) - X(t)$. Assume that $(t/n, X(t)/n) \in D_0$. For $0 \leq k \leq \omega$ we have from (P2) that

$$\left| \frac{X(t+k)}{n} - \frac{X(t)}{n} \right| \leq \frac{k\beta}{n} \leq 2\lambda,$$

so

$$\left\| \left(\frac{t+k}{n}, \frac{X(t+k)}{n} \right) - \left(\frac{t}{n}, \frac{X(t)}{n} \right) \right\|_\infty \leq 2\lambda,$$

and so $\left(\frac{t+k}{n}, \frac{X(t+k)}{n} \right)$ is in D.

Therefore, using (P3),

$$\begin{aligned}
&\mathbb{E}(X(t+k+1) - X(t+k) | H_{t+k}, \mathscr{E}) \\
&= f\left(\frac{t+k}{n}, \frac{X(t+k)}{n} \right) + \theta_k \\
&= f\left(\frac{t}{n}, \frac{X(t)}{n} \right) + \theta_k + \psi_k \\
&= f\left(\frac{t}{n}, \frac{X(t)}{n} \right) + \rho,
\end{aligned}$$

where $|\rho| \leq 2L\lambda$, since $|\theta_k| \leq \lambda$ (by (P3)) and $|\psi_k| \leq \frac{L\beta k}{n}$ (by (P4)).

Now, given H_t, let

$$Z_k = \begin{cases} X(t+k) - X(t) - kf\left(\frac{t}{n}, \frac{X(t)}{n} \right) - 2kL\lambda & \mathscr{E} \\ 0 & \neg\mathscr{E}. \end{cases}$$

Then

$$\mathbb{E}(Z_{k+1} - Z_k | Z_0, Z_1, \ldots, Z_k) \leq 0,$$

i.e. $Z_0, Z_1, \ldots, Z_\omega$ is a super-martingale.

Also

$$|Z_{k+1} - Z_k| \leq \beta + \left| f\left(\frac{t}{n}, \frac{X(t)}{n} \right) \right| + 2L\lambda \leq K_0\beta,$$

where $K_0 = O(1)$, since $f\left(\frac{t}{n}, \frac{X(t)}{n}\right) = O(1)$ by continuity and boundedness of D. So, using Theorem 21.15 we see that conditional on $H_t, \mathscr{E}$,

$$\mathbb{P}\left(X(t+\omega) - X(t) - \omega f(t/n, X(t/n)) \geq 2L\omega\lambda + K_0\beta\sqrt{2\alpha\omega}\right)$$
$$\leq \exp\left\{-\frac{2K_0^2\beta^2\alpha\omega}{2\omega K_0^2\beta^2}\right\} = e^{-\alpha}. \tag{22.3}$$

Similarly,

$$\mathbb{P}\left(X(t+\omega) - X(t) - \omega f(t/n, X(t/n)) \leq -2L\omega\lambda - K_0\beta\sqrt{2\alpha\omega}\right) \leq e^{-\alpha}. \tag{22.4}$$

Thus,

$$\mathbb{P}\left(|X(t+\omega) - X(t) - \omega f(t/n, X(t/n))| \geq 2L\omega\lambda + K_0\beta\sqrt{2\alpha\omega}\right) \leq 2e^{-\alpha}.$$

We have that $\omega\lambda$ and $\beta\sqrt{2\alpha\omega}$ are both $\Theta(n\lambda^2/\beta)$ giving

$$2L\omega\lambda + K_0\beta\sqrt{2\alpha\omega} \leq K_1\frac{n\lambda^2}{\beta}.$$

Now let $k_i = i\omega$ for $i = 0, 1, \ldots, i_0 = \lfloor \sigma n/\omega \rfloor$. We will show by induction that

$$\mathbb{P}\left(\exists j \leq i : |X(k_j) - z(k_j/n)n| \geq B_j\right) \leq 2ie^{-\alpha}, \tag{22.5}$$

where

$$B_j = B\left(\left(1 + \frac{L\omega}{n}\right)^{j+1} - 1\right)\frac{n\lambda}{L} \tag{22.6}$$

and where B is another constant.

The induction begins with $z(0) = \frac{X(0)}{n}$ and $B_0 = 0$. Note that

$$B_{i_0} \leq \frac{Be^{\sigma L}\lambda}{L}n = O(\lambda n).$$

Now write

$$|X(k_{i+1}) - z(k_{i+1}/n)n| = |A_1 + A_2 + A_3 + A_4|,$$
$$A_1 = X(k_i) - z(k_i/n)n,$$
$$A_2 = X(k_{i+1}) - X(k_i) - \omega f(k_i/n, X(k_i/n)),$$
$$A_3 = \omega z'(k_i/n) + z(k_i/n)n - z(k_{i+1}/n)n,$$
$$A_4 = \omega f(k_i/n, X(k_i/n)) - \omega z'(k_i/n).$$

We now bound each of these terms individually.

Our induction gives that with probability at most $2ie^{-\alpha}$,

$$|A_1| \le B_i.$$

Equations (22.3) and (22.4) give

$$|A_2| \le K_1 \frac{n\lambda^2}{\beta},$$

with probability $1 - 2e^{-\alpha}$.

$$A_3 = \omega z'(k_i/n) + z(k_i/n)n - z(k_{i+1}/n)n$$

Now

$$z(k_{i+1}/n) - z(k_i/n) = \frac{\omega}{n} z'(k_i/n + \hat{\omega}/n)$$

for some $0 \le \hat{\omega} \le \omega$ and so (P4) implies that

$$|A_3| = \omega |z'(k_i/n + \omega/n) - z'(k_i/n + \hat{\omega}/n)| \le L\frac{\omega^2}{n} \le 2L\frac{n\lambda^2}{\beta^2}.$$

Finally, (P4) gives

$$|A_4| \le \frac{\omega L |A_1|}{n} \le \frac{\omega L}{n} B_i.$$

Thus for some $B > 0$,

$$\begin{aligned} B_{i+1} &\le |A_1| + |A_2| + |A_3| + |A_4| \\ &\le \left(1 + \frac{\omega L}{n}\right) B_i + Bn\frac{\lambda^2}{\beta}. \end{aligned}$$

A little bit of algebra verifies (22.5) and (22.6).

Finally consider $k_i \le t < k_{i+1}$. From "time" k_i to t the change in X and nz is at most $\omega\beta = O(n\lambda)$. □

Remark 22.2 The above proof generalizes easily to the case where $X(t)$ is replaced by $X_1(t), X_2(t), \ldots, X_a(t)$ where $a = O(1)$.

The earliest mention of differential equations with respect to random graphs was in the paper by Karp and Sipser [476]. The paper by Ruciński and Wormald [650] was also influential. See Wormald [729] for an extensive survey on the differential equations method.

23
Branching Processes

In the Galton–Watson branching process, we start with a single particle comprising generation 0. In general, the nth generation consists of Z_n particles and each member x of this generation independently gives rise to a random number X of descendants in generation $n+1$. In this book we need the following theorem about the probability that the process continues indefinitely: let

$$p_k = \mathbb{P}(X = k), \qquad k = 0, 1, 2, \ldots.$$

Let

$$G(z) = \sum_{k=0}^{\infty} p_k z^k$$

be the probability generating function (p.g.f.) of X. Let $\mu = \mathbb{E}X$. Let

$$\eta = \mathbb{P}\left(\bigcup_{n \geq 0} \{Z_n = 0\}\right) \tag{23.1}$$

be the probability of ultimate extinction of the process.

Theorem 23.1 *η is the smallest non-negative root to the equation $G(s) = s$. Here $\eta = 1$ if $\mu < 1$.*

Proof If $G_n(z)$ is the p.g.f. of Z_n, then $G_n(z) = G(G_{n-1}(z))$. This follows from the fact that Z_n is the sum of Z_{n-1} independent copies of G. Let $\eta_n = \mathbb{P}(Z_n) = 0$. Then

$$\eta_n = G_n(0) = G(G_{n-1}(0)) = G(\eta_{n-1}).$$

It follows from (23.1) that $\eta_n \nearrow \eta$. Let ψ be any other non-negative solution to $G(s) = s$. We have

$$\eta_1 = G(0) \leq G(\psi) = \psi.$$

Now assume inductively that $\eta_n \leq \psi$ for some $n \geq 1$. Then

$$\eta_{n+1} = G(\eta_n) \leq G(\psi) = \psi.$$

□

24
Entropy

24.1 Basic Notions

Entropy is a useful tool in many areas. The entropy we talk about here was introduced by Shannon in [680]. We need some results on entropy in Chapter 13. We collect them here for convenience. For more on the subject we refer the reader to Cover and Thomas [227], or Gray [382] or Martin and England [572].

Let X be a random variable taking values in a finite set R_X. Let $p(x) = \mathbb{P}(X = x)$ for $x \in X$. Then the *entropy* of X is given by

$$h(X) = -\sum_{x \in R_X} p(x) \log p(x).$$

We have a choice for the base of the logarithm here. We use the natural logarithm, for use in Chapter 13.

Note that if X is chosen uniformly from R_X, i.e. $\mathbb{P}(X = x) = 1/|R_X|$ for all $x \in R_X$ then

$$h(X) = \sum_{x \in R_X} \frac{\log |R_X|}{|R_X|} = \log |R_X|.$$

We will see later that the uniform distribution maximizes entropy.

If Y is another random variable with a finite range then we define the *conditional entropy as*

$$h(X \mid Y) = \sum_{y \in R_Y} p(y) h(X_y) = -\sum_{x,y} p(x,y) \log \frac{p(x,y)}{p(y)}, \tag{24.1}$$

where X_y is the random variable with $\mathbb{P}(X_y = x) = \mathbb{P}(X = x \mid Y = y)$. Here $p(y) = \mathbb{P}(Y = y)$. The summation is over y such that $p(y) > 0$. We will use notation like this from now on, without comment.

Chain Rule:

Lemma 24.1

$$h(X_1,X_2,\ldots,X_m) = \sum_{i=1}^{m} h(X_i \mid X_1,X_2,\ldots,X_{i-1}). \tag{24.2}$$

Proof This follows by induction on m, once we have verified it for $m = 2$. For then

$$h(X_1,X_2,\ldots,X_m) \quad = \quad h(X_1,X_2,\ldots,X_{m-1}) \quad + \quad h(X_m \mid X_1,X_2,\ldots,X_{m-1}).$$

Now,

$$\begin{aligned} h(X_2 \mid X_1) &= -\sum_{x_1,x_2} p(x_1,x_2) \log \frac{p(x_1,x_2)}{p(x_1)} \\ &= -\sum_{x_1,x_2} p(x_1,x_2) \log p(x_1,x_2) + \sum_{x_1,x_2} p(x_1,x_2) \log p(x_1) \\ &= h(X_1,X_2) + \sum_{x_1} p(x_1) \log p(x_1) \\ &= h(X_1,X_2) + h(X_1). \end{aligned}$$

□

Inequalities:
Entropy is a measure of uncertainty and so we should not be surprised to learn that $h(X \mid Y) \leq h(X)$ for all random variables X, Y – here conditioning on Y represents providing information. Our goal is to prove this and a little more.

Let p, q be probability measures on the finite set X. We define the *Kullback–Liebler* distance

$$D(p||q) = \sum_{x \in A} p(x) \log \frac{p(x)}{q(x)},$$

where $A = \{x : p(x) > 0\}$.

Lemma 24.2

$$D(p||q) \geq 0$$

with equality if and only if $p = q$.

Proof Let

$$\begin{aligned} -D(p||q) &= \sum_{x \in A} p(x) \log \frac{q(x)}{p(x)} \\ &\leq \log \sum_{x \in A} p(x) \frac{q(x)}{p(x)} \\ &= \log 1 \\ &= 0. \end{aligned} \tag{24.3}$$

Inequality (24.3) follows from Jensen's inequality and the fact that log is a concave function. Because log is strictly concave, we will have equality in (24.3) if and only if $p = q$. □

It follows from this that

$$h(X) \leq \log |R_X|. \tag{24.4}$$

Indeed, let u denote the uniform distribution over R_X, i.e. $u(x) = 1/|R_X|$, then

$$0 \leq D(p||u) = \sum_x p(x)(\log p(x) + \log |R_X|) = -h(x) + \log |R_X|.$$

We can now show that conditioning does not increase entropy.

Lemma 24.3 *For random variables X, Y, Z,*

$$h(X \mid Y, Z) \leq h(X \mid Y).$$

Taking Z to be a constant, e.g. $Z = 1$ with probability one, we see

$$h(X \mid Y) \leq h(X).$$

Proof

$$\begin{aligned} h(X \mid Y) - h(X \mid Y,Z) &= -\sum_{x,y} p(x,y) \log \frac{p(x,y)}{p(y)} + \sum_{x,y,z} p(x,y,z) \log \frac{p(x,y,z)}{p(y,z)} \\ &= -\sum_{x,y,z} p(x,y,z) \log \frac{p(x,y)}{p(y)} + \sum_{x,y,z} p(x,y,z) \log \frac{p(x,y,z)}{p(y,z)} \\ &= \sum_{x,y,z} p(x,y,z) \log \frac{p(x,y,z)p(y)}{p(x,y)p(y,z)} \\ &= D(p_{x,y,z}||p(x,y)p(y,z)/p(y)) \\ &\geq 0. \end{aligned}$$

Note that $\sum_{x,y,z} p(x,y)p(y,z)/p(y) = 1$. □

Working through the above proof we see that $h(X) = h(X \mid Z)$ if and only if $p(x,z) = p(x)p(z)$ for all x,z, i.e. if and only if X and Z are independent.

24.2 Shearer's Lemma

The original proof is from Chung, Frank, Graham and Shearer [180]. The following proof is from Radakrishnan [639].

Lemma 24.4 *Let $X = (X_1, X_2, \dots, X_N)$ be a (vector) random variable and $\mathscr{A} = \{A_i : i \in I\}$ be a collection of subsets of a set B, where $|B| = N$, such that each element of B appears in at least k members of $\mathscr{A}$. For $A \subseteq B$, let $X_A = (X_j : j \in A)$. Then,*

$$h(X) \leq \frac{1}{k} \sum_{i \in I} h(X_{A_i}).$$

Proof We have, from Lemma 24.1 that

$$h(X) = \sum_{j \in B} h(X_j \mid X_1, X_2, \dots, X_{j-1}) \tag{24.5}$$

and

$$h(X_{A_i}) = \sum_{j \in A_i} h(X_j \mid X_\ell, \ell \in A_i, \ell < j). \tag{24.6}$$

We sum (24.6) for all $i = 1, 2, \dots, N$. Then

$$\begin{aligned}
\sum_{i \in I} h(X_{A_i}) &= \sum_{i \in I} \sum_{j \in A_i} h(X_j \mid X_\ell, \ell \in A_i, \ell < j) \\
&= \sum_{j \in B} \sum_{A_i \ni j} h(X_j \mid X_\ell, \ell \in A_i, \ell < j) && (24.7) \\
&\geq \sum_{j \in B} \sum_{A_i \ni j} h(X_j \mid X_1, X_2, \dots, X_{j-1}) && (24.8) \\
&\geq k \sum_{j \in B} h(X_j \mid X_1, X_2, \dots, X_{j-1}) && (24.9) \\
&= kh(X). && (24.10)
\end{aligned}$$

Here we obtain (24.8) from (24.7) by applying Lemma 24.3. We obtain (24.9) from (24.8) and the fact that each $j \in B$ appears in at least k A_is. We then obtain (24.10) by using (24.5). □

References

[1] M. Abdullah, C. Cooper and A.M. Frieze, Cover time of a random graph with given degree sequence, *Discrete Mathematics* 312 (2012) 3146–3163.

[2] D. Achlioptas, J-H. Kim, M. Krivelevich and P. Tetali, Two-coloring random hypergraphs, *Random Structures and Algorithms* 20 (2002) 249–259.

[3] D. Achlioptas and C. Moore, Almost all graphs with average degree 4 are 3-colorable, *Journal of Computer and System Sciences* 67 (2003) 441–471.

[4] D. Achlioptas and C. Moore, The chromatic number of random regular graphs, *Proceedings of RANDOM 2004*.

[5] D. Achlioptas and A. Naor, The two possible values of the chromatic number of a random graph, *Annals of Mathematics* 162 (2005) 1335–1351.

[6] R. Adamczak and P. Wolff, Concentration inequalities for non-Lipschitz functions with bounded derivatives of higher order, *Probability Theory and Related Fields* 161 (2014) 1–56.

[7] W. Aiello, A. Bonato, C. Cooper, J. Janssen and P. Pralat, A spatial web graph model with local influence regions, *Internet Mathematics* 5 (2009) 175–196.

[8] M. Ajtai, J. Komlós and E. Szemerédi, Largest random component of a k-cube, *Combinatorica* 2 (1982) 1–7.

[9] M. Ajtai, J. Komlós and E. Szemerédi, Topological complete subgraphs in random graphs, *Studia Scientiarum Mathematicarum Hungarica* 14 (1979) 293–297.

[10] M. Ajtai, J. Komlós and E. Szemerédi, The longest path in a random graph, *Combinatorica* 1 (1981) 1–12.

[11] M. Ajtai, J. Komlós and E. Szemerédi. The first occurrence of Hamilton cycles in random graphs, *Annals of Discrete Mathematics* 27 (1985) 173–178.

[12] D. Aldous, Exchangeability and related topics, *Lecture Notes in Mathematics*, 1117, Springer Verlag, New York (1985).

[13] D. Aldous, Asymptotics in the random assignment problem, *Probability Theory and Related Fields* 93 (1992) 507–534.

[14] D. Aldous, Asymptotic fringe distributions for general families of random trees, *Annals of Applied Probability* 1 (1991) 228–266.

[15] D. Aldous, Brownian excursions, critical random graphs and the multiplicative coalescent, *Annals of Probability* 25 (1997) 812–854.

[16] D. Aldous, The $\zeta(2)$ limit in the random assignment problem, *Random Structures and Algorithms* 4 (2001) 381–418.

[17] D. Aldous, G. Miermont and J. Pitman, Brownian bridge asymptotics for random p-mappings, *Electronic Journal of Probability* 9 (2004) 37–56.

[18] D. Aldous, G. Miermont and J. Pitman, Weak convergence of random p-mappings and the exploration process of inhomogeneous continuum random trees, *Probability Theory and Related Fields* 133 (2005) 1–17.

[19] P. Allen, J. Böttcher, Y. Kohayakawa and Y. Person, Tight Hamilton cycles in random hypergraphs, see arxiv.org.

[20] N. Alon, Choice numbers of graphs; a probabilistic approach, *Combinatorics, Probability and Computing* 1 (1992) 107–114.

[21] N. Alon, Restricted colorings of graphs, in *Surveys of Combinatorics 1993*, London Mathematical Society Lecture Notes Series 187 (K. Walker, ed.), Cambridge University Press (1993) 1–33.

[22] N. Alon, A note on network reliability, in *Discrete Probability and Algorithms* (Minneapolis, MN, 1993), IMA Vol. Math. Appl., 72, Springer, New York (1995) 11–14.

[23] N. Alon, M. Capalbo, Y. Kohayakawa, V. Rödl, A. Ruciński and E. Szemerédi, Universality and tolerance (extended abstract), in *41st Annual Symposium on Foundations of Computer Science* (Redondo Beach, CA, 2000), IEEE Computer Society Press, Los Alamitos, CA (2000) 14–21.

[24] N. Alon and Z. Füredi, Spanning subgraphs of random graphs, *Graphs and Combinatorics* 8 (1992) 91–94.

[25] N. Alon and M. Krivelevich, The concentration of the chromatic number of a random graph, *Combinatorica* 17 (1997) 303–313.

[26] N. Alon, M. Krivelevich and B. Sudakov, Finding a large hidden clique in a random graph, *Random Structures and Algorithm* 13 (1998) 457–466.

[27] N. Alon, M. Krivelevich and B. Sudakov, List coloring of random and pseudo random-graphs, *Combinatorica* 19 (1999) 453–472.

[28] N. Alon, M. Krivelevich and V. H. Vu, On the concentration of eigenvalues of random symmetric matrices, *Israel Journal of Mathematics* 131 (2002) 259–267.

[29] N. Alon, P. Pralat, and N. Wormald, Cleaning regular graphs with brushes, *SIAM Journal on Discrete Mathematics* 23 (2008) 233–250.

[30] N. Alon and J. Spencer, *The Probabilistic Method*, Third Edition, John Wiley and Sons (2008).

[31] N. Alon and B. Sudakov, Increasing the chromatic number of a random graph, *Journal of Combinatorics* 1 (2010) 345–356.

[32] N. Alon and R. Yuster, Threshold functions for H-factors, *Combinatorics, Probability and Computing* 2 (1993) 137–144.

[33] A. Amit and N. Linial, Random graph coverings I: general theory and graph connectivity, *Combinatorica* 22 (2002) 1–18.

[34] A. Amit and N. Linial, Random lifts of graphs II: edge expansion, *Combinatorics, Probability and Computing* 15 (2006) 317–332.

[35] A. Amit, N. Linial and J. Matoušek, Random lifts of graphs III: independence and chromatic number, *Random Structures and Algorithms* 20 (2002) 1–22.

[36] O. Angel, A. Flaxman and D. Wilson, A sharp threshold for minimum bounded-depth and bounded-diameter spanning trees and steiner trees in random networks, *Combinatorica* (2012) 1–33.

[37] S. Anoulova, J. Bennies, J. Lenhard, D. Metzler, Y. Sung and A. Weber, Six Ways of Looking at Burtin's Lemma, *The American Mathematical Monthly* 106 (1999) 345–351.

[38] J. Aronson, A.M. Frieze and B. Pittel, Maximum matchings in sparse random graphs: Karp–Sipser revisited, *Random Structures and Algorithms* 12 (1998) 111–178.

[39] R. Arratia, A.D. Barbour and S. Tavaré, A tale of three couplings: Poisson–Dirichlet and GEM approximations for random permutations, *Combinatorics, Probability and Computing* 15 (2006) 31–62.

[40] R. Arratia and S. Tavaré, The cycle structure of random permutations, *The Annals of Probability* 20 (1992) 1567–1591.

[41] C. Avin and G. Ercal, On the cover time of random geometric graphs, *Automata, Languages and Programming, Lecture Notes in Computer Science* 3580 (2005) 677–689.

[42] L. Babai, P. Erdős and S.M. Selkow, Random graph isomorphism, *SIAM Journal on Computing* 9 (1980) 628–635.

[43] L. Babai, M. Simonovits and J. Spencer, Extremal subgraphs of random graphs, *Journal of Graph Theory* 14 (1990) 599–622.

[44] A. Backhaus and T.F. Móri, Local degree distribution in scale free random graphs, *Electronic Journal of Probability* 16 (2011) 1465–1488.

[45] D. Bal, P. Bennett, A. Dudek and A.M. Frieze, The t-tone chromatic number of random graphs, see arxiv.org.

[46] D. Bal, P. Bennett, A.M. Frieze and P. Pralat, Power of k choices and rainbow spanning trees in random graphs, see arxiv.org.

[47] D. Bal and A.M. Frieze, Rainbow matchings and Hamilton cycles in random graphs, see arxiv.org.

[48] P. Balister and B. Bollobás, Percolation in the k-nearest neighbour graph. To appear.

[49] P. Balister, B. Bollobás, A. Sarkar and M. Walters, Connectivity of random k-nearest neighbour graphs, *Advances in Applied Probability* 37 (2005) 1–24.

[50] P. Balister, B. Bollobás, A. Sarkar and M. Walters, A critical constant for the k-nearest neighbour model, *Advances in Applied Probability* 41 (2009) 1–12.

[51] F. Ball, D. Mollison and G. Scalia-Tomba, Epidemics with two levels of mixing, *Annals of Applied Probability* 7 (1997) 46–89.

[52] J. Balogh, T. Bohman and D. Mubayi, Erdős–Ko–Rado in random hypergraphs, *Combinatorics, Probability and Computing* 18 (2009) 629–646.

[53] J. Balogh, B. Bollobás, M. Krivelevich, T. Müeller and M. Walters, Hamilton cycles in random geometric graphs, *Annals of Applied Probability* 21 (2011) 1053–1072.

[54] J. Balogh, R. Morris and W. Samotij, Independent sets in hypergraphs, see arxiv.org.

[55] C. Banderier, P. Flajolet, G. Schaeffer, and M. Soria, Random maps, coalescing saddles, singularity analysis, and Airy phenomena, *Random Structures and Algorithms* 19 (2001) 194–246.

[56] L. Barabási and R. Albert, Emergence of scaling in random networks, *Science* 286 (1999) 509–512.

[57] A.D Barbour, Poisson convergence and random graph, *Mathematical Proceedings of the Cambridge Philosophical Society* 92 (1982) 349–359.

[58] A.D. Barbour, L. Holst and S. Janson, *Poisson Approximation*, Oxford University Press, Oxford (1992).

[59] A.D. Barbour, S. Janson, M. Karoński and A. Ruciński, Small cliques in random graphs, *Random Structures and Algorithms* 1 (1990) 403–434.

[60] A.D. Barbour, M. Karoński and A. Ruciński, A central limit theorem for decomposable random variables with applications to random graphs, *Journal of Combinatorial Theory, B* 47 (1989) 125–145.

[61] A.D. Barbour and G. Reinert, The shortest distance in random multi-type intersection graphs, *Random Structures Algorithms* 39 (2011) 179–209.

[62] T. Bartnicki, J. Grytczuk, H.A. Kierstead and X. Zhu, The map-coloring game, *American Mathematical Monthly*, 114 (2007) 793–803.

[63] M. Bayati, D. Gamarnik and P. Tetali, Combinatorial approach to the interpolation method and scaling limits in sparse random graphs, *Annals of Probability* 41 (2013) 4080–4115.

[64] J. Beck, Combinatorial games: Tic-Tac-Toe theory, in *Encyclopedia of Mathematics and its Applications 114*, Cambridge University Press (2008).

[65] M. Bednarska and T. Łuczak, Biased positional games for which random strategies are nearly optimal, *Combinatorica* 20 (2000) 477–488.

[66] M. Behrisch, Component evolution in random intersection graphs, *The Electronic Journal of Combinatorics* 14 (2007) #R17.

[67] M. Behrisch, A. Taraz and M. Ueckerdt, Coloring random intersection graphs and complex networks, *SIAM Journal on Discrete Mathematics* 23 (2009) 288–299.

[68] I. Benjamini, G. Kozma and N. Wormald, The mixing time of the giant component of a random graph, *Random Structures and Algorithms* 45 (2014) 383–407.

[69] I. Benjamini, I. Shinkar and G. Tsur, Acquaintance time of a graph, see arxiv.org.

[70] E.A. Bender and E.R. Canfield, The asymptotic number of labelled graphs with given degree sequences, *Journal of Combinatorial Theory A* 24 (1978) 296–307.

[71] E.A. Bender and E.R. Canfield, The number of rooted maps on an orientable surface. *Journal of Combinatorial Theory, B* 53 (1991) 293–299.

[72] E.A. Bender, E.R. Canfield and B.D. McKay, The asymptotic number of labelled connected graphs, *Random Structures and Algorithms* 1 (1990) 127–169.

[73] A. Bender, Z. Gao and N.C. Wormald, The number of labeled 2-connected planar graphs, *Electronic Journal of Combinatorics* 9 (2002).

[74] P. Bennett and T. Bohman, A note on the random greedy independent set algorithm, see arxiv.org.

[75] S. Ben-Shimon, A. Ferber, D. Hefetz and M. Krivelevich, Hitting time results for Maker-Breaker games, *Random Structures and Algorithms* 41 (2012) 23–46.

[76] S. Ben-Shimon, M. Krivelevich and B. Sudakov, Local resilience and Hamiltonicity Maker-Breaker games in random regular graph, *Combinatorics, Probability and Computing* 20 (2011) 173–211.

[77] S. Ben-Shimon, M. Krivelevich and B. Sudakov, On the resilience of Hamiltonicity and optimal packing of Hamilton cycles in random graphs, *SIAM Journal of Discrete Mathematics* 25 (2011) 1176–1193.

[78] S. Berg, Random contact processes, snowball sampling and and factorial series distributions, *Journal of Applied Probability* 20 (1983) 31–46.

[79] C. Berge, *Graphs and Hypergraphs*, North-Holland Publishing Company, Amsterdam (1973).

[80] F. Bergeron, P. Flajolet and B. Salvy, Varieties of increasing trees, *Lecture Notes in Computer Science* 581 (1992) 24–48.

[81] J.Bertoin and G. Uribe Bravo, Super-critical percolation on large scale-free random trees, *The Annals of Applied Probability* 25 (2015) 81–103.

[82] S.N. Bernstein, *Theory of Probability*, (Russian), Moscow, (1927).

[83] J. Bertoin, Sizes of the largest clusters for supercritical percolation on random recursive trees, *Random Structures Algorithms* 44 (2014) 29–44.

[84] A. Beveridge, A. Dudek, A.M. Frieze, T. Müller and M. Stojaković, Maker-Breaker games on random geometric graphs, see arxiv.org in *Random Structures and Algorithms*.

[85] A. Beveridge, A.M. Frieze and C. McDiarmid, Random minimum length spanning trees in regular graphs, *Combinatorica* 18 (1998) 311–333.

[86] S. Bhamidi and R. van der Hofstad, Weak disorder asymptotics in the stochastic mean-field model of distance, *Annals of Applied Probability* 22 (2012) 29–69.

[87] S. Bhamidi, R. van der Hofstad and G. Hooghiemstra, Universality for first passage percolation on sparse uniform and rank-1 random graphs, see arxiv.org.

[88] S. Bhamidi, S. Sen and X. Wang, Continuum limit of inhomogeneous random graphs, see arxiv.org.

[89] Ph. Biane, J. Pitman and M. Yor, Probability laws related to the Jacobi theta and Riemann zeta functions, and brownian excursions, *Bulletin of the AMS* 38 (2001) 435–465.

[90] S.R. Blackbourn and S. Gerke, Connectivity of the uniform random intersection graph, *Discrete Mathematics* 309 (2009) 5130–5140.

[91] P. Blackwell, M. Edmonson-Jones and J. Jordan, Spectra of adjacency matrices of random geometric graphs, see arxiv.org.

[92] M. Bloznelis, Degree distribution of a typical vertex in a general random intersection graph, *Lithuanian Mathematical Journal* 48 (2008) 38–45.

[93] M. Bloznelis, Component evolution in general random intersection graphs, *SIAM Journal on Discrete Mathematics* 24 (2010) 639–654.

[94] M. Bloznelis, A random intersection digraph: indegree and outdegree distributions, *Discrete Mathematics* 310 (2010) 2560–2566.

[95] M. Bloznelis, The largest component in an inhomogeneous random intersection graph with clustering, *The Electronic Journal of Combinatorics* 17 (2010) #R110.

[96] M. Bloznelis and J. Damarackas, Degree distribution of an inhomogeneous random intersection graph, *The Electronic Journal of Combinatorics* 20(3) (2013) #P3.

[97] M. Bloznelis, E. Godehardt, J. Jaworski, V. Kurauskas and K. Rybarczyk, Recent progress in complex network analysis - models of random intersection graphs, in: B. Lausen, S. Krolak-Schwerdt, M. Bhmer, Eds., *Data Science, Learning by Latent Structures, and Knowledge Discovery*, Springer (2015) 59–68.

[98] M. Bloznelis, E. Godehardt, J. Jaworski, V. Kurauskas and K. Rybarczyk, Recent progress in complex network analysis - properties of random intersection graphs, in: B. Lausen, S. Krolak-Schwerdt and M. Bhmer, Eds., *Data Science, Learning by Latent Structures, and Knowledge Discovery*, Springer (2015) 69–78.

[99] M. Bloznelis and F. Goetze, Preferred attachment model of affiliation network, *Journal of Statistical Physics* 156 (2014) 800–821.

[100] M. Bloznelis, F. Göetze and J. Jaworski, Birth of a strongly connected giant in an inhomogeneous random digraph, *Journal of Applied Probability* 49 (2012) 601–611.

[101] M. Bloznelis, J. Jaworski and K. Rybarczyk, Component evolution in a secure wireless sensor network, *Networks* 53 (2009) 19–26.

[102] M. Bloznelis and M. Karoński, Random intersection graph process, to appear in *Internet Mathematics*.

[103] M. Bloznelis and T. Łuczak, Perfect matchings in random intersection graphs, *Acta Mathematica Hungarica*, 138 (2013) 15–33.

[104] M. Bloznelis and I. Radavicius, A note on Hamiltonicity of uniform random intersection graphs, *Lithuanian Mathematical Journal*, 51 (2011) 155–161.

[105] M. Bloznelis and K. Rybarczyk, k-connectivity of uniform s-intersection graphs, *Discrete Mathematics*, 333 (2014) 94–100.

[106] M. Bode, N. Fountoulakis and T. Müller, The probability that the hyperbolic random graph is connected, see arxiv.org.

[107] M. Bode, M. Fountoulakis and T. Müller, On the largest component of a hyperbolic model of complex networks, see arxiv.org.

[108] E. Bodine-Baron, B. Hassibi and A. Wierman, Generalizing Kronecker graphs in order to model searchable networks, in *47th Annual Allerton Conference on Communication, Control, and Computing* (2009) 194–201.

[109] E. Bodine-Baron, B. Hassibi and A. Wierman, Distance-dependent Kronecker graphs for modeling social networks, *IEEE Journal of Selected Topics in Signal Processing* 4 (2010) 718–731.

[110] T. Bohman, The triangle-free process, *Advances in Mathematics* 221 (2009) 1653–1677.

[111] T. Bohman, C. Cooper, A.M. Frieze, R. Martin and M. Ruszinko, On randomly generated intersecting hypergraphs, *Electronic Journal on Combinatorics* (2003) R29.

[112] T. Bohman and A.M. Frieze, Avoiding a giant component, *Random Structures and Algorithms* (2001) 75–85.

[113] T. Bohman and A.M. Frieze, Hamilton cycles in 3-out, *Random Structures and Algorithms* 35 (2009) 393–417.

[114] T. Bohman, A.M. Frieze and E. Lubetzky, A note on the random greedy triangle-packing algorithm, *Journal of Combinatorics* 1 (2010) 477–488.

[115] T. Bohman, A.M. Frieze and E. Lubetzky, Random triangle removal, see arxiv.org.

[116] T. Bohman, A.M. Frieze, R. Martin, M. Ruszinko and C. Smyth, On randomly generated intersecting hypergraphs II, *Random Structures and Algorithms* 30 (2007) 17–34.

[117] T. Bohman, A.M. Frieze, O. Pikhurko and C. Smyth, Anti-Ramsey properties of random graphs, *Journal of Combinatorial Theory, B* 100 (2010) 299–312.

[118] T. Bohman, A.M. Frieze and B. Sudakov, The game chromatic number of random graphs, *Random Structures and Algorithms* 32 (2008) 223–235.

[119] T. Bohman and P. Keevash, Dynamic concentration of the triangle-free process, see arxiv.org.

[120] T. Bohman and D. Kravitz, Creating a giant component, *Combinatorics, Probability and Computing* 15 (2006) 489–511.

[121] T. Bohman, D. Mubayi and M. Picolleil, The independent neighborhoods process, see arxiv.org.

[122] B. Bollobás, A probabilistic proof of an asymptotic formula for the number of labelled graphs, *European Journal on Combinatorics* 1(1980) 311–316.

[123] B. Bollobás, Random graphs, in *Combinatorics: Proceedings of the Eighth British Combinatorial Conference, University College, Swansea* Cambridge University Press (1981) 80–102.

[124] B. Bollobás, The diameter of random graphs, *Tranasactions of the American Mathematical Society* 267 (1981) 41–52.

[125] B. Bollobás, Degree sequence of random graphs, *Discrete Mathematics* 13 (1981) 1–19.

[126] B. Bollobás, Long paths in sparse random graphs, *Combinatorica* 2 (1982) 223–228.

[127] B. Bollobás, Vertices of given degree in a random graph, *Journal of Graph Theory* 6 (1982) 147–155.

[128] B. Bollobás, The evolution of random graphs, *Transactions of the Americam Mathematical Society* 286 (1984) 257–274.

[129] B. Bollobás, The evolution of sparse graphs, *Proceedings of a Cambridge Combinatorial Conference in honour of Paul Erős* (1984) 35–57.

[130] B. Bollobás, *Random Graphs*, First Edition, Academic Press, London (1985), Second Edition, Cambridge University Press (2001).

[131] B. Bollobás, The chromatic number of random graphs, *Combinatorica* 8 (1988) 49–56.

[132] B. Bollobás, *Modern Graph Theory*, Springer, New York (1998).

[133] B. Bollobás, Complete matchings in random subgraphs of the cube, *Random Structures and Algorithms* 1 (1990) 95–104.

[134] B. Bollobás, P. Catlin and P. Erdős, Hadwigers conjecture is true for almost every graph, *European Journal of Combinatorics* 1 (1980) 195–199.

[135] B. Bollobás, C. Cooper, T.I. Fenner and A.M. Frieze, On Hamilton cycles in sparse random graphs with minimum degree at least k, *Journal of Graph Theory* 34 (2000) 42–59.

[136] B. Bollobás and P. Erdős, Cliques in random graphs, *Mathematical Proceedings of the Cambridge Philosophical Society* 80 (1976) 419–427.

[137] B. Bollobás, T.I. Fenner and A.M. Frieze, An algorithm for finding hamilton paths and cycles in random graphs, *Combinatorica* 7 (1987) 327–341.

[138] B. Bollobás, T.I. Fenner and A.M. Frieze, Hamilton cycles in random graphs with minimal degree at least k, in A. Baker, B. Bollobas and A. Hajnal, Eds., *A tribute to Paul Erdos,* (1990) 59–96.

[139] B. Bollobás, T.I. Fenner and A.M. Frieze, Long cycles in sparse random graphs, in B. Bollobas, Ed., *Graph Theory and Combinatorics, Proceedings, Cambridge Combinatorial Conference in honour of Paul Erdős* (1984) 59–64.

[140] B. Bollobás and A. Frieze, On matchings and hamiltonian cycles in random graphs, *Annals of Discrete Mathematics* 28 (1985) 23–46.

[141] B. Bollobás and A. Frieze, Spanning maximal planar subgraphs of random graphs, *Random Structures and Algorithms* 2 (1991) 225–231.

[142] B. Bollobás, D. Gamarnik, O. Riordan and B. Sudakov, On the value of a random minimum length steiner tree, *Combinatorica* 24 (2004) 187–207.

[143] B. Bollobás, S. Janson and O. Riordan, The phase transition in inhomogeneous random graphs, *Random Structures and Algorithms* 31 (2007) 3–122.

[144] B. Bollobás, Y. Kohayakawa and T. Łuczak, The evolution of random subgraphs of the cube, *Random Structures and Algorithms* 3 (1992) 55–90.

[145] B. Bollobás, Y. Kohayakawa and T. Łuczak, Connectivity properties of random subgraphs of the cube, *Random Structures and Algorithms* 6 (1995) 221–230.

[146] B. Bollobás, R. Kozma and D. Miklós, *Handbook of Large-Scale Random Networks*, Springer and János Bolyai Mathematical Society, Budapest (2009).

[147] B. Bollobás and O. Riordan, Constrained graph processes, *Electronic Journal of Combinatorics* 7 (2000) R18.

[148] B. Bollobás and O. Riordan, The diameter of a scale free random graph, *Combinatorica* 24 (2004) 5–34.

[149] B. Bollobás, O. Riordan, J. Spencer and G. Tusnády, The degree sequence of a scale-free random graph process, *Random Structures and Algorithms* 18 (2001) 279–290.

[150] B. Bollobás and A. Thomason, Hereditary and monotone properties of graphs, in *The Mathematics of Paul Erdős, II, Algorithms and Combinatorics, 14,* Springer, Berlin (1997) 70–78.

[151] A. Bonato and P. Pralat, Probabilistic graph searching, see arxiv.org.

[152] C. Borgs, M. Brautbar, J. Chayes, S. Khanna and B. Lucier, The power of local information in social networks, see arxiv.org.

[153] J. Bourgain and G. Kalai, Influences of variables and threshold intervals under group symmetries, *Geometry and Functional Analysis* 7 (1997) 438–461.

[154] J. Bourgain, Appendix to [314].

[155] M. Bradonjić, R. Elsässer, T. Friedrich, T. Sauerwald and A. Stauffer, Efficient broadcast on random geometric graphs, *21st Annual ACM-SIAM Symposium on Discrete Algorithms* (2010) 1412–1421.

[156] M. Bradonjić, A. Hagberg, N.W. Hengartner and A.G. Percus, Component evolution in general random intersection graphs, in *Algorithms and Models for the Web-Graph, LNCS 6516*, Springer (2010) 36–43.

[157] G.R. Brightwell and M.J. Luczak, Vertices of high degree in the preferential attachment tree, *Electronic Journal of Probability* 17 (2012) 1–43.
[158] G.R. Brightwell, K. Panagiotou and A. Steger, Extremal subgraphs of random graphs, *Random Structures and Algorithms* 41 (2012) 147–178.
[159] T. Britton, M. Deijfen, A. Lageras and M. Lindholm, Epidemics on random graphs with tunable clustering *Journal of Applied Probability*, 45 (2008) 743–756.
[160] T. Britton, M. Deijfen and A. Martin-Löf, Generating simple random graphs with prescribed degree distribution, *Journal of Statistical Physics* 124 (2006) 1377–1397.
[161] A. Broder, A.M. Frieze and E. Shamir, Finding hidden Hamilton cycles, *Random Structures and Algorithms* 5 (1994) 395–410.
[162] A. Broder, A.M. Frieze, S. Suen and E. Upfal, An efficient algorithm for the vertex-disjoint paths problem in random graphs, *Proceedings of SODA '96*, 261–268.
[163] A. Broder, A.M. Frieze, S. Suen and E. Upfal, Optimal construction of edge-disjoint paths in random graphs, *SIAM Journal on Computing* 28 (1999) 541–574.
[164] T. Bui, S. Chaudhuri, T. Leighton and M. Sipser, Graph bisection algorithms with good average case behavior, *Combinatorica* 7 (1987) 171–191.
[165] Y.D. Burtin, Extremal metric characteristics of a random graph I, *Teoriya Veroyatnostei i ee Primeneniya* 19 (1974) 740–754.
[166] Y.D. Burtin, Extremal metric characteristics of a random graph II, *Teoriya Veroyatnostei i ee Primeneniya* 20 (1975) 82–99.
[167] Y.D. Burtin, On the probability of connectedness of a random subgraph of the n-cube, (in Russian) *Problemy Peredaci Informacii* 13 (1977) 90–95.
[168] Y.D. Burtin, On a simple formula for a random mappings and its applications, *Journal of Applied Probability* 17 (1980) 403–414.
[169] D.S. Callaway, J.E. Hopcroft, J.M. Kleinberg, M.E.J. Newman and S.H. Strogatz, Are randomly grown graphs really random? *Physical Review E* 64 (2001) 041902.
[170] E. Candellero and N. Fountoulakis, Clustering and the hyperbolic geometry of complex networks, see arxiv.org.
[171] Y. Caro, A. Lev, Y. Roditty, Z. Tuza and R. Yuster, On rainbow connection, *Electronic Journal of Combinatorics* 15 (2008) 11–13.
[172] J. Cerny and A. Teixeira, Critical window for the vacant set left by random walk on random regular graphs, see arxiv.org.
[173] D. Chakrabarti, Y. Zhan and C. Faloutsos, R-MAT: a recursive model for graph mining, in: *Proceedings of 4th International Conference on Data Mining* (2004) 442–446.
[174] S. Chatterjee and A. Dembo, Nonlinear large deviations, see arxiv.org.
[175] P. Chebolu and A.M. Frieze, Hamilton cycles in random lifts of complete directed graphs, *SIAM Journal on Discrete Mathematics* 22 (2008) 520–540.
[176] P. Chebolu, A.M. Frieze and P. Melsted, Finding a maximum matching in a sparse random graph in $O(n)$ expected time, *Journal of the ACM* 57 (2010) 1–27.

[177] L.H.Y. Chen, Poisson approximation for dependent trials, *Annals of Probability* 3 (1975) 534–545.
[178] W-C. Chen and W-C. Ni, Heap-ordered trees, 2-partitions and continued fractions, *European Journal of Combinatorics* 15 (1994) 513–517.
[179] W. Chen, W. Fang, G. Hu and M. Mahoney, On the hyperbolicity of small-world and tree-like random graphs, ISAAC 2012, LNCS 7676,(2012) 278–288.
[180] F. Chung, F. Frank, R. Graham and J. Shearer, Some intersection theorems for ordered sets and graphs, *Journal of Combinatorial Theory A* 43 (86) 23–37.
[181] F. Chung and L. Lu, Connected components in random graphs with given expected degree sequence *Annals of Combinatorics*, 6 (2002) 125–145.
[182] F. Chung and L. Lu. The average distances in random graphs with given expected degree, *Internet Math.* 1 (2003) 91–113.
[183] F. Chung and L. Lu, The volume of the giant component of a random graph with given expected degrees, *SIAM Journal on Discrete Mathematics* 20 (2006) 395–411.
[184] F. Chung, L. Lu and V. Vu, Eigenvalues of random power law graphs, *Annals of Combinatorics* 7 (2003) 21–33.
[185] F. Chung, L. Lu and V. Vu, The spectra of random graphs with expected degrees, *Proceedings of the National Academy of Sciences* 100 (2003) 6313–6318.
[186] V. Chvátal, Determining the stability number of a graph, *SIAM Journal on Computing* 6 (1977) 643–662.
[187] V. Chvátal, Almost all graphs with $1.44n$ edges are 3-colorable, *Random Structures and Algorithms* 2 (1991) 11–28.
[188] V. Chvátal and P. Erdős, Biased positional games, *Annals of Discrete Mathematics* 2 (1978) 221–228.
[189] A. Coja-Oghlan, Graph partitioning via adaptive spectral techniques, *Combinatorics, Probability and Computing* 19 (2010) 227–284.
[190] A. Coja-Oghlan, On the Laplacian eigenvalues of $G(n,p)$, *Combinatorics, Probability and Computing* 16 (2007) 923–946.
[191] A. Coja-Oghlan, Upper bounding the k-colorability threshold by counting covers, *Electronic Journal of Combinatorics* 20 (2013) P32.
[192] A. Coja-Oghlan, K. Panagiotou and A. Steger, On the chromatic number of random graphs, *Journal of Combinatorial Theory, B* 98 (2008) 980–993.
[193] A. Coja-Oghlan and D. Vilenchik, Chasing the k-colorability threshold, see arxiv.org.
[194] O. Cooley, M. Kang and C. Koch, The size of the giant component in random hypergraphs, see arxiv.org.
[195] D. Conlon and T. Gowers, Combinatorial theorems in sparse random sets, see arxiv.org.
[196] C. Cooper, Pancyclic Hamilton cycles in random graphs, *Discrete Mathematics* 91 (1991) 141–148.
[197] C. Cooper, On the thickness of sparse random graphs, *Combinatorics, Probability and Computing* 1 (1992) 303–309.
[198] C. Cooper, 1-pancyclic Hamilton cycles in random graphs, *Random Structures and Algorithms* 3 (1992) 277–287.

[199] C. Cooper, The size of the cores of a random graph with a given degree sequence, *Random Structures and Algorithms* (2004) 353–375.

[200] C. Cooper, The age specific degree distribution of web-graphs, *Combinatorics, Probability and Computing* 15 (2006) 637–661.

[201] C. Cooper and A.M. Frieze, On the number of hamilton cycles in a random graph, *Journal of Graph Theory* 13 (1989) 719–735.

[202] C. Cooper and A.M. Frieze, Pancyclic random graphs, in M. Karonski, J. Javorski and A. Rucinski, Eds., *Random Graphs,* Wiley (1990).

[203] C. Cooper and A.M. Frieze, The limiting probability that a-in, b-out is strongly connected, *Journal of Combinatorial Theory, B* 48 (1990) 117–134.

[204] C. Cooper and A.M. Frieze, Multicoloured Hamilton cycles in random graphs: an anti-Ramsey threshold, *Electronic Journal of Combinatorics* 2, (1995) R19.

[205] C. Cooper and A.M. Frieze, On the connectivity of random k-th nearest neighbour graphs, *Combinatorics, Probability and Computing* 4 (1996) 343–362.

[206] C. Cooper and A.M. Frieze, Hamilton cycles in random graphs and directed graphs, *Random Structures and Algorithms* 16 (2000) 369–401.

[207] C. Cooper and A.M. Frieze, Multi-coloured Hamilton cycles in randomly coloured random graphs, *Combinatorics, Probability and Computing* 11 (2002) 129–134.

[208] C. Cooper and A.M. Frieze, On a general model of web graphs, *Random Structures and Algorithms* 22 (2003) 311–335.

[209] C. Cooper and A.M. Frieze, The size of the largest strongly connected component of a random digraph with a given degree sequence, *Combinatorics, Probability and Computing* 13 (2004) 319–337.

[210] C. Cooper and A.M. Frieze, The cover time of random regular graphs, *SIAM Journal on Discrete Mathematics* 18 (2005) 728–740.

[211] C. Cooper and A.M. Frieze, The cover time of the preferential attachment graph, *Journal of Combinatorial Theory, B* 97 (2007) 269–290.

[212] C. Cooper and A.M. Frieze, The cover time of the giant component of a random graph, *Random Structures and Algorithms* 32 (2008) 401–439.

[213] C. Cooper and A.M. Frieze, The cover time of random geometric graphs, *Random Structures and Algorithms* 38 (2011) 324–349.

[214] C. Cooper and A.M. Frieze, Stationary distribution and cover time of random walks on random digraphs, *Journal of Combinatorial Theory, B* 102 (2012) 329–362.

[215] C. Cooper and A.M. Frieze, Component structure induced by a random walk on a random graph, *Random Structures and Algorithms* 42 (2013) 135–158.

[216] C. Cooper and A.M. Frieze, Long paths in random Apollonian networks, see arxiv.org.

[217] C. Cooper, A.M. Frieze, N. Ince, S. Janson and J. Spencer, On the length of a random minimum spanning tree, see arxiv.org.

[218] C. Cooper, A.M. Frieze and M. Krivelevich, Hamilton cycles in random graphs with a fixed degree sequence, *SIAM Journal on Discrete Mathematics* 24 (2010) 558–569.

[219] C. Cooper, A.M. Frieze and E. Lubetzky, Cover time of a random graph with given degree sequence II: Allowing vertices of degree two, *Random Structures and Algorithms* 45 (2014) 627–674.
[220] C. Cooper, A.M. Frieze and M. Molloy, Hamilton cycles in random regular digraphs, *Combinatorics, Probability and Computing* 3 (1994) 39–50.
[221] C. Cooper, A.M, Frieze and P. Pralat, Some typical properties of the Spatial Preferred Attachment model, *Internet Mathematics* 10 (2014), 27–47.
[222] C. Cooper, A.M. Frieze and T. Radzik, The cover time of random walks on random uniform hypergraphs, *Proceeedings of SIROCCO 2011*, 210–221.
[223] C. Cooper, A.M. Frieze, M. Molloy and B. Reed, Perfect matchings in random r-regular, s-uniform hypergraphs, *Combinatorics, Probability and Computing* 5 (1996) 1–15.
[224] C. Cooper, A.M. Frieze and B. Reed, Random regular graphs of non-constant degree: connectivity and Hamilton cycles, *Combinatorics, Probability and Computing* 11 (2002) 249–262.
[225] A. Collevecchio, A. Mehrabian and N. Wormald, Longest paths in random Apollonian networks and largest r-ary subtrees of random d-ary recursive trees, see arxiv.org.
[226] D. Coppersmith and G. Sorkin, Constructive bounds and exact expectations for the random assignment problem, *Random Structures and Algorithms* 15 (1999) 133–144.
[227] T. Cover and J. Thomas, Elements of Information Theory, 2nd Edition, Wiley-Interscience, (2006).
[228] V. Dani and C. Moore, Independent sets in random graphs from the weighted second moment method, *Proceedings of RANDOM 2011* (2011) 472–482.
[229] A. Darrasse, H-K. Hwang and M. Soria, Shape measures of random increasing k-trees, manuscript (2013).
[230] M. Deijfen and W. Kets, Random intersection graphs with tunable degree distribution and clustering, *Probability in the Engineering and Informational Sciences* 23 (2009) 661–674.
[231] D. Dellamonica, Jr., Y. Kohayakawa, V. Rödl and A. Ruciński, An improved upper bound on the density of universal random graphs, *Random Structures and Algorithms*, see arxiv.org.
[232] B. DeMarco and J. Kahn, Tight upper tail bounds for cliques, *Random Structures and Algorithms* 41 (2012) 469–487.
[233] B. DeMarco and J. Kahn, Mantel's theorem for random graphs, see arxiv.org.
[234] B. DeMarco and J. Kahn, Turán's theorem for random graphs, see arxiv.org.
[235] L. Devroye, Branching processes in the analysis of the heights of trees, *Acta Informatica* 24 (1987) 277–298.
[236] L. Devroye, O. Fawzi and N. Fraiman, Depth properties of scaled attachment random recursive tree, *Random Structures and Algorithms* 41 (2012) 66–98.
[237] L. Devroye and N. Fraiman, Connectivity in inhomogeneous random graphs, *Random Structures and Algorithms* 45 (2013) 408–420.
[238] L. Devroye and H-K. Hwang, Width and mode of the profile for some random trees of logarithmic height, *The Annals of Applied Probability* 16 (2006) 886–918.

[239] L. Devroye and S. Janson, Protected nodes and fringe subtrees in some random trees, *Electronic Communications in Probability* 19 (2014) 1–10.

[240] L. Devroy and J. Lu, The strong convergence of maximal degrees in uniform random recursive trees and dags, *Random Structure and Algorithms* 7 (1995) 1–14.

[241] J. Díaz, F. Grandoni and A. Marchetti-Spaccamela, Balanced Cut Approximation in Random Geometric Graphs, *Theoretical Computer Science* 410 (2009) 527–536.

[242] J. Díaz, D. Mitsche and X. Pérez-Giménez, Sharp threshold for hamiltonicity of random geometric graphs, *SIAM Journal on Discrete Mathematics* 21 (2007) 57–65.

[243] J. Díaz, X. Pérez-Giménez, J. Serna and N. Wormald, Walkers on the cycle and the grid, *SIAM Journal on Discrete Mathematics* 22 (2008) 747–775.

[244] J. Díaz, M.D. Penrose, J. Petit and M.J. Serna, Approximating layout problems on random geometric graphs, *Journal of Algorithms* 39 (2001) 78–116.

[245] J. Ding, J.H. Kim, E. Lubetzky and Y. Peres, Anatomy of a young giant component in the random graph, *Random Structures and Algorithms* 39 (2011) 139–178.

[246] J. Ding, E. Lubetzky and Y. Peres, Mixing time of near-critical random graphs, *Annals of Probability* 40 3 (2012) 979–1008.

[247] R. DiPietro, L.V. Mancini, A. Mei, A. Panconesi and J. Radhakrishnan, Redoubtable sensor networks, *ACM Transactions on Information and Systems Security* 11 (2008) 1–22.

[248] R.P. Dobrow and R.T. Smythe, Poisson approximations for functionals of random trees, *Random Structures and Algorithms* 9 (1996) 79–92.

[249] M. Dondajewski and J. Szymański, On a generalized random recursive tree, unpublished manuscript.

[250] M. Drmota, *Random Trees*, Springer, Vienna, (2009).

[251] M. Drmota and H-K. Hwang, Profiles of random trees: correlation and width of random recursive trees and binary search trees, *Advances in Applied Probability* 37 (2005) 321–341.

[252] A. Dudek and A.M. Frieze, Loose Hamilton cycles in random k-uniform hypergraphs, *Electronic Journal of Combinatorics* (2011) P48.

[253] A. Dudek and A.M. Frieze, Tight Hamilton cycles in random uniform hypergraphs, *Random Structures and Algorithms* 42 (2012) 374–385.

[254] A. Dudek, A.M. Frieze, P. Loh and S. Speiss, Optimal divisibility conditions for loose Hamilton cycles in random hypergraphs, *Electronic Journal of Combinatorics* 19 (2012).

[255] A. Dudek, A.M. Frieze, A. Ruciński and M. Šilekis, Loose Hamilton cycles in regular hypergraphs, *Combinatorics, Probability and Computing* 24 (2015) 179–194.

[256] A. Dudek, A.M. Frieze, A. Ruciński and M. Šilekis, Embedding the Erdős–Rényi hypergraph into the random regular hypergraph and hamiltonicity, see arxiv.org.

[257] A. Dudek, A.M. Frieze and C. Tsourakakis, Rainbow connection of random regular graphs, see arxiv.org.

[258] A. Dudek, D. Mitsche and P. Pralat, The set chromatic number of random graphs, see arxiv.org.

[259] A. Dudek and P. Pralat, An alternative proof of the linearity of the size-Ramsey number of paths, to appear in *Combinatorics, Probability and Computing.*

[260] A. Dudek and P. Pralat, Acquaintance time of random graphs near connectivity threshold, see arxiv.org.

[261] R. Durrett, Rigorous result for CHKNS random graph model, in: *Proceedings of Discrete Random Walks (Paris 2003), Discrete Mathematics and Theoretical Computer Science AC* (2003) 95–104.

[262] R. Durrett, *Random Graph Dynamics*, Cambridge University Press (2007).

[263] R. Durrett, *Probability: Theory and Examples*, 4th Edition, Cambridge University Press (2010).

[264] M. Dwass, The total progeny in a branching process, *Journal of Applied Probability* 6 (1969) 682–686.

[265] M. Dyer, A. Flaxman, A.M. Frieze and E. Vigoda, Random colouring sparse random graphs with fewer colours than the maximum degree, *Random Structures and Algorithms* 29 (2006) 450–465.

[266] M.E. Dyer, A.M. Frieze and L.R. Foulds, On the strength of connectivity of random subgraphs of the n-cube, in P. Karonski, Ed., *Random Graphs '85, Annals of Discrete Mathematics 33*, North-Holland (1987) 17–40.

[267] M.E. Dyer, A.M. Frieze and C. Greenhill, On the chromatic number of a random hypergraph, see arxiv.org.

[268] M.E. Dyer, A.M. Frieze and C.J.H. McDiarmid, Linear programs with random costs, *Mathematical Programming* 35 (1986) 3–16.

[269] M.E. Dyer, A.M. Frieze and B. Pittel, The average performance of the greedy matching algorithm, *The Annals of Applied Probability* 3 (1993) 526–552.

[270] E. Ebrahimzadeh, L. Farczadi, P. Gao, A. Mehrabian, C. Sato, N. Wormald and J. Zung, On the longest paths and the diameter in random apollonian networks. *Electronic Notes in Discrete Mathematics* 43(5) (2013) 355-365.

[271] J. Edmonds, Paths, trees and flowers, *Canadian Journal of Mathematics* 17 (1965) 449–467.

[272] C. Efthymiou, MCMC sampling colourings and independent sets of $G(n,d/n)$ near uniqueness threshold, *Proceedings of SODA 2014* 305–316.

[273] G.P. Egorychev, A solution of the Van der Waerden's permanent problem, *Preprint IFSO-L3 M Academy of Sciences SSSR, Krasnoyarks* (1980).

[274] P. Erdős, Graph theory and probability, *Canadian Journal of Mathematics* 11 (1959) 34–38.

[275] P. Erdős and A. Rényi, On random graphs I, *Publicationes Mathematicae* 6 (1959) 290–297.

[276] P. Erdős and A. Rényi, On the evolution of random graphs, *Publications of the Mathematical Institute of the Hungarian Academy of Sciences* 5 (1960) 17–61.

[277] P. Erdős and A. Rényi, On the strength of connectedness of a random graph, *Acta Mathematica Academiae Scientiarum Hungaricae* 8 (1961) 261–267.

[278] P. Erdős and A. Rényi, On random matrices, *Publications of the Mathematical Institute of the Hungarian Academy of Sciences* 8 (1964) 455–461.

[279] P. Erdős and A. Rényi, On the existence of a factor of degree one of a connected random graph, *Acta Mathematica Academiae Scientiarum Hungaricae* 17 (1966) 359–368.

[280] P. Erdős and M. Simonovits, Supersaturated graphs and hypergraphs, *Combinatorica* 3 (1983) 181–192.

[281] P. Erdős, M. Simonovits and V. T. Sós, Anti-Ramsey theorems, *Colloquia Mathematica Societatis János Bolya* 10, Infinite and Finite Sets, Keszethely, (1973).

[282] P. Erdős and J. Spencer, Evolution of the n-cube, *Computers & Mathematics with Applications* 5 (1979) 33–39.

[283] P. Erdős, S. Suen and P. Winkler, On the size of a random maximal graph, *Random Structures and Algorithms* 6 (1995) 309–318.

[284] L. Eschenauer and V.D. Gligor, A key managment scheme for distributed sensor networks, in *Proceedings of the 9th ACM Conference on Computer and Communication Security* (2002) 41–47.

[285] H. van den Esker, A geometric preferential attachment model with fitness, see arxiv.org.

[286] R. Fagin, Probabilities in finite models, *Journal of Symbolic Logic* 41 (1976) 50–58.

[287] D.I. Falikman, The proof of the Van der Waerden's conjecture regarding to doubly stochastic matrices, *Matematicheskie Zametki* 29 (1981).

[288] M. Faloutsos, P. Faloutsos and C. Faloutsos, On power-law relationships of the internet topology, *ACM SIGCOMM, Boston* (1999) 251–262.

[289] V. Feldman, E. Grigorescu, L. Reyzin, S. Vempala and Y. Xiao, Statistical algorithms and a lower bound for detecting planted cliques, *STOC 2013*, 655–664.

[290] W. Feller, *An Introduction to Probability Theory and its Applications*, 3rd Edition, Wiley, New York (1968).

[291] Q. Feng, H.M. Mahmoud and A. Panholzer, Phase changes in subtree varieties in random recursive and binary search trees, *SIAM Journal of Discrete Mathematics* 22 (2008) 160–184.

[292] A. Ferber, Closing gaps in problems related to Hamilton cycles in random graphs and hypergraphs, see arxiv.org.

[293] A. Ferber, R. Glebov, M. Krivelevich and A. Naor, Biased games on random boards, to appear in *Random Structures and Algorithms*.

[294] A. Ferber, G. Kronenberg and E. Long, packing, covering and counting hamilton cycles in random directed graphs, see arxiv.org.

[295] A. Ferber, G. Kronenberg, F. Mousset and C. Shikhelman, Packing a randomly edge-colored random graph with rainbow k-outs, see arxiv.org.

[296] A. Ferber, R. Nenadov, A. Noever, U. Peter and N. Škorić, Robust hamiltonicity of random directed graphs, see arxiv.org.

[297] A. Ferber, R. Nenadov and U. Peter, Universality of random graphs and rainbow embedding, see arxiv.org.

[298] T.I. Fenner and A.M. Frieze, On the connectivity of random m-orientable graphs and digraphs, *Combinatorica* 2 (1982) 347–359.

[299] D. Fernholz and V. Ramachandran, The diameter of sparse random graphs, *Random Structures and Algorithms* 31 (2007) 482–516.

[300] J.A. Fill, E.R. Scheinerman and K.B. Singer-Cohen, Random intersection graphs when $m = \omega(n)$: an equivalence theorem relating the evolution of the $G(n,m,p)$ and $G(n,p)$ models, *Random Structures and Algorithms* 16 (2000) 156–176.

[301] G. Fiz Pontiveros, S. Griffiths and R. Morris, The triangle-free process and $R(3,k)$, see arxiv.org.

[302] A. Flaxman, The lower tail of the random minimum spanning tree, *The Electronic Journal of Combinatorics* 14 (2007) N3.

[303] A. Flaxman, A.M. Frieze and T. Fenner, High degree vertices and eigenvalues in the preferential attachment graph, *Internet Mathematics* 2 (2005) 1–20.

[304] A. Flaxman, A.M. Frieze and J. Vera, A geometric preferential attachment model of networks, *Internet Mathematics* 3 (2007) 187–205.

[305] A.Flaxman, A.M. Frieze and J. Vera, A geometric preferential attachment model of networks, *Internet Mathematics* 4 (2007) 87–112.

[306] A. Flaxman, D. Gamarnik and G.B. Sorkin, Embracing the giant component, *Random Structures and Algorithms* 27 (2005) 277–289.

[307] J.E. Folkert, The distribution of the number of components of a random mapping function, PhD Dissertation, Michigan State University, (1955).

[308] C.M. Fortuin, P.W. Kasteleyn and J. Ginibre, Correlation inequalities on some partially ordered sets, *Communications in Mathematical Physics* 22 (1971) 89–103.

[309] N. Fountoulakis, On the evolution of random graphs on spaces of negative curvature, see arxiv.org.

[310] N. Fountalakis and K. Panagiotou, Sharp load thresholds for cuckoo hashing, *Random Structures and Algorithms* 41 (2012) 306–333.

[311] N. Fountalakis, M. Khosla and K. Panagiotou, The multiple-orientability thresholds for random hypergraphs, see arxiv.org.

[312] N. Fountoulakis, D. Kühn and D. Osthus, The order of the largest complete minor in a random graph, *Random Structures and Algorithms* 33 (2008) 127–141.

[313] N. Fountoulakis and B. A. Reed, The evolution of the mixing rate of a simple random walk on the giant component of a random graph, *Random Structures Algorithms* 33 (2008) 68–86.

[314] E. Friedgut, Sharp thresholds of graph properties, and the k-sat problem, *Journal of the American Mathematical Society* 12 (1999) 1017–1054.

[315] E. Friedgut, Hunting for sharp thresholds, *Random Structures Algorithms* 26 (2005) 37–51.

[316] E. Friedgut and G. Kalai, Every monotone graph property has a sharp threshold, *Proceedings of the American Mathematical Society* 124 (1996) 2993–3002.

[317] J. Friedman, A proof of Alon's second eigenvalue conjecture and related problems, *Memoirs of the American Mathematical Society* (2008).

[318] T. Friedrich and A. Krohmer, Cliques in hyperbolic random graphs, *IEEE INFOCOM* (2015).

[319] T. Friedrich, T. Sauerwald and A. Stauffer, Diameter and broadcast time of random geometric graphs in arbitrary dimensions, *Algorithmica* 67 (2013) 65–88.

[320] A.M. Frieze, On the value of a random minimum spanning tree problem, *Discrete Applied Mathematics* 10 (1985) 47–56.

[321] A.M. Frieze, On large matchings and cycles in sparse random graphs, *Discrete Mathematics* 59 (1986) 243–256.

[322] A.M. Frieze, Maximum matchings in a class of random graphs, *Journal of Combinatorial Theory, B* 40 (1986) 196–212.

[323] A.M. Frieze, An algorithm for finding hamilton cycles in random digraphs, *Journal of Algorithms* 9 (1988) 181–204.

[324] A.M. Frieze, On the independence number of random graphs, *Discrete Mathematics* 81 (1990) 171–176.

[325] A.M. Frieze, Perfect matchings in random bipartite graphs with minimal degree at least 2, *Random Structures and Algorithms* 26 (2005) 319–358.

[326] A.M. Frieze, Loose Hamilton cycles in random 3-uniform hypergraphs, *Electronic Journal of Combinatorics* 17 (2010) N28.

[327] A.M. Frieze, On a greedy 2-matching algorithm and hamilton cycles in random graphs with minimum degree at least three, *Random Structures and Algorithms* 45 (2014) 443–497.

[328] A.M. Frieze, N. Goyal, L. Rademacher and S. Vempala, Expanders via random spanning trees, *SIAM Journal on Computing* 43 (2014) 497–513.

[329] A.M. Frieze and S. Haber, An almost linear time algorithm for finding Hamilton cycles in sparse random graphs with minimum degree at least three, to appear in *Random Structures and Algorithms*.

[330] A.M. Frieze, S. Haber and M. M. Lavrov, On the game chromatic number of sparse random graphs, *SIAM Journal of Discrete Mathematics* 27 (2013) 768–790.

[331] A.M. Frieze, M.R. Jerrum, M. Molloy, R. Robinson and N.C. Wormald, Generating and counting Hamilton cycles in random regular graphs, *Journal of Algorithms* 21 (1996) 176–198.

[332] A.M. Frieze and T. Johansson, On random k-out sub-graphs of large graphs, see arxiv.org.

[333] A.M. Frieze and R. Kannan, A new approach to the planted clique problem, *Proceedings of Foundations of Software Technology and Theoretical Computer Science*, Bangalore, India (2008).

[334] A.M. Frieze and M. Krivelevich, On two Hamilton cycle problems in random graphs, *Israel Journal of Mathematics* 166 (2008) 221–234.

[335] A.M. Frieze, M. Krivelevich and C. Smyth, On the chromatic number of random graphs with a fixed degree sequence, *Combinatorics, Probability and Computing* 16 (2007) 733–746.

[336] A.M. Frieze, S. Haber and M. Lavrov, On the game chromatic number of sparse random graphs, *SIAM Journal of Discrete Mathematics* 27 (2013) 768–790.

[337] A.M. Frieze and P. Loh, Rainbow Hamilton cycles in random graphs, *Random Structures and Algorithms* 44 (2014) 328–354.

[338] A.M. Frieze and T. Łuczak, Hamiltonian cycles in a class of random graphs: one step further, in M.Karonski, J. Jaworski and A. Rucinski, Eds., *Proceedings of Random Graphs '87,* (1990) 53–59.

[339] A.M. Frieze and T. Łuczak, Edge disjoint trees in random graphs, *Periodica Mathematica Hungarica* 21 (1990) 28–30.

[340] A.M. Frieze and T. Łuczak, On the independence and chromatic numbers of random regular graphs, *Journal of Combinatorial Theory* 54 (1992) 123–132.

[341] A.M. Frieze and C.J.H. McDiarmid, On random minimum length spanning trees, *Combinatorica* 9 (1989) 363–374.

[342] A.M. Frieze and B.D. McKay, Multicoloured trees in random graphs, *Random Structures and Algorithms* 5 (1994) 45–56.

[343] A.M. Frieze and P. Melsted, Maximum matchings in random bipartite graphs and the space utilization of cuckoo hashtables, *Random Structures and Algorithms* 41 (2012) 334–364.

[344] A.M. Frieze, D. Mitsche, X. Pérez-Giménez and P. Pralat, On-line list colouring of random graphs, see arxiv.org.

[345] A.M. Frieze and W. Pegden, Between 2- and 3-colorability, see arxiv.org.

[346] A.M. Frieze and W. Pegden, Looking for vertex number one, see arxiv.org.

[347] A.M. Frieze and B. Pittel, Perfect matchings in random graphs with prescribed minimal degree, *Trends in Mathematics*, Birkhauser Verlag, Basel (2004) 95–132.

[348] A.M. Frieze, M. Ruszinko and L. Thoma, A note on random minimum length spanning trees, *Electronic Journal of Combinatorics* 7 (2000) R41.

[349] A.M. Frieze and G. Sorkin, Efficient algorithms for three-dimensional axial and planar random assignment problems, *Random Structures and Algorithms* 46 (2015) 160–196.

[350] A.M. Frieze and C. Tsourakakis, On certain properties of random apollonian networks, *WAW2012* (2012) 93–112.

[351] A.M. Frieze and C.E. Tsourakakis, Rainbow connectivity of sparse random graphs, *Electronic Journal of Combinatorics* 19 (2012) 1–17.

[352] A.M. Frieze and L. Zhao, Optimal construction of edge-disjoint paths in random regular graphs, *Combinatorics, Probability and Computing* 9 (2000) 241–264.

[353] M. Fuchs, The subtree size profile of plane-oriented recursive trees, manuscript, *Proceedings of the Eighth Workshop on Analytic Algorithmics and Combinatorics (ANALCO)* (2011) 85–92.

[354] M. Fuchs, H-K. Hwang and R. Neininger, Profiles of random trees: limit theorems for random recursive trees and binary search trees, *Algorithmica* 46 (2006) 367–407.

[355] Z. Füredi and J. Komlós, The eigenvalues of random symmetric matrics, *Combinatorica* 1 (1981) 233–241.

[356] D. Gale and L.S. Shapley, College admissions and the stability of marriage, *American Mathematical Monthly* 69 (1962) 9–14.

[357] P. Gao and M. Molloy, The stripping process can be slow, see arxiv.org.

[358] P. Gao, X. Pérez-Giménez and C.M. Sato, Arboricity and spanning-tree packing in random graphs with an application to load balancing, see arxiv.org.

[359] P. Gao and N. Wormald, Orientability thresholds for random hypergraphs, see arxiv.org.

[360] J. Gastwirth, A probability model of a pyramid scheme, *American Statistician* 31 (1977) 79–82.

[361] H. Gebauer and T. Szabó, Asymptotic random graph intuition for the biased connectivity game, *Random Structures and Algorithms* 35 (2009) 431–443.

[362] S. Gerke, H-J. Prömel, T. Schickinger, A. Steger and A. Taraz, K_4-free subgraphs of random graphs revisited, *Combinatorica* 27 (2007) 329–365.

[363] S. Gerke, H-J. Prömel and T. Schickinger, K_5-free subgfraphs of random graphs, *Random Structures and Algorithms* 24 (2004) 194–232.

[364] S. Gerke, D. Schlatter, A. Steger and A. Taraz, The random planar graphs process, *Random Structures and Algorithms* 32 (2008) 236–261.

[365] I.B. Gertsbakh, Epidemic processes on random graphs: Some preliminary results, *Journal of Applied Probability* 14 (1977) 427–438.

[366] I. Gessel and R.P. Stanley, Stirling polynomials, *Journal of Combinatorial Theory, A* 24 (1978) 24–33.

[367] E.N. Gilbert, Random graphs, *Annals of Mathematical Statistics* 30 (1959) 1141–1144.

[368] O. Giménez and M. Noy, Asymptotic enumeration and limit laws of planar graphs, *Journal of the American Mathematical Society* 22 (2009) 309–329.

[369] R. Glebov and M. Krivelevich, On the number of Hamilton cycles in sparse random graphs, *SIAM Journal on Discrete Mathematics* 27 (2013) 27–42.

[370] R. Glebov, H. Naves and B. Sudakov, The threshold probability for cycles, see arxiv.org.

[371] Y.V. Glebskii, D.I. Kogan, M.I. Liagonkii and V.A. Talanov, Range and degree of realizability of formulas in the restricted predicate calculus, *Cybernetics 5* (1969) 142–154.

[372] E. Godehardt and J. Jaworski, On the connectivity of a random interval graph, *Random Structures and Algorithms* 9 (1996) 137–161.

[373] E. Godehardt and J. Jaworski, Two models of random intersection graphs for classification, in *Data Analysis and Knowledge Organization*, Springer, Berlin, 22 (2003) 67–82.

[374] E. Godehardt, J. Jaworski and K. Rybarczyk, On the connectivity of random intersection graphs, in H.J. Lenz and R. Decker R., Eds., *Studies in Classification, Data Analysis and Knowledge Organization 33,* Springer Verlag, Heidelberg - Berlin (2007) 67–74.

[375] A. Goel, S. Rai and B. Krishnamachari, Monotone properties of random geometric graphs have sharp thresholds, *Annals of Applied Probability* 15 (2005) 2535–2552.

[376] W. Goh and E. Schmutz, Limit distribution for the maximum degree of a random recursive tree, *Journal of Computing and Applied Mathematics* 142 (2002) 61–82.

[377] Ch. Goldschmidt and J.B. Martin, Random recursive trees and the Bolthausen–Sznitman coalescent, *Electronic Journal of Probability* 10 (2005) 718–745.

[378] S.W. Golomb, Random permutations, *Bulletin of the American Mathematical Society* 70 (1964) 747.

[379] V.L. Goncharov, Sur la distribution des cycles dans les permutations, *Proceedings of the USSR Academy of Sciences* 35, (1942) 267–269.

[380] V.L. Goncharov, Some facts in combinatorics, *Izv. Acad. Nauk SSSR, Ser. Mat.* 8 (1944) 3–48 (in Russian). [See also: On the field of combinatory analysis, *Transactions of the American Mathematical Society* 19 (1962) 1–46.]

[381] D.A. Grable, On random greedy triangle packing, *Electronic Journal on Combinatorics* 4 (1997) R11.

[382] R. Gray, *Entropy and Information Theory*, Springer, (2011).

[383] C. Greenhill, S. Janson and A. Ruciński, On the number of perfect matchings in random lifts, *Combinatorics, Probability and Computing* 19 (2010) 791–817.

[384] G.R. Grimmett, Random labelled trees and their branching networks, *Journal of the Australian Mathematical Society* 30 (1980) 229–237.

[385] G. Grimmett and C. McDiarmid, On colouring random graphs, *Mathematical Proceedings of the Cambridge Philosophical Society* 77 (1975) 313–324.

[386] C. Groër, B.D. Sullivan and S. Poole, A mathematical analysis of the R-MAT random graph generator, *Networks*, 58 (2011) 159–170.

[387] L. Gugelmann, K. Panagiotou and U. Peter, Random hyperbolic graphs: degree sequence and clustering, in *Proceedings of the 39th International Colloquium Conference on Automata, Languages, and Programming (ICALP)* (2012) 573–585.

[388] P. Gupta and P.R. Kumar, Critical power for asymptotic connectivity in wireless networks, in W. M. McEneany, G. Yin and Q. Zhang, Eds., *Stochastic Analysis, Control, Optimization and Applications: A Volume in Honor of W. H. Fleming*, Boston, MA: Birkhauser, (1998) 547–566.

[389] Y. Gurevich and S. Shelah, Expected computation time for Hamiltonian path problem, *SIAM Journal on Computing* 16 (1987) 486–502.

[390] E. Győri, B. Rorthchild and A. Ruciński, Every graph is contained in a sparest possible balanced graph, *Mathematical Proceedings of the Cambridge Philosophical Society* 98 (1985) 397–401.

[391] S. Haber and M. Krivelevich, The logic of random regular graphs, *Journal of Combinatorics* 1 (2010) 389–440.

[392] A. Hamm and J. Kahn, On Erdős–Ko–Rado for random hypergraphs I, see arxiv.org.

[393] A. Hamm and J. Kahn, On Erdős–Ko–Rado for random hypergraphs II, see arxiv.org.

[394] J.C. Hansen and J. Jaworski, Random mappings with exchangeable in-degrees, *Random Structures and Algorithms* 33 (2008) 105–126.

[395] J.C. Hansen and J. Jaworski, Predecessors and successors in random mappings with exchangeable in-degrees, *Journal of Applied Probability* 50 (2013) 721–740.

[396] T. Harris, A lower bound for the critical probability in a certain percolation, *Proceedings of the Cambridge Philosophical Society* 56 (1960) 13–20.

[397] H. Hatami, Random cubic graphs are not homomorphic to the cycle of size 7, *Journal of Combinatorial Theory, B* 93 (2005) 319–325.

[398] H. Hatami and M. Molloy, The scaling window for a random graph with a given degree sequence, *Random Structures and Algorithms* 41 (2012) 99–123.

[399] P. Haxell, Y. Kohayakawa and T. Łuczak, Turáns extremal problem in random graphs: forbidding even cycles, *Journal of Combinatorial Theory, B* 64 (1995) 273–287.

[400] P. Haxell, Y. Kohayakawa and T. Łuczak, Turáns extremal problem in random graphs: forbidding odd cycles, *Combinatorica* 16 (1995) 107–122.

[401] J. He and H. Liang, On rainbow-k-connectivity of random graphs, *Information Processing Letters* 112 (2012) 406–410.

[402] A. Heckel and O. Riordan, The hitting time of rainbow connection number two, *Electronic Journal on Combinatorics* 19 (2012) 1–16.

[403] D. Hefetz, M. Krivelevich, M. Stojaković and T. Szabó, *Positional Games* (Oberwolfach Seminars), Birkhauser (2014).

[404] D. Hefetz, M. Krivelevich and T. Szabo, Sharp threshold for the appearance of certain spanning trees in random graphs, *Random Structures and Algorithms* 41 (2012) 391–412.

[405] D. Hefetz, A. Steger and B. Sudakov, Random directed graphs are robustly hamiltonian, see arxiv.org.

[406] W. Hoeffding, Probability inequalities for sums of bounded random variables, *Journal of the American Statistical Association* 58 (1963) 13–30.

[407] R. van der Hofstad, Random graphs and complex networks: Volume 1, see arxiv.org.

[408] R. van der Hofstad, Critical behaviour in inhomogeneous random graphs, *Random Structures and Algorithms* 42 (2013) 480–508.

[409] R. van der Hofstad, G. Hooghiemstra and P. Van Mieghem, On the covariance of the level sizes in random recursive trees, *Random Structures and Algorithms* 20 (2002) 519–539.

[410] R. van der Hofstad, S. Kliem and J. van Leeuwaarden, Cluster tails for critical power-law inhomogeneous random graphs, see arxiv.org

[411] C. Holmgren and S. Janson, Limit laws for functions of fringe trees for binary search trees and recursive trees, see arxiv.org

[412] P. Horn and M. Radcliffe, Giant components in Kronecker graphs, *Random Structures and Algorithms* 40 (2012) 385–397.

[413] H-K. Hwang, Profiles of random trees: plane-oriented recursive trees, *Random Structures and Algorithms* 30 (2007) 380–413.

[414] G.I. Ivchenko, On the asymptotic behaviour of the degrees of vertices in a random graph, *Theory of Probability and Applications* 18 (1973) 188–195.

[415] S. Janson, Poisson convergence and Poisson processes with applications to random graphs, *Stochastic Processes and their Applications* 26 (1987) 1–30.

[416] S. Janson, Normal convergence by higher semi-invariants with applications to random graphs, *Annals of Probability* 16 (1988) 305–312.

[417] S. Janson, A functional limit theorem for random graphs with applications to subgraph count statistics *Random Structures and Algorithms* 1 (1990) 15–37.

[418] S. Janson, Poisson approximation for large deviations, *Random Structures and Algorithms* 1 (1990) 221–230.

[419] S. Janson, Multicyclic components in a random graph process, *Random Structures and Algorithms* 4 (1993) 71–84.

[420] S. Janson, The numbers of spanning trees, Hamilton cycles and perfect matchings in a random graph, *Combinatorics, Probability and Computing* 3 (1994) 97–126.
[421] S. Janson, The minimal spanning tree in a complete graph and a functional limit theorem for trees in a random graph, *Random Structures Algorithms* 7 (1995) 337–355.
[422] S. Janson, Random regular graphs: Asymptotic distributions and contiguity, *Combinatorics, Probability and Computing* 4 (1995) 369–405.
[423] S. Janson, One, two and three times $\log n/n$ for paths in a complete graph with random weights, *Combinatorics, Probability and Computing* 8 (1999) 347–361.
[424] S. Janson, Asymptotic degree distribution in random recursive trees, *Random Structures and Algorithms*, 26 (2005) 69–83.
[425] S. Janson, Monotonicity, asymptotic normality and vertex degrees in random graphs, *Bernoulli* 13 (2007) 952–965.
[426] S. Janson, Plane recursive trees, Stirling permutations and an urn model, in *Fifth Colloquium on Mathematics and Computer Science, Nancy*, (2008) 541–547.
[427] S. Janson, Asymptotic equivalence and contiguity of some random graphs, *Random Structures Algorithms* 36 (2010) 26–45.
[428] S. Janson, M. Kuba and A. Panholzer, Generalized Stirling permutations, families of increasing trees and urn models, *Journal of Combinatorial Theory, A* 118 (2010) 94–114.
[429] S. Janson and M.J. Luczak, A simple solution to the k-core problem, *Random Structures Algorithms* 30 (2007) 50–62.
[430] S. Janson, K. Oleszkiewicz and A. Ruciński, Upper tails for subgraph counts in random graphs, *Israel Journal of Mathematics* 142 (2004) 61–92.
[431] S. Janson, T. Łuczak and A. Ruciński, An exponential bound for the probability of nonexistence of a specified subgraph in a random graph, in M. Karoński, J. Jaworski and A. Ruciski, Eds., *Random Graphs '87*, Wiley (1990) 73–87.
[432] S. Janson, T. Łuczak and A. Ruciński, *Random Graphs*, John Wiley and Sons, New York, (2000).
[433] S. Janson and K. Nowicki, The asymptotic distributions of generalized U-statistics with applications to random graphs, *Probability Theory and Related Fields* 90 (1991) 341–375.
[434] S. Janson and O. Riordan, Susceptibility in inhomogeneous random graphs, *The Electronic Journal of Combinatorics* 19 (2012) #P31
[435] S. Janson and N. Wormald, Rainbow Hamilton cycles in random regular graphs, *Random Structures Algorithms* 30 (2007) 35–49.
[436] S. Janson, D.E. Knuth, T. Łuczak and B.G. Pittel, The birth of the giant component, *Random Structures and Algorithms* 4 (1993) 233–358.
[437] J. Janssen, P. Pralat, and R. Wilson, Geometric graph properties of the spatial preferred attachment model, *Advances in Applied Mathematics* 50 (2013) 243–267.
[438] J. Janssen, P. Pralat, and R. Wilson, Asymmetric distribution of nodes in the spatial preferred attachment model, in *Proceedings of the 10th Workshop on*

Algorithms and Models for the Web Graph (WAW 2013), Lecture Notes in Computer Science 8305, Springer (2013) 1–13.

[439] J. Jaworski, On a random mapping (T, P_j), *Journal of Applied Probability* 21 (1984) 186–191.

[440] J. Jaworski, Random mappings with independent choices of images, in *Random Graphs'87*, Wiley, New York (1990) 89–101.

[441] J. Jaworski, Predecessors in a random mapping, *Random Structures and Algorithms* 13 (1998) 501–519.

[442] J. Jaworski, Epidemic processes on digraphs of random mappings, *Journal of Applied Probability* 36 (1999) 780–798.

[443] J. Jaworski and T. Łuczak, Cycles in a uniform graph process, *Combinatorics, Probability and Computing* 1 (1992) 223–239.

[444] J. Jaworski and Z. Palka, Remarks on a general model of a random digraph, *Ars Combinatoria* 65 (2002) 135–144.

[445] J. Jaworski and I.H. Smit, On a random digraph, *Annals of Discrete Mathematics* 33 (1987) 111–127.

[446] J. Jaworski, M. Karoński and D. Stark, The degree of a typical vertex in generalized random intersection graph model, *Discrete Mathematics* 30 (2006) 2152–2165.

[447] J. Jaworski and D. Stark, The vertex degree distribution of passive random intersection graph models, *Combinatorics, Probability and Computing* 17 (2008) 549–558.

[448] M.R. Jerrum, Large cliques elude the metropolis process, *Random Structures and Algorithms* 3 (1992) 347–359.

[449] J. Jiang, M. Mitzenmacher and J. Thaler, Parallel peeling algorithms, see arxiv.org.

[450] J. Jonasson, On the cover time of random walks on random graphs, *Combinatorics, Probability and Computing* 7 (1998), 265–279.

[451] A. Johansson, J. Kahn and V. Vu, Factors in random graphs, *Random Structures and Algorithms* 33 (2008) 1–28.

[452] J. Jordan, Geometric preferential attachment in non-uniform metric spaces, *Electronic Journal of Probability* 18 (2013) 1–15.

[453] J. Jordan and A. Wade, Phase transitions for random geometric preferential attachment graphs, *Advances in Applied Probability* 47 (2015).

[454] F. Juhász, On the spectrum of a random graph, in L. Lovasz and V.T. Sos, Eds., *Algebraic Methods in Graph Theory, Colloquia Mathematica Societatis Janos Bolyai, North Holland* 25 (1981) 313–316.

[455] M. Kahle, Topology of random simplicial complexes: a survey, *AMS Contemporary Volumes in Mathematics*, to appear.

[456] J. Kahn, E. Lubetzky and N. Wormald, The threshold for combs in random graphs, see arxiv.org.

[457] J. Kahn, E. Lubetzky and N. Wormald, Cycle factors and renewal theory, see arxiv.org.

[458] J. Kahn and E. Szemerédi, On the second eigenvalue in random regular graphs - Section 2, in *Proceedings of the 21st Annual ACM Symposium on Theory of Computing* (1989) 587–598.

[459] S. Kalikow and B. Weiss, When are random graphs connected? *Israel Journal of Mathematics* 62 (1988) 257–268.

[460] N. Kamčev, M. Krivelevich and B. Sudakov, Some remarks on rainbow connectivity, see arxiv.org.

[461] M. Kang, M. Karoński, C. Koch and T. Makai, Properties of stochastic Kronecker graphs, to appear in *Journal of Combinatorics*.

[462] M. Kang and T. Łuczak, Two critical periods in the evolution of random planar graphs, *Transactions of the American Mathematical Society* 364 (2012) 4239–4265.

[463] M. Kang, W. Perkins and J. Spencer, The Bohman–Frieze process near criticality, see arxiv.org.

[464] D. Karger and C. Stein, A new approach to the minimum cut problem, *Journal of the ACM* 43 (1996) 601–640.

[465] M. Karoński, A review of random graphs, *Journal of Graph Theory* 6 (1982) 349–389.

[466] M. Karoński, Random graphs, in *Handbook of Combinatorics*, Elsevier Science B.V. (1995) 351–380.

[467] M. Karoński and T. Łuczak, The phase transition in a random hypergraph, *Journal of Computational and Applied Mathematics* 1 (2002) 125–135.

[468] M. Karoński and B. Pittel, Existence of a perfect matching in a random $(1 + \varepsilon^{-1})$–out bipartite graph, *Journal of Combinatorial Theory, B* (2003) 1–16.

[469] M. Karoński and A. Ruciński, On the number of strictly balanced subgraphs of a random graph, in *Graph Theory, Proc. Łagów, 1981*, Lecture Notes in Mathematics 1018, Springer, Berlin (1983) 79–83.

[470] M. Karoński and A. Ruciński, Problem 4, in *Graphs and other Combinatorial Topics, Proceedings, Third Czech. Symposium on Graph Theory, Prague* (1983).

[471] M. Karoński and A. Ruciński, Poisson convergence and semi-induced properties of random graphs, *Mathematical Proceedings of the Cambridge Philosophical Society* 101 (1987) 291–300.

[472] M. Karoński, E. Scheinerman and K. Singer-Cohen, On random intersection graphs: the subgraph problem, *Combinatorics, Probability and Computing* 8 (1999) 131–159.

[473] R.M. Karp, An upper bound on the expected cost of an optimal assignment, in D. Johnson, Ed., *Discrete Algorithms and Complexity: Proceedings of the Japan-US Joint Seminar*, Academic Press, New York, (1987), 1–4.

[474] R.M. Karp, The transitive closure of a random digraph, *Random Structures and Algorithms* 1 (1990) 73–93.

[475] R.M. Karp, A. Rinnooy-Kan and R.V. Vohra, Average case analysis of a heuristic for the assignment problem, *Mathematics of Operations Research* 19 (1994) 513–522.

[476] R.M. Karp and M. Sipser, Maximum matchings in sparse random graphs, in *Proceedings of the 22nd IEEE Symposium on the Foundations of Computer Science* (1981) 364–375.

[477] Zs. Katona, Width of scale-free tree, *Journal of Applied Probability* 42 (2005) 839–850.

[478] Zs. Katona, Levels of scale-free tree, manuscript

[479] L. Katz, Probability of indecomposability of a random mapping function, *Annals of Mathematical Statistics* 26 (1955) 512–517.
[480] G. Kemkes, X. Perez-Gimenez and N. Wormald, On the chromatic number of random d-regular graphs, *Advances in Mathematics* 223 (2010) 300–328.
[481] R. Keusch and A. Steger, The game chromatic number of dense random graphs, *Electronic Journal of Combinatorics* 21 (2014) 1–14.
[482] J-H. Kim, The Ramsey number $R(3,t)$ has order of magnitude $t^2/\log t$, *Random Structures and Algorithms* (1995) 173–207.
[483] J-H. Kim and V. Vu, Sandwiching random graphs: universality between random graph models, *Advances in Mathematics* 188 (2004) 444–469.
[484] J.H. Kim and N.C. Wormald, Random matchings which induce Hamilton cycles, and hamiltonian decompositions of random regular graphs, *Journal of Combinatorial Theory* B 81 (2001) 20–44.
[485] J.F.C. Kingman, The population structure associated with the Ewens sampling formula, *Theoretical Population Biology* 11 (1977) 274–283.
[486] W. Kinnersley, D. Mitsche and P. Pralat, A note on the acquaintance time of random graphs, *Electronic Journal of Combinatorics* 20(3) (2013) #P52.
[487] M. Kiwi and D. Mitsche, A bound for the diameter of random hyperbolic graphs, see arxiv.org.
[488] J. Kleinberg, The small-world phenomenon: An algorithmic perspective, in *Proceedings of the 32nd ACM Symposium on Theory of Computing* (2000) 163–170.
[489] D.E. Knuth, R. Motwani and B. Pittel, Stable husbands, *Random Structures and Algorithms* 1 (1990) 1–14.
[490] F. Knox, D. Kühn and D. Osthus, Edge-disjoint Hamilton cycles in random graphs, to appear in *Random Structures and Algorithms*.
[491] L.M. Koganov, Universal bijection between Gessel–Stanley permutations and diagrams of connections of corresponding ranks, *Mathematical Surveys* 51 (1996) 333–335.
[492] Y. Kohayakawa, B. Kreuter and A. Steger, An extremal problem for random graphs and the number of graphs with large even girth, *Combinatorica* 18 (1998) 101–120.
[493] Y. Kohayakawa, T. Łuczak and V. Rödl, On K_4-free subgraphs of random graphs, *Combinatorica* 17 (1997) 173–213.
[494] V.F. Kolchin, A problem of the allocation of particles in cells and cycles of random permutations, *Theory of Probability and its Applications* 16 (1971) 74–90.
[495] V.F. Kolchin, Branching processes, random trees, and a generalized scheme of arrangements of particles, *Mathematical Notes* 21 (1977) 386–394.
[496] V.F. Kolchin, *Random Mappings*, Optimization Software, Inc., New York (1986).
[497] V.F. Kolchin, *Random Graphs*, Cambridge University Press, Cambridge (1999).
[498] J. Komlós and E. Szemerédi, Limit distributions for the existence of Hamilton circuits in a random graph, *Discrete Mathematics* 43 (1983) 55–63.
[499] W. Kordecki, On the connectedness of random hypergraphs, *Commentaciones Mathematicae* 25 (1985) 265–283.

[500] W. Kordecki, Normal approximation and isolated vertices in M. Karonski, J. Jaworski and A. Rucinski, Eds., *Random Graphs'87*, John Wiley and Sons, Chichester (1990) 131–139.

[501] I.N. Kovalenko, Theory of random graphs, *Cybernetics and Systems Analysis* 7 (1971) 575–579.

[502] D. Krioukov, F. Papadopoulos, M. Kitsak, A. Vahdat and M. Boguná, Hyperbolic geometry of complex networks, *Physics Review, E* (2010) 82:036106.

[503] M. Krivelevich, The choice number of dense random graphs, *Combinatorics, Probability and Computing* 9 (2000) 19–26.

[504] M. Krivelevich, C. Lee and B. Sudakov, Long paths and cycles in random subgraphs of graphs with large minimum degree, see arxiv.org.

[505] M. Krivelevich, P.-S. Loh and B. Sudakov, Avoiding small subgraphs in Achlioptas processes, *Random Structures and Algorithms* 34 (2009) 165–195.

[506] M. Krivelevich, E. Lubetzky and B. Sudakov, Hamiltonicity thresholds in Achlioptas processes, *Random Structures and Algorithms* 37 (2010) 1–24.

[507] M. Krivelevich, E. Lubetzky and B. Sudakov, Longest cycles in sparse random digraphs, *Random Structures and Algorithms* 43 (2013) 1–15.

[508] M. Krivelevich, E. Lubetzky and B. Sudakov, Cores of random graphs are born Hamiltonian, *Proceedings of the London Mathematical Society* 109 (2014) 161–188.

[509] M. Krivelevich and W. Samotij, Optimal packings of Hamilton cycles in sparse random graphs, *SIAM Journal on Discrete Mathematics* 26 (2012) 964–982.

[510] M. Krivelevich and B. Sudakov, The chromatic numbers of random hypergraphs, *Random Structures and Algorithms* 12 (1998) 381–403.

[511] M. Krivelevich and B. Sudakov, Minors in expanding graphs, *Geometric and Functional Analysis* 19 (2009) 294–331.

[512] M. Krivelevich and B. Sudakov, The phase transition in random graphs - a simple proof, *Random Structures and Algorithms* 43 (2013) 131–138.

[513] M. Krivelevich, B. Sudakov, V. Vu and N. Wormald, Random regular graphs of high degree, *Random Structures and Algorithms* 18 (2001) 346–363.

[514] M. Krivelevich, B. Sudakov, V.H. Vu and N.C. Wormald, On the probability of independent sets in random graphs, *Random Structures and Algorithms* 22 (2003) 1–14.

[515] M. Krivelevich and V.H. Vu, Choosability in random hypergraphs, *Journal of Combinatorial Theory, B* 83 (2001) 241–257.

[516] M. Kuba and A. Panholzer, On the degree distribution of the nodes in increasing trees, *Journal of Combinatorial Theory (A)* 114 (2007) 597–618.

[517] L. Kucera, Expected complexity of graph partitioning problems, *Discrete Applied Mathematics* 57 (1995) 193–212.

[518] V. Kurauskas, On small subgraphs in random intersection digraphs, *Discrete Mathematics* 313 (2013) 872–885.

[519] V. Kurauskas and M. Bloznelis, Large cliques in sparse random intersection graphs, see arxiv.org.

[520] A.N. Lageras and M. Lindholm, A note on the component structure in random intersection graphs with tunable clustering, *The Electronic Journal of Combinatorics* 15 (2008) #N10.

[521] C. Lee and B. Sudakov, Dirac's theorem for random graphs, *Random Structures and Algorithms* 41 (2012) 293–305.

[522] M. Lelarge, A new approach to the orientation of random hypergraphs, see arxiv.org.

[523] C. Lennon and B. Pittel, On a likely number of solutions for a stable marriage problem, *Combinatorics, Probability and Computing* (2009) 371–421.

[524] J. Leskovec, Dynamics of large networks, PhD in Computer Science, Carnegie Mellon University, (2008).

[525] J. Leskovec, D. Chakrabarti, J. Kleinberg and C. Faloutsos, Realistic, mathematically tractable graph generation and evolution, using Kronecker multiplication, in *PKDD '05: Proceedings of the 9th European Conference on Principles and Practice of Knowledge Discovery in Databases* (2005) 133–145.

[526] J. Leskovec, J. Kleinberg and C. Faloutsos, Graphs over time: densification laws, shrinking diameters and possible explanations, in *Proceedings of ACM SIGKDD Conference on Knowledge Discovery in Data Mining* (2005) 177187.

[527] J. Leskovec, D. Chakrabarti, J. Kleinberg, C. Faloutsos and Z. Ghahramani, Kronecker graphs: an approach to modeling networks, *Journal of Machine Learning Research* 11 (2010) 985–1042.

[528] K. Lin and G. Reinert, Joint vertex degrees in the inhomogeneous random graph model $\mathscr{G}(n,\{p_{ij}\})$, *Advances in Applied Probability* 44 (2012) 139–165.

[529] N. Linial and R. Meshulam, Homological connectivity of random 2-complexes, *Combinatorica* 26 (2006) 475–487.

[530] N. Linial and E. Rozenman, Random lifts of graphs: perfect matchings, *Combinatorica* 25 (2005) 407–424.

[531] S. Linusson and J. Wästlund, A proof of Parisi's conjecture on the random assignment problem, *Probability Theory and Related Fields* 128 (2004) 419–440.

[532] J. Lu, and Q. Feng, (1998). Strong consistency of the number of vertices of given degrees in nonuniform random recursive trees, *Yokohama Mathematical Journal* 45 (1998) 61–69.

[533] E. Lubetzky and A. Sly, Cutoff phenomena for random walks on random regular graphs, *Duke Mathematical Journal* 153 (2010) 475–510.

[534] E. Lubetzky and Y. Zhao, On the variational problem for upper tails in sparse random graphs, see arxiv.org.

[535] T. Łuczak, On the equivalence of two basic models of random graphs, in M. Karonski, J. Jaworski and A. Rucinski, Eds., *Proceedings of Random Graphs'87,* Wiley, Chichester (1990) 151–158.

[536] T. Łuczak, On the number of sparse connected graphs, *Random Structures and Algorithms* 1 (1990) 171–174.

[537] T. Łuczak, Component behaviour near the critical point, *Random Structures and Algorithms* 1 (1990) 287–310.

[538] T. Łuczak, Cycles in a random graph near the critical point, *Random Structures and Algorithms* 2 (1991) 421–440.

[539] T. Łuczak, Size and connectivity of the k-core of a random graph, *Discrete Mathematics* 91 (1991) 61–68.

[540] T. Łuczak, The chromatic number of random graphs, *Combinatorica* 11 (1991) 45–54.

[541] T. Łuczak, A note on the sharp concentration of the chromatic number of random graphs, *Combinatorica* 11 (1991) 295–297.

[542] T. Łuczak, The phase transition in a random graph, in D. Miklos, V.T. Sos and T. Szonyi, Eds., *Combinatorics, Paul Erdős is Eighty, Vol. 2*, Bolyai Society of Mathematical Studies 2, Journal of the Bolyai Mathematical Society, Budapest (1996) 399–422.

[543] T. Łuczak, Random trees and random graphs, *Random Structures and Algorithms* 13 (1998) 485–500.

[544] T. Łuczak, On triangle free random graphs, *Random Structures and Algorithms* 16 (2000) 260–276.

[545] T. Łuczak, B. Pittel and J.C. Wierman, The structure of a random graph at the point of the phase transition, *Transactions of the American Mathematical Society* 341 (1994) 721–748

[546] T. Łuczak and P. Pralat, Chasing robbers on random graphs: zigzag theorem, *Random Structures and Algorithms* 37 (2010) 516–524.

[547] T. Łuczak and A. Ruciński, Tree-matchings in graph processes, *SIAM Journal on Discrete Mathematics* 4 (1991) 107–120.

[548] T. Łuczak, A. Ruciński and B. Voigt, Ramsey properties of random graphs, *Journal of Combinatorial Theory, B* 56 (1992) 55–68.

[549] T. Łuczak, Ł. Witkowski and M. Witkowski, Hamilton cycles in random lifts of graphs, see arxiv.org.

[550] M. Mahdian and Y. Xu. Stochastic Kronecker graphs, *Random Structures and Algorithms* 38 (2011) 453–466.

[551] H. Mahmoud, *Evolution of Random Search Trees*, John Wiley & Sons, Inc., New York, (1992).

[552] H. Mahmoud, The power of choice in the construction of recursive trees, *Methodology and Computing in Applied Probability* 12 (2010) 763–773.

[553] H.M. Mahmoud and R.T. Smythe, A survey of recursive trees, *Theory of Probability and Mathematical Statistics* 51 (1995) 1–27.

[554] H.M. Mahmoud, R.T. Smythe and J. Szymański, On the structure of random plane-oriented recursive trees and their branches, *Random Structures and Algorithms* 4 (1993) 151–176.

[555] E. Marczewski, Sur deux propriétés des classes d'ensembles, *Fundamenta Mathematicae* 33 (1945) 303–307.

[556] D. Matula, The largest clique size in a random graph, *Technical Report*, Department of Computer Science, Southern Methodist University, (1976).

[557] D. Matula, Expose-and-merge exploration and the chromatic number of a random graph, *Combinatorica* 7 (1987) 275–284.

[558] R. Mauldin, *The Scottish Book: mathematics from the Scottish Cafe*, Birkhäuser Boston, (1981).

[559] G. McColm, First order zero-one laws for random graphs on the circle, *Random Structures and Algorithms* 14 (1999) 239–266.

[560] G. McColm, Threshold functions for random graphs on a line segment, *Combinatorics, Probability and Computing* 13 (2004) 373–387.

[561] C. McDiarmid, Determining the chromatic number of a graph, *SIAM Journal on Computing* 8 (1979) 1–14.
[562] C. McDiarmid, Clutter percolation and random graphs, *Mathematical Programming Studies* 13 (1980) 17–25.
[563] C. McDiarmid, On the method of bounded differences, in J. Siemons, Ed., *Surveys in Combinatorics, London Mathematical Society Lecture Notes Series* 141, Cambridge University Press, (1989).
[564] C. McDiarmid, Expected numbers at hitting times, *Journal of Graph Theory* 15 (1991) 637–648.
[565] C. McDiarmid, Random channel assignment in the plane, *Random Structures and Algorithms* 22 (2003) 187–212.
[566] C. McDiarmid and T. Müller, On the chromatic number of random geometric graphs, *Combinatorica* 31 (2011) 423–488.
[567] C. McDiarmid and B. Reed, Colouring proximity graphs in the plane, *Discrete Mathematics* 199 (1999) 123–137.
[568] C. McDiarmid, A. Steger and D. Welsh, Random planar graphs, *Journal of Combinatorial Theory, B* 93 (2005) 187–205.
[569] B.D. McKay, Asymptotics for symmetric 0-1 matrices with prescribed row sums, *Ars Combinatoria* 19 (1985) 15–25.
[570] B.D. McKay and N.C. Wormald, Asymptotic enumeration by degree sequence of graphs with degrees $o(n^{1/2})$, *Combinatorica* 11 (1991) 369–382.
[571] B.D. McKay and N.C. Wormald, The degree sequence of a random graph I, The models, *Random Structures and Algorithms* 11 (1997) 97–117.
[572] N. Martin and J. England, *Mathematical Theory of Entropy*. Cambridge University Press (2011).
[573] A. Mehrabian, Justifying the small-world phenomenon via random recursive trees, manuscript (2014).
[574] A. Meir and J.W. Moon, The distance between points in random trees, *Journal of Combinatorial Theory* 8 (1970) 99–103.
[575] A. Meir and J.W. Moon, Cutting down random trees, *Journal of the Australian Mathematical Society* 11 (1970) 313–324.
[576] A. Meir and J.W. Moon, Cutting down recursive trees, *Mathematical Biosciences* 21 (1974) 173–181.
[577] A. Meir and J.W. Moon, On the altitude of nodes in random trees, *Canadian Journal of Mathematics* 30 (1978) 997–1015.
[578] S. Micali and V. Vazirani, An $O(|V|^{1/2}|E|)$ algorithm for finding maximum matching in general graphs, *Proceedings of the 21st IEEE Symposium on Foundations of Computing* (1980) 17–27.
[579] M. Mihail and C. Papadimitriou, On the eigenvalue power law, *Randomization and Approximation Techniques, 6th International Workshop* (2002) 254–262.
[580] S. Milgram, The small world problem, *Psychology Today* 2 (1967) 60–67.
[581] I.M. Milne-Thomson, *The Calculus of Finite Differences,* Macmilan, London (1951).
[582] M. Molloy, Cores in random hypergraphs and Boolean formulas, *Random Structures and Algorithms* 27 (2005) 124–135.

[583] M. Molloy and B. Reed, The size of the largest component of a random graph on a fixed degree sequence, *Combinatorics, Probability and Computing* 7 (1998) 295–306.

[584] M. Molloy, H. Robalewska, R.W. Robinson and N.C. Wormald, 1-factorisations of random regular graphs, *Random Structures and Algorithms* 10 (1997) 305–321.

[585] R. Montgomery, Embedding bounded degree spanning trees in random graphs, see arxiv.org.

[586] R. Montgomery, Sharp threshold for embedding combs and other spanning trees in random graphs, see arxiv.org.

[587] J.W. Moon, The distance between nodes in recursive trees, *London Mathematical Society Lecture Notes Series* 13 (1974) 125–132.

[588] T.F. Móri, On random trees, *Studia Scientiarum Mathematicarum Hungarica* 39 (2003) 143–155.

[589] T.F. Móri The maximum degree of the Barabási–Albert random tree, *Combinatorics, Probability and Computing* 14 (2005) 339–348.

[590] K. Morris, A. Panholzer and H. Prodinger, On some parameters in heap ordered trees, *Combinatorics, Probability and Computing* 13 (2004) 677–696.

[591] E. Mossel and A. Sly, Gibbs rapidly samples colorings of $G_{n,d/n}$, *Journal of Probability Theory and Related Fields* 148 (2010) 37–69.

[592] T. Müller, Two point concentration in random geometric graphs, *Combinatorica* 28 (2008) 529–545.

[593] T. Müller, Private communication.

[594] T. Müller, X. Pérez and N. Wormald, Disjoint Hamilton cycles in the random geometric graph, *Journal of Graph Theory* 68 (2011) 299–322.

[595] T. Müller and M. Stokajović, A threshold for the Maker–Breaker clique game, *Random Structures and Algorithms* 45 (2014) 318–341.

[596] L. Mutafchiev, A limit distribution related to a random mappings and its application to an epidemic process, *Serdica* 8 (1982) 197–203.

[597] H.S. Na and A. Rapoport, Distribution of nodes of a tree by degree, *Mathematical Biosciences* 6 (1970) 313–329.

[598] A. Nachmias and Y. Peres, Critical random graphs: diameter and mixing time, *Annals of Probability* 36 (2008) 1267–1286.

[599] A. Nachmias and Y. Peres, The critical random graph, with martingales, *Israel Journal of Mathematics* 176 (2010) 29–41.

[600] C. Nair, B. Prabhakar and M. Sharma, Proofs of the Parisi and Coppersmith–Sorkin random assignment conjectures, *Random Structures and Algorithms* 27 (2005) 413–444.

[601] D. Najock and C.C. Heyde (1982). On the number of terminal vertices in certain random trees with an application to stemma construction in philology, *Journal of Applied Probability* 19 (1982) 675–680.

[602] C. St. J. Nash-Williams, Edge-disjoint spanning trees of finite graphs, *Journal of the London Mathematical Society* (1961) 445–450.

[603] R. Nenadov, Y. Person, N. Škoric and A. Steger, An algorithmic framework for obtaining lower bounds for random Ramsey problems, see arxiv.org.

[604] R. Nenadov and A. Steger, A short proof of the random Ramsey theorem, to appear in *Combinatorics, Probability and Computing*.

[605] R. Nenadov, A. Steger and M. Stojacović, On the threshold for the Maker–Breaker H-game, see arxiv.org.

[606] S. Nicoletseas, C. Raptopoulos and P.G. Spirakis, The existence and efficient construction of large independent sets in general random intersection graph, in *ICALP 2004, LNCS 3142*, Springer (2004) 1029–1040.

[607] S. Nikoletseas, C. Raptopoulos and P.G. Spirakis, On the independence number and Hamiltonicity of uniform random intersection graphs, *Theoretical Computer Science* 412 (2011) 6750–6760.

[608] I. Norros and H. Reittu, On a conditionally Poissonian graph process, *Advances in Applied Probability* 38 (2006) 59–75.

[609] D. Osthus, H.J. Prömel, and A. Taraz, On random planar graphs, the number of planar graphs and their triangulations, *Journal of Combinatorial Theory, B* 88 (2003) 119–134.

[610] Z. Palka, On the number of vertices of given degree in a random graph, *Journal of Graph Theory* 8 (1984) 167–170.

[611] Z. Palka, Extreme degrees in random graphs, *Journal of Graph Theory* 11 (1987) 121–134.

[612] E. Palmer and J. Spencer, Hitting time for k edge disjoint spanning trees in a random graph, *Periodica Mathematica Hungarica* 91 (1995) 151–156.

[613] A. Panholzer and H. Prodinger, The level of nodes in increasing trees revisited, *Random Structures and Algorithms* 31 (2007) 203–226.

[614] A. Panholzer and G. Seitz, Ordered increasing k-trees: introduction and analysis of a preferential attachment network model, *Discrete Mathematics and Theoretical Computer Science (AofA 2010)* (2010) 549–564.

[615] [PKBnV10] F. Papadopoulos, D. Krioukov, M. Boguná and A. Vahdat, Greedy forwarding in dynamic scale-free networks embedded in hyperbolic metric spaces, in *Proceedings of the 29th Conference on Information Communications, INFOCOM10* (2010) 2973–2981.

[616] G. Parisi, A conjecture on random bipartite matching, Physics e-Print archive (1998).

[617] E. Peköz, A. Röllin and N. Ross, Degree asymptotics with rates for preferential attachment random graphs, *The Annals of Applied Probability* 23 (2013) 1188–1218.

[618] M. Penrose, On k-connectivity for a geometric random graph, *Random Structures and Algorithms* 15 (1999) 145–164.

[619] M. Penrose, *Random Geometric Graphs*, Oxford University Press (2003).

[620] N. Peterson and B. Pittel, Distance between two random k-out graphs, with and without preferential attachment, see arxiv.org.

[621] J. Pitman, Random mappings, forest, and subsets associated with the Abel–Cayley–Hurwitz multinomial expansions, *Seminar Lotharingien de Combinatoire* 46 (2001) 46.

[622] B. Pittel, On distributions related to transitive closures of random finite mappings, *The Annals of Probability* 11 (1983) 428–441.

[623] B. Pittel, On tree census and the giant component in sparse random graphs, *Random Structures and Algorithms* 1 (1990) 311–342.

[624] B. Pittel, On likely solutions of a stable marriage problem, *Annals of Applied Probability* 2 (1992) 358–401.

[625] B. Pittel, Note on the heights of random recursive trees and random m-ary search trees, *Random Structures and Algorithms* 5 (1994) 337–347.

[626] B. Pittel, On a random graph evolving by degrees, *Advances in Mathematics* 223 (2010) 619–671.

[627] B. Pittel, L. Shepp and E. Veklerov, On the number of fixed pairs in a random instance of the stable marriage problem, *SIAM Journal on Discrete Mathematics* 21 (2008) 947–958.

[628] B. Pittel, J. Spencer and N. Wormald, Sudden emergence of a giant k-core in a random graph, *Journal of Combinatorial Theory, B* 67 (1996) 111–151.

[629] B. Pittel and R. Weishar, The random bipartite nearest neighbor graphs, *Random Structures and Algorithms* 15 (1999) 279–310.

[630] D. Poole, On weak hamiltonicity of a random hypergraph, see arxiv.org.

[631] L. Pósa, Hamiltonian circuits in random graphs, *Discrete Mathematics* 14 (1976) 359–364.

[632] P. Pralat, Cleaning random graphs with brushes, *Australasian Journal of Combinatorics* 43 (2009) 237–251.

[633] P. Pralat, Cleaning random d-regular graphs with brooms, *Graphs and Combinatorics* 27 (2011) 567–584.

[634] P. Pralat and N. Wormald, Meyniel's conjecture holds for random d-regular graphs, see arxiv.org.

[635] V. M. Preciado and A. Jadbabaie, Spectral analysis of virus spreading in random geometric networks, *Proceedings of the 48th IEEE Conference on Decision and Control* (2009) 4802–4807.

[636] H. Prodinger, Depth and path length of heap ordered trees, *International Journal of Foundations of Computer Science* 7 (1996) 293–299.

[637] H. Prodinger and F.J. Urbanek, On monotone functions of tree structures, *Discrete Applied Mathematics* 5 (1983) 223–239.

[638] H. Prüfer, Neuer Beweis eines Satzes über Permutationen, *Archives of Mathematical Physics* 27 (1918) 742–744.

[639] J. Radakrishnan, Entropy and Counting, www.tcs.tifr.res.in/~jaikumar/Papers/EntropyAndCounting.pdf.

[640] M. Radcliffe and S.J. Young, Connectivity and giant component of stochastic Kronecker graphs, see arxiv.org.

[641] S. Rai, The spectrum of a random geometric graph is concentrated, *Journal of Theoretical Probability* 20 (2007) 119–132.

[642] A. Rényi, On connected graphs I, *Publ. Math. Inst. Hungar. Acad. Sci.* 4 (1959) 385–387.

[643] A. Rényi and G. Szekeres, On the height of trees, *Journal of the Australian Mathematical Society* 7 (1967) 497–507.

[644] O. Riordan, Spanning subgraphs of random graphs, *Combinatorics, Probability and Computing* 9 (2000) 125–148.

[645] O. Riordan, The small giant component in scale-free random graph, *Combinatorics, Probability and Computing* 14 (2005) 897–938.

[646] O. Riordan, Long cycles in random subgraphs of graphs with large minimum degree, see arxiv.org.

[647] O. Riordan and A. Selby, The maximum degree of a random graph, *Combinatorics, Probability and Computing* 9 (2000) 549–572.

[648] O. Riordan and L. Warnke, Achlioptas process phase transitions are continuous, *Annals of Applied Probability* 22 (2012) 1450–1464.

[649] O. Riordan and N. Wormald, The diameter of sparse random graphs, *Combinatorics, Probability and Computing* 19 (2010) 835–926.

[650] R.W. Robinson and N.C. Wormald, Almost all cubic graphs are Hamiltonian, *Random Structures and Algorithms* 3 (1992) 117–126.

[651] R.W. Robinson and N.C. Wormald, Almost all regular graphs are Hamiltonian, *Random Structures and Algorithms* 5 (1994) 363–374.

[652] V. Rödl and A. Ruciński, Lower bounds on probability thresholds for Ramsey properties, in *Combinatorics, Paul Erdős is Eighty, Vol.1*, Bolyai Society of Mathematical Studies (1993), 317–346.

[653] V. Rödl and A. Ruciński, Threshold functions for Ramsey properties, *Journal of the American Mathematical Society* 8 (1995) 917–942.

[654] D. Romik, *The Surprising Mathematics of Longest Increasing Subsequences*, Cambridge University Press, New York (2014).

[655] S.M. Ross, A random graph, *Journal of Applied Probability* 18 (1981) 309–315.

[656] H. Rubin and R. Sitgreaves, Probability distributions related to random transformations on a finite set, *Technical Report 19A*, Applied Mathematics and Statistics Laboratory, Stanford University (1954).

[657] A. Ruciński, When a small subgraphs of a random graph are normally distributed, *Probability Theory and Related Fields* 78 (1988) 1–10.

[658] A. Ruciński, Matching and covering the vertices of a random graph by copies of a given graph, *Discrete Mathematics* 105 (1992) 185–197.

[659] A. Ruciński and A. Vince, Strongly balanced graphs and random graphs, *Journal of Graph Theory* 10 (1986) 251–264.

[660] A. Ruciński and N. Wormald, Random graph processes with degree restrictions, *Combinatorics, Probability and Computing* 1 (1992) 169–180.

[661] A. Rudas, B. Toth and B. Való, Random trees and general branching processes, *Random Structures and Algorithms* 31 (2007) 186–202.

[662] K. Rybarczyk, Equivalence of the random intersection graph and $G(n,p)$, *Random Structures and Algorithms* 38 (2011) 205–234.

[663] K. Rybarczyk, Sharp threshold functions for random intersection graphs via a coupling method, *The Electronic Journal of Combinatorics* 18(1) (2011) #P36.

[664] K. Rybarczyk, Diameter, connectivity, and phase transition of the uniform random intersection graph, *Discrete Mathematic* 311 (2011) 1998–2019.

[665] K. Rybarczyk, Constructions of independent sets in random intersection graphs, *Theoretical Computer Science* 524 (2014) 103–125.

[666] K. Rybarczyk, The coupling method for inhomogeneous random intersection graphs, see arxiv.org.

[667] K. Rybarczyk and D. Stark, Poisson approximation of the number of cliques in a random intersection graph, *Journal of Applied Probability* 47 (2010) 826–840.

[668] K. Rybarczyk and D. Stark, Poisson approximation of counts of subgraphs in random intersection graphs, submitted.

[669] V.N. Sachkov, *Combinatorial Methods in Discrete Mathematics*, Cambridge University Press (1996).

[670] A.A. Saposhenko, Metric properties of almost all boolean functions, (in Russian), *Diskretny Analiz* 10 (1967) 91–119.

[671] A.A. Saposhenko, Geometric structure of almost all boolean functions, (in Russian), *Problemy Kibernetiki* 30 (1975) 227–261.

[672] D. Saxton and A. Thomason, Hypergraph containers, see arxiv.org.

[673] M. Schacht, Extremal results for random discrete structures, see arxiv.org.

[674] J. Schmidt and E. Shamir, A threshold for perfect matchings in random d-pure hypergraphs, *Discrete Mathematics* 45 (1983) 287–295.

[675] J. Schmidt–Pruzan and E. Shamir, Component structure in the evolution of random hypergraphs, *Combinatorica* 5 (1985) 81–94.

[676] R. Schneider, *Convex Bodies: The Brunn–Minkowski Theory*, Cambridge University Press (2013).

[677] A. Scott, On the concentration of the chromatic number of random graphs (2008), see arxiv.org.

[678] C. Seshadhri, A. Pinar, and T.G. Kolda, An in-depth analysis of stochastic Kronecker graphs, *The Journal of the ACM*, see arxiv.org.

[679] E. Shamir and J. Spencer, Sharp concentration of the chromatic number of random graphs $G_{n,p}$, *Combinatorica* 7 (1987) 121–129.

[680] C. Shannon, A mathematical theory of communication, *Bell System Technical Journal* 27 (1948) 379423.

[681] S. Shelah and J. Spencer, Zero-one laws for sparse random graphs, *Journal of the American Mathematical Society* 1 (1988) 97–115.

[682] L.A. Shepp and S.P. Lloyd, Ordered cycle length in a random permutation, *Transactions of the American Mathematical Society* 121 (1966) 340–357.

[683] L. Shi and N. Wormald, Coloring random regular graphs, *Combinatorics, Probability and Computing* 16 (2007) 459–494.

[684] L. Shi and N. Wormald, Coloring random 4-regular graphs, *Combinatorics, Probability and Computing* 16 (2007) 309–344.

[685] B. Söderberg, General formalism for inhomogeneous random graphs, *Physical Review, E* 66 (2002) 066121.

[686] J. Spencer, Private Communication.

[687] J. Spencer, Enumerating graphs and Brownian motion, *Communications in Pure and Applied Mathematics* 50 (1997) 291–294.

[688] J. Spencer, *The Strange Logic of Random Graphs*, Springer, Berlin (2001).

[689] J. Spencer and N. Wormald, Birth control for giants, *Combinatorica* 27 (2007) 587–628.

[690] D. Stark, The vertex degree distribution of random intersection graphs, *Random Structures Algorithms* 24 (2004) 249–258.

[691] J. M. Steele, On Frieze's $\zeta(3)$ limit forlengths of minimal spanning trees, *Discrete Applied Mathematics* 18 (1987) 99–103.

[692] C. Stein, A bound for the error in the normal approximation to the distribution of a sum of dependent random variables, in *Proceedings of the Sixth Berkeley Symposium on Mathematical Statistics and Probability , 1970*, University of California Press, Berkeley, Vol. II (1972) 583–602.

[693] V.E. Stepanov, On the distribution of the number of vertices in strata of a random tree, *Theory of Probability and Applications* 14 (1969) 65–78.

[694] V.E. Stepanov, Limit distributions of certain characteristics of random mappings, *Theory of Probability and its Applications* 14 (1969) 612–636.

[695] M. Stojaković and T. Szabó, Positional games on random graphs, *Random Structures and Algorithms* 26 (2005) 204–223.

[696] Ch. Su, J. Liu and Q. Feng, A note on the distance in random recursive trees, *Statistics & Probability Letters* 76 (2006) 1748–1755

[697] B. Sudakov and V. Vu, Local resilience of graphs, *Random Structures and Algorithms* 33 (2008) 409–433.

[698] J. Szymański, On the nonuniform random recursive tree, *Annals of Discrete Mathematics* 33 (1987) 297–306.

[699] J. Szymański, On the maximum degree and the height of random recursive tree, in *Random Graphs'87*, Wiley, New York (1990) 313–324.

[700] J. Szymański, Concentration of vertex degrees in a scale free random graph process, *Random Structures and Algorithms* 26 (2005) 224–236.

[701] J. Tabor, On k-factors in stochastic Kronecker graphs, (2014) unpublished manuscript.

[702] M. Talagrand, Concentration of measures and isoperimetric inequalities in product spaces, *Publications Mathematiques de IT.H.E.S.* 81 (1996) 73–205.

[703] M.A. Tapia and B.R. Myers, Generation of concave node-weighted trees, *IEEE Transactions on Circuits and Systems* 14 (1967) 229–230.

[704] S-H. Teng and F.F. Yao, k-nearest-neighbor clustering and percolation theory, *Algorithmica* 49 (2007) 192–211.

[705] A. Thomason, A simple linear expected time algorithm for Hamilton cycles, *Discrete Mathematics* 75 (1989) 373–379.

[706] P. Turán, On an extremal problem in graph theory, *Matematikai és Fizikai Lapok (in Hungarian)* 48 (1941) 436–452.

[707] T.S. Turova, Phase transition in dynamical random graphs, *Journal of Statistical Physics* 123 (2006) 1007–1032.

[708] T.S. Turova, Diffusion approximation for the componentsin critical inhomogeneous random graphsof rank 1, *Random Structures Algorithms* 43 (2013) 486–539.

[709] T.S. Turova, Asymptotics for the size of the largest component scaled to "$\log n$" in inhomogeneous random graphs, *Arkiv fur Matematik* 51 (2013) 371–403.

[710] W.T. Tutte, A census of planar triangulations, *Canadian Journal of Mathematics* 14 (1962) 21–38.

[711] W.T. Tutte, A census of planar maps, *Canadian Journal of Mathematics* 15 (1963) 249–271.

[712] W.F. de la Vega, Long paths in random graphs, *Studia Scientiarum Mathematicarum Hungarica* 14 (1979) 335–340.

[713] A. M. Vershik and A. A. Shmidt, Limit measures arising in the asymptotic theory of symmetric groups, I, *Theory of Probability and its Applications* 22 (1977) 70–85.

[714] E. Vigoda, Improved bounds for sampling colorings, *Journal of Mathematical Physics* 41 (2000) 1555–1569.

[715] O.V. Viskov, Some comments on branching process, *Mathematical Notes* 8 (1970) 701–705.

[716] V. Vu. On some degree conditions which guarantee the upper bound of chromatic (choice) number of random graphs, *Journal of Graph Theory* 31 (1999) 201–226.

[717] V. Vu, Spectral norm of random matrices, *Combinatorica* 27 (2007) 721–736.

[718] D.W. Walkup, Matchings in random regular bipartite graphs, *Discrete Mathematics* 31 (1980) 59–64.

[719] D.W. Walkup, On the expected value of a random asignment problem, *SIAM Journal on Computing* 8 (1979) 440–442.

[720] J. Wástlund, An easy proof of the $\zeta(2)$ limit in the random assignment problem, *Electronic Communications in Probability* 14 (2009) 261–269.

[721] K. Weber, Random graphs - a survey, *Rostocker Mathematisches Kolloquium* (1982) 83–98.

[722] D. West, *Introduction to Graph Theory*, Second edition, Prentice Hall (2001).

[723] E.P. Wigner, On the distribution of the roots of certain symmetric matrices, *Annals of Mathematics* 67 (1958) 325–327.

[724] L.B. Wilson, An analysis of the stable assignment problem, *BIT* 12 (1972) 569–575.

[725] E.M. Wright, The number of connected sparsely edged graphs, *Journal of Graph Theory* 1 (1977) 317–330.

[726] E.M. Wright, The number of connected sparsely edged graphs II, *Journal of Graph Theory* 2 (1978) 299–305.

[727] E.M. Wright, The number of connected sparsely edged graphs III, *Journal of Graph Theory* 4 (1980) 393–407.

[728] N.C. Wormald, Models of random regular graphs, in J.D. Lamb and D.A. Preece, editors, *Surveys in Combinatorics, 1999, volume 267 of London Mathematical Society Lecture Note Series*, Cambridge University Press (1999) 239–298.

[729] N.C. Wormald, The differential equation method for random graph processes and greedy algorithms, in M. Karonski and H.J. Promel, Eds., *Lectures on approximation and randomized algorithms, Advanced Topics in Mathematics*, Polish Scientific Publishers PWN, Warsaw (1999) 73–155.

[730] O. Yagan and A.M. Makowski, Zero-one laws for connectivity in random key graphs, *IEEE Transations on Information Theory* 58 (2012) 2983–2999.

[731] B. Ycart and J. Ratsaby, The VC dimension of k-uniform random hypergraphs, *Random Structures and Algorithms* 30 (2007) 564–572.

[732] S.J. Young, *Random Dot Product Graphs: A Flexible Model of Complex Networks*, ProQuest (2008).

[733] S.J. Young and E.R. Scheinerman, Random dot product graph models for social networks, in *Algorithms and Models for the Web-Graph, LNCS 4863*, Springer (2007) 138–149.

[734] Z. Zhang, L. Rong and F. Comellas, High dimensional random apollonian networks, *Physica A: Statistical Mechanics and its Applications* 364 (2006) 610–618.

[735] J. Zhao, O. Yagan and V. Gligor, On k-connectivity and minimum vertex degree in random s-intersection graphs, *SIAM Meeting on Analytic Algorithmics and Combinatorics*, (2015), see arxiv.org.

[736] T. Zhou, G. Yan and B.H. Wang, Maximal planar networks with large clustering coefficient and power-law degree distribution, *Physics Reviews, E* 71 (2005) 046141.

Author Index

Main Index

Printed in the United States
by Baker & Taylor Publisher Services